ETHER LIPIDS

CHEMISTRY AND BIOLOGY

CONTRIBUTORS

Wolfgang J. Baumann
Hildegard Debuch
Roy Gigg
Howard Goldfine
Per-Otto Hagen
Donald J. Hanahan
Lloyd A. Horrocks
Morris Kates
Donald C. Malins
Helmut K. Mangold
Claude Piantadosi
Peter Seng
Fred Snyder
Guy A. Thompson, Jr.
Usha Varanasi

ETHER LIPIDS

CHEMISTRY AND BIOLOGY

Edited by FRED SNYDER

MEDICAL DIVISION
OAK RIDGE ASSOCIATED UNIVERSITIES
OAK RIDGE, TENNESSEE

ACADEMIC PRESS New York and London 1972

ACADEMIC PRESS, INC.
111 Fifth Avenue, New York, New York 10003

United Kingdom Edition published by
ACADEMIC PRESS, INC. (LONDON) LTD.
24/28 Oval Road, London NW1

LIBRARY OF CONGRESS CATALOG CARD NUMBER: 72-182644

PRINTED IN THE UNITED STATES OF AMERICA

For Cathy

CONTENTS

Chapter VII The Enzymic Pathways of Ether-Linked Lipids and Their Precursors

Fred Snyder

Chapter VIII Biological Effects and Biomedical Applications of Alkoxylipids

Helmut K. Mangold

Chapter IX Content, Composition, and Metabolism of Mammalian and Avian Lipids That Contain Ether Groups

Lloyd A. Horrocks

Chapter XIV Bacterial Plasmalogens

Howard Goldfine and Per-Otto Hagen

Chapter XV Ether-Linked Lipids in Extremely Halophilic Bacteria

Morris Kates

Chapter XVI The Search for Alkoxylipids in Plants

Helmut K. Mangold

LIST OF CONTRIBUTORS

Numbers in parentheses indicate the pages on which the authors' contributions begin.

WOLFGANG J. BAUMANN (51), The Hormel Institute, University of Minnesota, Austin, Minnesota

HILDEGARD DEBUCH (1), Physiologisch-Chemisches Institut der Universität, Köln-Lindenthal, Germany

ROY GIGG (87), Division of Chemistry, National Institute for Medical Research, Mill Hill, London, England

HOWARD GOLDFINE (329), Department of Microbiology, University of Pennsylvania School of Medicine, Philadelphia, Pennsylvania

PER-OTTO HAGEN (329), Department of Surgery, Duke Medical Center, Durham, North Carolina

DONALD J. HANAHAN (25), Department of Biochemistry, University of Arizona College of Medicine, Tucson, Arizona

LLOYD A. HORROCKS (177), Department of Physiological Chemistry, The Ohio State University College of Medicine, Columbus, Ohio

MORRIS KATES (351), Department of Biochemistry, University of Ottawa, Ottawa, Canada

DONALD C. MALINS (297), U.S. Department of Commerce, National Oceanic and Atmospheric Administration, National Marine Fisheries Service, Seattle, Washington

HELMUT K. MANGOLD (157, 399), Institut für Technologie und Biochemie, H. P. Kaufmann-Institut der Bundesanstalt für Fettforschung, Münster (Westf.), Germany

Claude Piantadosi (81, 109), Department of Medicinal Chemistry, School of Pharmacy, University of North Carolina, Chapel Hill, North Carolina

Peter Seng (1), Physiologisch-Chemisches Institut der Universität, Köln-Lindenthal, Germany

Fred Snyder (121, 273), Medical Division, Oak Ridge Associated Universities, Oak Ridge, Tennessee

Guy A. Thompson, Jr. (313, 321), Department of Botany, University of Texas, Austin, Texas

Usha Varanasi (297), U.S. Department of Commerce, National Oceanic and Atmospheric Administration, National Marine Fisheries Service, Seattle, Washington

PREFACE

In the half century that has elapsed since investigators first reported the existence of lipids containing ether bonds, a vast amount of literature on the subject has been compiled. The tempo of research in the field accelerated greatly a few years ago after the discovery of cell-free systems capable of synthesizing both *O*-alkyl and *O*-alk-1-enyl moieties in lipids. This rapid escalation has made it difficult for "Ether Lipids: Chemistry and Biology" to include the most recent breakthroughs in glyceryl ether research, but all the chapters thoroughly cover the ether lipids to 1971, and in some instances it has been possible to include descriptions of more recent publications. We hope that this volume, being the first book published on this subject, will fill a need for many investigators and students.

The work includes one chapter on the history of the studies of ether lipids and others on the chemical syntheses, analytical procedures, biological effects, and metabolic pathways of these lipids. In general, the biochemical topics are organized according to the biological source of the data, such as mammals, birds, marine organisms, molluscs, protozoa, bacteria, and plants. I believe that this organization is helpful in focusing on the suitability and usefulness of the various tissues that are being used to elucidate the role of ether lipids in living systems. All the contributors have themselves done valuable research in the ether field, and a number of them give personal accounts of their own work that would not appear in the usual scientific paper.

"Ether Lipids: Chemistry and Biology" is written primarily for persons engaged in lipid research. But since the chapters cover chemistry, biology, and biomedical applications, they should also serve as an important reference to workers who are highly specialized in any of these fields. At the same time, the broader concepts are adequately presented for the general audience, especially graduate students who are studying lipids and membranes. This treatise should be an indispensable source of otherwise unavailable information on a group of lipids now evoking considerable interest in the world of biochemical research.

I express many thanks to the various authors for their excellent coverage and critical review of the material assigned. Moreover, a very special debt of gratitude is owed my wife, Cathy, for her outstanding help in editing this book.

Fred Snyder

NOMENCLATURE

There are two basic types of ether-linked aliphatic moieties in glycerolipids: alkyl (I) and alk-1-enyl (II).

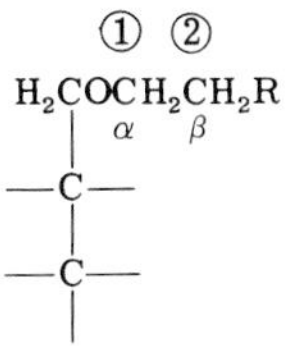

Alkyl

(I)

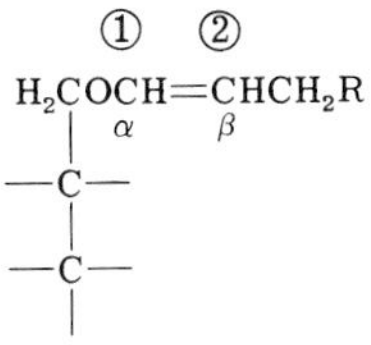

Alk-1-enyl

(II)

The terms alkyl and alk-1-enyl merely describe the type of bond between the 1 and 2 (α,β) carbon atoms in the aliphatic side chain that is adjacent to the ether bond and are not meant to reflect any other meaning about the molecule, such as unsaturation or substituted groupings further out along the aliphatic chains. Except for di- and triethers, the *O*-alkyl and *O*-alk-1-enyl moieties have only been found at the 1-position of the glycerol moiety in the natural state. This book is primarily concerned with the *O*-ethers, although *S*-alkyl moieties have also been studied in biological systems. The literature contains several synonyms for *O*-alkyl and *O*-alk-1-enyl, and these are given to provide a background to a newcomer to this field:

O-Alkyl—glyceryl ethers, saturated, alkoxy
O-Alk-1-enyl—plasmalogen, alkenyl, vinyl, enol, unsaturated

Rules for the specific naming of ether-linked glycerolipids are described under The Nomenclature of Lipids [see *The Journal of Lipid Research*

8, 523 (1967)] as proposed by the Commission on Biochemical Nomenclature, which is sponsored by the International Union of Pure and Applied Chemistry (IUPAC) and the International Union of Biochemistry (IUB). Unfortunately, the rules do not spell out general names to describe the ether-linked glycerolipids and their individual classes, although such usage has always varied with authors and with history. In this book, I have attempted to keep the use of general terms as close as possible to that recommended by the IUPAC–IUB Commission on Biochemical Nomenclature, i.e., the terms alkyl and alk-1-enyl have been used to describe the ether moieties. The prefix *O* has not been used if the aliphatic moiety is clearly on glycerol, e.g., alkylglycerols, alk-1-enylglycerols, alkyl glycerolipids, or alk-1-enyl glycerolipids, but the *O* has been used if glycerol is not clearly designated, e.g., *O*-alkyl lipids, *O*-alk-1-enyl lipids, *O*-alkyl moiety, or *O*-alk-1-enyl moiety, or if the point of attachment of the substituted ether group(s) is ambiguous. The general names also reflect the order of substituents at the three positions of glycerol in that the first prefix refers to position 1, the second to position 2, etc. For example, alk-1-enylacylglycerophosphorylethanolamine is equivalent to 1-alk-1′-enyl-2-acyl-*sn*-glycero-3-phosphorylethanolamine.

I suggest the following terminology for naming the general classes of ether-linked glycerolipids:

General Terms

ether-linked glycerolipids
alkyl glycerolipids
alk-1-enyl glycerolipids
O-alkyl lipids
O-alk-1-enyl lipids
O-alkyl moieties
O-alk-1-enyl moieties
alkyl phospholipids
alk-1-enyl phospholipids
alkyl choline phospholipids
alk-1-enyl choline phospholipids
alkyl ethanolamine phospholipids
alk-1-enyl ethanolamine phospholipids

*Lipid Classes**

trialkylglycerol
dialkylacylglycerol

* Numbers can be used to designate the position on glycerol if required.

alk-1-enyldiacylglycerol
alkyldiacylglycerol
dialkylglycerol
alk-1-enylacylglycerol
alkylacylglycerol
alk-1-enylglycerol
alkylglycerol
alkylacylglycerophosphate or alkylacyl-GP
alk-1-enylacylglycerophosphate or alk-1-enylacyl-GP
alkylglycerophosphate or alkyl-GP
alk-1-enylglycerophosphate or alk-1-enyl-GP
alkylacylglycerophosphorylcholine or alkylacyl-GPC
alk-1-enylacylglycerophosphorylcholine or alk-1-enylacyl-GPC
alkylglycerophosphorylcholine or alkyl-GPC
alk-1-enylglycerophosphorylcholine or alk-1-enyl-GPC
alkylacylglycerophosphorylethanolamine or alkylacyl-GPE
alk-1-enylacylglycerophosphorylethanolamine or alk-1-enylacyl-GPE
alkylglycerophosphorylethanolamine or alkyl-GPE
alk-1-enylglycerophosphorylethanolamine or alk-1-enyl-GPE
alkylacylglycerophosphorylserine or alkylacyl-GPS
alk-1-enylacylglycerophosphorylserine or alk-1-enylacyl-GPS
alkylglycerophosphorylserine or alkyl-GPS
alk-1-enylglycerophosphorylserine or alk-1-enyl-GPS

CHAPTER I

THE HISTORY OF ETHER-LINKED LIPIDS THROUGH 1960

Hildegard Debuch and Peter Seng

I. Plasmalogens

A. Detection and Investigation of Structure

It was by accident that Robert Feulgen,* together with K. Voit, detected plasmalogens in 1924. At that time he had already described the well-known Feulgen reaction, i.e., the staining of cell nuclei with fuchsin-sulfurous acid. The procedure consisted of fixing tissue slices in the usual way by dehydrat-

* Robert Feulgen was born in Werden (Ruhr), Germany, on September 2, 1884. After studying medicine, he became a research assistant at Berlin in 1912, and at Giessen/Lahn in 1919. He obtained a professorship at Giessen/Lahn and stayed there until he died on October 24, 1955.

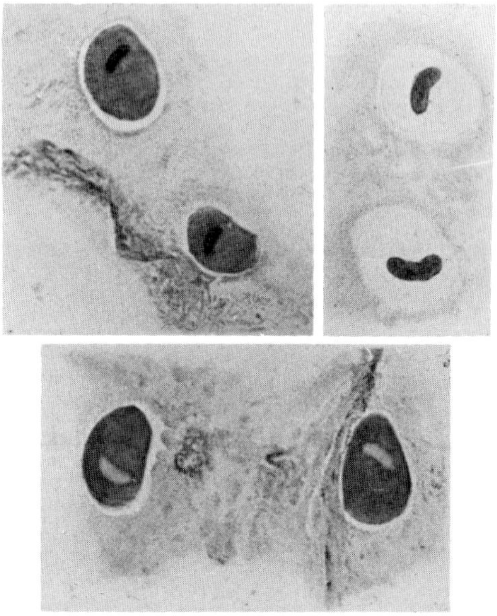

FIG. 1. Different stainings of *Nyctotherus*. Reproduction from Feulgen's original paper (Feulgen and Voit, 1924). Bottom, plasmal reaction; top right, nucleal reaction; top left, both reactions on the same organism.

ing with alcohol, then treating them with acid to hydrolyze the deoxyribonucleic acid, and finally exposing the deoxyribose thus formed to fuchsin-sulfurous acid, producing a red-violet complex.

The following story, never before published, was told to me (H.D.) by Feulgen himself. It is reported here to emphasize once more the importance of accidental observations.

It was a very hot summer day, and people in the lab decided to leave, and to enjoy themselves in the woods. But what to do with the many fresh tissue slices they had already prepared, how to preserve them? They put them into a solution of $HgCl_2$ and went away. The next day, when they came back to the lab a little tired, they put some of these slices by an error into the fuchsin-sulfurous acid without fixation and without acid treatment. When they looked through the microscope, they found to their surprise that the plasma of the cells was stained, while the nuclei were not [Fig. 1]. When Robert Feulgen saw this, he concluded that an aldehyde must be present in the plasma of the cell. Since glycolysis was under investigation in many biochemical labs at that time, he first thought the plasmal stain was caused by acetaldehyde. However, he soon realized that the fuchsin-sulfurous acid-sensitive material was extracted from the tissue by alcohol and, furthermore, that it did not seem to be in a free state in the cell. Therefore he called the aldehyde—because of its presence in the plasma of the cell—"plasmal," and the unknown mother substance releasing this aldehyde "plasmalogen."

Since Feulgen and his co-workers had discovered a new substance, they naturally looked for its presence in other tissues. Indeed, they found "plasmalogen" in the whole animal kingdom, from protozoa to human beings, in nearly every kind of tissue, although in varying quantities (Stepp *et al.*, 1927; Imhäuser, 1927). We now know that particularly muscle, adrenal glands, cerebellum, and cerebrum (especially white matter) contain high levels of plasmalogens. However, Feulgen and Voit (1924) found them not only in these tissues, but in nearly every cell, although some organs, like liver, produced only a weak coloration.

Of course, it was not very long before Feulgen worked out a method for the quantitative estimation of "plasmal," the so-called plasmal reaction (for blood: Feulgen and Imhäuser, 1927; for lipid fractions from tissues: Feulgen and Grünberg, 1939). In this reaction, the aldehydes are liberated from the plasmalogen by acid treatment. After reaction with fuchsin-sulfurous acid, the colored complex is extracted from the water-soluble products by an alcohol immiscible with water, such as amyl or capryl alcohols. The color reaction follows Lambert-Beer's law, and the red color absorbs maximally at 530 nm. A slight variation, easier to use in a routine analytical setup, has been described by Klenk and Debuch (1963). Feulgen and Imhäuser's original method for the determination of plasmalogens in blood was a "direct" method, in that it did not call for prior extraction of the plasmalogens from the serum, but several workers have since modified this procedure. Some of them are Leupold (1950b), Feulgen *et al.* (1951, 1953a, 1953b), Waelsch (1951), Leupold and Büttner (1952, 1953a), Leupold *et al.* (1954), and Thiele (1955c). The histochemical technique for aldehyde-containing lipids has also been modified (Hack, 1952). In 1956 Wittenberg *et al.* devised a stoichiometric method for determining aldehydes as *p*-nitrophenylhydrazone derivatives (Korey and Wittenberg, 1954; Wittenberg *et al.*, 1956). Another chemical method should also be mentioned here, the one described by Rapport and Franzl (1957a). It is based on the increased uptake of iodine by the double bond of α, β-unsaturated ethers which is much more reactive than other kinds of olefinic bonds.

Surprisingly enough, it was 15 years (1939) before Feulgen and Bersin isolated an aldehyde-containing phosphatide from beef muscle. We know today that the product they obtained was an artifact, but in the light of our present knowledge, that is very understandable. Even nowadays, 30 years later, there is no method available that separates plasmalogens from other phospholipids without changing—at least partially—one or the other. But in 1939 Feulgen and Bersin obtained their "plasmalogen" after treating the total phospholipids extracted from tissues with alkali in fairly high concentrations. Thus the ester linkages of the phospholipids containing choline, ethanolamine, or serine were hydrolyzed, while the acetal linkages formed from the plasmalogens were stable. Under these conditions, the fatty acids had been saponified to soaps and neither they nor the choline or ethanolamine esters of glycerophosphoric acid were soluble in organic solvents, but the aldehyde-containing phosphatides were. Since analysis of the isolated product revealed an aldehyde to phosphorus ratio of 1:1, Feulgen and Bersin (1939) called the new substance "acetalphosphatide," and gave it the following formula (**1**):

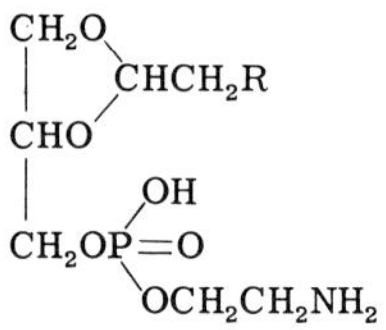

(**1**)

Acetalphosphatide (R designates a fatty aldehyde residue)

It can be seen that the isolated acetalphosphatide contained ethanolamine· It had no distinct melting point and decomposed at about 150°C. It was not soluble in ether, petroleum ether, or acetone, slightly soluble in benzene or cold ethanol, much more soluble in methanol, and easily soluble in chloroform. The analysis of the product crystallized from ethanol was in good agreement with theoretical values for a mixture of palmityl- and stearyl-acetalphosphatide, $C_{21}H_{44}O_6NP$ or $C_{23}H_{48}O_6NP$, respectively. Carbon: 57.61–59.30% (found, 58.22%); hydrogen: 10.14–10.39% (found, 10.28%); nitrogen: 3.20–3.01% (found, 3.10%); phosphorus: 7.10–6.67% (found, 6.67%). Feulgen and Bersin (1939) also isolated the ethanolamine ester of glycerophosphoric acid from the acetalphosphatide (**1**) by liberating the aldehydes with $HgCl_2$ in aqueous suspension. They obtained the ester in a crystallized form from water/ethanol; it had a melting point of 86°–87°C. It probably was a mixture of the α- and β-derivatives, such as the one identified by Ansell and Norman (1956) in rat brain extracts. In 1951, Thannhauser *et al.* (1951a, b) isolated a product similar to Feulgen's (**1**) from beef brain.

Feulgen and Bersin (1939), describing their product, emphasized that they could find it only after treating the phospholipids with alkali (33% KOH, 100°C, 20 min), and therefore already suggested the possibility that Feulgen's acetalphosphatide was not identical to the original plasmalogen. It bothered them that their isolated product was not soluble in ether when the total phospholipids were, and that it reacted with fuchsin-sulfurous acid at a rate slower than did the plasmalogen in the original phospholipid extract. The increase in fuchsin-sulfurous acid reaction time was also noted by Anchel and Waelsch (1944), who compared the reaction time of glyceryl aldehyde to that of synthetic acetals. However, the validity of drawing conclusions from this comparison has long been open to question, since Ehrlich *et al.* (1948) later observed that the presence of other naturally occurring lipids seems to inhibit the color development of higher fatty aldehydes by fuchsin-sulfurous acid. But Anchel's observation did throw some doubt on the identification of plasmalogen as Feulgen's acetalphosphatide. Independently, some investigators (Klenk and Böhm, 1951; Schmidt *et al.*, 1951, 1953) suspected that there could be another lipid group in the plasmalogen molecule, since both Feulgen's product and that isolated by Thannhauser *et al.* (1951a, b) from beef brain, using Feulgen's methods, were insoluble in ether. Yet no insoluble phosphatide could be found. Klenk and Böhm (1951), who had a plasmalogen-containing phospholipid fraction in which the aldehyde accounted for 20%, could not confirm that the theoretical amount of Feulgen's acetalphosphatide was present. Therefore, Klenk (1952) postulated a formula for plasmalogen in which one of the hydroxyl groups of the glycerophosphoric acid residue was esterified with a fatty acid (**2**).

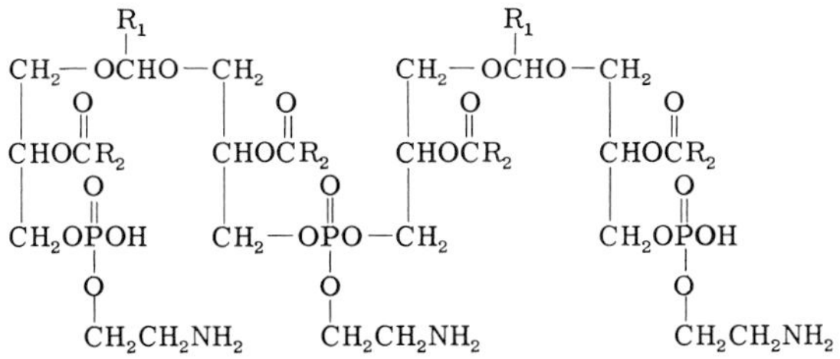

(**2**)

R_1's represent fatty aldehyde residues; R_2's represent fatty acid residues

At that time, there was virtually no method available for separating the different phospholipid groups on a preparative scale, at least not those of the "cephalin" type. But Klenk and Böhm (1951), using Folch's (1942) fractionation method (based on precipitation in organic solvents) for brain lipids, also showed that a serine plasmalogen exists in the human brain's so-called "serine-cephalin" fraction, since it contained approximately 5% plasmal and since 98% of its nitrogen was in serine's amino groups.

Two years later, Klenk and Gehrmann (1953) showed that choline derivatives of plasmalogen must also exist. They isolated a lecithin fraction from beef heart muscle on alumina; this fraction had an aldehyde content of about 10%, and contained nearly all its nitrogen in the choline moiety. In the same year, Hack (1953) independently drew the same conclusion using paper chromatography. By 1960 choline plasmalogen had also been found in bovine muscle (Rapport *et al.*, 1954; Rapport and Alonzo, 1955), pig liver (Marinetti and Erbland, 1957; Marinetti *et al.*, 1958), serum (Böhm and Richarz, 1957), egg yolk (Thiele, 1955a), and ram spermatozoa (Lovern *et al.*, 1957; Hartree and Mann, 1958), and had been isolated in a pure lyso state from beef heart (Hartree and Mann, 1960a).

Finally, Klenk and Debuch (1954) demonstrated that the ethanolamine plasmalogens of human brain contained a fatty acid moiety as well as an aldehyde moiety in the same molecule. Different analytical results brought them to this conclusion: Again using Folch's (1942) procedure for separating brain lipids, they prepared an "ethanolamine-cephalin"-containing fraction corresponding to Folch's "Fraction V." The aldehydes accounted for more than 20% of this fraction and, according to Feulgen's formula (**1**), should have represented something over half the acetalphosphatide. The ratio of fatty acids to aldehydes was 2:1, so that if Feulgen's formula were correct, one would expect to find approximately 66% diester and 33% acetalphosphatide, and mild acid treatment or alkaline hydrolysis should both render about half the total phosphorus water-soluble. However, after hydrogenation of the lipids, alkali treatment solubilized only 30–40% of the total phosphorus. In addition, the phosphorus content of the fraction was less than 4% rather than the theoretical 5.5%. These findings could only be explained by assuming that plasmalogen contains a fatty acid group in addition to the aldehyde, and that the theoretical molar ratio of diester-phospholipid to plasmalogen is 1:2 in the ethanolamine fraction of human brain lipids. Klenk and Debuch (1954) hence postulated three formulas, two of which, (**3**) and (**4**), are given here.

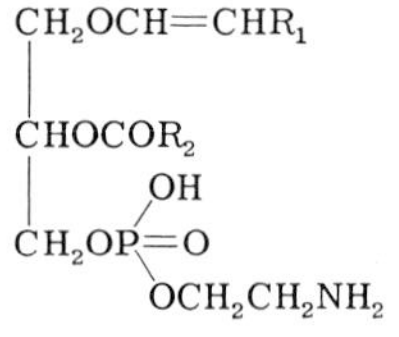

(**3**)

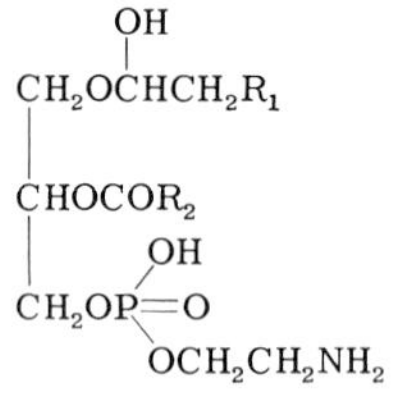

(**4**)

R_1's represent fatty aldehyde residues; R_2's represent fatty acid residues

When one assumes that the fraction Klenk and Debuch were working with was 66% phospholipid of the type shown in (**3**) or (**4**) and 33% phospholipid of the well-known diester type (**5**), the theoretical values match those

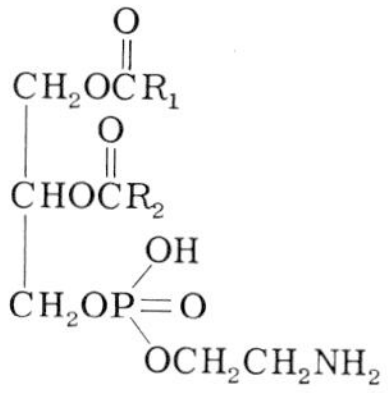

(**5**)

Phosphatidylethanolamine (R_1 and R_2 represent fatty acid residues)

obtained by the investigators to a high degree. A little later in the same year, Rapport *et al.* (1954), after determining the phosphorus and nitrogen content of plasmalogen-rich phospholipid fractions, also postulated that two fatty chains might exist in the plasmalogens.

After Klenk and Gehrmann (1953) had demonstrated the existence of choline plasmalogen, Klenk and Debuch (1955) showed that it, too, contains a fatty acid residue. They used a lecithin fraction of beef heart, which contained almost 40% choline plasmalogen and 60% phosphatidylcholine, and split the aldehydes off by treatment with acetic acid at 37°C. A lysolecithin was produced, and it could be separated from the unreacted lecithin by countercurrent distribution. The amount of lysophosphatidylcholine found corresponded to the theoretical value of plasmalogen in the lecithin fraction used. In the United States some months later, Rapport and Alonzo (1955) used a photometric method to estimate the ester groups in a lecithin fraction, and they also concluded that there was a fatty acid in the choline plasmalogen.

Conclusive evidence for the presence of a fatty acid in the ethanolamine plasmalogen of brain was published by Debuch (1956). She treated a plasmalogen-rich so-called "cephalin" fraction with acid and separated the lysophosphatidylethanolamine from the unchanged corresponding diester phospholipid. In this way, it was now possible to investigate the fatty acids of the plasmalogens. Since a method for getting the aldehydes was already available (see below), one could analyze the plasmalogen structure without separating it from the phosphatidylethanolamine.

As early as 1929, Feulgen *et al.* had extracted phospholipids from horse muscle, and after acid treatment, liberated the aldehydes and distilled them with steam. The thiosemicarbazones obtained were derived from palmitaldehyde and stearaldehyde in a ratio of 9:1. Later on, Behrens (1930) changed Feulgen's method in that he prepared the thiosemicarbazones by adding thiosemicarbacid directly to an aqueous emulsion of phospholipids previously acidified and neutralized. Anchel and Waelsch (1942) also isolated C_{16}- and C_{18}-fatty aldehydes from brain and liver; they made *p*-carboxyphenylhydrazone and carboxymethoxime derivatives. In 1939, Feulgen and Bersin had suspected that there was an unsaturated aldehyde group in brain, which Klenk (1945) identified as *n*-octadecenal.

Klenk (1944) was also the first to develop a method that allowed us to obtain aldehydes in a pure state, even when working with smaller amounts of lipid. He used methanolysis to get the aldehydes as dimethylacetal derivatives along with fatty acid methyl esters. Since the acetal linkage is stable in alkali, the aldehyde derivatives could be separated from those of the fatty acids after saponification. Klenk (1945) then converted the dimethylacetals to the oximes; with these derivatives, he was able to separate the saturated from the unsaturated derivatives by crystallization from ethanol and determine the amounts of palmitic, stearic, and octadecenoic acids after converting the oximes into the cyanides and finally into the acids. The ratio of C_{16}- to C_{18}-saturated fatty acids—derived from the corresponding aldehydes—was 39 to 61 in brain tissue. Leupold (1950a) used a similar method to investigate the unsaturated aldehyde moieties, and found that they accounted for more than 50% of the total plasmal of brain phosphatides. They were mostly *cis*-Δ-9-octadecenal, but he detected some *cis*-Δ-11-octadecenals. Leupold also found traces of C_{14}-aldehydes in brain phosphatides. Additional work on the total phospholipid fractions of different organs, such as the study of Klenk *et al.* (1952) on muscles from horse and beef, showed that the ratio between the different chain lengths in the aldehyde mixture varied from tissue to tissue.

After separating the monoacyl compounds resulting from mild acid treatment of brain phospholipids from phosphatidylethanolamine, Debuch (1956) found that these ethanolamine plasmalogens did not contain any saturated fatty acids. Similar results for choline plasmalogens were later obtained by other authors (Klenk and Krickau, 1957; Gray, 1958).

Debuch (1956) also hydrogenated the plasmalogen-rich phospholipid fraction (**6**) and subsequently hydrolyzed the fatty acid esters to isolate, for the first time, a mixture of the phosphoric acid esters of chimyl and batyl alcohol (**7**).

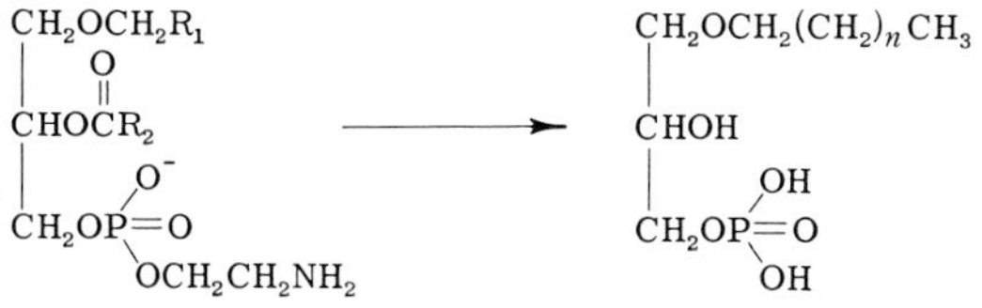

(6) (7)

Hydrogenated ethanolamine plasmalogen
R_1 represents fatty alcohol residue
R_2 represents fatty acid residue

Chimyl alcohol (n = 14)
Batyl alcohol (n = 16)
Phosphoric acid ester

Although the isolation of glyceryl ether derivatives made formula (**3**) (discussed by Klenk and Debuch in 1954) very probable for plasmalogen, it could not be considered conclusive proof. But Rapport *et al.* (1957) described the preparation of a lysoplasmalogen of the ethanolamine type that had a bromine uptake different from that of the usual olefinic bonds and concluded (Rapport and Franzl, 1957a) that the aldehydogenic moiety exists in its enol form. Independently, Debuch (1957, 1958a, b) showed that formula (**3**) was also correct with respect to the linkage of the aldehyde. Since the enol form of an aldehyde has a double bond between the α- and β-carbon atom, it should be vulnerable to attack by ozone (**8**).

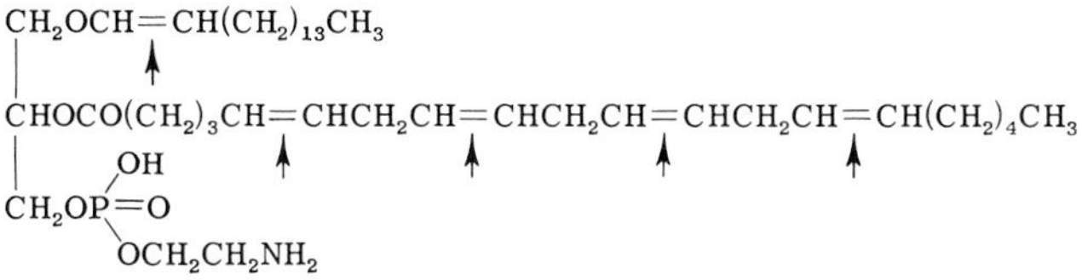

(**8**)

Ethanolamine plasmalogen (↑ represents site of attack by ozone)

After hydrolysis and further oxidation of the ozonides, one should obtain three final products: (1) C_{15}- and C_{17}-n-fatty acids, derived from the enol ether of hexadecanal and octadecanal, respectively, (2) mono- and dicarboxylic acids with chain lengths of nine carbon atoms or less derived from the various parts of the unsaturated fatty acids and fatty aldehydes, and (3) the phospholipid residue. If the aldehydes were originally bound through a hemiacetal linkage, such as in Feulgen's original formula, one should obtain palmitic and stearic acids. The results actually obtained by Debuch (1957, 1958a) identified 75% of all the fatty acids as penta- and heptadecanoic acids, indicating that the hemiacetal form of the plasmalogen, if indeed there was any at all, existed only at a very low level. Blietz

(1958) confirmed the presence of the enol ether structure in ethanolamine plasmalogens after cleaving the intact plasmalogen in tritiated water.

The puzzle of plasmalogen structure was now nearly solved, but one question remained: Was the aldehyde situated at the α- or β-position of the glycerol? Rapport and Franzl (1957b; Franzl and Rapport, 1955) treated a plasmalogen-rich choline phosphatide fraction from beef heart with snake phospholipase A. At that time, every lipid chemist was certain that this enzyme specifically splits the fatty acid–ester linkage in the α-position (Hanahan, 1954). Since Rapport and Franzl's work (1957b) showed that phospholipase A reacted with a plasmalogen-containing phospholipid fraction in which the ether-insoluble lysophosphatide gave a positive plasmal reaction, it was evident that the plasmalogen—or at least a part of it—had also acted as a substrate for phospholipase A. Thus it was only logical for the authors to conclude that the aldehydes were in the β-position. Hack and Ferrans (1958, 1959) also showed that plasmalogen is hydrolyzed by phospholipase A to lysoplasmalogen.

Gray (1957, 1958) was able to use chemical methods to prove the position of the aldehydes of both choline and ethanolamine plasmalogens. He liberated the aldehydes by acid, and after isolating and hydrogenating the resulting lyso compounds, he oxidized these with permanganate. Subsequent acid hydrolysis removed the fatty acids, and the remaining compounds were identified as methylglyoxals. This indicated that the starting material was an α-lysolecithin or α-lyso-"cephalin," and therefore was additional evidence for the postulate that the β-position was occupied by aldehydes. Some authors, like Marinetti and Erbland (1957), concluded from their own results that the aldehydes of pig heart plasmalogens were predominantly in the α-position. They (Marinetti *et al.*, 1958) prepared glyceryl ether phosphates after hydrogenation and mild acid or basic hydrolysis of the lecithin fraction from pig heart in a manner similar to the one Klenk and Debuch employed for brain lipids (1954) and for beef heart lipids (1955). After prolonged acid hydrolysis, Marinetti *et al.* (1958) obtained glyceryl ethers and used periodic acid oxidation to get aldehydes containing the *o*-alkyl chain [reaction (1)].

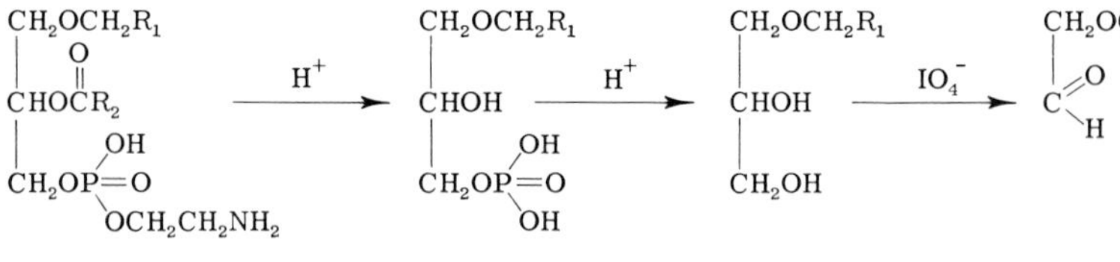

(1)

Products obtained after acid hydrolysis and periodic acid oxidation of hydrogenated plasmalogens
R_1 represents fatty alcohol residue; R_2 represents fatty acid residue

Later on, Marinetti *et al.* (1958) showed that the ethers obtained from pig heart were mainly α-, while those from beef heart were 87% β-derivatives. In 1959, he (Marinetti *et al.*, 1959a) stated that the presence of diethers had complicated the beef heart determinations, and that beef heart plasmalogens, as well as those from pig heart, were mainly linked to the α-position of glycerol. Marinetti *et al.* (1959b) also threw some doubt on the position of the attack of phospholipase A on lecithin, since they concluded from their experiments that this enzyme could also attack β-linked fatty acids from the plasmalogens.

Experiments with ethanolamine plasmalogens by Debuch (1959a, b) proved beyond a doubt that the aldehyde group was in the α-position of the glycerol moiety. Chimyl and batyl alcohol phosphoric acid esters (a mixture identical to the one obtained in 1956 from brain) were isolated as a pure lipid class. Without any fractionation these products were oxidized by periodate after hydrolysis of the phosphoric acid esters. Both the periodate uptake and the resulting formaldehyde indicated that the glyceryl ethers were totally α-derivatives, and it was concluded that the plasmalogens, at least the ethanolamine plasmalogens from brain, possess formula (**3**). This was confirmed by the infrared spectrum of the glyceryl ethers.

B. Occurrence and Physiology

Since 1924 there have been many histochemical studies made on the occurrence of plasmalogens, more or less semiquantitative in nature. It was shown that plasmalogens exist all over the animal kingdom (Imhäuser, 1927). Some investigators even reported their occurrence in plant tissue, for example, in soy beans (Feulgen *et al.*, 1949), peanuts (Lovern, 1952), green peas (Feulgen *et al.*, 1949; Wagenknecht, 1957), olive oil (Thiele, 1953a), and in tubercle bacteria (Böhm, 1952). If indeed there are any plasmalogens in plants, the amount is very small. (See Chapter XVI.)

When one compares the different modern estimations of plasmalogen content to the old figures, one must realize that the values found depend somewhat on the method of estimation. Furthermore, the old values were usually calculated on the basis of a molecular weight far too low for plasmalogens, namely the 451 called for by the structure of Feulgen's acetalphosphatide (**1**). The plasmalogens as we now know them to be have a molecular weight of approximately 750.

In some animal tissues, like egg yolk, the plasmalogens comprise a very small part of the total phospholipids. Feulgen *et al.* (1949) found that the plasmalogen content of egg yolk, originally about 1%, rises during the development of the embryonic chicken. This was confirmed by Thiele (1955a). On the other hand, some cells, like beef heart muscle (Hartree and

TABLE I

Distribution of Plasmalogens in Human Brain[a,b]

Ganglion-cell nuclei	Striatum	Cortex	Hypo-thalamus	Thalamus	White matter	Myelin sheaths	Pallidum
0.2	2.62	2.65	2.97	3.98	5.86	10.13[c]	4.59

[a] From Stammler and Debuch (1954) and Debuch and Stammler (1956).
[b] Values are given in percent of dry weight.
[c] Calculated according to Brante (1949); see also Stammler and Debuch (1954).

Mann, 1960a) and ram spermatozoa (Boguth, 1952; Lovern *et al.*, 1957; Hartree and Mann, 1958), are rich in plasmalogens. Hartree and Mann (1960b) thought that plasmalogens take a special part in the energy source of spermatozoa. One tissue that is particularly rich in plasmalogens is brain (Imhäuser, 1927; Feulgen and Bersin, 1939). This has been further brought out by histochemical studies (Stammler, 1952), as well as by methods estimating the amount of plasmalogens in different parts of human brain (Stammler and Debuch, 1954; Debuch and Stammler, 1956; Erickson and Lands, 1959; Webster, 1960) (Table I).

Investigators have long tried to find physiological properties for plasmalogens. The reader is referred to some older publications on tissue aldehydes and their relation to drugs (Oster and Pellett, 1955), the influence of hormones on plasmalogen levels in rat kidney (Oster and Oster, 1946; Oster, 1947), and the reactions of plasmalogens with amines (Oster and Mulinos, 1944). Hajdu *et al.* (1957) identified β-palmitoyl lysolecithin in mammalian blood. It had digitalis-like activity and they believed it originated from a choline plasmalogen precursor. Vogt's (1956) data indicated that the so-called "Darmstoff," a smooth muscle-stimulating substance from gastrointestinal tissue, was a phosphatidic acid-like compound containing an acetal group. Do plasmalogens perhaps play a role in pathological conditions? Buddecke and Andresen (1959) investigated the plasmal content in human aortas and found that arteriosclerotic aortas contained much less than normals, and Hack and Ferrans (1960) observed a nitrogen-free plasmalogen in infarcted myocardium of dog.

Ever since plasmalogens had been detected, clinicians were looking for their biological function in man. Everyone hoped to find a relationship between plasmalogen levels in blood or serum and disease, and there was much discussion and investigation along this line. For those readers who are interested, publications related to this matter are listed in the Appendix to this chapter. Since the aldehyde group in plasmalogens is acid-labile, it is not surprising that Voit (1925) found that gastric juice acted on these lipids, and that Thiele (1955b) ascribed this function to the hydrogen ion concentration. There were some investigations of the plasmalogen levels in serum after feeding meals rich in plasmalogen (Feulgen *et al.*, 1928; Portwich *et al.*, 1954) and after feeding other diets (Schäfer and Taubert, 1951) and in patients with liver diseases (Thiele, 1953b). Injections of plasmal directly into the veins seemed to be metabolized quite quickly (Feulgen *et al.*, 1928; Leupold and Büttner, 1953b). However, no clear picture of the role played by plasmalogens in the organism emerged before 1960.

II. Alkoxylipids

A. Detection, Occurrence, and Investigation of Structure

In 1909 Dorée gave what was possibly the first hint that, in addition to the higher alcohols such as cetyl alcohol ($C_{16}H_{33}OH$), another type of nonsterol alcohol occurs in the unsaponifiable lipids of the starfish *Asterias rubens* (Dorée, 1909). Although he published no real proof for the existence of alkoxy lipids, one can assume that the author had a crude mixture of the glyceryl ethers in his hands. However, Kossel and Edlbacher (1915) obviously obtained a rather pure preparation of batyl alcohol. They extracted pieces of testes and intestinal sacs of the starfish *Astropecten aurantiacus* twice, first with ethanol and then with ether. The lipid extracted by ether was saponified by refluxing with ethanolic KOH. The unsaponifiable fraction was extracted from the soaps by ether. Purification was achieved by recrystallization from hot ethanol after addition of some hot water; these crystals melted at 71°C. Heating with acetic anhydride produced an acetate, with a lower melting point. The nonacetylated compound produced a faint precipitate with digitonin and gave no positive Liebermann-Burchard or Salkowski reactions. A molecular weight of 372 and a molecular formula $C_{23}H_{48}O_3$ were calculated from the elemental organic analysis and the effect of the compound on the boiling point of acetone. Kossel and Edlbacher (1915) called this compound "astrol" and assumed that it was an alcohol originally esterified in the starfish tissues and barely soluble in ethanol. For many years no one noticed Kossel and Edlbacher's communication in connection with alkoxy lipids. Bergmann and Stansbury (1943), however, prepared batyl alcohol from starfish and compared its properties to the data of Kossel and Edlbacher as well as to a sample of *d*-α-(*n*-octadecyl)glyceryl ether synthesized by Baer and Fischer (1941). There is really no doubt that "astrol" and batyl alcohol are identical.

In the 1920s, Tsujimoto (who had detected squalene in 1916) and Toyama were among those who had the most intimate knowledge of the unsaponifiable compounds of liver oils in the world. They (Tsujimoto and Toyama, 1922a, b, c) started a thorough investigation of some alcoholic compounds, different from sterols, that were said to occur in the liver oils of elasmobranch fishes. They saponified the liver oil of the Kagurazame (*Hexanchus corinus* Jordan and Gilbert) with ethanolic KOH and extracted the unsaponifiables with ether. No squalene was detected there, and this semiliquid fraction contained only 1.5% cholesterol, as determined by the digitonin method. From the acid number of the liberated fatty acids and from the amount of liberated glycerol they calculated that roughly 30% of the fatty acids must have been esterified to an unknown substance, now present in the unsaponifiable matter. Consequently, there must be higher alcohols with either one or two hydroxyl groups. The high saponification number determined for acetylated unsaponifiable matter made the latter possibility more likely. The unsaponifiable matter was further fractionated by recrystallization from acetone. Finally, they obtained two fractions, one solid and the other liquid at room temperature.

The authors assigned two possible molecular formulas, $C_{20}H_{42}O_3$ and $C_{21}H_{44}O_3$, to the purified solid material, based on elemental organic analysis and the saponification numbers of its acetylated derivative. A molecular weight of 412 was calculated for this derivative from its effect on the freezing point of benzene. The purified liquid material was acetylated. After distillation of the acetylated products, the main fraction was saponified and proved to be a liquid alcohol. From their results, Tsujimoto and Toyama deduced that its molecular formula must be either $C_{20}H_{40}O_3$ or $C_{21}H_{42}O_3$, and the uptake of hydrogen catalyzed by platinum oxide indicated that it had one double bond per mole. The hydrogenated product had properties identical to those of the solid alcohol.

In their first study, Tsujimoto and Toyama found these two dihydroxy alcohols in liver oils from quite a number of different species of the elasmobranch families of rays (Batoidei) and sharks (Selachoidei). They therefore called them batyl alcohol ($C_{20}H_{42}O_3$ or $C_{21}H_{44}O_3$) and selachyl alcohol ($C_{20}H_{40}O_3$ or $C_{21}H_{42}O_3$). The authors could neither decide which were the right formulas nor prove whether the third oxygen atom was contained in an ether bond or in some other form that could not be acetylated.

Two years later, Toyama (1924b) stated that all available data gave better support to 21 than to 20 carbon atoms per mole of batyl or selachyl alcohol. He also separated a third alcohol from crude batyl alcohol preparations; it had a melting point of 61°C and analyses indicated that its formula was $C_{19}H_{40}O_3$. The liver oils of the ratfish (*Chimaera monstrosa*) and some other related fishes contained even higher amounts of this homolog than of batyl alcohol, so Toyama called it chimyl alcohol. He could not detect a lower homolog of selachyl alcohol. Drawing on his knowledge of the roughly quantitative estimates of the distribution of cetyl, octadecyl, and oleyl alcohols in a number of fish liver oils (Toyama, 1924a), he discussed the possibility that batyl alcohol, for example, might be derived from a condensation reaction of octadecyl alcohol and glycerol that formed an ether bond (Toyama 1924b). Similar reactions could be hypothesized from cetyl alcohol and glycerol to chimyl alcohol, and from oleyl alcohol and glycerol to selachyl alcohol. As further support for his view, Toyama (1924b) showed that nonylic acid was produced by the dry distillation (or by oxidative degradation with permanganate) of selachyl acetate as well as of oleyl acetate. Dry distillation of selachyl alcohol yielded some pungent gaseous products at 300°–320°C. However, the author did not mention the possible production of acrolein, although its presence would have indicated the glyceryl ether nature of these alcohols (compare André and Bloch, 1935).

This was a time of great activity in the field of vitamins. While investigating the antirachitic properties of the unsaponifiable lipids from shark livers (the Greenland shark, *Somniosus microcephalus*), Weidemann (1926) obtained highly purified preparations of batyl and selachyl alcohol (taking advantage of the physical properties possessed by the lead salts of selachyl alcohol's diphtalic acid esters after hydrolization), but he erroneously reported the presence of a methoxy group in batyl alcohol.

The true nature of the crucial oxygen atom was described by Heilbron and Owens (1928). They refluxed batyl alcohol with hydroiodic acid and obtained octadecyl iodide. This result was conclusive evidence of a glyceryl ether structure, and these authors did not see the slightest evidence of a methoxy group. Knight (1930) compared chimyl, batyl, and selachyl alcohols [isolated by Drummond and Baker (1929) from cod and shark liver oils] to α-monoglycerides. He used experiments on surface films on water of varying temperatures, and obtained data that favored an α-glyceryl ether structure. G. G. Davies *et al.* (1930) synthesized batyl alcohol and confirmed Heilbron's earlier results with respect to the glyceryl ether structure, but it was not until 1933 that a clear decision was possible on the question of the ether-linked alcohols' position on the glycerol moiety; Adam's (1933) investigation of surface films merely confirmed Knight's (1930) results. However, a new approach to the analysis of compounds with neighboring hydroxyl groups had been developed by Crigee (1931): the oxidative treatment of batyl alcohol with lead tetraacetate to yield "glycolaldehyde octadecylether" and formaldehyde. This reaction was used by W. H. Davies *et al.* (1933) to provide the final proof of the α-glyceryl ether structure [reaction (2)].

$$CH_2OR{-}CHOH{-}CH_2OH \xrightarrow{\text{lead tetraacetate or periodic acid}} CH_2OR{-}CHO + HCHO \qquad (2)$$

Oxidation of neighboring hydroxyl groups

(Crigee, 1931; W. H. Davies *et al.*, 1933; Karnovsky and Rapson, 1946a; Emmerie, 1953; Karnovsky and Brumm, 1955; Malins, 1960)

Contradictory measurements of the optical rotation of glyceryl ethers puzzled several investigators (Toyama, 1924b; W. H. Davies *et al.*, 1933), but the steric configuration of the glycerol moiety remained unexplained. A systematic investigation of the problem was first reported by Toyama and Ishikawa (1938, cited according to Baer and Fischer, 1941). Their results, confirmed by Baer and Fischer (1941) and Fischer and Baer (1941), were that the natural batyl alcohol in chloroform was either weakly dextro- or levorotatory, or neither, depending on its concentration. High concentrations had negative rotation properties, a 10% solution had no rotatory property, and more diluted solutions exhibited positive rotation. Only the derivatives had clear-cut optical activities. In Fischer and Baer's own words (1941): "... the optical activity can be augmented, or perhaps we should say, made apparent, by further substitution, e.g., acetylation and acetonation."

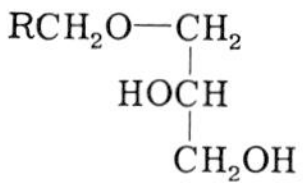

(9)

Steric configuration of naturally occurring glyceryl ethers (corr. after Fischer and Baer, 1941)

For *d*-chimyl alcohol, R represents $C_{15}H_{31}$; for *d*-selachyl alcohol, R represents $C_{17}H_{33}$; and for *d*-batyl alcohol, R represents $C_{17}H_{35}$

Starting with their synthetic, optically active, "acetone glycerol" (Baer and Fischer, 1939), they succeeded in synthesizing pure enantiomeric forms of *d*(+)- and *l*(−)-α-octadecyl- and hexadecylglycerols. By comparing the optical activities of these well-defined synthetic compounds with those of pure biological samples (batyl, chimyl, and selachyl alcohols), Baer and Fischer (1941) and Fischer and Baer (1941) obtained conclusive evidence that the naturally occurring α-glyceryl ethers belong to the *d*-series. Baer (1963) also reported more detailed physical data on α-glyceryl ether.

André and Bloch (1932), like Tsujimoto and Toyama (1922a) before them, were particularly interested in naturally occurring fats, i.e., those present in the tissue before saponification. After calculating acetylation numbers and glycerol:fatty acid ratios of the liver oil of *Scimnorrhinus lichia* Bonnaterre, they suggested that the naturally occurring alcoholic compounds are esterified with fatty acids, and they postulated a diester form. By means of a solvent partition technique, André and Bloch (1935) succeeded in obtaining a fraction from the liver oil mentioned above that contained 74% diester ethers of glycerol, as they had predicted. However, it took 25 years before Mangold and Malins (1960) isolated the diester ethers as an intact class of lipids, using thin-layer chromatographic procedures.

Saturated ether lipids were identified in mammalian tissues after a report on the use of yellow bone marrow in the treatment of leukopenia (agranulocytosis) was published by Giffin and Watkins (1930). Marberg and Wiles (1937) showed that the therapeutic principle could be concentrated in the unsaponifiable fraction of this tissue, and this led to the isolation and identification of batyl alcohol from the unsaponifiable lipid fraction of yellow bone marrow of cattle by Holmes *et al.* (1941). Prelog *et al.* obtained batyl alcohol from pig spleen (1943) and chimyl alcohol from bull and pig testes (1944), while Hardegger *et al.* (1943) isolated batyl alcohol from arteriosclerotic human aortas. These investigators were the first to use column chromatography on activated alumina to fractionate unsaponifiable matter.

Karnovsky and Rapson (1946a) used the reaction initially introduced by Crigee (1931) to develop a practical method for measuring the α-glyceryl ethers in the unsaponifiable matter. The neighboring hydroxy groups of the glycerol moiety were oxidized by periodic acid, producing an ether of glycolaldehyde with a long-chain alcohol and additional formaldehyde, the amount of which was measured gravimetrically as its dimedone derivative. Karnovsky and Brumm (1955) later improved this method by measuring the formaldehyde colorimetrically after reaction with chromotropic acid. The door to future analytical methods in the field of saturated ethers was later opened by Blomstrand and Gürtler (1959), who reported successful gas chromatographic analysis of glyceryl ether acetates, and by Malins

(1960) who succeeded in gas chromatographic analyses of aldehydes derived from the α-glyceryl ethers of dogfish (*Squalus acanthias*) liver lipids by periodic acid oxidation. But in 1946, reliable quantitative measurements had already been obtained for a great variety of animal lipids of marine origin, especially from elasmobranch and teleost fishes (Karnovsky and Rapson, 1946a, b). One of Karnovsky and Rapson's interesting findings was that older animals (sharks) contained higher concentrations of glyceryl ether lipids than did younger individuals, and embryos contained the least of all. However, they were not able to detect measurable amounts of ether lipids in vegetable oils or in human liver. A further series of determinations in a number of animal species was published in 1947 and 1948 (Karnovsky *et al.*, 1947, 1948a, b).

Emmerie (1953) was the first investigator who managed to separate derivatives of the naturally occurring α-glyceryl ether lipids. He oxidized the isolated ether fraction with excess periodic acid and separated the three resulting glycolaldehydes by paper chromatography. However, the method was laborious and difficult.

Karnovsky and Brumm (1955) made the first attempt to investigate the metabolism of the saturated ether lipids. Pieces of diverticula from living starfish (*Asterias rubens*) were incubated with ^{14}C-labeled acetate or glycerol. The precursors were incorporated into the glyceryl ether lipids at a rate comparable to that of other lipid classes. ^{14}C-Radioactivity from acetate appeared only in the fatty alcohol moiety, whereas that from labeled glycerol was found in both glycerol and fatty alcohol. Two Swedish investigators, Bergström and Blomstrand (1956), demonstrated an approach to studying the intestinal absorption and metabolism of chimyl alcohol in the rat. Since about 5% of the radioactivity of 1-^{14}C-labeled hexadecyl glyceryl ether given by stomach tube was recovered in the feces, the intestinal absorption of this compound must be 95%. About 50% of the absorbed activity appeared in the lymph (thoracic duct). Only 3–4% of the total lymph activity was in phospholipids, while about 45% was in the fatty acid moiety of triglycerides and diester ethers, and the remaining 50% was found in the chimyl alcohol, half of it esterified with two fatty acids per mole and half of it free. These results and those of Karnovsky and Brumm (1955) gave the first important indication that the glyceryl ethers are rapidly metabolized in animals.

After the report by Brante (1949) on "cephalin B," additional data were supplied by Dawson (1954), Edgar (1956), and Weiss (1956) on the occurrence of alkali- and acid-stable phospholipids in mammalian tissues. Following mild alkali hydrolysis of egg yolk phosphatides, Carter *et al.* (1958) obtained intact sphingomyelin and a phosphorylethanolamine derivative of batyl alcohol. Svennerholm and Thorin (1960) hydrolyzed polar lipid fractions from calf and human brains, human liver, kidney, and spleen in the same way and isolated the phosphorylethanolamine derivative and traces of the phosphorylcholine derivative of batyl alcohol by Florisil column chromatography. From the chromatographic data, the authors concluded that originally—at least in the phosphorylethanolamine derivative—an alkyl acyl phosphatide group must have been present among the tissue lipids. However, they did not succeed in separating the intact alkyl acyl phosphatide from its diacyl or alkenyl acyl analogs, either by chromatographic or other procedures.

B. Biological Effects

The biological function or the biosynthesis of alkyl glyceryl ether lipids in their two naturally occurring forms, diester ethers and alkyl acyl phosphatides, was not elucidated during the time covered by this chapter, but a considerable number of biological effects were successively ascribed to these lipids during the years 1925 to 1960 (see Chapter VIII). In 1926, Weidemann had already proven that they did *not* possess any antirachitic properties, but there were various reports of other therapeutic effects, such as an antileukopenic effect (Giffin and Watkins, 1930; Marberg and Wiles, 1937), a central depressant action (Berger, 1948), an erythropoietic effect (Sandler, 1949), a tuberculostatic effect (Emmerie *et al.*, 1952), a wound-healing property (Bodman and Maisin, 1958), and a radioprotective effect (Brohult, 1960). These effects were neither proved nor finally refuted before 1960.

But by 1960 the occurrence of alkyl glyceryl ether lipids in animals had been shown to be widespread, one could almost say universal (Deuel, 1951; Hilditch, 1956; Bodman and Maisin, 1958). Plant lipids were also investigated (Karnovsky and Rapson, 1946b), and by 1952 Harrison and Hawke (1952) had found glyceryl ethers in the seeds of *Acacia giraffae* (African Camel's-Thorn) (see Chapter XVI). The first data of Karnovsky and Brumm (1955) on *in vitro* incorporation of precursors and of Bergström and Blomstrand (1956) on *in vivo* absorption of *O*-alkyl lipids gave strong support to the hypothesis that the ether lipids must somehow play a role of physiological significance, but definition of this role and theories on the biosynthesis of ether lipids remained merely speculative (Lovern, 1937; Baer and Fischer, 1941; Fischer and Baer, 1941; Hilditch, 1956). Current knowledge of enzymic systems that synthesize ether bonds in lipids is summarized in Chapter VII.

REFERENCES

Adam, N. K. (1933). *J. Chem. Soc., London* p. 164.
Anchel, M., and Waelsch, H. (1942). *J. Biol. Chem.* **145**, 605.
Anchel, M., and Waelsch, H. (1944). *J. Biol. Chem.* **152**, 501.
André, E., and Bloch, A. (1932). *C. R. Acad. Sci.* **195**, 627.
André, E., and Bloch, A. (1935). *Bull. Soc. Chim. Fr.* [5] **2**, 789.
Ansell, G. B., and Norman, J. M. (1956). *J. Neurochem.* **1**, 32.
Baer, E. (1963). *Progr. Chem. Fats Other Lipids* **6**, 31–86.
Baer, E., and Fischer, H. O. L. (1939). *J. Biol. Chem.* **128**, 463.
Baer, E., and Fischer, H. O. L. (1941). *J. Biol. Chem.* **140**, 397.
Behrens, M. (1930). *Hoppe-Seyler's Z. Physiol. Chem.* **191**, 183.
Berger, F. M. (1948). *J. Pharmacol. Exp. Ther.* **93**, 470 (cited after Karnovsky and Brumm, 1955).
Bergmann, W., and Stansbury, H. A. (1943). *J. Org. Chem.* **8**, 283.
Bergström, S., and Blomstrand, R. (1956). *Acta Physiol. Scand.* **38**, 166.
Blietz, R. (1958). *Hoppe-Seyler's Z. Physiol. Chem.* **310**, 120.
Blomstrand, R., and Gürtler, J. (1959). *Acta Chem. Scand.* **13**, 1466.
Bodman, J., and Maisin, J. H. (1958). *Clin. Chim. Acta* **3**, 253.
Böhm, P. (1952). *Hoppe-Seyler's Z. Physiol. Chem.* **291**, 155.
Böhm, P., and Richarz, G. (1957). *Hoppe-Seyler's Z. Physiol. Chem.* **306**, 201.
Boguth, W. (1952). *Naturwissenschaften* **18**, 432.
Brante, G. (1949). *Acta Physiol. Scand.* **18**, Suppl. 63, 189.
Brohult, A. (1960). *Nature (London)* **188**, 591.
Buddecke, E. A., and Andresen, G. (1959). *Hoppe-Seyler's Z. Physiol. Chem.* **314**, 38.
Carter, H. E., Smith, D. B., and Jones, D. N. (1958). *J. Biol. Chem.* **232**, 681.
Crigee, R. (1931). *Ber. Deut. Chem. Ges.* **64**, 260.
Davies, G. G., Heilbron, I. M., and Owens, W. M. (1930). *J. Chem. Soc., London* p. 2542.
Davies, W. H., Heilbron, I. M., and Jones, W. E. (1933). *J. Chem. Soc., London* p. 165.
Dawson, R. M. C. (1954). *Biochem. J.* **56**, 621.
Debuch, H. (1956). *Hoppe-Seyler's Z. Physiol. Chem.* **304**, 109.
Debuch, H. (1957). *Biochem. J.* **67**, 27P.
Debuch, H. (1958a). *Hoppe-Seyler's Z. Physiol. Chem.* **311**, 266.
Debuch, H. (1958b). *J. Neurochem.* **2**, 243.
Debuch, H. (1959a). *Hoppe-Seyler's Z. Physiol. Chem.* **314**, 49.
Debuch, H. (1959b). *Hoppe-Seyler's Z. Physiol. Chem.* **317**, 182.
Debuch, H., and Stammler, A. (1956). *Hoppe-Seyler's Z. Physiol. Chem.* **305**, 111.
Deuel, H. J., Jr. (1951). "The Lipids. Their Chemistry and Biochemistry," Vol. I. Wiley (Interscience), New York.
Dorée, C. (1909). *Biochem. J.* **4**, 72.
Drummond, J. C., and Baker, L. C. (1929). *Biochem. J.* **23**, 274.
Edgar, G. W. F. (1956). *Acta Anat.* **27**, 240.
Ehrlich, G., Taylor, H. E., and Waelsch, H. (1948). *J. Biol. Chem.* **173**, 547.
Emmerie, A. (1953). *Rec. Trav. Chim. Pays-Bas* **72**, 893.
Emmerie, A., Engel, C., and Klip, W. (1952). *J. Sci. Food Agr.* **3**, 264.
Erickson, N. E., and Lands, W. E. M. (1959). *Proc. Soc. Exp. Biol. Med.* **102**, 512.
Feulgen, R., and Bersin, T. (1939). *Hoppe-Seyler's Z. Physiol. Chem.* **260**, 217.
Feulgen, R., and Grünberg, H. (1939). *Hoppe-Seyler's Z. Physiol. Chem.* **257**, 90.
Feulgen, R., and Imhäuser, K. (1927). *Biochem. Z.* **181**, 30.

Feulgen, R., and Voit, K. (1924). *Pfluegers Arch. Gesamte Physiol. Menschen Tiere* **206,** 389.
Feulgen, R., Imhäuser, K., and Westhues, M. (1928). *Biochem. Z.* **193,** 251.
Feulgen, R., Imhäuser, K., and Behrens, M. (1929). *Hoppe-Seyler's Z. Physiol. Chem.* **180,** 161.
Feulgen, R., Feller, P., and Andresen, G. (1949). *Ber. Oberhess. Ges. Nat.-Heilk.* **24,** 270.
Feulgen, R., Boguth, W., and Andresen, G. (1951). *Hoppe-Seyler's Z. Physiol. Chem.* **287,** 90.
Feulgen, R., Boguth, W., and Andresen, G. (1953a). *Hoppe-Seyler's Z. Physiol. Chem.* **293,** 79.
Feulgen, R., Boguth, W., Seydl, G., and Andresen, G. (1953b). *Hoppe-Seyler's Z. Physiol. Chem.* **295,** 271.
Fischer, H. O. L., and Baer, E. (1941). *Schweiz. Med. Wochenschr.* **71,** 321.
Folch, J. (1942). *J. Biol. Chem.* **146,** 35.
Franzl, R. E., and Rapport, M. M. (1955). *Fed. Proc., Fed. Amer. Soc. Exp. Biol.* **14,** 213.
Giffin, H. Z., and Watkins, C. H. (1930). *J. Amer. Med. Ass.* **95,** 587.
Gray, G. M. (1957). *Biochem. J.* **67,** 25P.
Gray, G. M. (1958). *Biochem. J.* **70,** 425.
Hack, M. H. (1952). *Anat. Rec.* **112,** 275.
Hack, M. H. (1953). *Biochem. J.* **54,** 602.
Hack, M. H., and Ferrans, V. J. (1958). *Fed. Proc., Fed. Amer. Soc. Exp. Biol.* **17,** 235.
Hack, M. H., and Ferrans, V. J. (1959). *Hoppe-Seyler's Z. Physiol. Chem.* **315,** 157.
Hack, M. H., and Ferrans, V. J. (1960). *Circ. Res.* **4,** 738.
Hajdu, S., Weiss, H., and Titus, E. (1957). *J. Pharmacol. Exp. Ther.* **120,** 99.
Hanahan, D. J. (1954). *J. Biol. Chem.* **207,** 879.
Hardegger, E., Ruzicka, L., and Tagman, E. (1943). *Helv. Chim. Acta* **26,** 2205.
Harrison, G. S., and Hawke, F. (1952). *J. S. Afr. Chem. Inst.* **5,** 1.
Hartree, E. F., and Mann, T. (1958). *Biochem. J.* **69,** 50.
Hartree, E. F., and Mann, T. (1960a). *Biochem. J.* **75,** 251.
Hartree, E. F., and Mann, T. (1960b). *J. Reprod. Fert.* **1,** 23.
Heilbron, I. M., and Owens, W. M. (1928). *J. Chem. Soc., London* p. 942.
Hilditch, T. P. (1956). "The Chemical Constitution of Natural Fats." Chapman & Hall, London.
Holmes, H. N., Corbet, R. E., Geiger, W. B., Kornblum, N., and Alexander, W. (1941). *J. Amer. Chem. Soc.* **63,** 2607.
Imhäuser, K. (1927). *Biochem. Z.* **186,** 360.
Karnovsky, M. L., and Brumm, A. F. (1955). *J. Biol. Chem.* **216,** 689.
Karnovsky, M. L., and Rapson, W. S. (1946a). *J. Soc. Chem. Ind., London* **65,** 138.
Karnovsky, M. L., and Rapson, W. S. (1946b). *J. Soc. Chem. Ind., London* **65,** 425.
Karnovsky, M. L., Rapson, W. S., and Black, M. (1947). *J. Soc. Chem. Ind., London* **66,** 95.
Karnovsky, M. L., Rapson, W. S., Schwartz, H. M., Black, M., and van Rensburg, N. J. (1948a). *J. Soc. Chem. Ind., London* **67,** 104.
Karnovsky, M. L., Lategan, A. W., Rapson, W. S., and Schwartz, H. M. (1948b). *J. Soc. Chem. Ind., London* **67,** 193.
Klenk, E. (1944). *Hoppe-Seyler's Z. Physiol. Chem.* **281,** 25.
Klenk, E. (1945). *Hoppe-Seyler's Z. Physiol. Chem.* **282,** 18.
Klenk, E. (1952). *Colloq. Ges. Physiol. Chem.* **3,** 27.

Klenk, E., and Böhm, P. (1951). *Hoppe-Seyler's Z. Physiol. Chem.* **288,** 98.
Klenk, E., and Debuch, H. (1954). *Hoppe-Seyler's Z. Physiol. Chem.* **296,** 179.
Klenk, E., and Debuch, H. (1955). *Hoppe-Seyler's Z. Physiol. Chem.* **299,** 66.
Klenk, E., and Debuch, H. (1963). *Progr. Chem. Fats Other Lipids* **6,** 3–29.
Klenk, E., and Gehrmann, G. (1953). *Hoppe-Seyler's Z. Physiol. Chem.* **292,** 110.
Klenk, E., and Krickau, G. (1957). *Hoppe-Seyler's Z. Physiol. Chem.* **308,** 98.
Klenk, E., Stoffel, W., and Eggers, H. J. (1952). *Hoppe-Seyler's Z. Physiol. Chem.* **290,** 246.
Knight, B. C. J. G. (1930). *Biochem. J.* **24,** 257.
Korey, S. R., and Wittenberg, J. (1954). *Fed. Proc., Fed. Amer. Soc. Exp. Biol.* **13,** 244.
Kossel, A., and Edlbacher, S. (1915). *Hoppe-Seyler's Z. Physiol. Chem.* **94,** 277.
Leupold, F. (1950a). *Hoppe-Seyler's Z. Physiol. Chem.* **285,** 182.
Leupold, F. (1950b). *Hoppe-Seyler's Z. Physiol. Chem.* **285,** 216.
Leupold, F., and Büttner, H. (1952). *Hoppe-Seyler's Z. Physiol. Chem.* **291,** 178.
Leupold, F., and Büttner, H. (1953a). *Hoppe-Seyler's Z. Physiol. Chem.* **292,** 13.
Leupold, F., and Büttner, H. (1953b). *Verh. Deut. Ges. Inn. Med.* **59,** 210.
Leupold, F., Büttner, H., and Ranniger, K. (1954). *Hoppe-Seyler's Z. Physiol. Chem.* **294,** 107.
Lovern, J. A. (1937). *Biochem. J.* **31,** 755.
Lovern, J. A. (1952). *Nature (London)* **169,** 969.
Lovern, J. A., Olley, J., Hartree, E. F., and Mann, T. (1957). *Biochem. J.* **67,** 630.
Malins, D. (1960). *Chem. Ind. (London)* p. 1359.
Mangold, H. K., and Malins, D. C. (1960). *J. Amer. Oil Chem. Soc.* **37,** 383.
Marberg, C. M., and Wiles, H. O. (1937). *J. Amer. Med. Ass.* **109,** 1965.
Marinetti, G. V., and Erbland, J. (1957). *Biochim. Biophys. Acta* **26,** 429.
Marinetti, G. V., Erbland, J., and Stotz, E. (1958). *J. Amer. Chem. Soc.* **80,** 1624.
Marinetti, G. V., Erbland, J., and Stotz, E. (1959a). *J. Amer. Chem. Soc.* **81,** 861.
Marinetti, G. V., Erbland, J., and Stotz, E. (1959b). *Biochim. Biophys. Acta* **33,** 403.
Oster, K. A. (1947). *Exp. Med. Surg.* **5,** 219.
Oster, K. A., and Mulinos, M. G. (1944). *J. Pharmacol. Exp. Ther.* **80,** 132.
Oster, K. A., and Oster, J. G. (1946). *J. Pharmacol. Exp. Ther.* **87,** 306.
Oster, K. A., and Pellett, C. A. (1955). *Exp. Med. Surg.* **13,** 385.
Portwich, F., Leupold, F., and Büttner, H. (1954). *Deut. Z. Verdau.-Stoffwechselkr.* **14,** 174.
Prelog, V., Ruzicka, L., and Steinmann, F. (1943). *Helv. Chim. Acta* **26,** 2222.
Prelog, V., Ruzicka, L., and Steinmann, F. (1944). *Helv. Chim. Acta* **27,** 674.
Rapport, M. M., and Alonzo, N. (1955). *J. Biol. Chem.* **217,** 199.
Rapport, M. M., and Franzl, R. E. (1957a). *J. Neurochem.* **1,** 303.
Rapport, M. M., and Franzl, R. E. (1957b). *J. Biol. Chem.* **225,** 851.
Rapport, M. M., Lerner, B., and Alonzo, N. (1954). *Fed. Proc., Fed. Amer. Soc. Exp. Biol.* **13,** 278.
Rapport, M. M., Lerner, B., Alonzo, N., and Franzl, R. E. (1957). *J. Biol. Chem.* **225,** 859.
Sandler, O. E. (1949). *Acta Med. Scand.* **133,** Suppl. 225, 72.
Schäfer, G., and Taubert, M. (1951). *Arzneim.-Forsch.* **5,** I 97.
Schmidt, G., Ottenstein, B., Zöllner, N., and Thannhauser, S. J. (1951). *Abstr. Pap., Int. Congr. Pure Appl. Chem., 12th, 1951* p. 104.
Schmidt, G., Ottenstein, B., and Bessman, M. J. (1953). *Fed. Proc., Fed. Amer. Soc. Exp. Biol.* **12,** 265.

Stammler, A. (1952). *Deut. Z. Nervenheilk.* **168,** 305.
Stammler, A., and Debuch, H. (1954). *Hoppe-Seyler's Z. Physiol. Chem.* **296,** 80.
Stepp, W., Feulgen, R., and Voit, K. (1927). *Biochem. Z.* **181,** 284.
Svennerholm, L., and Thorin, H. (1960). *Biochim. Biophys. Acta* **41,** 371.
Thannhauser, S. J., Boncoddo, N. F., and Schmidt, G. (1951a). *J. Biol. Chem.* **188,** 417.
Thannhauser, S. J., Boncoddo, N. F., and Schmidt, G. (1951b). *J. Biol. Chem.* **188,** 423.
Thiele, O. W. (1953a). *Z. Exp. Med.* **121,** 246.
Thiele, O. W. (1953b). *Klin. Wochenschr.* **31,** 907.
Thiele, O. W. (1955a). *Hoppe-Seyler's Z. Physiol. Chem.* **299,** 151.
Thiele, O. W. (1955b). *Hoppe-Seyler's Z. Physiol. Chem.* **300,** 131.
Thiele, O. W. (1955c). *Hoppe-Seyler's Z. Physiol. Chem.* **300,** 157.
Toyama, Y. (1924a). *Chem. Umsch. Geb. Fette, Oele, Wachse Harze* **31,** 13.
Toyama, Y. (1924b). *Chem. Umsch. Geb. Fette, Oele, Wachse Harze* **31,** 61.
Toyama, Y., and Ishikawa, T. (1938). *J. Chem. Soc. Jap.* **59,** 1367.
Tsujimoto, M. (1916). *J. Chem. Ind. Jap.* p. 227.
Tsujimoto, M., and Toyama, Y. (1922a). *Chem. Umsch. Geb. Fette, Oele, Wachse Harze* **29,** 27.
Tsujimoto, M., and Toyama, Y. (1922b). *Chem. Umsch. Geb. Fette, Oele, Wachse Harze* **29,** 37.
Tsujimoto, M., and Toyama, Y. (1922c). *Chem. Umsch. Geb. Fette, Oele, Wachse Harze* **29,** 43.
Vogt, W. (1956). *J. Physiol. (London)* **133,** 64P.
Voit, K. (1925). *Z. Biol.* **83,** 223.
Waelsch, H. (1951). *Hoppe-Seyler's Z. Physiol. Chem.* **288,** 123.
Wagenknecht, A. C. (1957). *Science* **126,** 1288.
Webster, G. R. (1960). *Biochim. Biophys. Acta* **44,** 109.
Weidemann, G. (1926). *Biochem. J.* **20,** 685.
Weiss, B. (1956). *J. Biol. Chem.* **223,** 523.
Wittenberg, J. B., Korey, S. R., and Swenson, F. H. (1956). *J. Biol. Chem.* **219,** 31.

APPENDIX—BIOMEDICAL APPLICATIONS

Braun-Falco, O., Theisen, H., and Seckfort, H. (1958). *Klin. Wochenschr.* **36,** 763.
Christl, H. (1953). *Hoppe-Seyler's Z. Physiol. Chem.* **293,** 83.
Franke, R., Thiele, F. H., and Keller, N. (1953). *Z. Exp. Med.* **120,** 422.
Graffi, A. (1944). *Arch. Exp. Zellforsch. Besonders Gewebezuecht.* **25,** 127.
Graffi, A. (1949). *Arch. Geschwulstforsch.* **1,** 69.
Hornykiewytsch, T. (1952). *Strahlentherapie* **86,** 175.
Hornykiewytsch, T., and Seydl, G. (1952). *Strahlentherapie* **88,** 129.
Hornykiewitsch, T., Seydl, G., and Thiele, O. W. (1954). *Strahlentherapie* **95,** 523.
Klenk, E. (1945). *Hoppe-Seyler's Z. Physiol. Chem.* **282,** 18.
Klenk, E., and Friedrichs, E. (1952). *Hoppe-Seyler's Z. Physiol. Chem.* **290,** 169.
Knick, B. (1953). *Verh. Deut. Ges. Inn. Med.* **59,** 304.
Knick, B., Severin, G., and Tilling, W. (1953). *Arzneim.-Forsch.* **7,** I 168.
Kroczek, H. (1940). *Z. Mikrosk.-Anat. Forsch.* **50,** 511.
Lapp, H. (1951). *Verh. Deut. Pathol. Ges.* **35,** 145.
Leupold, F., and Büttner, H. (1953). *Verh. Deut. Ges. Inn. Med.* **59,** 210.

Leupold, F., Büttner, H., and Ranniger, K. (1954a). *Hoppe-Seyler's Z. Physiol Chem.* **294,** 107.

Leupold, F., Büttner, H., and Ranniger, K. (1954b). *Klin. Wochenschr.* **32,** 745.

Schäfer, G., and Taubert, M. (1950). *Arzneim.-Forsch.* **4,** I 593.

Seckfort, H. (1953). *Verh. Deut. Ges. Inn. Med.* **59,** 212.

Seckfort, H. (1954). *Verh. Deut. Ges. Inn. Med.* **60,** 967.

Seckfort, H., Busanny-Caspari, W., and Andres, E. (1956). *Klin. Wochenschr.* **34,** 464.

Seckfort, H., Busanny-Caspari, W., and Andres, E. (1957a). *Klin. Wochenschr.* **35,** 295.

Seckfort, H., Busanny-Caspari, W., and Andres, E. (1957b). *Klin. Wochenschr.* **35,** 980.

Seckfort, H., Busanny-Caspari, W., and Andres, E. (1958). *Klin. Wochenschr.* **36,** 434.

Thiele, O. W. (1953). *Klin. Wochenschr,* **31,** 907.

Thiele, O. W. (1954). *Z. Exp. Med.* **123,** 65.

Thiele, O. W. (1955). *Z. Exp. Med.* **125,** 136.

Tonutti, E. (1941). *Klin. Wochenschr.* **19,** 1196.

CHAPTER II

ETHER-LINKED LIPIDS: CHEMISTRY AND METHODS OF MEASUREMENT

Donald J. Hanahan

I. Introduction

The ether bond, C—O—C, is widespread in nature and occurs in many biologically important compounds. Some representative ether-containing

compounds are

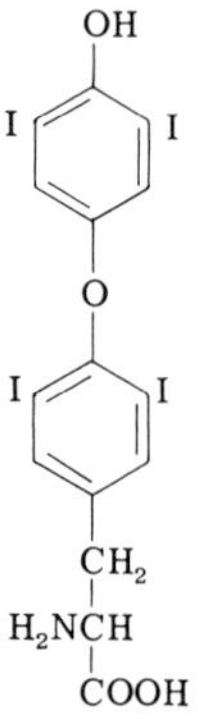

Thyroxine

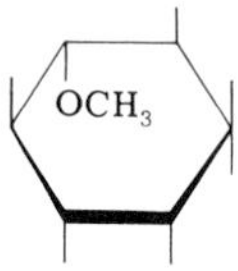

5-*O*-Methyl-*myo*-inositol (sequoyitol)

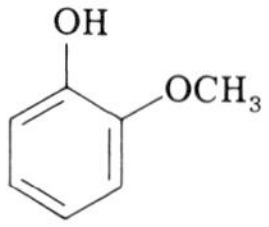

O-Methoxyphenol (guaicol)

In the past several years, increasing attention has been paid to one group of C—O—C-containing compounds, the ether-containing lipids. These ethers may be divided into two major classes, the alkyl glyceryl ethers, (**1**) and (**2**), and the alk-1-enyl glyceryl ethers, (**3**):

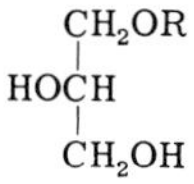

(**1**)

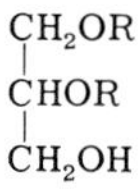

(**2**)

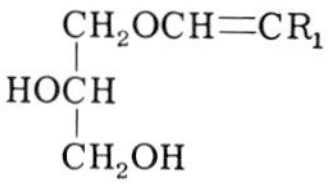

(**3**)

R designates long-chain hydrocarbon

The isolation (and identification) of the alkyl glyceryl ethers (**1**) and dialkyl glyceryl ethers (**2**), which are regarded chemically as saturated ethers, from their more complex parent phospholipids and glycerides, has been accomplished with relative ease. On the other hand, the isolation of the alk-1-enyl glyceryl ethers (**3**) has posed some problems because of the instability of the molecule, but this difficulty has now been largely circumvented.

On the basis of present evidence we think these ethers do not exist naturally in the free form, but that they are found as esterified derivatives:

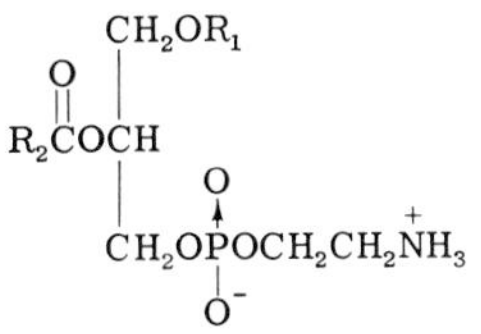

(4)

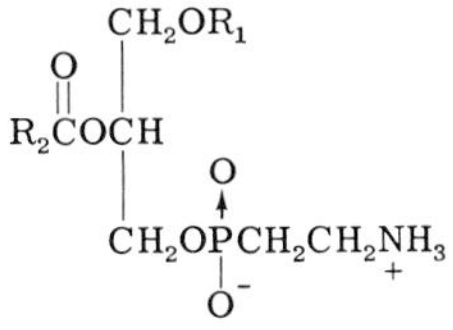

(5)

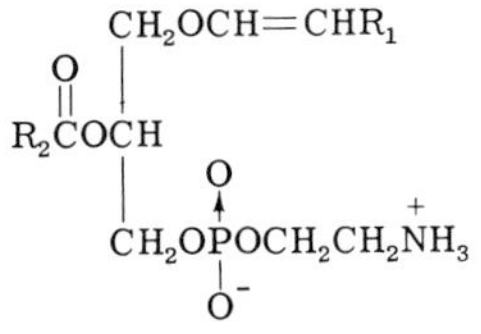

(6)

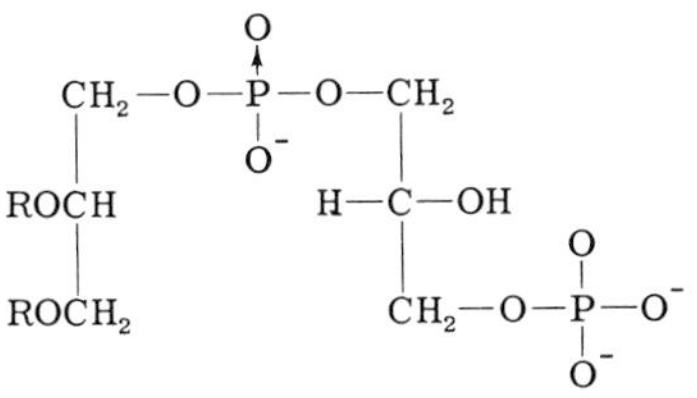

(7)

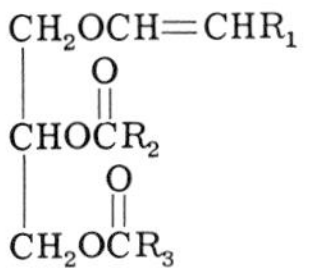

(8)

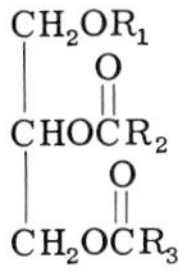

(9)

Structures (**4**) and (**6**) represent the widespread phosphodiester (C—O—P) derivatives, and (**8**) and (**9**) are the neutral lipid analogs. Structure (**5**) describes the recently discovered ether-containing phosphonolipid (C—P) found in protozoa, and structure (**7**) is the unusual diether phospholipid found in an extremely halophilic bacterium. It is of considerable interest that in some tissues the glyceryl ethers are found almost exclusively in the ethanolamine-containing phospholipids, whereas the alk-1-enyl glyceryl ethers are found in the choline- as well as in the ethanolamine-containing fractions.

Glyceryl ethers are quite widespread and have been isolated from diverse locations. A few representative sources of alkyl and alk-1-enyl glyceryl ethers are given in Table I. Investigations have indicated the natural occurrence of polyols other than glycerol, such as the ethanediols, propanediols, and related derivatives. Although a true saturated ether derivative of the diols has not been detected, it is probably only a matter of time before this novel compound is found. The occurrence of alk-1-enyl ether derivatives

TABLE I

Typical Sources of Alkyl and Alk-1-enyl Glyceryl Ethers

Class of ether	Source	References
Alkyl glyceryl ether (**1**)	Bovine erythrocytes	Hanahan *et al.*, 1963
	Bovine red marrow	Thompson and Hanahan, 1962
	Tetrahymena pyriformis	Thompson, 1967
	Neoplasms	Snyder and Wood, 1968, 1969
Dialkyl glyceryl ether (**2**)	*Halobacterium cutirubrum*	Sehgal *et al.*, 1962
Alk-1-enyl glyceryl ether (**3**)	Cardiac muscle	Thiele *et al.*, 1960
	Mammalian erythrocytes	Dawson *et al.*, 1960
	Shark liver	Schmid *et al.*, 1967
	Human perinephric fat	Schmid and Mangold, 1966
	Brain	Wells and Dittmer, 1967
		Horrocks, 1968
	Neoplasms	Snyder and Wood, 1968, 1969

of diols is noted by Bergelson (1969) in his excellent review on the diol lipids. Similarly, Snyder (1969) has reviewed in a detailed manner the biochemistry of the glyceryl ether-containing lipids identified before that date.

A further examination of ether-containing compounds found in natural sources reveals the presence of thioethers. Representative structures of this class of important compounds are

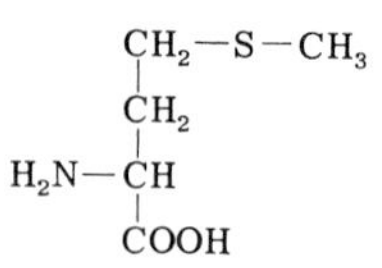

Methionine

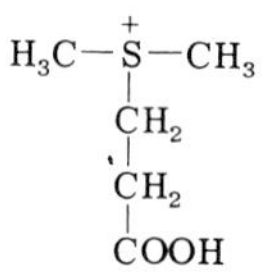

Dimethyl propiothetin

$CH_3C{-}S{-}CH_3$

Dimethyl sulfide

Recently, glyceryl thioethers were found in the lipids of bovine heart (Ferrell and Radloff, 1970). Prior to this report, glyceryl thioethers had been synthesized in the laboratory (Piantodosi *et al.*, 1969), analyzed (Wood *et al.*, 1969b), and used in metabolic studies (Snyder *et al.*, 1969). Quantitative analysis of these thioethers can be achieved by gas–liquid chromatography (Wood *et al.*, 1969a), if a preliminary separation of the glyceryl thio from the glyceryl (*O*-alkyl) ethers is done by thin-layer chromatography on Silica Gel G. The nonlipid thioethers above are of considerable chemical and biochemical interest; reference is made to a review article by Wallenfels and Diekmann (1967).

It is the intent in this chapter to dwell mainly on the chemistry of the *O*-ether bonds as found in glycerolipids and on the possible influence of the contiguous hydroxyl groups on their reactivity and then to translate this information to methods applicable for their estimation. Primary attention will be devoted to the alkyl glyceryl ethers (**1**) and the alk-1-enyl glyceryl ethers (**2**). Chapters IV and XV discuss the dialkyl glyceryl diether (**3**).

II. Alkyl Glyceryl Ethers (1-Alkylglycerols)

The alkyl glyceryl ethers are comparable in their reactivity to the classic aliphatic ether, R_1OR_2. Although the presence of the contiguous hydroxyl groups does provide a convenient function with which to further characterize the ether, there is little or no evidence suggesting that the hydroxyl function modifies the chemical activity of the ether bond per se.

The methods for isolating the alkyl glyceryl ethers from either phospholipids or neutral lipids include (a) lithium aluminum hydride cleavage of carboxylate and phosphate esters (Thompson, 1965); (b) acetolysis, wherein the phosphate ester is replaced by an acetoxy group but carboxylate esters are unaffected (Bevan *et al.*, 1953; Renkonen, 1967); and (c) alkaline methanolysis, wherein the carboxylate esters are readily attacked but the phosphate esters remain intact (Hanahan *et al.*, 1963). Depending on the nature of the experiment and the problem at hand one could obviously employ any one or all of these techniques. However, the best general method for the release of glyceryl monoethers is the lithium aluminum hydride reaction. The reaction is rapid, smooth, and has no apparent effect on the ether bond in either the alkyl glyceryl ethers or the alk-1-enyl glyceryl ethers (Wood and Snyder, 1967). However, it is not quantitative for alk-1-enyl lipids (Wykle *et al.*, 1970); see Chapter IX.

A. General Reactivity

The C—O—C bond in alkyl glyceryl ethers is comparatively unreactive, as would be expected from its aliphatic structure. This bond is stable to bases, and to most oxidizing and reducing agents. Further, it is well known that it undergoes only one type of reaction, namely, cleavage by acid:

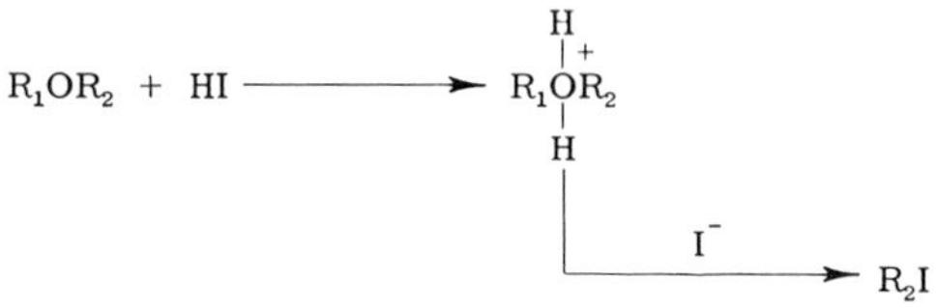

Thus, nucleophilic attack by a halide ion on the protonated ether occurs with formation of an alkyl halide. The alkyl glyceryl ethers, like simpler ethers, are more easily and readily cleaved by hydroiodic acid (HI) than by hydrobromic acid (HBr) or hydrochloric acid (HCl). The formation of the alkyl iodides provides a basis for the identification of alkyl glyceryl ethers containing unsaturated (olefinic) hydrocarbon chains (Guyer *et al.*, 1963; Hanahan, 1965), since hydrogen iodide can attack either side of the double bond with formation of isomeric products (Hanahan, 1965).

B. Qualitative Characterization

Inasmuch as chemical reactions of the C—O—C bond in the alkyl glyceryl ethers are limited almost entirely to the acid-cleavage reaction cited above, physical characterization and derivative formation utilizing the hydroxyl functions are of particular value. Perhaps the most important approach to proof of structure of the naturally occurring alkyl glyceryl ethers is through comparison with chemically synthesized compounds (see Chapter III). Briefly, basically all synthetic approaches reported to date have been modifications of the classic Williamson reaction involving an alkyl halide and a metal alkoxide. This technique was utilized by Baer and his associates in their synthetic program designed to prepare optically active alkyl glyceryl ethers (Baer *et al.*, 1944; Baer and Fischer, 1947). Whereas the L-, or 3-*O*-, form of the ethers can be synthesized from commonly available starting materials, synthesis of the D-, or 1-*O*-, configuration requires starting material that is quite rare or difficult to obtain. However, this problem was circumvented by utilizing a basic modification of the procedure described by Lands and Zschocke (1965), which involved the conversion of 3-*O*-benzylglycerol to 1-*O*-benzylglycerol. L-(3-*O*)-Alkyl glyceryl ethers were used as the starting material (Chacko and Hanahan, 1968). The ditosylates of the L-(3-*O*) ethers were prepared and then subjected to a displacement reaction with freshly fused potassium acetate in anhydrous ethanol. Subsequent alkaline hydrolysis allowed release of the desired D-(1-*O*)-alkyl glyceryl ether in good yields. Certain characteristics of the synthetic and naturally occurring alkyl glyceryl ethers are illustrated in Table II. Further detailed characterization of the alkyl glyceryl ethers has been published by Wood *et al.* (1969b).

As mentioned, the free hydroxyl functions on the alkyl glyceryl ethers allow a number of derivative possibilities. These derivatives (Table III) have been of most use for identification purposes, especially in gas–liquid chromatography (GLC). Of the reactions presented in Table III, the periodic acid reaction has been particularly useful in establishing the position of the ether linkage on the glycerol backbone. The primary product, alkoxyacetaldehyde ($ROCH_2CHO$), can be subjected directly to GLC or can first be converted by lithium aluminum hydride to the alkoxyethanols, $ROCH_2CH_2OH$. W. J. Baumann and co-workers (1967) synthesized a series of the alkoxyethanols (ethanediol derivatives) with the hydrocarbon side chain of the ether varying from 12 to 20 carbons. They also prepared the *cis*- and *trans*-9-octadecenyloxy and the *cis-cis*-9,12-octadecadienyloxy derivatives, and described various physical characteristics of all these compounds in detail. In addition, they presented the synthesis and properties of several long-chain acyl derivatives of the ethers mentioned above. In a comparable study, Wood and Baumann (1968) reported the

TABLE II

Physical Characteristics of Certain Alkylglycerols

Compound	Source	MP (°C)	D	References
D-1-Octadecylglycerol	Chemical synthesis (*de novo*)	71°–72°	−12.6°[a]	Baer and Fischer, 1947
L-3-Octadecylglycerol	Chemical synthesis (*de novo*)	71°–72°	+12.4°[a]	
D-1-Octadecylglycerol	Hydrogenation of naturally occurring selachyl alcohol	70.5°–71.0°	−15.2[b]	Chacko and Hanahan, 1968
L-3-Octadecylglycerol	Chemical synthesis	70.5°–71.5°	+16.0°[b]	Chacko and Hanahan, 1968
D-1-Octadecylglycerol	Inversion of the L-3-form	71°–72°	−16.2°	Chacko and Hanahan, 1968

[a] As isopropylidene derivative in substance (neat).

[b] As isopropylidene derivative in *n*-hexane. These data show the importance of the isopropylidene derivatives of these ethers in enhancing the optical rotation value of the 1- and 3-alkylglycerols. Obviously, this reaction would be of no value with the 2-alkylglycerols.

TABLE III

Typical Derivatives of Alkylglycerols

Reagent	Derivative	Reference
Acetone	—OR —O —O	Hanahan *et al.*, 1963
Acetic anhydride	—OR —OAc —OAc	Blomstrand and Gürtler, 1959
Trimethylsilyl chloride	—OR $—OSiMe_3$ $—OSiMe_3$	Wood and Snyder, 1966
Periodic acid	—OR =O H	W. J. Baumann *et al.*, 1967

synthesis of saturated and unsaturated monoethers and monoesters of 1,2-ethanediol, with hydrocarbon side chains ranging from 12 to 20 carbons. These compounds were analyzed by GLC as the acetate, trifluoroacetate, and trimethyl silyl ether derivatives.

Recently, Ramachandran *et al.* (1969) have described the preparation of a series of derivatives of the alkyl glyceryl ethers that could prove most useful and satisfactory for qualitative and quantitative identification purposes:

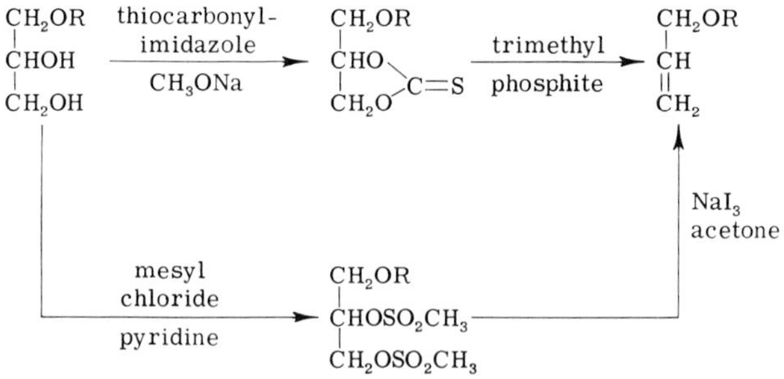

According to the authors, the cyclic thiocarbonate is obtained in quantitative yield, has significant absorption in the ultraviolet region of the spectrum, and shows satisfactory behavior on GLC columns. The allyl alkyl ether was also chromatographed.

A number of other approaches have been successfully employed in qualitatively characterizing or identifying the C—O—C bond. Prominent among these techniques are infrared spectroscopy, nuclear magnetic resonance spectroscopy (NMR), mass spectrometry, and thin-layer chromatography (TLC). As knowledge of these techniques has progressed, there have been reports of quantitative applications, which will be noted where pertinent.

Infrared spectroscopy of the alkyl glyceryl ethers has been employed by a number of investigators in establishing certain structural features of these compounds. It was soon apparent that the saturated ether bond in a native glyceryl ether phospholipid such as structure (4) could not be identified by infrared spectroscopy. The C—O—C absorption at 1100 cm^{-1} in these lipid ethers was simply swamped out by the intense phosphate absorption in 1110–1060 cm^{-1} region. Consequently, it has been necessary to isolate the alkyl glyceryl ether from the parent lipid molecule and then study the free ethers. It is possible to use the absorption band at 1100 cm^{-1} for quantitative assay (Hanahan, 1963) but unfortunately the reproducibility and range is limited. A detailed report on the infrared spectra of the 1- and 2-isomers of alkyl glyceryl ethers and of their mono- and diesters has been presented by Oswald *et al.* (1967). They published evidence of polymorphism in crystalline forms of these compounds. The most stable crystalline form of 1-alkyl glyceryl ethers, which exist in two forms, was the A form of hexagonal packing. The second form, B, was triclinic crystals. The 2-alkyl glyceryl ethers exhibited only the hexagonal packing form. A number of distinctive absorption bands were noted for these forms in the infrared spectral region. In a provocative and detailed study, W. J. Baumann and Ulshöfer (1968) examined the infrared spectra of long-chain ethers of glycerol, 1,2-ethanediol, and propanediols. A wide variety of diethers, triethers, and monoethers were used. Certain specific characteristics of the infrared absorption pattern for the alkyl glyceryl ethers and certain of their derivatives are worthwhile noting, e.g., the 1132–1110 cm^{-1} absorption region so characteristic of the alkyl ether bond, the asymmetrical C—O stretching vibration. Its particular frequency is dependent on the environment of the ether group and hence on the degree of polarization of the C—O bond. W. J. Baumann and Ulshöfer (1968) give a detailed discussion of the influence of adjacent groups, such as esters and ethers, on the spectra.

Nuclear magnetic resonance spectrometry has been employed to a limited extent in identification of the alkyl glyceryl ethers, but has not lent itself to any quantitative evaluation to date. Perhaps the most pertinent

TABLE IV

Major Fragments Obtained from the Dimethyl Ether of Octadecenylglycerol (Selachyl Alcohol)[a] *Mass Spectrometry*

Mass No.	Fragment
370	Molecular ion
$M - 90$	[$H\dot{C}OCH_3$ | H_2COCH_3 + $H\cdot$] fragment lost
$M - 120$	$CH_2{=}CH{-}CH{=}CH(CH_2)_{13}CH_3$
121	$CH_2O\cdot$ | $CHOCH_3$ + 2 $H\cdot$ | CH_2OCH_3
103	$\cdot CH_2$ | $CHOCH_3$ | CH_2OCH_3
89	$\dot{C}HOCH_3$ | CH_2OCH_3
45	$\cdot CH_2OCH_3$

[a] Taken from Hallgren and Larsson (1962).

studies on this topic have emanated from the laboratories of Wood and Snyder (1967), who showed that the 1- and 2-isomers of alkyl glyceryl ethers have considerably different NMR spectra, and that this can be used to distinguish these species. Further, they showed that mono- and di-unsaturated species had different spectra and that qualitative checks on fractionation techniques, syntheses, etc., could be based on these differences. Carter *et al.* (1958) reported on the NMR spectra of alkyl glyceryl ether acetates. These spectra (for the diacetate of octadecylglycerol and the alkyl glyceryl ether isolated from eggs) showed that one can differentiate the four equivalent hydrogens in 2-octadecylglycerol from the nonequivalent hydrogens (two doublets) on carbon-3 of 1-octadecylglycerol. Serdarevich and Carroll (1966) published NMR spectra of 1- and 2-alkyl glyceryl ethers prepared by conventional methods using fatty alcohols derived from D(+)-12-methyltetradecanoic acid and D(+)-14-methylhexadecanoic acid. Specifically, they recorded the spectra of the isopropylidene and 1,3-benzylidene derivatives and also of the free glyceryl ethers. Perhaps the most important characteristic for distinguishing between 1- and 2-alkyl glyceryl ethers is the protons attached to glycerol carbons. The spectra displayed by these two ethers were distinctly different.

Mass spectrometry coupled with GLC is of considerable advantage in establishing the structure of glyceryl ethers and their derivatives. An early study employing this combined technique was that of Hallgren and Larsson (1962), who investigated the structure of the dimethyl ethers of 1-octadecylglycerol (batyl alcohol) and 1-*cis*-9-octadecenylglycerol (selachyl alcohol). The major fragments of the latter derivative are listed in Table IV.

Thin-layer chromatography has been an invaluable aid in the lipid field in general, and its usefulness in studies on the glyceryl ethers has proven no exception. The alkyl and alk-1-enyl glyceryl ethers can easily be separated by TLC, and can then be estimated quantitatively by a photodensitometric technique described in Section II, C. An extensive treatment of various facets of the chromatographic examination of the ether lipids (as well as other lipids) is presented by experts in the field in a recent book edited by Marinetti (1967); see also reviews on this subject by Renkonen (1971) and Snyder (1971).

C. Quantitative Measurement

Any quantitative (or for that matter, qualitative) assay for glyceryl ethers usually requires prior separation of the ether-containing compounds and subsequent isolation of the free glyceryl ethers per se. The quantitative measurement of an ether can now be achieved. Although a particular technique by itself may give a quantitative response on a weight basis, the problems associated with the proper identification of lipids makes it imperative for any investigator to employ more than one procedure for meaningful results. The methods outlined below are useful and reproducible and can be of value if used in adjunct with other techniques such as TLC and infrared and mass spectrometry. The procedures outlined here were developed mainly for assay of the 1-alkylglycerols, since these are the most abundant natural types. However, if the 2-isomers are present, modified procedures must be employed. The reader is referred to the technique developed by Wood and Snyder (1967) for the separation of long-chain saturated and unsaturated 1- and 2-alkylglycerols by the difference in their mobility on Silica Gel G impregnated with sodium arsenite or boric acid.

1. *Periodic Acid Uptake*

The use of periodic acid for the estimation of vicinal hydroxyl groups is a well-established technique. Its application to the 1-alkylglycerols was first described by Karnovsky and Brumm (1955) and more recently outlined by Thompson and Kapoulas (1969).

The vicinal hydroxyls of the 1-alkylglycerols (and also 3-alkylglycerols) react smoothly and quantitatively with periodic acid at room temperature and neutral pH.

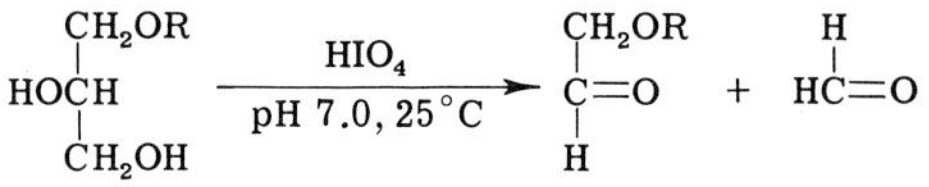

The uptake of periodic acid can be monitored by a direct titrimetric procedure measuring the unused periodic acid (Siggia, 1949), by a colorimetric assay based on detection of the formaldehyde released (Karnovsky and Brumm, 1955; Thompson and Kapoulas, 1969), or by a spectrophotometric assay (Dixon and Lipkin, 1954). These methods, used in proper perspective, can be of considerable value. However, periodic acid does react with other glycols (and also with amino alcohols), so the uniqueness of the sample under assay must be convincingly proven. Further, the presence of the 2-isomers of glyceryl ethers must be considered in any interpretation of the quantitativeness of these reactions.

As regards identification of the length and nature of the side chain of glyceryl monoethers, one of the products of the reaction with periodic acid, the alkoxy acetaldehyde, can be examined by GLC as outlined by Gelman and Gilbertson (1969). They employed a colorimetric assay for these alkoxy aldehydes through the use of the *p*-nitrophenylhydrazone derivative, which was extracted with hexane and ultimately examined in ethanol at 380 nm. Although the authors did not discuss the point, it has been noted by Rapport and Norton (1962) that the hydrazone method gives low values for fatty aldehydes for various reasons such as sensitivity to light and solubility problems. Obviously some care must be exercised in the application of this particular technique.

2. *Isopropylidene Formation*

One of the more useful procedures for estimating the side-chain length of the alkyl glyceryl ethers and at the same time providing quantitative information on the level of alkyl glyceryl ethers is that of isopropylidene formation.

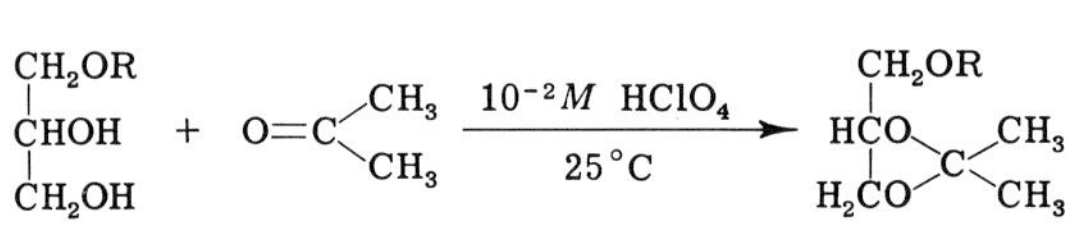

The isopropylidene derivatives, which can be obtained in quantitative yields, are then subjected to GLC. Further information on the properties of these derivatives can be obtained in articles by Hanahan *et al.* (1963) and Wood *et al.* (1969a). The technique does not apply to the 2-isomers of glyceryl ethers, since they do not react with acetone under these conditions. A particular advantage of the isopropylidene derivative is that it can easily be isolated by silicic acid chromatography (or by GLC) in any desired amount and then converted back to the free form, without racemization, by treatment with dilute acid.

3. *Combined TLC-Photodensitometric Technique*

Wood and Snyder (1968) have described a procedure that can be used for the simultaneous quantitative determination of the alkylglycerols (**1**) and alk-1-enylglycerols (**3**). The main feature of this method is the release of the free ethers from the native neutral lipid or phospholipid by the action of lithium aluminum hydride. The ethers are quantitatively isolated by solvent extraction and then applied in known amounts to a Silica Gel G TLC plate. After development in a proper solvent, the plates are charred and subjected to quantitative densitometric assay as described by Privett *et al.* (1965). The separation of the two types of ethers [(**1**) and (**3**)] on TLC was adequate and the relative error of the procedure was calculated to be less than 10% for samples with a level of 1 μg of ethers per mg of total lipid. Inasmuch as these ethers are easily extractable from a chromatography plate, it is possible to examine the separated components by several other techniques and make a more precise determination of their composition. Additional data obtained with ^{14}C-labeled plasmalogens have demonstrated that the use of $LiAlH_4$ gives low values for alk-1-enyl glycerolipids (Wykle *et al.*, 1970; Snyder *et al.*, 1971); Vitride,* rather than $LiAlH_4$, is a better reducing agent in the procedure described in this section.

4. *Other Methods*

Other quantitative methods for assay of glyceryl ethers have been reported. Wood and Snyder (1966) showed that the alkylglycerols can be quantitatively assayed by use of trifluoracetyl and trimethyl silyl ether derivatives, and that it is also possible to resolve the 1- and 2-isomers of the alkylglycerols as trifluoracetyl derivatives. Ramachandran *et al.* (1968) reported that it is possible to identify and quantify the alkylglycerols by GLC of their iodide, isopropylidene, and allyl derivatives.

* Sodium bis(2-methoxyethoxy)aluminum hydride (Eastman Chem. Co.).

III. Dialkyl Glyceryl Ethers

The presence of glyceryl diethers in naturally occurring lipids has been substantiated only within the past few years. The significant breakthroughs in this area of investigation were reports by Sehgal *et al.* (1962) and Kates *et al.* (1965a) on the high levels of these substances in an extremely halophilic bacteria, *Halobacterium cutirubrum*. The ether was found as part of the bacterium's major phospholipid, for which the suggested structure was 2,3-di-*O*-(3′,7′,11′,15-tetramethylhexadecyl)glyceryl-1-phosphoryl-([1″-(3″)-glyceryl-3″-(1″)-phosphate]) [see formula (7), Section I]. Methanolysis of the halophile's lipids yielded dialkylglycerols as a major component and alkylglycerols as a minor component; these ethers contained the phytanyl moiety (3′,7′,11′,15-tetramethylhexadecyl) as the side chain. The diethers were synthesized as follows:

(a)

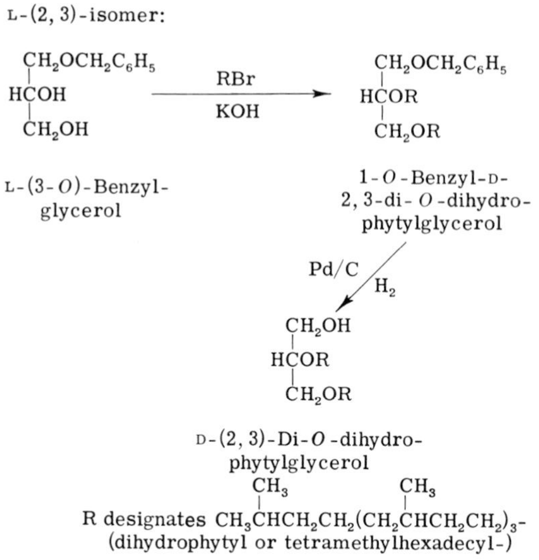

(b)

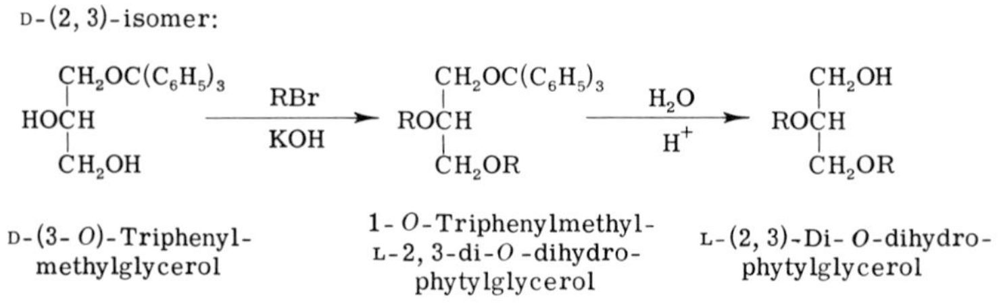

On the basis of the characteristics of these two isomers, the naturally occurring diether was established as L-(2,3)-di-*O*-3′,7′,11′,15′-tetrahydrophytylglycerol. Later studies by Kates *et al.* (1967) showed the absolute stereochemical configuration of the alkyl side chain to be (3*R*, 7*R*, 11*R*)-15. The diether is unique in that it has the L-configuration, whereas diglycerides have the D-configuration. The small amount of alkylglycerol was separated by TLC on boric acid-impregnated silicic acid into equal amounts of α- and β-isomers. The α-isomer was 3-*O*-(3′*R*,7′*R*,11′*R*,15-tetramethylhexadecyl)glycerol, and the β-isomer was the 2-*O*-(3′*R*,7′*R*, 11′*R*,15-tetramethylhexadecyl)glycerol.

The infrared spectrum of the diphytanyl-*sn*-glycerol had the following major peaks: *OH*, 3450 cm^{-1}; *CH_2 and CH_3*, 2960, 2930, 2860, 1465, and 730 cm^{-1}; *$C(CH_3)_2$*, 1370–1380 cm^{-1}; *C—O—C*, 1110 cm^{-1}; and *C—O*, 1045 cm^{-1}. The specific optical rotation of this diether was $[\alpha]_D + 7.6°$ (2.95, in $CHCl_3$). The monoethers had the following specific rotation values in $CHCl_3$: *α-isomer*, −0.95° and *β-isomer*, +2.4°. Through stereospecific synthesis of monoethers, Joo *et al.* (1968) were able to show that the α- and β-isomers of the alkyl glyceryl ethers from *H. cutrirubrum* had the 3-*sn*-glycerol or L-configuration. This is opposite to what has been found for the alkyl glyceryl ethers isolated from fish liver lipids or from mammalian erythrocyte lipids. The chemical reactivity of the ether bond in these diethers and monoethers is very similar to that described for the alkyl bond in a previous section (II, A). It is possible to analyze the dialkylglycerols by GLC of their trimethylsilyl or acetate derivatives (Wood *et al.*, 1969a). The monophytanylglycerol, in the form of its di-*O*-methyl ether derivative, can also be examined by GLC. Essentially, these dialkyl glyceryl ethers present the same problems in analysis and identification as are encountered with the alkyl glyceryl ethers (see Section II), and comparable approaches must be used in solving their quantitative assay.

IV. Alk-1-enyl Glyceryl Ethers

The glyceryl vinyl ether, $HOCH_2HCOHCH_2OCH{=}CHR_1$, is unique in that it does not obey the usual rule of stability associated with ethers. Its lability under mild acid conditions immediately demonstrates a different type of structure than the one noted for the alkyl glyceryl ethers described in Section II. This lability has caused considerable difficulty in the elucidation of the structure of alk-1-enyl ether-containing phospholipids, and accounts in the main for the recent proof of their structure as compared to that of the alkyl glyceryl ethers.

The alk-1-enyl glyceryl ethers exist mainly as phospholipids (plasmalogens), and to a limited extent as glycerides. Much of the early work on these ethers was accomplished with the intact plasmalogens, but as experience and knowledge accumulated it was possible to isolate the alk-1-enylglycerols per se and conduct studies on their physical and chemical reactivity. Certain facets of the chemical and physical characteristics of the alk-1-enyl glyceryl ethers are described in the following section. As will be apparent, despite rather extensive chemical studies, primary consideration of the chemical reactivity of the alk-1-enyl glyceryl ethers has been limited to hydrolysis, iodination, hydrogenation, and ozonolysis.

A. General Reactivity and Proof of Structure

As might be expected, the alk-1-enyl glyceryl ethers exhibit considerably more reactivity than the alkyl glyceryl ethers, and this reactivity has been of value in determining their structure and configuration.

1. *Iodoacetal Formation*

A particularly useful reaction in establishing the nature of the ether bond involves the use of iodine and methanol (Siggia and Edsberg, 1968). This reaction, which forms the basis of a quantitative analytical procedure for alk-1-enyl ethers, proceeds as follows:

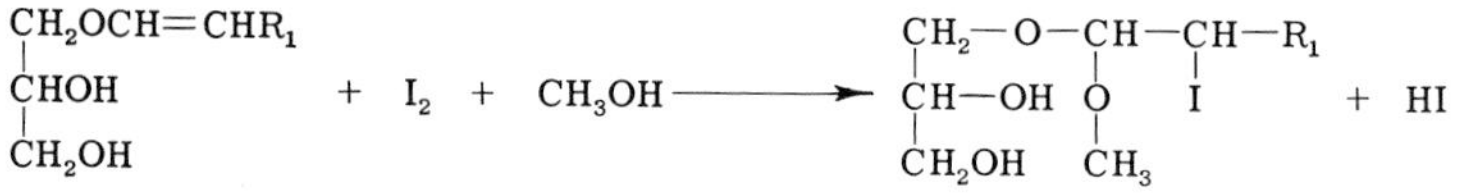

The reaction is fast and provides a facile approach to determination of the level of the ether bond in lipid mixtures.

2. *Hydrolysis*

Although there is some question as to the lability of the alk-1-enyl glyceryl ether to alkali (Pietruszko and Gray, 1962), there is no doubt that it can be cleaved quite easily and smoothly by mineral acid. An early method (Siggia, 1949) for determination of alk-1-enyl ethers was based on this fact:

$$ROCH{=}CH_2 + H_2O \xrightarrow{H^+} ROH + CH_3\overset{\overset{O}{\|}}{C}H$$

After acid cleavage, the acetaldehyde, in this instance, was assayed by its combination with bisulfite. Such general behavior of alk-1-enyl ethers with acid applies equally well to the alk-1-enylglycerols, which react as follows:

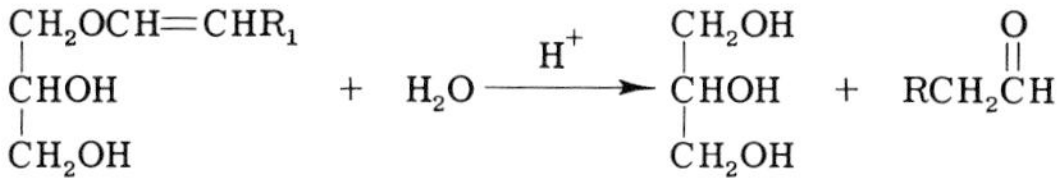

This also forms the basis of a reaction leading to the quantitative determination of long-chain fatty aldehydes (see methods Section II, B).

Recently Frosolono and Rapport (1969) reported the kinetics of the acid-catalyzed hydrolysis of plasmalogens and certain model compounds. This process is now thought to involve an initial protonation of the unsaturated linkage to form a carbonium ion

$$CH_2{=}CHOR + H^+ \underset{}{\overset{\text{slow}}{\rightleftharpoons}} CH_3\overset{+}{C}HOR$$

then reaction with water and ultimate cleavage to yield an aldehyde and alcohol:

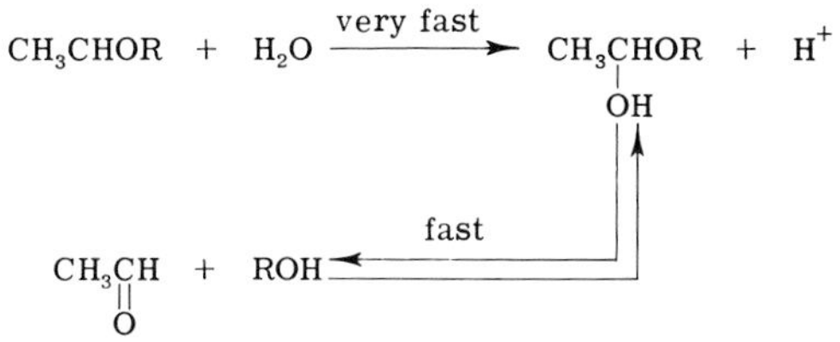

True second-order rate constants were determined for the plasmalogen hydrolysis as well as for the model compounds, such as isobutyl vinyl ether, 1-butenyl ethyl ether, and dihydropyran. It was concluded that the bulky hydrocarbon unit attached to the alk-1-enyl ether group on plasmalogens did not affect the chemical reactivity of the alk-1-enyl ether in these complex phospholipids. Evidence was presented in support of a faster hydrolysis of the *cis*-alk-1-enyl ether than of the *trans*-alk-1-enyl ether.

3. *Derivative Formation*

A number of derivatives that are useful for identification and assay can be prepared from alk-1-enylglycerols (see Section IV, B). Under proper conditions (Warner and Lands, 1963), it is possible to hydrogenate the double bond of the alk-1-enyl ether with the formation of an alkylglycerol. This particular reaction has been of value, as shall be outlined in Section IV, A, 4, in establishing the basic structure of the alk-1-enyl ether. It is interesting that hydrogenation occurs without inversion of configuration (Cymerman-Craig *et al.*, 1966). The reaction is quantitative and the resulting alkylglycerol is easily isolated and identified in the manner previously discussed (Section II).

Ozonolysis has been useful in the proof of structure of the alk-1-enyl ether linkage. In particular, Debuch (1958) has shown that ozone attacks this double bond (and also other unsaturated bonds) with the release of odd-numbered fatty aldehydes. Inasmuch as the chain length of the hydrocarbon must be one carbon longer than that of the released aldehydes, Debuch justifiably concluded that the plasmalogen structure is an "enol ether" and not a hemiacetal. After ozonolysis of a hemiacetal, one would expect to find the even-numbered long-chain fatty acids. See Chapter I for additional discussion.

4. *Geometric Configuration of the Double Bond*

After elucidation of the chemical nature of the ether linkage in "plasmalogens," the question naturally arose as to whether the geometric configuration of this double bond was *cis* or *trans*. Evidence supporting a *cis*- configuration was reported almost simultaneously by Norton *et al.* (1962) and Warner and Lands (1963). In each of these studies, the approach was similar in that a native choline-containing plasmalogen was treated with phospholipase C yielding an alk-1-enylacylglycerol that could be subjected to alkaline treatment to yield the alk-1-enylglycerol. The free compound could then be hydrogenated to give the alkylglycerol or subjected to periodate oxidation to yield an alkenylglycolaldehyde. These reactions are illustrated in the following scheme:

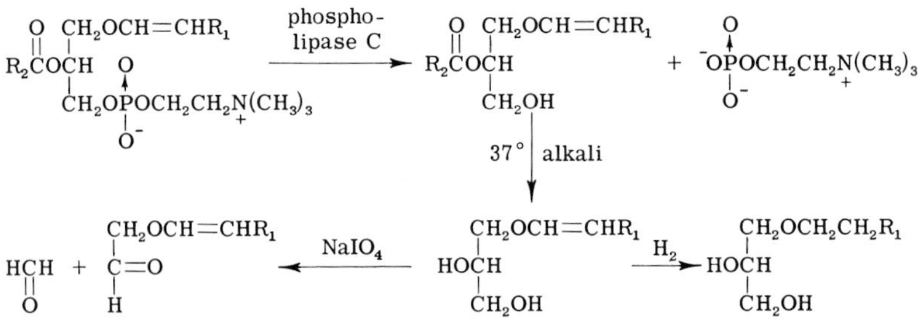

The proof of the configuration of the double bond by these two laboratories rested essentially on the infrared spectral characteristics of the various compounds noted above. Norton *et al.* (1962) compared the spectra of the alk-1-enyl glyceryl ether preparation with that of synthetic 1-butenyl

TABLE V

Major Infrared Absorption Bands of Alk-1-enyl Glyceryl Ethers

Wavelength (cm^{-1})	Group
1667–1670	H—C=C—H stretching
1110	C—O—C
738	=C—H deformation (*cis*)
725	$(CH_2)_n$ rocking

ether, whereas Warner and Lands (1963) compared the spectra of their preparations with that of synthetic methyl 1-dodec-1-enyl ether. The latter was separated by alumina chromatography into the *cis* and *trans* forms, a particularly excellent approach, since the spectra of naturally occurring alk-1-enylglycolaldehyde and those of *cis*- or *trans*-methyl 1-dodec-1-enyl ether could thus be closely compared. It was found that the spectra of *cis*-methyl 1-dodec-1-enyl ether ($CH_3OCH{=}CHR$) and of the alk-1-enylglycolaldehyde ($HOCCH_2OCH{=}CHR_1$) differed only by the presence of a 1750 cm^{-1} band caused by the carbonyl group in the latter. It was unequivocally shown by Warner and Lands (1963) that a *cis* configuration could be assigned to the double bond in the alk-1-enyl linkage found in phospholipids. The major infrared absorption bands found are shown in Table V. There is no doubt that the 1667 cm^{-1} band is most useful for ascertaining the presence of a *cis*-α, β-unsaturated group in alk-1-enylglycerols (assuming no amide, such as a sphingomyelin, is present). The double bond "conjugated" to an ether intensifies the absorption in 1667 cm^{-1} region to a very significant degree over that observed with saturated linkages present in glycerides. Norton *et al.* (1962) showed that, in a substituted vinyl ether, the *trans* bond is shifted from 970 cm^{-1} (10.3 μ) to 932 cm^{-1} (10.72 μ), whereas *cis*-alk-1-enyl ethers absorbed at 733 cm^{-1} (13.6 μ). The latter compounds do not have a significant absorbance in the 1030 cm^{-1} (9.7 μ) to 854 cm^{-1} (11.7 μ) region, and the *trans* isomer had no absorbance in the 833 cm^{-1} (12 μ) to 714 cm^{-1} (14 μ) region.

Schmid *et al.* (1967) described the geometric and optical configuration of the neutral "plasmalogens" (alk-1-enyldiacylglycerols) of ratfish liver. On the basis of infrared data and chemical behavior they concluded that the ether group of the diacyl glycerolipids was an alk-1-enyl ether possessing a *cis* configuration, and on the basis of optical activity they considered these compounds to be D(+)-1-*cis*-alk-1′-enyldiacylglycerols.

TABLE VI

Optical Rotatory Dispersion Values of the Diacetates of Alk-1-enylglycerols and Alkylglycerols[a]

Compound	$[\alpha]_{322}$	$[\alpha]_{250}$	$[\alpha]_{238}$	$[\alpha]_{233}$	$[\alpha]_{204}$
Diacetate of natural alk-1-enylglycerol					
(a) original	−17.6°	−40.6° (trough)	−15.5° (peak)	−27.7°	—
(b) hydrogenated	−23.4°	—	—	−41.5° (peak)	−238°
Diacetate of octadecylglycerol (batyl alcohol)	−43.7°	—	—	−65.7° (peak)	−270°
Diacetate of hexadecylglycerol (chimyl alcohol)	—	—	—	−42.1° (peak)	−254°

[a] Taken from Cymerman-Craig *et al.* (1966).

5. *Absolute Configuration*

Since synthetic plasmalogens were not available for comparison with naturally occurring alk-1-enyl glyceryl ethers, the correct stereochemical configuration of these ethers was always assumed to be similar to that of other phosphoglycerides for which the optical configuration had been elucidated. However, Cymerman-Craig *et al.* (1966) clarified this matter in their report on the absolute configuration of naturally occurring alk-1-enyl glyceryl ethers. Pig heart choline plasmalogens were hydrolyzed by the procedure outlined by Warner and Lands (1963; and see Section IV, A, 4), and the alk-1-enyl glyceryl ethers were purified by chromatography on silicic acid. These ethers were then converted in part to the diacetates and in part hydrogenated, and then acetylated. The optical rotatory dispersion patterns of these derivatives were then obtained down to 232 nm. It was not possible to proceed further because of the rapidly increasing ultraviolet absorption of the alk-1-enyl ether chromophore. The rotary dispersion values (R.D.) in ethanol are shown in the Table VI. The results showed that the alk-1-enylglycerol and its hydrogenated product, alkylglycerol, have the same absolute configuration as naturally occurring 1-hexadecylglycerol (chimyl alcohol) and 1-octadecylglycerol (batyl alcohol). Cymerman-Craig *et al.* concluded that these alk-1-enyl ethers were all of the L-(1)-configuration.

B. Quantitative Measurement

Several methods have been described for assay of alk-1-enyl ethers in lipids. Prominent among these are the iodine uptake reaction (see Section

IV, A, 1), which has been the only direct analysis applicable to the intact plasmalogens, and the *p*-nitrophenylhydrazine reaction with long-chain fatty aldehydes liberated by acid cleavage of the alk-1-enyl ether linkage. However, these assays are not unequivocally applicable to all tissues or substances and have given values as much as 8–10% lower than the actual value. Renkonen (1968) has described a thin-layer chromatographic technique for the separation of the methylated dinitrophenylated derivatives of ethanolamine glycerophospholipids. The multiple unidimensional development system employed in this procedure gave an apparent good separation of alkenyl (plasmalogens), alkylacyl, and diacylglycerophospholipids. However, on the basis of evidence presented in this report, it is not possible to evaluate the usefulness of this approach as a quantitative technique. At the present time, three methods appear most suited for quantitative estimation of the alk-1-enyl ether group. It is again strongly suggested that certain adjunct systems for characterization, such as infrared and mass spectrometry, be employed as well for total assay.

1. *Acetal Formation*

In the technique described by Gray (1969), the alk-1-enyl ether linkage in plasmalogens is cleaved with acidified methanol, leading to the quantitative formation of the fatty aldehyde dimethyl acetals. Although this approach is a reasonable and effective one, the presence of methyl esters in the resulting product necessitates chromatographic purification of the dimethyl acetals. Perhaps the more effective procedure is to selectively release the aldehyde with 90% acetic acid.

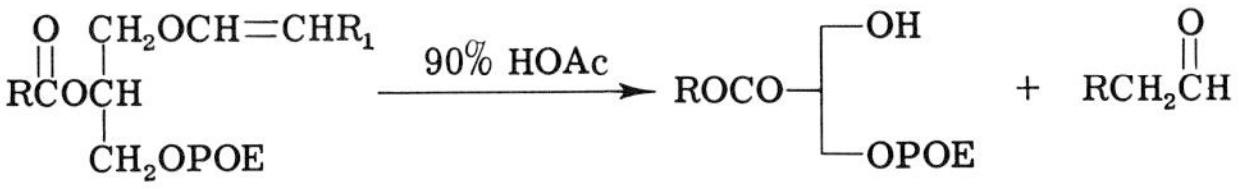

Subsequently, the aldehydes can be quantitatively separated from the lysophosphatidylethanolamine by chromatography on silicic acid, and then assayed by the fuchsin color reaction. The fuchsin reaction is a time-honored reaction and has served us well in the past, but it does tend to give results approximately 10% higher than actual values. An alternate procedure is to convert the fatty aldehydes to the dimethyl acetals.

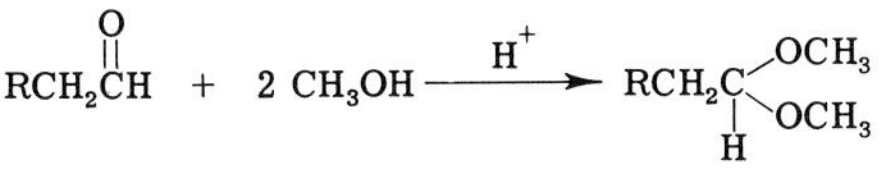

This product can be weighed and then analyzed by GLC.

Although the dimethyl acetals appear to be of considerable value for assay of alk-1-enyl ethers, numerous investigators (Gray, 1960; Marcus, 1962; Morrison and Smith, 1964; N. A. Baumann *et al.*, 1965; Stein and Slawson, 1966; Mahadevan *et al.*, 1967) have reported alterations in dimethyl acetals on certain types of GLC columns. Such anomalous behavior led Stein and Slawson (1966) to investigate the GLC behavior of a representative acetal, the dimethyl acetal of octadecanal. This compound underwent a significant alteration on certain column phases to yield methyl octadec-1-enyl ether or octadecanal. The presence of acid in the sample accelerated the conversion. The column effluent was collected and found to consist of a mixture of *cis* and *trans* isomers, with a ratio of *cis/trans* of 0.90 from *trans*-alk-1-enyl ether and 1.0 from *cis*-alk-1-enyl ether. On the basis of their observations, Stein and Slawson proposed the following reaction to explain the elimination of a molecule of methanol:

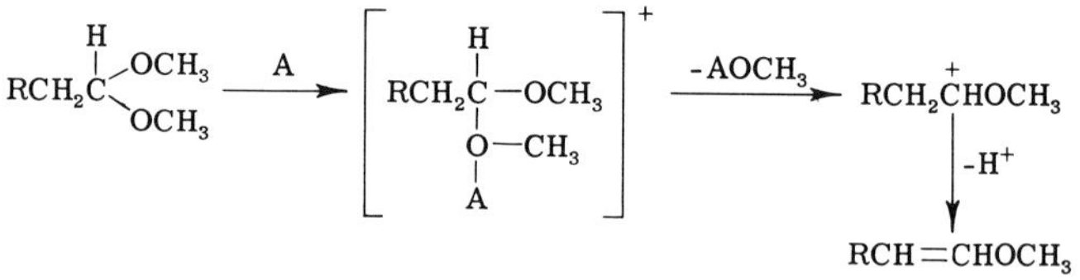

Initial attack by a Lewis acid, A, occurs on the acetal oxygen with subsequent formation of a carbonium ion, which then loses a proton. One possible source of the Lewis acid might have been the Al^{3+} in the silicate solid support used for GLC. According to Stein and Slawson, the isomerization reaction resulting in *cis–trans* isomers could have involved a reversible addition of a proton to the alk-1-enyl ether to give a carbonium ion, which would have with no stereospecific memory:

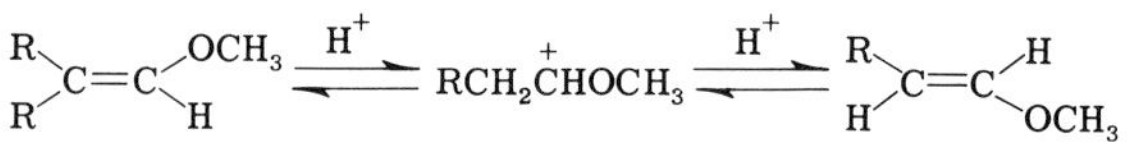

Mahadevan *et al.* (1967) describe the thermal degradation of the dimethyl acetal of hexadecyl aldehyde to the corresponding alk-1-enyl methyl ether during GLC.

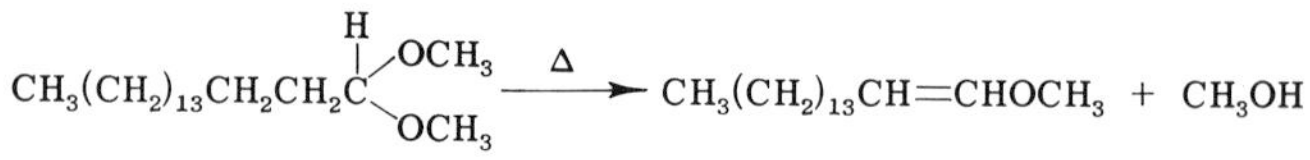

The effluents from preparative GLC columns (polar as well as nonpolar columns, ethylene glycol succinate polyester–phosphoric acid columns, and β-cyclodextrin acetate columns) were collected and analyzed by TLC, and infrared and mass spectrometry. The evidence obtained supported the fact that the dimethyl acetal was completely converted to a derivative with a less polar behavior. Free aldehyde was unaltered. Infrared spectral examination of the column effluent revealed no aldehyde or acetal bands, but instead a strong doublet at 1667 cm^{-1} (6 μ) indicative of –O—CH═CH–, and in addition, a 934 cm^{-1} (10.7 μ) band indicative of enol ethers. The mass spectra had one peak at $m/e = 254$, corresponding to hexadec-1-enyl methyl ether, and one at $m/e = M - 32$, possibly a hydrocarbon formed by loss of methanol from the ether. The hexadec-1-enyl methyl ether was easily hydrolyzed to yield hexadecanal and hydrogenated to form hexadecyl methyl ether.

As noted by Stein and Slawson (1966), the difficulties associated with the GLC of dimethyl acetals can be circumvented by qualitative and quantitative calibrations of the GLC columns, together with some attempt to prove that the substance emerging from the column is the same as that placed on it.

2. *Dioxolane Derivatives*

Inasmuch as the dimethyl acetals are not stable to acid and undergo degradation during GLC, Rao *et al.* (1967) and Panganamala *et al.* (1969) proposed the use of another type of acetal in analysis of the fatty aldehydes, namely, the 1,3-dioxolane derivative. These cyclic ethers are reportedly much more stable than the dimethyl acetals and can be separated by silver nitrate TLC into unsaturated and saturated components, eluted, and then separated by GLC. In this procedure, the plasmalogens were hydrolyzed with 90% acetic acid, and the fatty aldehydes were extracted with chloroform and/or diethyl ether. The fatty aldehydes were purified by silicic acid chromatography and then converted into 1,3-dioxolanes by the following reaction:

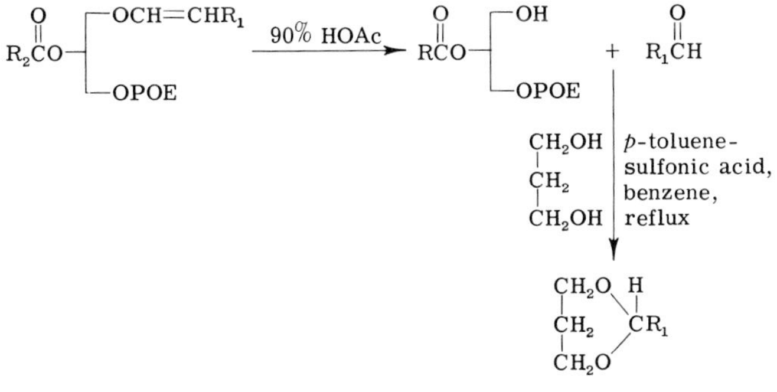

If this reaction is conducted on the intact phospholipids, artifacts are often formed.

A note of caution must be introduced regarding the use of acid in the cleavage of the alk-1-enyl ether group. Recently, Wood and Healy (1970) have shown through the use of labeled alk-1-enylglycerols that the ether bond can be completely cleaved by acid treatment. Although long-chain fatty aldehydes were the primary products, other compounds, such as cyclic acetals, were formed. This would cause low results in any quantitative measurement of the aldehydes, for example, by TLC.

3. *Diacetates*

Albro and Dittmer (1968) have described a procedure which appears to lend itself to assay of the alk-1-enyl glyceryl ethers. The ether-containing lipid mixture is subjected to hydrogenolysis with lithium aluminum hydride as described by Thompson (1965), and the alk-1-enylglycerols are isolated quantitatively in intact form by chromatography on silicic acid. The alk-1-enylglycerols can be analyzed by weight and by periodate oxidation, and then converted to the diacetyl derivative, which can be assayed quantitatively for length of the side chain by GLC on a 5% SE-30 liquid phase.

4. *Combined TLC-Densitometric Method*

The method described by Wood and Snyder (1968) and outlined in Section II, C, 3, is applicable to the detection of the alk-1-enylglycerols.

REFERENCES

Albro, P. W., and Dittmer, J. C. (1968). *J. Chromatogr.* **38,** 230.
Baer, E., and Fischer, H. O. L. (1947). *J. Biol. Chem.* **170,** 337.
Baer, E., Rubin, L. J., Fischer, H. O. L. (1944). *J. Biol. Chem.* **155,** 447.
Baumann, N. A.; Hagen, P-O., and Goldfine, H. (1965). *J. Biol. Chem.* **240,** 1559.
Baumann, W. J., and Ulshöfer, H. W. (1968). *Chem. Phys. Lipids* **2,** 114.
Baumann, W. J., Schmid, H. H. O., Ulshöfer, H. W., and Mangold, H. K. (1967). *Biochim. Biophys. Acta* **144,** 355.
Bergelson, L. D. (1969). *Progr. Chem. Fats Other Lipids* **10,** 239.
Bevan, T. H., Brown, D. A., Gregory, G. I., and Malkin, T. (1953). *J. Chem. Soc., London* p. 127.
Blomstrand, R., and Gürtler, J. (1959). *Acta Chem. Scand.* **13,** 1466.
Carter, H. E., Smith, D. B., and Jones, D. N. (1958). *J. Biol. Chem.* **232,** 681.
Chacko, G. K., and Hanahan D. J. (1968). *Biochim. Biophys. Acta* **164,** 252.
Cymerman-Craig, J., Hamon, D. P. G., Purushothaman, K. K., Roy, S. K., and Lands, W. E. M. (1966). *Tetrahedron* **22,** 175.
Dawson, R. M. C., Hemington, N., and Lindsay, D. B. (1960). *Biochem. J.* **77,** 226.

Debuch, H. (1958). *Hoppe-Seyler's Z. Physiol. Chem.* **311,** 266.
Dixon, J. S., and Lipkin, D. (1954). *Anal. Chem.* **26,** 1092.
Ferrell, W. J., and Radloff, D. M. (1970). *Physiol. Chem. Phys.* **2,** 551.
Frosolono, M. F., and Rapport, M. M. (1969). *J. Lipid Res.* **10,** 504.
Gelman, R. A., and Gilbertson, J. R. (1969). *Anal. Biochem.* **31,** 463.
Gray, G. M. (1960). *J. Chromatogr.* **4,** 52.
Gray, G. M. (1969). *Methods Enzymol.* **14,** 678–684.
Guyer, K. E., Hoffman, W. A., Horrocks, L. A., and Cornwell, D. G. (1963). *J. Lipid Res.* **4,** 385.
Hallgren, B., and Larsson, S. (1962). *J. Lipid Res.* **3,** 31.
Hanahan, D. J. (1963). Unpublished observations.
Hanahan, D. J. (1965). *J. Lipid Res.* **6,** 350.
Hanahan, D. J., Ekholm, J., and Jackson, C. M. (1963). *Biochemistry* **2,** 630.
Horrocks, L. A. (1968). *J. Lipid Res.* **9,** 469.
Joo, C. N., Shier, T., and Kates, M. (1968). *J. Lipid Res.* **9,** 782.
Karnovsky, M. L., and Brumm, A. F. (1955). *J. Biol. Chem.* **216,** 689.
Kates, M., Yengoyan, L. S., and Sastry, P. S. (1965a). *Biochim. Biophy. Acta* **98,** 252.
Kates, M., Palameta, B., and Yengoyan, L. S. (1965b). *Biochemistry* **4,** 1595.
Kates, M., Joo, C. N., Palameta, B., and Shier, T. (1967). *Biochemistry* **6,** 3329.
Lands, W. E. M., and Zschocke, A. (1965). *J. Lipid Res.* **6,** 324.
Mahadevan, V., Viswanathan, C. V., and Phillips, F. (1967). *J. Lipid Res.* **8,** 2.
Marcus, A. J. (1962). *J. Clin. Invest.* **41,** 2198.
Marinetti, G. V., ed. (1967). "Lipid Chromatographin Analysis," Vol. 1. Dekker, New York.
Morrison, W. R., and Smith, L. M. (1964). *J. Lipid Res.* **5,** 600.
Norton, W. T., Gottfried, E. L., and Rapport, M. M. (1962). *J. Lipid Res.* **3,** 456.
Oswald, E. O., Piantadosi, E. E., Anderson, C. E., and Snyder, F. (1967). *Chem. Phys. Lipids* **1,** 270.
Panganamala, R. V., Buntine, D. W., Geer, J. C., and Cornwell, D. G. (1969). *Chem. Phys. Lipids* **3,** 401.
Piantadosi, C., Snyder, F., and Wood, R. (1969). *J. Pharm. Sci.* **58,** 1028.
Pietruszko, R., and Gray, G. M. (1962). *Biochim. Biophys. Acta* **56,** 232.
Privett, O. S., Blank, M. L., Codding, D. W., and Nickell, E. E. (1965). *J. Amer. Oil Chem. Soc.* **42,** 381.
Ramachandran, S., Sprecher, H. W., and Cornwell, D. G. (1968). *Lipids* **3,** 511.
Ramachandran, S., Panganamala, R. V., and Cornwell, D. G. (1969). *J. Lipid Res.* **10,** 465.
Rao, P. V., Ramachandran, S., and Cornwell, D. G. (1967). *J. Lipid Res.* **8,** 380.
Rapport, M. M., and Norton, W. T. (1962). *Annu. Rev. Biochem.* **31,** 103.
Renkonen, O. (1967). *Advan. Lipid Res.* **5,** 329.
Renkonen, O. (1968). *J. Lipid Res.* **9,** 34.
Renkonen, O. (1971). *In* "Progress in Thin-Layer Chromatography and Related Methods" (A. Niederwieser and G. Pataki, eds.), Vol. II, pp. 143–181. Ann Arbor Sci. Publ., Ann Arbor, Michigan.
Schmid, H. H. O., and Mangold, H. K. (1966). *Biochem. Z.* **346,** 13.
Schmid, H. H. O., Baumann, W. J., and Mangold, H. K. (1967). *J. Amer. Chem. Soc.* **89,** 4797.
Sehgal, S. N., Kates, M., and Gibbons, N. E. (1962). *Can. J. Biochem. Physiol.* **40,** 69.
Serdarevich, B., and Carroll, K. K. (1966). *Can. J. Biochem. Physiol.* **44,** 743.

Siggia, S. (1949). "Quantitative Organic Analysis via Functional Groups." Wiley, New York.
Siggia, S., and Edsberg, R. L. (1948). *Anal. Chem.* **20,** 762.
Snyder, F. (1969). *Progr. Chem. Fats Other Lipids* **10,** 289–335.
Snyder, F. (1971). *In* "Progress in Thin-Layer Chromatography and Related Methods" (A. Niederwieser and G. Pataki, eds.), Vol. II, pp. 105–141. Ann Arbor Sci. Publ., Ann Arbor, Michigan.
Snyder, F., and Wood, R., (1968). *Cancer Res.* **28,** 992.
Snyder, F., and Wood, R., (1969). *Cancer Res.* **29,** 251.
Snyder, F., Piantadosi, C., and Wood, R. (1969). *Proc. Soc. Exp. Biol. Med.* **130,** 1170.
Snyder, F., Blank, M. L., and Wykle, R. L. (1971). *J. Biol. Chem.* **246,** 3639.
Stein, R. A., and Slawson, V. (1966). *J. Chromatogr.* **25,** 204.
Thiele, O. W., Schröder, H., and Berg, W. (1960). *Hoppe-Seyler's Z. Physiol. Chem.* **322,** 147.
Thompson, G. A., Jr. (1965). *J. Biol. Chem.* **240,** 1912.
Thompson, G. A., Jr. (1967). *Biochemistry* **6,** 2015.
Thompson, G. A., Jr., and Hanahan, D. J. (1962). *Arch. Biochem. Biophys.* **96,** 671.
Thompson, G. A., Jr., and Kapoulas, V. M. (1969). *Methods Enzymol.* **14,** 671–673.
Wallenfels, K., and Diekmann, H. (1967). *In* "The Chemistry of the Ether Linkage" (S. Patai, ed.), pp. 207–242. Wiley (Interscience), New York.
Warner, H. R., and Lands, W. E. M. (1963). *J. Amer. Chem. Soc.* **85,** 60.
Wells, M. A., and Dittmer, J. C. (1967). *Biochemistry* **6,** 3169.
Wood, R., and Baumann, W. J. (1968). *J. Lipid Res.* **9,** 733.
Wood, R., and Healy, K. (1970). *Lipids* **5,** 661.
Wood, R., and Snyder, F. (1966). *Lipids* **1,** 62.
Wood, R., and Snyder, F. (1967). *Lipids* **2,** 161.
Wood, R., and Snyder, F. (1968). *Lipids* **3,** 129.
Wood, R., Baumann, W. J., Snyder, F., and Mangold, H. K. (1969a). *J. Lipid Res.* **10,** 128.
Wood, R., Piantadosi, C., and Snyder, F. (1969b). *J. Lipid Res.* **10,** 370.
Wykle, R. L., Blank, M. L., and Snyder, F. (1970). *FEBS Lett.* **12,** 57.

CHAPTER III

THE CHEMICAL SYNTHESES OF ALKOXYLIPIDS

Wolfgang J. Baumann

I. Introduction

During the past decade the most significant progress in lipid science was probably made in the development of analytical techniques. Advanced analytical methodology has not only provided the basis for modern lipid biochemistry but has also made possible and inspired research on the chemical synthesis of lipids.

Application of refined analytical methods to synthetic problems has many times revealed the insufficiency of standard procedures and has shown that traditional synthetic routes, intermediates, and reaction conditions suitable for the preparation of short-chain or aromatic compounds are not necessarily applicable to the preparation of lipids. Generally, reactions of long-chain intermediates are relatively nonspecific, and may lead to a number of products having similar physical properties. The low specificity may be explained by the fact that the reaction rates for long-chain intermediates are significantly lower than those for the corresponding short-chain homologs, although the reactivity of a given functional group is largely independent of the chain length.

Evidently, *lipid syntheses require tailormade intermediates of very specific reactivities* in order to secure maximum yields of the desired product. Specifically for ether synthesis, the reactivity of an alkylating agent should be sufficiently high to secure reaction with the alkoxide within a reasonable period of time. On the other hand, an alkylating agent of very high reactivity may decompose or may participate in a side reaction before the site of alkylation becomes accessible.

This chapter has been written to serve the chemist and biochemist involved in various areas of research on ether lipids as a guide to convenient and efficient synthetic procedures. Naturally, a chapter on a topic as comprehensive as "The Chemical Syntheses of Alkoxylipids" requires emphasis on certain aspects and the omission of others. From the vast amount of literature papers were selected for their basic or practical interest or for their presentation of new or essentially improved synthetic procedures. Synthetic routes that are of historical interest will be mentioned briefly.

II. Syntheses of Long-Chain Alkyl Glycerol Ethers

Long-chain alkyl glycerol ethers are the most important representatives of the alkoxylipids. They are constituents of a large variety of naturally occurring neutral and polar lipids (for reviews, see Mangold and Baumann, 1967; Snyder, 1968). Asymmetrical glycerol ethers occur in optically active forms. α-Alkyl glycerol ethers usually possess the D-configuration (Baer and Fischer, 1941), as do the constituent alk-1-enyl ethers of neutral and polar plasmalogens (Baumann *et al.*, 1966b; Schmid *et al.*, 1967; Craig *et al.*, 1966), i.e., they occur as 1-alkyl-*sn*-glycerols. The α,β-dialkyl glycerol ethers found in the lipids of extremely halophilic bacteria (Sehgal *et al.*, 1962) exist in the L-form, i.e., they are 2,3-dialkyl-*sn*-glycerols (Kates *et al.*, 1965a).

In this chapter the system of stereospecific numbering (*sn*-system; Hirschman, 1960) will be used throughout; it has been recommended by the IUPAC-IUB Commission on Biochemical Nomenclature (1967). In several instances, however, the D,L-nomenclature according to Baer and Fischer (1939b) will also be used, not only because of its historical merits, but primarily because in original publications the Baer nomenclature is used almost exclusively. In order to minimize confusion Greek letters will be used to designate the glycerol oxygens of racemic compounds or of enantiomorphic forms of unknown configuration.

α-Alkyl glycerol ethers are of special importance because they are conveniently used as key intermediates in the synthesis of most other alkoxylipids, such as di- and triethers of glycerol, glycol ethers, alkylacylglycerols, and alkoxyphospholipids. Therefore, the synthesis of α-alkyl glycerol ethers will be dealt with in greatest detail.

Formation of the ether linkage between a long-chain moiety and a specific hydroxy group of glycerol usually involves reaction of a suitable glycerol derivative with a long-chain alkylating agent. In principle, the same glycerol derivatives can be used for the synthesis of glycerol ethers as for the preparation of partial glycerides (for reviews, see Malkin and Bevan, 1957; Hartman, 1958; Mattson and Volpenhein, 1962). Protection of one or two of the hydroxy groups of glycerol can be achieved by ketalation, acetalation, or etherification with bulky, acid-labile substituents. After alkylation in basic medium, the protective groups are easily removed by acid-catalyzed hydrolysis or by hydrogenolysis. In contrast to the preparation of glycerides, no special precautions need to be taken when removing the blocking groups, because long-chain aliphatic ethers, unlike esters (E. Fischer, 1920), do not suffer hydrolysis or other alterations in dilute acidic medium. On the other hand, alkylation requires more drastic reaction conditions than does acylation, and this can lead to the formation of a number of side products. Special care has to be taken in selecting optimum alkylation conditions in order to suppress side reactions.

A. α-Alkyl Glycerol Ethers

More than a century ago, Berthelot (1854) and Reboul (1860) reported syntheses of alkyl glycerol ethers. Their procedures, however, were found unsuitable for the preparation of long-chain homologs (G. G. Davies *et al.*, 1930). The first long-chain α-alkyl glycerol ethers (**3**), namely hexadecyl and octadecyl glycerol ethers, were synthesized by alkylation of the sodium salt of allyl alcohol (**1**) with hexadecyl or octadecyl chloride, respectively, followed by oxidation of the allyl ether (**2**) by means of hydrogen peroxide in glacial acetic acid (G. G. Davies *et al.*, 1930). The procedure, which, of course, is suitable only for the preparation of saturated and racemic glycerol ethers, was improved upon later by other investigators who used alkyl iodides for alkylation, and peracetic acid (Kornblum and Holmes, 1942) or performic acid (Stegerhoek and Verkade, 1956a) for oxidation of the allyl group. [1-^{14}C]-Hexadecylglycerol was also prepared by this procedure (Bergström and Blomstrand, 1956).

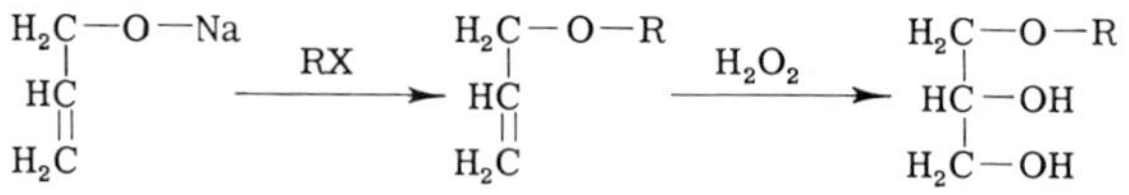

(**1**) (**2**) (**3**)

Most modern procedures for the synthesis of asymmetrical monoalkyl glycerol ethers involve alkylation of the acetone ketal of glycerol. Isopropylideneglycerol (**4**) was first utilized by E. Fischer *et al.* (1920) for the

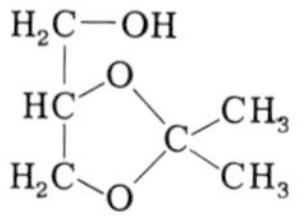

(**4**)

preparation of α-monoglycerides, and is easily synthesized by the procedures used by Renoll and Newman (1955) or Hartman (1960). The ketal (**4**) can be prepared free of the six-ring isomer—which is quite unstable because of axial methyl substitution—and following alkylation, the blocking group can easily be removed by acid-catalyzed hydrolysis. Isopropylideneglycerol can be prepared in optically active form (H. O. L. Fischer and Baer, 1941), as the D-isomer from D-mannitol (Baer and Fischer, 1939a; Baer, 1952) or as the L-isomer from L-arabinose of mesquite gum (Baer and Fischer, 1939c), and thus can serve for the preparation of both enantiomorphic forms of glycerol ethers.

Condensation of isopropylideneglycerol with a long-chain alkylating agent is commonly achieved via the potassium or sodium salt. Earlier methods involved the formation of the sodium salt of D- and L-isopropylideneglycerol from sodium naphthalene in dimethyl glycol ether, alkylation with long-chain alkyl iodides (Baer and Fischer, 1941) or tosylates (Baer *et al.*, 1944), and subsequent hydrolysis in acetic acid. Baer *et al.* (1944) also attempted the synthesis of D-selachyl alcohol, the common naturally occurring 1-*cis*-9′-octadecenyl-*sn*-glycerol, and of its L-isomer by reacting oleyl tosylate with the sodium salt of the suitable isopropylideneglycerol. However, the procedure led largely to the *trans* isomer (Baer *et al.*, 1947). In order to avoid isomerization, Baer *et al.* (1947) reacted the tosylate of isopropylideneglycerol with the sodium salt of *cis*-9-octadecenol and obtained *cis*-9′-octadecenylglycerol, but in very low yield. Baylis *et al.* (1958) prepared hexadecyl and octadecyl glycerol ethers by condensation of isopropylideneglycerol and long-chain alkyl iodides in the presence of sodium in boiling xylene, whereas S. C. Gupta and Kummerow (1959) preferred alkyl tosylates for alkylation of the potassium salt of isopropylideneglycerol. Condensation can also conveniently be achieved by the use of sodium amide in toluene (Anatol *et al.*, 1964; Berecoechea and Anatol, 1965; Berecoechea, 1966). The same authors also described a synthesis of dodecyl- and tetradecylglycerols from the corresponding long-chain alcohols and epichlorohydrin in the presence of boron trifluoride followed by hydrolysis in sodium formate solution; however, they did not sufficiently establish the purity of the final products.

The large number of synthetic procedures that have been developed (Table I) may be taken as an indication of the difficulties which are generally encountered in the preparation of pure α-alkyl glycerol ethers, especially unsaturated ones. In 1964, Baumann and Mangold reinvestigated various synthetic routes and showed that most difficulties arise from unsuitable alkylating agents (Baumann *et al.*, 1966a). They described two convenient methods for the synthesis of asymmetrical alkyl glycerol ethers (**6**) with high overall yields, utilizing long-chain methanesulfonates for alkylation of isopropylideneglycerol (**4**). Potassium in benzene or potassium hydroxide in xylene were used to achieve condensation, and aqueous hydrochloric acid in methanol for hydrolysis of intermediate (**5**) (Baumann and Mangold, 1964).

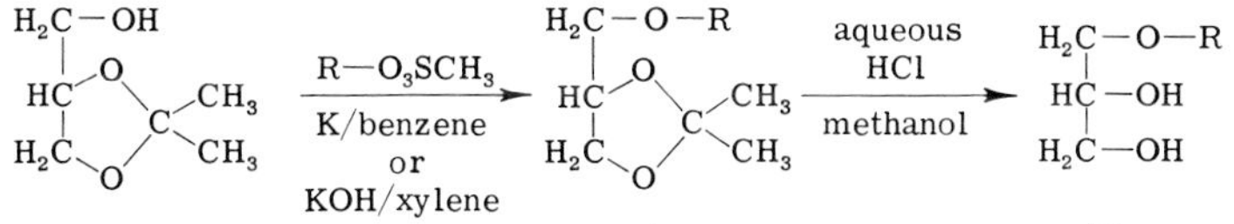

(**4**) (**5**) (**6**)

When potassium hydroxide is used for condensation, the water formed is removed by azeotropic distillation (Kates *et al.*, 1963a). Both procedures are suitable for the preparation of sterically pure *cis*- and *trans*-unsaturated ethers, as well as for the synthesis of optically active isomers (Baumann *et al.*, 1966b) such as 3-alkyl-*sn*-glycerols (the L-isomers) from 1,2-isopropylidene-*sn*-glycerol.

TABLE I

Long-Chain α-Alkyl Glycerol Ethers

α-Alkyl	Formula	Molecular weight	Yield (%)	MP (°C)	CST[a] (°C)	Ref.
Decyl	$C_{13}H_{28}O_3$	232.4	63,89	38.5[b]	55	*c*
Undecyl	$C_{14}H_{30}O_3$	246.4	75	37	61.5	*d*
Dodecyl	$C_{15}H_{32}O_3$	260.4	73,75	49.5	66.5	*c*
			76	48	—	*e*
Tridecyl	$C_{16}H_{34}O_3$	274.5	79	47.5	70.5	*d*
Tetradecyl	$C_{17}H_{36}O_3$	288.5	80,81	58.5	74	*c*
			50	55	—	*e*
Pentadecyl	$C_{18}H_{38}O_3$	302.5	69	58	77.5	*d*
12′-Methyltetradecyl	$C_{18}H_{38}O_3$	302.5	59	43–44.3	—	*f*
Hexadecyl	$C_{19}H_{40}O_3$	316.5	—	64–65	—	*g*
			—	62–63	—	*h*
			64	62–63	—	*i*
			69,79	65.5	81	*c*
			55	59,64	—	*e*
			73,77	65.5–66.5	—	*j*
Heptadecyl	$C_{20}H_{42}O_3$	330.6	74	64.5	84	*d*
14′-Methylhexadecyl	$C_{20}H_{42}O_3$	330.6	62	48.5–50	—	*f*
Octadecyl	$C_{21}H_{44}O_3$	344.6	55,67	70–71	—	*g,k*
			—	71	—	*h*
			87	71–72	—	*l*
			61	71–72	—	*i*
			90	69–70	—	*m*
			78	71–71.5	87.5	*c*
			66,72	65,72	—	*e*
			75	71–72	—	*n*
			60,64	70.5–72	—	*j*
Nonadecyl	$C_{22}H_{46}O_3$	358.6	71	71.5	90.5	*d*
Eicosyl	$C_{23}H_{48}O_3$	372.6	79,85	76.5–77	93	*c*
			93	75–76	93	*o*
Phytanyl	$C_{23}H_{48}O_3$	372.6	22	—	—	*p*
Heneicosyl	$C_{24}H_{50}O_3$	386.7	89	76.5	95.5	*d*
cis-9-Hexadecenyl	$C_{19}H_{38}O_3$	314.5	57	12.5	68	*c*
cis-9-Octadecenyl	$C_{21}H_{42}O_3$	342.6	14	17.6–19	—	*q*
			45,49	18–19	75	*c*
			87	—	—	*n*
trans-9-Octadecenyl	$C_{21}H_{42}O_3$	342.6	45	50–51	—	*q*
			81	49	77	*c*
cis,cis-9,12-Octadecadienyl	$C_{21}H_{40}O_3$	340.6	68	8	62	*c*
cis,cis,cis-9,12,15-Octadecatrienyl[r]	$C_{21}H_{38}O_3$	338.5	66	13	52.5	*c*

The procedure was also used for the preparation of ^{14}C-labeled α-alkyl glycerol ether of high specific activity. Figure 1 shows a thin-layer chromatogram of several intermediates and products as they are used or obtained in the synthesis of α-[1'-^{14}C]-hexadecyl glycerol ether from [1-^{14}C]-palmitic acid via mesylates (Mangold *et al.*, 1965).

3-*cis*-9'-Octadecenyl-*sn*-glycerol (the L-form) can also be prepared (Palameta and Kates, 1966) by alkylation of 1,2-isopropylidene-*sn*-glycerol with oleyl bromide, if the latter is available as the pure geometric isomer (Baumann and Mangold, 1966c). The only naturally occurring L-monoalkyl glycerol ether, found in the lipids of *Halobacterium cutirubrum*, was recently synthesized by Joo *et al.* (1968) by condensation of phytanyl bromide with 1,2-isopropylideneglycerol in the presence of potassium hydroxide in benzene.

Chacko and Hanahan (1968) have described a very elegant route to the synthesis of the common 1-alkyl-*sn*-glycerols (**8**), the D-isomers, from the more readily accessible L-isomers (**6**). The procedure involves a Walden inversion of the ditosylate (**7**), by tosylate displacement with potassium acetate in ethanol followed by alkaline hydrolysis.

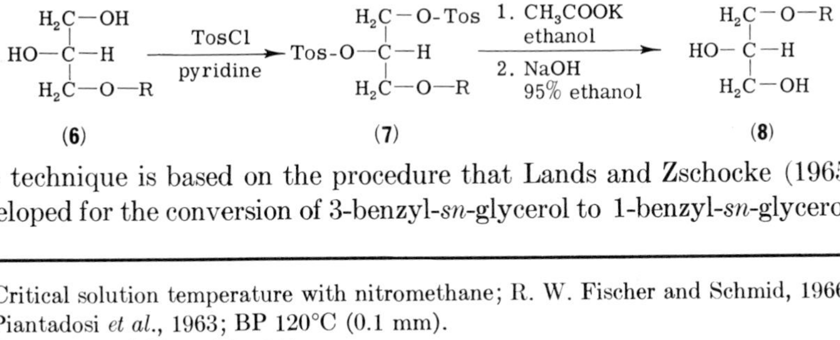

(**6**) (**7**) (**8**)

The technique is based on the procedure that Lands and Zschocke (1965) developed for the conversion of 3-benzyl-*sn*-glycerol to 1-benzyl-*sn*-glycerol.

---

[a] Critical solution temperature with nitromethane; R. W. Fischer and Schmid, 1966.
[b] Piantadosi *et al.*, 1963; BP 120°C (0.1 mm).
[c] Baumann and Mangold, 1964.
[d] Baumann *et al.*, 1965.
[e] Anatol *et al.*, 1964; Berecoechea and Anatol, 1965; Berecoechea, 1966.
[f] Serdarevich and Carroll, 1966.
[g] G. G. Davies *et al.*, 1930.
[h] Baer and Fischer, 1941; D- and L-isomers.
[i] Baylis *et al.*, 1958.
[j] Chacko and Hanahan, 1968; D- and L-isomers.
[k] Kornblum and Holmes, 1942.
[l] Stegerhoek and Verkade, 1956a.
[m] S. C. Gupta and Kummerow, 1959.
[n] Palameta and Kates, 1966; L-isomer.
[o] Baumann *et al.*, 1966b; L-isomer.
[p] Joo *et al.*, 1968; L-isomer.
[q] Baer *et al.*, 1947; D-isomer.
[r] Different batches of starting material contained up to 20% of *trans* isomer.

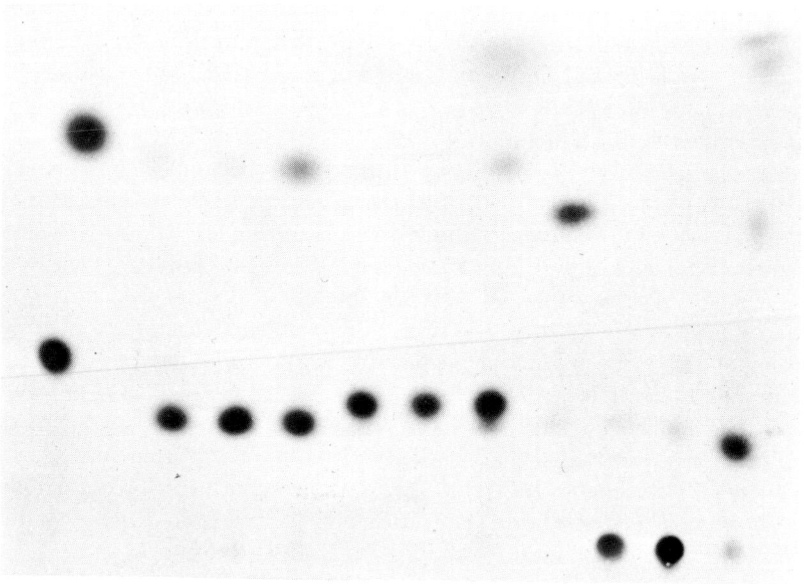

FIG. 1. Thin-layer autoradiogram of intermediates and products of the synthesis of racemic α-[1-^{14}C]-hexadecyl glycerol ether from [1-^{14}C]-palmitic acid. Adsorbent, Silica Gel G. Developing solvent, hexane:diethyl ether:acetic acid, 80:20:1, v/v/v. The fractions are: (1) palmitic acid, (2) methyl palmitate, (3) hexadecanol, crude, (4) hexadecanol, recrystallized, (5) filtrate of recrystallization, (6) hexadecyl methanesulfonate, crude, (7) hexadecyl methanesulfonate, recrystallized, (8) filtrate of recrystallization, (9) α-hexadecylisopropylideneglycerol, crude, (10) α-hexadecylglycerol, crude, (11) α-hexadecylglycerol, recrystallized, (12) filtrate of recrystallization.

Serdarevich and Carroll (1966) prepared α-alkyl glycerol ethers having 12′-methyltetradecyl and 14′-methylhexadecyl side chains by reaction of the bromides with isopropylideneglycerol in the presence of potassium hydroxide in benzene.

2′-Methoxy-substituted α-alkylglycerols, which were found in liver oil of Greenland shark, were synthesized by Hallgren and Ställberg (1967). Bromination of hexadecanoyl chloride and subsequent methanolysis gave methyl 2-bromohexadecanoate, which was reacted with sodium methoxide. Lithium aluminum hydride reduction, tosylation of the alcohol, and etherification with isopropylideneglycerol according to S. C. Gupta and Kummerow (1959) yielded 1-(2′-methoxyhexadecyl)-*rac*-glycerol.

B. β-Alkyl Glycerol Ethers

Symmetrical alkyl glycerol ethers have been found in lipid extracts of *Halobacterium cutirubrum* (Joo *et al.*, 1968), but they may be artifacts, produced during methanolysis of dialkyl glycerol ethers, rather than natural constituents (Kates *et al.*, 1971). Numerous syntheses of β-alkyl glycerol ethers have been reported. All procedures involve alkylation of 1,3-benzylideneglycerol (**9**), followed by removal of the acetal blocking group.

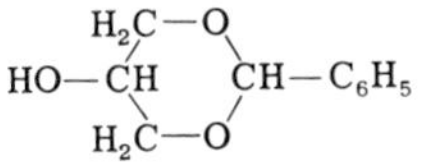

(**9**)

1,3-Benzylideneglycerol was first described by Hibbert and Carter (1929). It is commonly used as starting material for the preparation of β-monoglycerides. Its synthesis has been improved by Verkade and van Roon (1942), and it was found that the crystalline 1,3-benzylideneglycerol usually used in syntheses consists of a mixture of *cis*- and *trans*-2-phenyl-5-hydroxy-1,3-dioxanes (Baggett *et al.*, 1960; Dobinson and Foster, 1961).

Alkylation of 1,3-benzylideneglycerol (**9**) was achieved by W. H. Davies *et al.* (1934) by reaction of its potassium salt with hexadecyl or octadecyl iodide in benzene solution. The 2-phenyl-5-alkoxy-1,3-dioxanes (**10**) were hydrolyzed in 75% aqueous HCl/ethanol to yield β-alkyl glycerol ethers. S. C. Gupta and Kummerow (1959) prepared symmetrical glycerol ethers via the saturated alkyl tosylates, whereas Bevan and Malkin (1960) achieved condensation of alkyl iodides with the potassium salt of 1,3-benzylideneglycerol in xylene solution, and found that higher yields of β-alkyl glycerol ethers (**11**) can be obtained when the acetal blocking group is removed by hydrogenolysis rather than acidic hydrolysis. Oswald *et al.* (1966) used radioactively labeled alkyl bromides for alkylation, and acidic hydrolysis for removal of the benzylidene group from (**10**). Arnold *et al.* (1967) prepared 2-hexadecylglycerol by condensation of (**9**) with alkyl iodide in aqueous KOH and subsequent hydrolysis of (**10**) in H_2SO_4 ethanol/water.

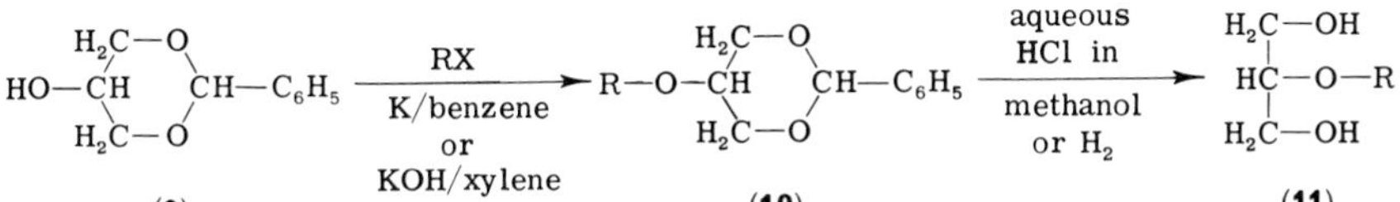

(**9**) (**10**) (**11**)

Syntheses of symmetrical monoethers of glycerol with branched side chains were reported by Serdarevich and Carroll (1966) as well as by Joo *et al.* (1968). Both used potassium hydroxide in benzene for condensation of the long-chain alkyl halides with benzylideneglycerol and achieved removal of the acetal group by acidic hydrolysis. The preparation of 2-*O*-3′*R*,7′*R*,-11′*R*,15′-tetramethylhexadecylglycerol is of special interest (Joo *et al.*, 1968), as this 2-phytanylglycerol ether was also obtained by methanolysis of the lipids of *Halobacterium cutirubrum*.

The physical characteristics of β-alkyl ethers of glycerol are very similar to those of the corresponding α-alkyl ethers. Distinction of the two isomers by melting points, infrared spectra (Baumann and Ulshöfer, 1968), or migration rate in adsorption chromatography is not reliable (Wood and Snyder, 1967). However, both isomers can easily be distinguished by chromatography on boric acid-impregnated silicic acid layers, since they have different complexing properties (Prey *et al.*, 1961; Serdarevich and Carroll, 1966; Wood and Snyder, 1967).

C. α, β-Dialkyl Glycerol Ethers

The first preparations of long-chain dialkyl glycerol ethers were undertaken to aid in establishing the structures of glycerol diethers from bacterial lipids. Such unusual lipid constituents had been found in *Halobacterium cutirubrum* (Sehgal *et al.*, 1962; Faure *et al.*, 1963), but their presence was also suspected in other organisms (see review by Snyder, 1968).

Kates *et al.* (1963a) prepared α, β-dihexadecyl- and dioctadecylglycerols (**14**) by alkylation of α-benzylglycerol (**12**) (Sowden and Fischer, 1941; Howe and Malkin, 1951) with the corresponding long-chain alkyl bromides in boiling benzene using potassium hydroxide for condensation. The benzyl group was removed from (**13**) by hydrogenolysis over a palladium catalyst.

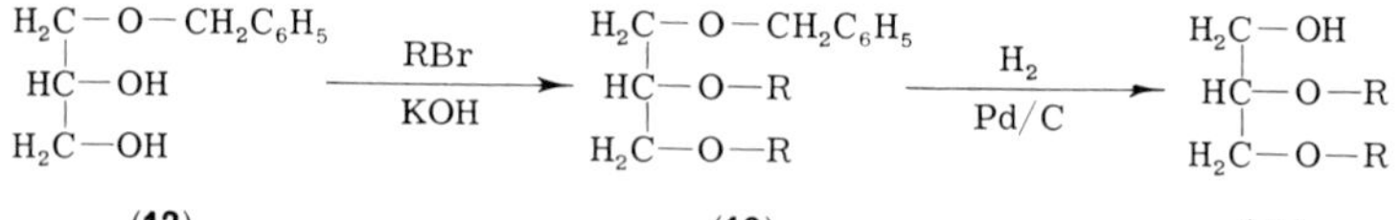

(**12**) (**13**) (**14**)

The procedure was also successfully applied to the preparation of optically active 1,2-dialkyl-*sn*-glycerols (the D-isomers; Kates *et al.*, 1963a, 1965a; Baer and Stanacev, 1965) from 3-benzyl-*sn*-glycerol (Sowden and Fischer, 1941). 2,3-Dialkyl-*sn*-glycerol, the L-isomer, was prepared by Joo *et al.* (1968) from 1-benzyl-*sn*-glycerol (Lands and Zschocke, 1965). This route of synthesis does not permit, of course, the preparation of unsaturated dialkyl ethers of glycerol, or of ethers having two different alkyl groups.

Helferich and Sieber (1927, 1928) were the first investigators to describe another valuable method for protecting primary hydroxy groups of glycerol: reaction with triphenylchloromethane (trityl chloride). Trityl ethers are stable in basic medium, i.e., under conditions of alkylation, and following alkylation the trityl group can conveniently be removed by acid-catalyzed hydrolysis. Thus, diethers of glycerol, even those with unsaturated aliphatic chains, can be prepared via trityl intermediates. Verkade *et al.* (1937) have thoroughly reviewed the use of trityl derivatives in lipid syntheses.

α, β-Dialkyl glycerol ethers with identical alkyl groups can be directly synthesized (Kates *et al.*, 1965a; Palameta and Kates, 1966; Thomas and Law, 1966) by alkylation of α-tritylglycerol (Helferich *et al.*, 1923; Jackson and Lundberg, 1963) with alkyl bromides. Following alkylation, the trityl ether linkage is cleaved with aqueous hydrochloric acid in methanol (Kates *et al.*, 1965a) or anhydrous hydrogen chloride in petroleum hydrocarbon (Thomas and Law, 1966). The optically active diethers of glycerol in L-form that have two phytanyl groups, as they do in halophilic bacteria (Kates *et al.*, 1966), were first prepared by these procedures. 1-Triphenylmethyl-*sn*-glycerol (**15**) (D-tritylglycerol, Baer and Fischer, 1945) was reacted with phytanyl bromide, and intermediate (**16**) was hydrolyzed to yield 2,3-di-*O*-(3′*R*,7′*R*,11′*R*,15′-tetramethylhexadecyl)-*sn*-glycerol (**17**) (Kates *et al.*, 1965a).

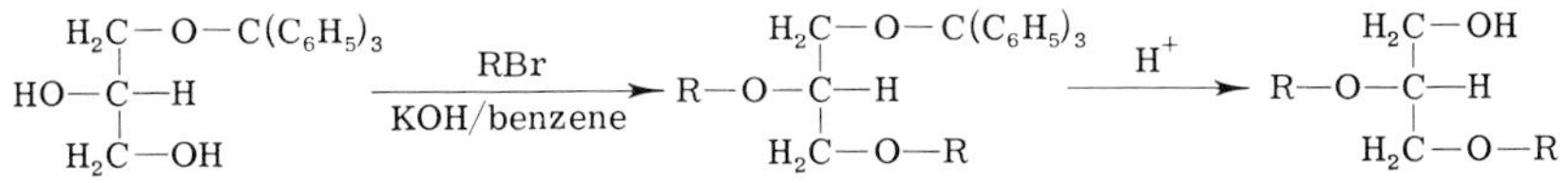

(**15**) (**16**) (**17**)

R designates 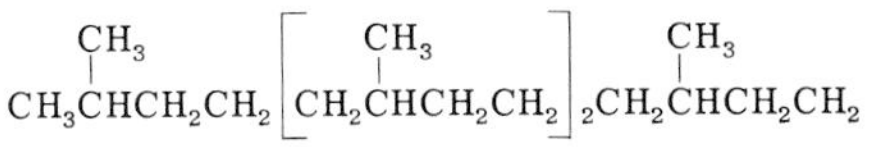

In analogy to the tritylation of glycerol, α-alkyl glycerol ethers can be reacted with trityl chloride to prepare specifically substituted, mixed α, β-dialkyl glycerol ethers (Table II).

Baumann and Mangold (1966a) have described a synthesis applicable to the preparation of saturated and unsaturated, as well as mixed, dialkyl ethers, which is based on the β-alkylation of α-alkyl α'-trityl glycerol with alkyl methanesulfonates. Optically active 2,3-dialkyl-*sn*-glycerols can be prepared by a similar sequence of reactions (Baumann *et al.*, 1966b; Palameta and Kates, 1966). First, the primary hydroxy group of an alkyl glycerol ether (**18**) is reacted with triphenylchloromethane in pyridine. Further alkylation of the tritylated glycerol ether (**19**) with alkyl methanesulfonate using potassium hydroxide in xylene for condensation, leads to

TABLE II

Long-Chain α,β-Dialkyl Glycerol Ethers

α-Alkyl	β-Alkyl	Formula	Molecular weight	Yield (%)	MP (°C)	CST[a] (°C)	Ref.
Octyl	Octyl	$C_{19}H_{40}O_3$	316.5	52	*b*	—	*c*
Dodecyl	Dodecyl	$C_{27}H_{56}O_3$	428.6	54	38	—	*c*
				40	37	—	*d*
Dodecyl	Tetradecyl	$C_{29}H_{60}O_3$	456.8	66	42–43	—	*d*
Tetradecyl	Dodecyl	$C_{29}H_{60}O_3$	456.8	60	42	—	*d*
Dodecyl	Hexadecyl	$C_{31}H_{64}O_3$	484.9	55	43–44	—	*d*
Tetradecyl	Tetradecyl	$C_{31}H_{64}O_3$	484.9	60	51–52	—	*d*
				49[e]	37–38	—	*f*
Hexadecyl	Dodecyl	$C_{31}H_{64}O_3$	484.9	65	43–44	—	*d*
Dodecyl	Octadecyl	$C_{33}H_{68}O_3$	512.9	58	49–50	—	*d*
Tetradecyl	Hexadecyl	$C_{33}H_{68}O_3$	512.9	60	52	—	*d*
Hexadecyl	Tetradecyl	$C_{33}H_{68}O_3$	512.9	53	54	—	*d*
Octadecyl	Dodecyl	$C_{33}H_{68}O_3$	512.9	76	41.5–42	123	*g*
				25	48	—	*d*
Tetradecyl	Octadecyl	$C_{35}H_{72}O_3$	541.0	70	55	—	*d*
Hexadecyl	Hexadecyl	$C_{35}H_{72}O_3$	541.0	50	54.5–55.5	—	*h*
				55[e]	48.5–49.5	—	*h*
				60	59.5–60	132	*g*
				78	59	—	*c*
				54	55	—	*d*
Octadecyl	Tetradecyl	$C_{35}H_{72}O_3$	541.0	78	52–52.5	131.5	*g*
				50	54	—	*d*
Hexadecyl	Octadecyl	$C_{37}H_{76}O_3$	569.0	91	61.5–62	139.5	*g*
				42	60–61	—	*d*
Octadecyl	Hexadecyl	$C_{37}H_{76}O_3$	569.0	88	58.5–59	139.5	*g*
				50	58–59	—	*d*
Octadecyl	Octadecyl	$C_{39}H_{80}O_3$	597.1	50	63–64	—	*h*
				77[e]	53.5–54.5	—	*h*
				89	64–65	147	*g*
				65	64–65	—	*d*
Octadecyl	Eicosyl	$C_{41}H_{84}O_3$	625.1	87	66.5–67	154	*g*
Eicosyl	Eicosyl	$C_{43}H_{88}O_3$	653.1	70[i]	63–63.5	159	*j*
Phytanyl	Phytanyl	$C_{43}H_{88}O_3$	653.1	31[e]	—	—	*k*
				46,20[i]	—	—	*k,l*
Octadecyl	*cis*-9-Octadecenyl	$C_{39}H_{78}O_3$	595.1	69	23–24	136	*g*
				57[i]	—	—	*m*
Octadecyl	*trans*-9-Octadecenyl	$C_{39}H_{78}O_3$	595.1	91	48–48.5	138.5	*g*
cis-9-Octadecenyl	Octadecyl	$C_{39}H_{78}O_3$	595.1	80[n]	20–21	136.5	*g*
				64[i]	—	—	*m*

TABLE II (*Continued*)

α-Alkyl	β-Alkyl	Formula	Molecular weight	Yield (%)	MP (°C)	CST[a] (°C)	Ref.
cis-9-Octa-decenyl	*cis*-9-Octa-decenyl	$C_{39}H_{76}O_3$	593.0	83[n]	—	126	*g*
				69[i]	—	—	*m*
				64	—	—	*o*
				52	—	—	*c*
				84[e]	—	—	*p*

[a] Critical solution temperature with nitromethane; R. W. Fischer and Schmid, 1966.
[b] BP (0.2 mm) 150°–155°C.
[c] Paltauf and Spener, 1968.
[d] Anatol *et al.*, 1964; Berecoechea and Anatol, 1965; Berecoechea, 1966.
[e] 1,2-Dialkyl-*sn*-glycerol (D-isomer).
[f] Baer and Stanacev, 1965.
[g] Baumann and Mangold, 1966a.
[h] Kates *et al.*, 1963a.
[i] 2,3-Dialkyl-*sn*-glycerol (L-isomer).
[j] Baumann *et al.*, 1966b.
[k] Kates *et al.*, 1965a.
[l] Joo *et al.*, 1968.
[m] Palameta and Kates, 1966.
[n] Yield of crude material.
[o] Thomas and Law, 1966.
[p] Chacko and Hanahan, 1968.

1-trityl-2,3-dialkyl-*sn*-glycerol (**20**) without formation of significant amounts of dialkyl ether, *R—O—R*. Final hydrolytic cleavage of the trityl ether bond in methanolic hydrochloric acid gives 2,3-dialkyl-*sn*-glycerol (**21**) in good yields.

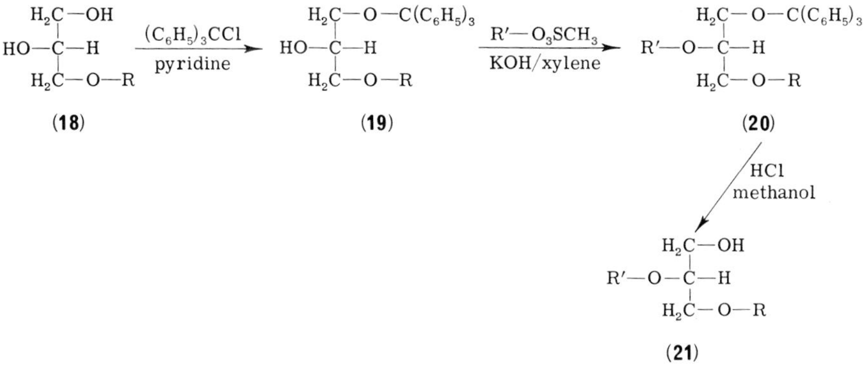

Use of alkyl methanesulfonates instead of bromides permits the synthesis of pure *cis*- and *trans*-unsaturated glycerol diethers without isomerization or other alteration of double bonds, and without formation of a significant amount of by-products. Esters of methanesulfonic acid (mesylates) are easily prepared (Baumann and Mangold, 1964), and have been versatile intermediates in the synthesis of various other types of lipids (for reviews, see Mangold, 1968; Lundberg, 1969).

Anatol *et al.* (1964) described a number of saturated dialkyl ethers of glycerol that they prepared via α-alkyl-α′-tritylglycerol. Alkylation was achieved with long-chain alkyl *p*-toluenesulfonates in the presence of sodium amide in toluene. Detritylation with HCl/isopropanol led to the asymmetrical glycerol ethers (Berecoechea and Anatol, 1965; Berecoechea, 1966). Chacko and Hanahan (1968) prepared 1,2-di-*cis*-9′-octadecenyl-*sn*-glycerol by a similar procedure, but used oleyl bromide for alkylation of the tritylated monoether and glacial acetic acid for removal of the protective group.

In 1968, Paltauf and Spener described an excellent synthesis of dialkyl glycerol ethers, including unsaturated ones, based on the alkylation of α-tetrahydropyranylglycerol (**22**) with long-chain alkyl methanesulfonates. Intermediate (**23**) was hydrolyzed at room temperature in a mixture of concentrated hydrochloric acid, methanol and diethyl ether to give diether (**24**) in good yield and high purity.

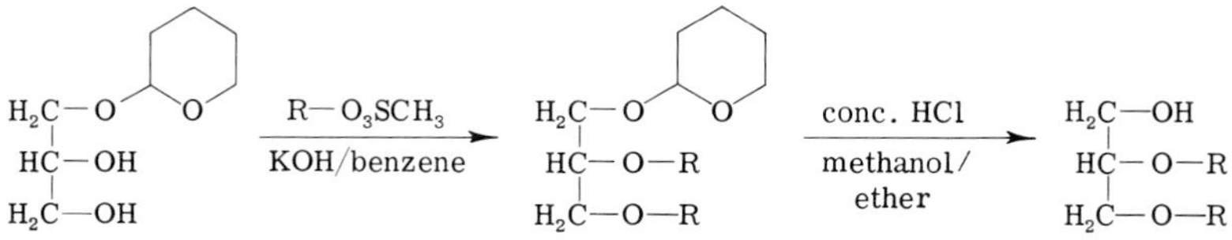

(**22**) (**23**) (**24**)

α-Tetrahydropyranylglycerol is easily prepared in its racemic form according to Barry and Craig (1955). Its use in place of tritylated glycerol significantly facilitates acid-catalyzed removal of the blocking group as well as purification of the final products. Krabisch and Borgström (1965) had previously utilized α-tetrahydropyranylglycerol for the synthesis of diglycerides. They had shown that the pyranyl "ether" linkage, actually an acetal linkage, is cleaved at least ten times more easily by acid than is the trityl ether linkage.

D. α, α′-Dialkyl Glycerol Ethers

Symmetrical dialkyl glycerol ethers have not as yet been detected as natural lipid constituents. They can be prepared by partial alkylation of α-alkyl glycerol ethers with long-chain alkyl methanesulfonates (Baumann and Mangold, 1966a). It was found that formation of the α, α′-diether is favored when potassium in benzene instead of potassium hydroxide in xylene was used for condensation. The symmetrical glycerol ether was then isolated by adsorption chromatography.

A procedure leading specifically to α, α′-dialkyl ethers of glycerol was described by Damico *et al.* (1967), but due to the hydrogenation step required, the synthesis is limited to the preparation of ethers having saturated alkyl groups. 1,3-Benzylideneglycerol (**9**) was reacted with benzyl chloride in benzene in the presence of potassium hydroxide to yield benzyl ether (**25**) which upon acid-catalyzed hydrolysis gave β-benzyl

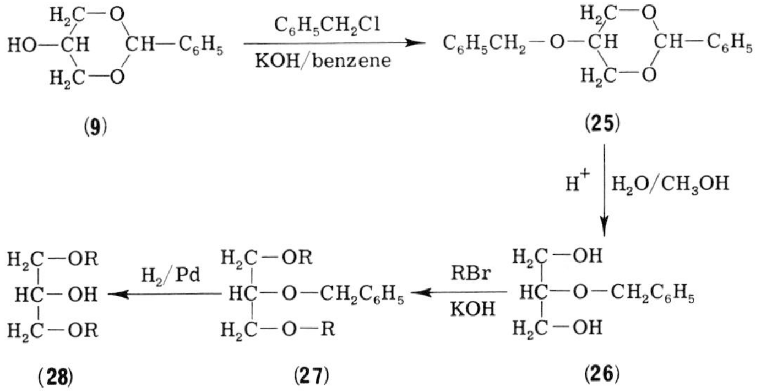

glycerol ether (**26**). Alkylation of (**26**) was achieved by reaction with alkyl bromide, and triether (**27**) was finally subjected to hydrogenation to give the α, α′-dialkyl glycerol ether (**28**). A number of medium-chain α, α′-dialkylglycerols were prepared by Fauran *et al.* (1964) by condensation of 1-chloro-3-alkoxy-2-propanols with sodium alkoxide; however, the compounds were not characterized satisfactorily.

E. Trialkyl Glycerol Ethers

More recently, trialkyl ethers of glycerol have received considerable attention, despite the fact that these compounds have not as yet been detected in biological materials. Unlike glycerides, trialkyl glycerol ethers are neither hydrolyzed nor absorbed when fed to rats (Spener *et al.*, 1968), and thus serve as ideal lipid markers in fat absorption studies since they are structurally similar to triglycerides. Morgan and Hofmann (1970a, b) used tritium-labeled 1-hexadecyl-2, 3-didodecylglycerol in feeding experiments and found that less than 0.2% of the triether was absorbed, and that the radioactivity was present exclusively as triether in feces. Because trialkyl glycerol ethers do not appear to be toxic and do not significantly affect digestion, these compounds may play a future role not only as markers in observing malabsorption syndromes but also as noncaloric fat substitutes in diets.

Trialkyl glycerol ethers (**29**) can be prepared by complete alkylation of alkyl glycerol ethers or of dialkyl glycerol ethers (Table III).

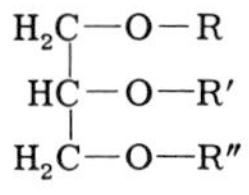

(**29**)

Baumann and Mangold (1966a) synthesized triethers of glycerol having identical, or two or three different, alkyl chains by alkylation of mono- or diethers of glycerol with long-chain alkyl methanesulfonates in the presence of potassium hydroxide in xylene. Direct alkylation of glycerol was unsuccessful under the conditions employed. Schmid *et al.* (1966) showed that the method is also suitable for the preparation of triethers having unsaturated alkyl chains. Selection of optimum alkylation conditions is extremely important for the synthesis of glycerol triethers in order to retard formation of long-chain dialkyl ethers, R—O—R, that can only be efficiently removed from the final product by chromatography.

Fauran *et al.* (1964) prepared medium-chain trialkyl glycerol ethers by alkylation of diethers of glycerol with alkyl halides in toluene in the presence of sodium, whereas Berecoechea (1966) used tosylate for alkylation and sodium amide for condensation. Paltauf and Spener (1968) reported the synthesis of several $1'$-^{14}C-labeled trialkyl glycerol ethers. They used long-chain alkyl methanesulfonates for alkylation and potassium hydroxide in benzene to achieve condensation. Lutton and Stewart (1970) have recently studied the phase behavior of a number of trialkyl glycerol ethers by thermal and diffraction methods and have shown that triethers are dimorphic.

III. Syntheses of Long-Chain Ether Esters of Glycerol

André and Bloch (1935) pioneered the investigation of alkyldiacylglycerols which they had found in marine oils. It has been demonstrated since that these compounds occur in a large number of human (Schmid and Mangold, 1966) and animal tissues (for review, see Snyder, 1968), and that alkyldiacylglycerols can be separated from triglycerides by thin-layer adsorption chromatography (Mangold and Malins, 1960). Dialkyl-

TABLE III

Long-Chain Trialkyl Glycerol Ethers

α-Alkyl	β-Alkyl	α′-Alkyl	Formula	Molecular weight	Yield (%)	MP (°C)	CST[a] (°C)	Ref.
Octyl	Octyl	Octyl	$C_{27}H_{56}O_3$	428.7	66	—	—	*b*
Dodecyl	Decyl	Decyl	$C_{35}H_{72}O_3$	541.0	—	−5	146	*c*
Dodecyl	Dodecyl	Dodecyl	$C_{39}H_{80}O_3$	597.1	71	—	—	*b*
Dodecyl	Tetradecyl	Tetradecyl	$C_{43}H_{88}O_3$	653.2	—	17.5–18	168	*c*
Dodecyl	Dodecyl	Hexadecyl	$C_{43}H_{88}O_3$	653.2	>30	23	—	*d*
Dodecyl	Hexadecyl	Hexadecyl	$C_{47}H_{96}O_3$	709.3	33	36.5–37	178	*e*
Dodecyl	Hexadecyl	Octadecyl	$C_{49}H_{100}O_3$	373.3	46	34–35; 27.5–28	182.5	*e*
Dodecyl	Octadecyl	Octadecyl	$C_{51}H_{104}O_3$	765.4	43	43–44	185.5	*e*
Tetradecyl	Hexadecyl	Octadecyl	$C_{51}H_{104}O_3$	765.4	80	43	—	*f*
Hexadecyl	Hexadecyl	Hexadecyl	$C_{51}H_{104}O_3$	765.4	39	47–48	186	*e*
						47.8;42.1	—	*g*
					75	47	—	*b*
Octadecyl	Octadecyl	Octadecyl	$C_{57}H_{116}O_3$	849.6	36	57.5–58	196	*e*
						57.3;51	—	*g*
Octadecyl	Octadecyl	*cis*-9-Octadecenyl	$C_{57}H_{114}O_3$	847.5	68	23–25	188.5	*c,h*
Octadecyl	*cis*-9-Octadecenyl	*cis*-9-Octadecenyl	$C_{57}H_{112}O_3$	845.5	61	5–7	183	*c,h*
cis-9-Octadecenyl	*cis*-9-Octadecenyl	*cis*-9-Octadecenyl	$C_{57}H_{110}O_3$	843.5	57	−11	176.5	*c,h*
					59	—	—	*b*

[a] Critical solution temperature with nitromethane; R. W. Fischer and Schmid, 1966.
[b] Paltauf and Spener, 1968.
[c] Baumann and Mangold, 1965.
[d] Morgan and Hofmann, 1970a.
[e] Baumann and Mangold, 1966a.
[f] Berecoechea, 1966.
[g] Lutton and Stewart, 1970.
[h] Schmid *et al.*, 1966.

acylglycerols have not been conclusively identified as natural lipid constituents.

Alkyldiacylglycerols and dialkylacylglycerols are conveniently prepared by acylation of the corresponding glycerol ethers. As a curiosity it may be mentioned that the alkylation of a 1,2-diglyceride has also been investigated as a route for the preparation of alkyldiacylglycerols (Karpova *et al.*, 1966). The problems involved in the acylation of glycerol ethers are usually the same as those encountered in syntheses of glycerides. Therefore, the preparation of ether esters of glycerol will be dealt with in brief and general terms. Reference is made to the excellent reviews on glyceride syntheses by Hartman (1958) and by Mattson and Volpenhein (1962).

Stegerhoek and Verkade (1956a) prepared alkylacylglycerols and alkyldiacylglycerols derived from α-octadecyl and β-octadecyl glycerol ethers by acylation with acyl chlorides in pyridine/chloroform solution. α-Monoether monoesters of glycerol were prepared via α-alkyl-α′-tritylglycerol (**30**). Acylation of (**30**) in the β-position gave α-alkyl-β-acyl-

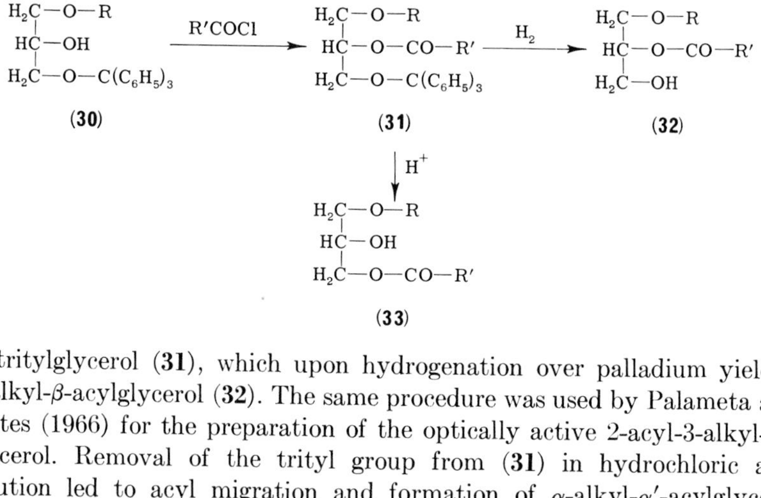

(**30**) (**31**) (**32**) (**33**)

α′-tritylglycerol (**31**), which upon hydrogenation over palladium yielded α-alkyl-β-acylglycerol (**32**). The same procedure was used by Palameta and Kates (1966) for the preparation of the optically active 2-acyl-3-alkyl-*sn*-glycerol. Removal of the trityl group from (**31**) in hydrochloric acid solution led to acyl migration and formation of α-alkyl-α′-acylglycerol (**33**). The partially acylated glycerol ethers were used for the preparation of mixed-acid alkyldiacylglycerols (Stegerhoek and Verkade, 1956a). Lawson and Getz (1961) prepared alkyldiacylglycerols, including some with unsaturated carbon chains.

Baumann and Mangold (1966b) reported syntheses of series of saturated and unsaturated alkyldiacylglycerols and dialkylacylglycerols from the corresponding glycerol ethers by reaction with acyl chlorides in benzene/pyridine solution, using an acylation procedure similar to that described by A. Gupta and Malkin (1952). Optically active 1,2-diacyl-3-alkyl-*sn*-glycerols and 1-acyl-2,3-dialkyl-*sn*-glycerols, the L-isomers, were prepared in a similar way (Baumann *et al.*, 1966b). Oswald *et al.* (1966) described the preparation of a number of ^{14}C-labeled α-alkyl- and β-alkyl-diacylglycerols; they achieved acylation with a mixture of fatty acid and trifluoroacetic anhydride. Lutton and Stewart (1970) have recently studied some of the physical characteristics of alkyl glycerides.

The only synthesis of an unsaturated α-alkyl-β-acylglycerol was described by Chacko and Hanahan (1968). 1-*cis*-9′-Octadecenyl-2-octadecanoyl-*sn*-glycerol was synthesized from 1-*cis*-9′-octadecenyl-2-octadecanoyl-3-trityl-*sn*-glycerol by removal of the trityl group with boric acid in triethyl borate. The 1,2-ether ester could be isomerized in the presence of catalytic amounts of perchloric acid (Martin, 1953) to yield 1-*cis*-9′-octadecenyl-3-octadecanoyl-*sn*-glycerol. Chacko and Hanahan (1968) mentioned that acylation of 1-alkylglycerols preferentially occurs in position 3 unless a large excess of acyl chloride is used. The synthesis of unsaturated α-alkyl-β-acylglycerols by the method used by Krabisch and Borgström (1965) for the preparation of unsaturated 1,2-diglycerides via the tetrahydropyranyl ether appears feasible but has not been investigated.

IV. Syntheses of Other Neutral Alkoxylipids

Recently, it has become apparent that the lipids of a variety of biological materials may contain alcohol moieties other than glycerol, and that short-chain diols are widespread lipid constituents. Diol lipids have been detected in animal and plant tissues, as well as in microorganisms (for a review, see Bergelson, 1969) It has been shown that diol lipids not only occur as esters (Demarteau-Ginsburg, 1958) and alk-1-enyl ether esters (Bergelson *et al.*, 1966), but also as alkyl ethers (Varanasi and Malins, 1969).

Saturated and unsaturated monoalkyl ethers of glycol can be prepared (Baumann *et al.*, 1967) in high purity by oxidative cleavage of α-alkyl glycerol ethers (**34**) with sodium metaperiodate in pyridine (Baumann *et al.*, 1969) and subsequent reduction of the aldehyde (**35**) with lithium aluminum hydride to yield (**36**).

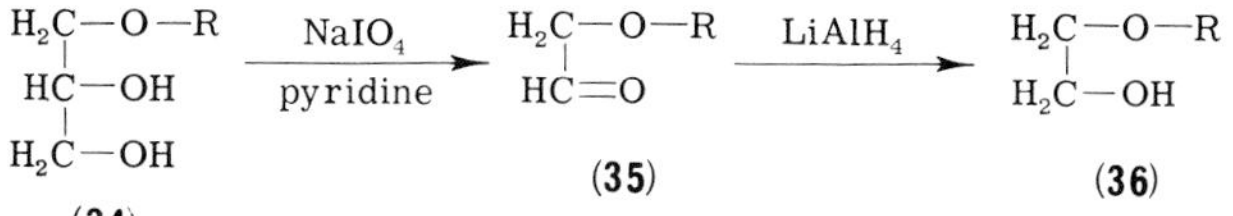

(**34**) (**35**) (**36**)

This method is also applicable to the preparation of the alk-1-enyl glycol ethers from the corresponding glycerol derivatives (Baumann *et al.*, 1968). Earlier procedures for the preparation of glycol ethers involved alkylation of sodium glycolate (Cooper and Partridge, 1950; Chakhovskoy *et al.*, 1956;

Wrigley *et al.*, 1960), oxyethylation of alcohols (Wrigley *et al.*, 1960), or direct alkylation of glycol (Shirley *et al.*, 1953; Fieser *et al.*, 1956).

Monoethers of other diols, such as 1,3-propanediol (Baumann *et al.*, 1967), can be prepared via the monotrityl derivatives using essentially the same procedures of tritylation, alkylation with long-chain methanesulfonates, and hydrolysis as described for the syntheses of glycerol ethers. Diethers of glycol have been prepared by condensation of chloromethyl alkyl ethers via a Wurtz synthesis (Teramura and Oda, 1951) or by direct alkylation of glycol (Cooper and Partridge, 1950; Shirley *et al.*, 1953; Fieser *et al.*, 1956). Baumann *et al.* (1967) have prepared diethers of glycol and 1,3-propanediol by alkylation of the corresponding monoethers with long-chain alkyl methanesulfonates in xylene in the presence of potassium hydroxide. Ether esters of diols are conveniently obtained from monoethers using standard acylation procedures (Baumann *et al.*, 1967).

Methods for the detection (Bergelson, 1969; Kramer *et al.*, 1971a, b) as well as isolation of diol lipids (Calderon and Baumann, 1970a, b) have been developed. However, the separation of intact, natural diol lipids still creates some difficulties because of the similarity of these minor lipid constituents to the bulk of glycerol-derived lipids.

Another unusual ether lipid, namely *O*-alkylcholesterol, has been detected by Funasaki and Gilbertson (1968) in the lipids of bovine cardiac muscle. Paltauf (1968) described the synthesis of a number of cholesterol ethers having saturated and unsaturated aliphatic chains, and determined the properties of these lipids. The synthesis involved reaction of cholesterol with long-chain alkyl methanesulfonates in the presence of potassium hydroxide in boiling benzene. The final product was purified by adsorption chromatography and recrystallization.

V. Syntheses of Alkoxyphospholipids

Alkoxyphospholipids occur in nature as analogs of phosphatidylethanolamine (Carter *et al.*, 1958; Svennerholm and Thorin, 1960; Hanahan and Watts, 1961) and phosphatidylcholine (Renkonen, 1962; Pietruszko and Gray, 1962). In these alkoxylipids the α-acyl group is replaced by an alkyl function. Alkoxyphospholipids containing two ether groups are also known (Sehgal *et al.*, 1962). They were shown to possess structures analogous to those of phosphatidylglycerophosphate and phosphatidylglycerol (Kates *et al.*, 1963b; Faure *et al.*, 1963, 1964).

Most synthetic routes developed for the preparation of ester phospholipids from diacylglycerols are also suitable for the preparation of ether phospholipids from alkylacylglycerols or dialkylglycerols. Due to the relative stability of the ether bond in basic, and to a large extent also in acidic media, the preparation of ether phospholipids is often simpler than that of the ester analogs.

A. Alkylglycerophosphates

In 1956, Stegerhoek and Verkade described the synthesis of an alkylacylglycerophosphate using the procedure reported by Baer (1951) for the preparation of the diacyl analog (Stegerhoek and Verkade, 1956b). α-Octadecyl-β-acylglycerol (**37**) was reacted with diphenylphosphoryl chloride (Brigl and Müller, 1959) in the presence of pyridine, and the corresponding

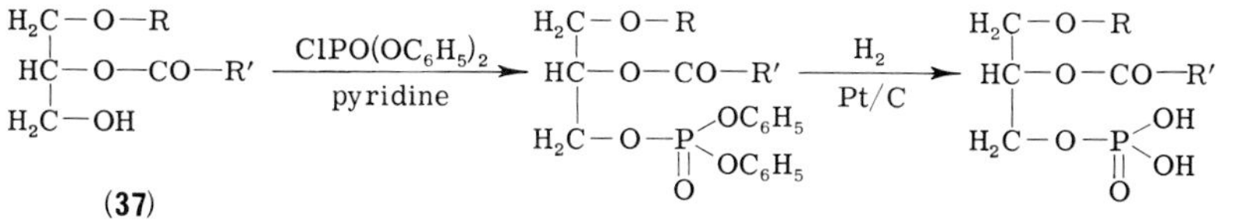

(**37**) (**38**) (**39**)

α′-phosphoric acid diphenyl ester (**38**) was obtained in high yields. Finally, hydrogenolysis was effected in the presence of platinum on active carbon. By this procedure α-octadecyl-β-hexadecanoyl- and α-octadecyl-β-octadecanoylglycerophosphates (**39**) were prepared. With α-alkyl-α′-acylglycerol as starting material the corresponding β-phosphoric acids were similarly obtained. Removal of the acyloxy group with sodium methoxide yielded the corresponding α-octadecylglycerophosphate. The procedure, of course, is suitable only for the preparation of saturated phosphatidic acid analogs.

Chen and Barton (1970) prepared the D-isomer of a diether analog of phosphatidic acid, namely, 1,2-dioctadecyl-*sn*-glycero-3-phosphate, by applying Baer's phosphorylation procedure (Baer, 1951) to 1,2-dioctadecyl-*sn*-glycerol. 1,2-Dioctadecenylglycerophosphate tritium-labeled in 9,10-positions was prepared by Paltauf (1969). Kates *et al.* (1971) described two synthetic routes to an L-isomer, 2,3-diphytanyl-*sn*-1-glycerophosphate. Again, phosphorylation of 2,3-diphytanyl-*sn*-glycerol (**40**) with diphenylphosphoryl chloride in anhydrous pyridine followed by catalytic hydrogenolysis of the phenyl ester bonds led to the dialkylglycerophosphate (**44**). A second procedure involved tosylation of diether (**40**) and conversion of tosylate (**41**) to 1-iodo-2,3-diphytanyl-*sn*-glycerol (**42**). Reaction of (**42**) with silver di-*p*-nitrobenzyl phosphate (Zervas and Dilaris, 1956) in anhydrous benzene, and catalytic hydrogenolysis of the resulting *p*-nitrobenzyl ester (**43**) yielded 2,3-diphytanyl-*sn*-1-glycerophosphate (**44**) isolated as sodium or potassium salt.

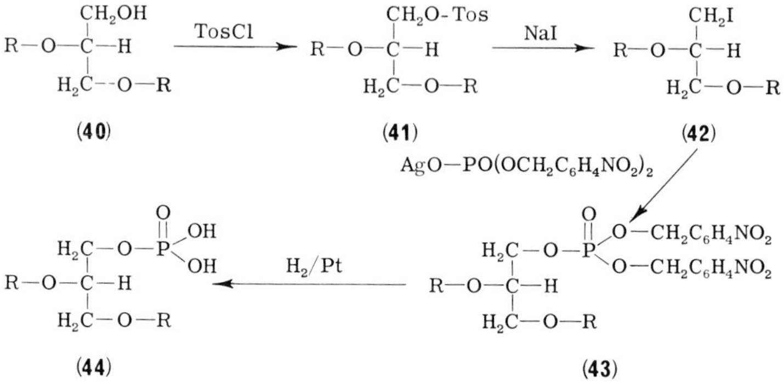

(40) (41) (42) (43) (44)

The diether glycerophosphate was condensed with cytidine monophosphate morpholidate in pyridine to yield 2, 3-diphytanyl-*sn*-glycerol-1-(cytidine 5′-diphosphate) (**45**), the diether analog of CDP-diglyceride.

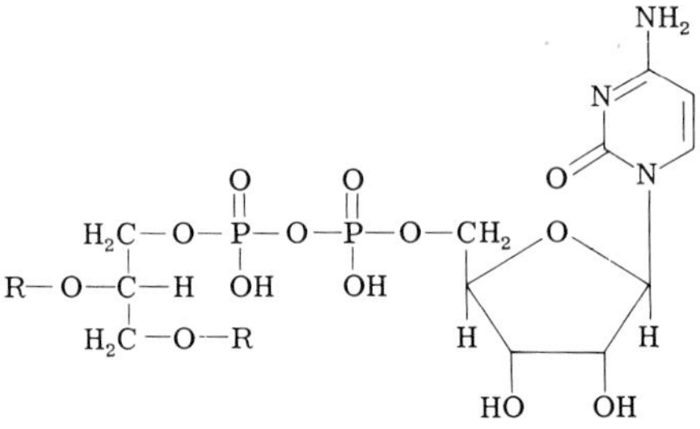

(45)

R designates 3*R*,7*R*,11*R*,15-tetramethylhexadecyl

B. Alkylglycerophosphorylethanolamines

Baylis *et al.* (1958) prepared the α-monoether analogs of cephalin with the intention of making standards available for establishing the structures of naturally occurring alkyl (Carter *et al.*, 1958) and alk-1-enyl ether phospholipids (Klenk and Debuch, 1954). They synthesized C_{16}- and C_{18}-substituted cephalin analogs by condensation of the iodides derived from α-alkyl-β-acylglycerol with silver 2-(benzyloxycarbonylamino)-ethylphenylphosphate in boiling xylene. α-Alkyl-β-acylglycerophosphorylethanolamines were obtained after hydrogenation in glacial acetic acid over Adam's catalyst. The β-alkyl analogs of cephalin were prepared similarly via the iodide of α-acyl-β-alkylglycerol (Bevan and Malkin, 1960).

Unsaturated monoether monoester glycerophosphorylethanolamines can also be prepared by applying standard procedures of cephalin synthesis

(Rose, 1947; Baer and Buchnea, 1959) to alkylacylglycerols in place of diacylglycerols. Chacko and Hanahan (1968) treated 1-*cis*-9′-octadecenyl-2-octadecanoylglycerol with phosphorus oxychloride in quinoline, and the resulting product was reacted with *N*-phthaloylethanolamine in pyridine. Hydrolysis followed by chromatographic purification of the crude *N*-phthaloyl phospholipid and final removal of the phthaloyl blocking group with hydrazine in ethanol (Daemen *et al.*, 1962) gave the desired 1-*cis*-9′-octadecenyl-2-octadecanoyl-3-glycerophosphorylethanolamine (**46**).

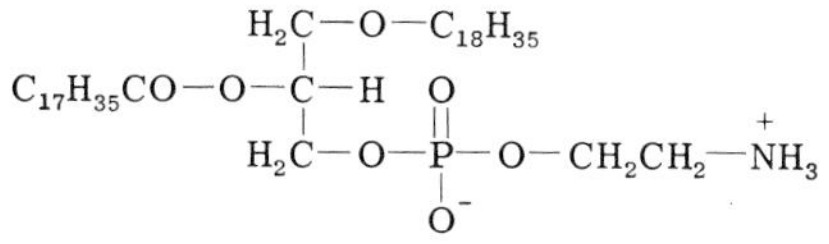

(**46**)

The 1,2-diether analogs of phosphatidylethanolamine were similarly synthesized from 1,2-dialkyl-*sn*-glycerols, both saturated and unsaturated ones (Chacko and Hanahan, 1968). Thomas and Law (1966) had previously prepared the racemic dioleylglycerophosphorylethanolamine by condensing the glycerol diether with 2-phthalimidoethyl dichlorophosphate in pyridine (Hirt and Berchtold, 1957) and subsequent reaction with hydrazine in ethanol. Paltauf (1969) prepared the racemic 1,2-dioctadecenylglycerophorphorylethanolamine labeled with tritium in 9,10 positions.

C. Alkylglycerophosphorylcholines

A procedure for the synthesis of monoether analogs of lecithin was described by Chacko and Hanahan (1968). The reaction of 1-*cis*-9′-octadecenyl-2-octadecanoyl-*sn*-glycerol with phosphorus oxychloride in quinoline (Baer and Kindler, 1962) gave the dichlorophosphate. The latter was reacted with choline iodide in pyridine, the remaining P—Cl bond was hydrolyzed by addition of water, and the final product, 1-*cis*-9′-octadecenyl-2-octadecanoyl-*sn*-glycero-3-phosphorylcholine (**47**), was recovered after treatment with silver carbonate and cation–anion exchange resin.

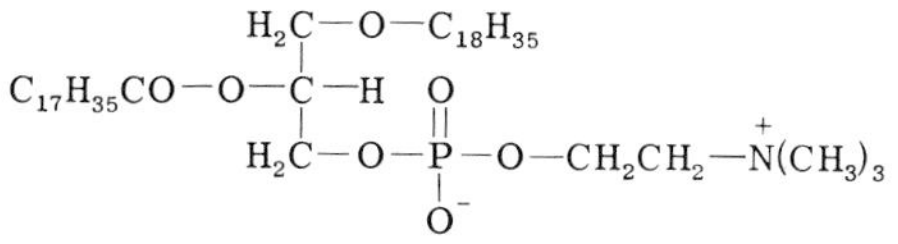

(**47**)

The first synthesis of a diether analog of lecithin was reported by Stanacev *et al.* (1964). 1,2-Dioctadecyl-*sn*-glycero-3-phosphorylcholine (**51**) was obtained by phosphorylation of 1,2-dioctadecyl-*sn*-glycerol (**48**) with monophenylphosphoryl dichloride in pyridine, esterification of the

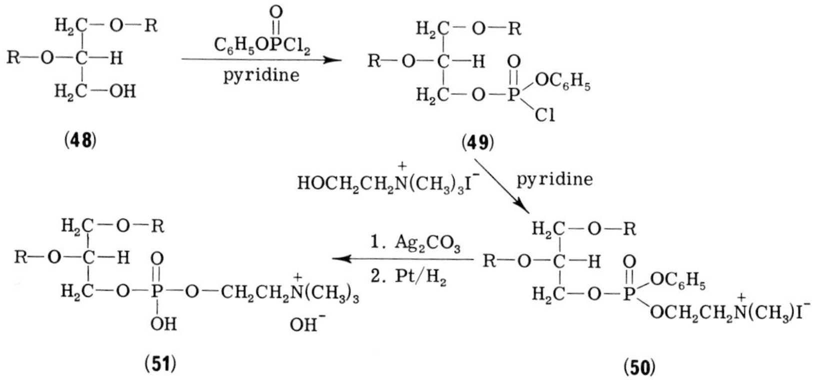

resulting dioctadecylglycerolphenylphosphoryl chloride (**49**) with choline iodide, conversion of dioctadecylglycerolphenylphosphorylcholine iodide (**50**) into the corresponding carbonate, and removal of the phenyl group by catalytic hydrogenation with platinum as catalyst. Eibl *et al.* (1967) described a somewhat modified procedure for the preparation of monoether and diether analogs of lecithin, having the choline ester group in position 3 or 2. Chacko and Hanahan (1968) prepared 1,2-di-*cis*-9′-octadecenyl-*sn*-glycero-3-phosphorylcholine from the glycerol diether by the sequence of reactions given above for the preparation of the corresponding monoether monoester phospholipid. Paltauf (1969) described the racemic 9,10-tritiated analog.

The synthesis of lysolecithin analogs, via the corresponding benzyl derivatives, was reported by Arnold *et al.* (1967).

D. Dialkylglycerophosphorylglycerol and Dialkylglycerophosphorylglycerophosphates

Joo and Kates (1968, 1969) have described the syntheses of the phytanyl diether analogs of phosphatidylglycerol and of phosphatidylglycerophosphate, the major lipid constituents of extremely halophilic bacteria, in particular *H. cutirubrum* (Kates *et al.*, 1965b). The major lipid of this bacterium, 2,3-diphytanyl-*sn*-glycero-1-phosphoryl-3′-*sn*-glycero-1′-phosphate, was prepared by condensation of the silver salt of 2,3-diphytanyl-*sn*-glycerol-1-(*O*-*p*-nitrobenzyl)phosphate (Joo and Kates, 1969) (**52**) with 1-iodo-2-*tert*-butyl-3-diphenylphosphorylglycerol (**53**).

Condensation product (**54**) was hydrogenated to remove the nitrobenzyl

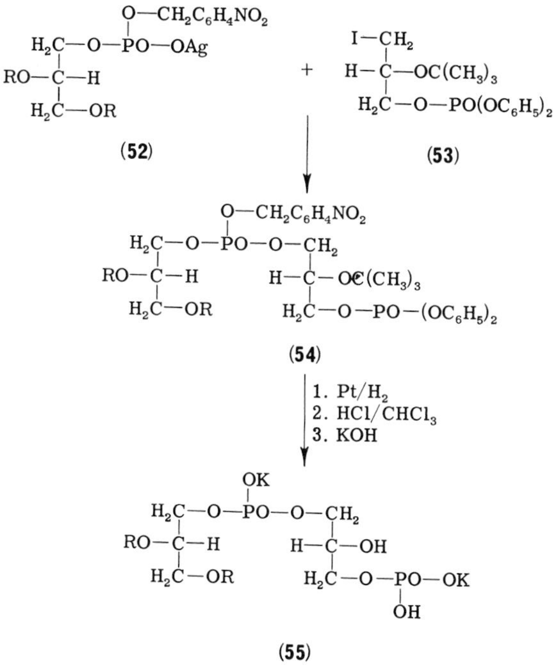

(**52**) (**53**) (**54**) (**55**)

and phenyl groups, and treated with anhydrous hydrogen chloride in chloroform for removal of the *tert*-butyl group. For comparison with natural preparations the potassium salt (**55**) of the phosphatidylglycerophosphate analog was prepared.

The phytanyl ether analog of phosphatidylglycerol was prepared by condensation of the silver salt (**52**) with 1-iodo-2-*tert*-butyl-3-benzylglycerol

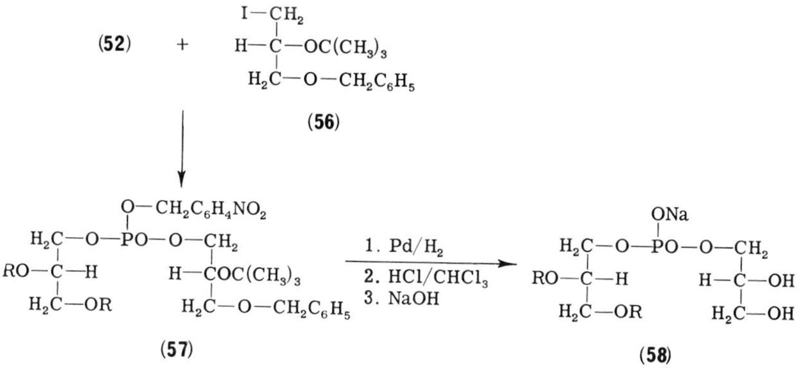

(**56**) (**57**) (**58**)

(**56**), hydrogenolysis of condensation product (**57**) and treatment with HCl/chloroform. Neutralization led to the sodium salt of 2,3-diphytanyl-*sn*-glycero-1-phosphoryl-3′-*sn*-glycerol (**58**).

VI. Syntheses of Some Phospholipid Analogs

Synthetic analogs of naturally occurring glycerophospholipids, in which the acyl groups are replaced by alkyl moieties, and in which the phosphate group is simultaneously modified, have been used as model compounds in enzymic investigations and also in therapeutic studies. Phosphonic acid monoether analogs of phosphatidylethanolamine and phosphatidylcholine were described by Chacko and Hanahan (1969). Baer and Stanacev (1965) and Rosenthal and Pousada (1965) prepared a number of cephalin-related diether glycerophosphonates. Syntheses of nonhydrolyzable analogs of phosphatidic acid (Rosenthal *et al.*, 1964; Rosenthal and Pousada, 1965) having the structures of dialkoxy propylphosphonates were also reported. More recently, Rosenthal (1966) described the synthesis of some other diether phosphonates related to lecithin (Rosenthal and Pousada, 1969), cephalin, and phosphatidic acid, and of a diether phosphinate analog of lecithin (Rosenthal *et al.*, 1969).

REFERENCES

Anatol, J., Berecoechea, J., and Giraud, D. (1964). *C. R. Acad. Sci.* **258,** 6466.

André, E., and Bloch, A. (1935). *Bull. Soc. Chim. Fr.* [5] **2,** 789.

Arnold, D., Weltzien, H. U., and Westphal, O. (1967). *Justus Liebigs Ann. Chem.* **709,** 234.

Baer, E. (1951). *J. Biol. Chem.* **189,** 235.

Baer, E. (1952). *Biochem. Prep.* **2,** 31.

Baer, E., and Buchnea, D. (1959). *J. Amer. Chem. Soc.* **81,** 1758.

Baer, E., and Fischer, H. O. L. (1939a). *J. Biol. Chem.* **128,** 463.

Baer, E., and Fischer, H. O. L. (1939b). *J. Biol. Chem.* **128,** 475.

Baer, E., and Fischer, H. O. L. (1939c). *J. Amer. Chem. Soc.* **61,** 761.

Baer, E., and Fischer, H. O. L. (1941). *J. Biol. Chem.* **140,** 397.

Baer, E., and Fischer, H. O. L. (1945). *J. Amer. Chem. Soc.* **67,** 944.

Baer, E., and Kindler, A. (1962). *Biochemistry* **1,** 518.

Baer, E., and Stanacev, N. Z. (1965). *J. Biol. Chem.* **240,** 44.

Baer, E., Rubin, L. J., and Fischer, H. O. L. (1944). *J. Biol. Chem.* **155,** 447.

Baer, E., Fischer, H. O. L., and Rubin, L. J. (1947). *J. Biol. Chem.* **170,** 337.

Baggett, N., Brimacombe, J. S., Foster, A. B., Stacey, M., and Whiffen, D. H. (1960). *J. Chem. Soc., London* p. 2574.

Barry, P. J., and Craig, B. M. (1955). *Can. J. Chem.* **33,** 716.

Baumann, W. J., and Mangold, H. K. (1964). *J. Org. Chem.* **29,** 3055.

Baumann, W. J., and Mangold, H. K. (1965). Unpublished results.

Baumann, W. J., and Mangold, H. K. (1966a). *J. Org. Chem.* **31,** 498.

Baumann, W. J., and Mangold, H. K. (1966b). *Biochim. Biophys. Acta* **116,** 570.
Baumann, W. J., and Mangold, H. K. (1966c). *J. Lipid Res.* **7,** 568.
Baumann, W. J., and Ulshöfer, H. W. (1968). *Chem. Phys. Lipids* **2,** 114.
Baumann, W. J., Kammereck, R., and Mangold, H. K. (1965). Unpublished results.
Baumann, W. J., Jones, L. L., Barnum, B. E., and Mangold, H. K. (1966a). *Chem. Phys. Lipids* **1,** 63.
Baumann, W. J., Mahadevan, V., and Mangold, H. K. (1966b). *Hoppe-Seyler's Z. Physiol. Chem.* **347,** 52.
Baumann, W. J., Schmid, H. H. O., Ulshöfer, H. W., and Mangold, H. K. (1967). *Biochim. Biophys. Acta* **144,** 355.
Baumann, W. J., Schmid, H. H. O., Kramer, J. K. G., and Mangold, H. K. (1968). *Hoppe-Seyler's Z. Physiol. Chem.* **349,** 1677.
Baumann, W. J., Schmid, H. H. O., and Mangold, H. K. (1969). *J. Lipid Res.* **10,** 132.
Baylis, R. L., Bevan, T. H., and Malkin, T. (1958). *J. Chem. Soc., London* p. 2962.
Berecoechea, J. (1966). Ph.D. Thesis, University of Reims, France.
Berecoechea, J., and Anatol, J. (1965). *C. R. Acad. Sci.* **260,** 3700.
Bergelson, L. D. (1969). *Progr. Chem. Fats Other Lipids* **10,** 239.
Bergelson, L. D., Vaver, V. A., Prokazova, N. V., Ushakov, A. N., and Popkova, G. A. (1966). *Biochim. Biophys. Acta* **116,** 511.
Bergström, S., and Blomstrand, R. (1956). *Acta Physiol. Scand.* **38,** 166.
Berthelot, M. (1854). *Ann. Chim. Phys.* [3] **41,** 216.
Bevan, T. H., and Malkin, T. (1960). *J. Chem. Soc., London* p. 350.
Brigl, P., and Müller, H. (1939). *Chem. Ber.* **72,** 2121.
Calderon, M., and Baumann, W. J. (1970a). *J. Lipid Res.* **11,** 167.
Calderon, M., and Baumann, W. J. (1970b). *Biochim. Biophys. Acta* **210,** 7.
Carter, H. E., Smith, D. B., and Jones, D. N. (1958). *J. Biol. Chem.* **232,** 681.
Chacko, G. K., and Hanahan, D. J. (1968). *Biochim. Biophys. Acta* **164,** 252.
Chacko, G. K., and Hanahan, D. J. (1969). *Biochim. Biophys. Acta* **176,** 190.
Chakhovskoy, N., Martin, R. H., and van Nechel, R. (1956). *Bull. Soc. Chim. Belg.* **65,** 453.
Chen, J.-S., and Barton, P. G. (1970). *Can. J. Biochem.* **48,** 585.
Cooper, F. C., and Partridge, M. W. (1950). *J. Chem. Soc., London* p. 459.
Craig, J. C., Hamon, D. P. G., Purushothaman, K. K., Roy, S. K., and Lands, W. E. M. (1966). *Tetrahedron* **22,** 175.
Daemen, F. J. M., De Haas, D. H., and Van Deenen, L. L. M. (1962). *Rec. Trav. Chim. Pays-Bas* **81,** 348.
Damico, R., Callahan, R. C., and Mattson, F. H. (1967). *J. Lipid Res.* **8,** 63.
Davies, G. G., Heilbron, I. M., and Owens, W. M. (1930). *J. Chem. Soc., London* p. 2542.
Davies, W. H., Heilbron, I. M., and Jones, W. E. (1934). *J. Chem. Soc., London* p. 1232.
Demarteau-Ginsburg, H. (1958). Ph.D. Thesis, University of Paris, France.
Dobinson, B., and Foster, A. B. (1961). *J. Chem. Soc., London* p. 2338.
Eibl, H., Arnold, D., Weltzien, H. U., and Westphal, O. (1967). *Justus Liebigs Ann. Chem.* **709,** 226.
Fauran, C., Miocque, M., and Gautier, J.-A. (1964). *Bull. Soc. Chim. Fr.* **4,** 831.
Faure, M., Marechal, J., and Troestler, J. (1963). *C. R. Acad. Sci.* **257,** 2187.
Faure, M., Marechal, J., and Troestler, J. (1964). *C. R. Acad. Sci.* **259,** 941.
Fieser, M., Fieser, L. F., Toromanoff, E., Hirata, Y., Heymann, H., Tefft, M., and Bhattacharya, S. (1956). *J. Amer. Chem. Soc.* **78,** 2825.

Fischer, E. (1920). *Ber. Deut. Chem. Ges.* **53,** 1621.
Fischer, E., Bergmann, M., and Bärwind, H. (1920). *Ber. Deut. Chem. Ges.* **53,** 1589.
Fischer, H. O. L., and Baer, E. (1941). *Chem. Rev.* **29,** 287.
Fischer, R. W., and Schmid, H. H. O. (1966). *In* "Standard Methods of Chemical Analysis" (F. J. Welcher, ed.), Vol. IIIA, p. 669. Van Nostrand Co., Princeton, New Jersey.
Funasaki, H., and Gilbertson, J. R. (1968). *J. Lipid Res.* **9,** 766.
Gupta, A., and Malkin, T. (1952). *J. Chem. Soc., London* p. 2405.
Gupta, S. C., and Kummerow, F. A. (1959). *J. Org. Chem.* **24,** 409.
Hallgren, B., and Ställberg, G. (1967). *Acta Chem. Scand.* **21,** 1519.
Hanahan, D. J., and Watts, R. (1961). *J. Biol. Chem.* **236,** PC 59.
Hartman, L. (1958). *Chem. Rev.* **58,** 845.
Hartman, L. (1960). *Chem. Ind. (London)* p. 711.
Helferich, B., and Sieber, H. (1927). *Hoppe-Seyler's Z. Physiol. Chem.* **170,** 31.
Helferich, B., and Sieber, H. (1928). *Hoppe-Seyler's Z. Physiol. Chem.* **175,** 311.
Helferich, B., Speidel, P. E., and Toeldte, W. (1923). *Ber. Deut. Chem. Ges.* **56,** 766.
Hibbert, H., and Carter, N. M. (1929). *J. Amer. Chem. Soc.* **51,** 1601.
Hirschman, H. (1960). *J. Biol. Chem.* **235,** 2762.
Hirt, R., and Berchtold, R. (1957). *Helv. Chim. Acta* **40,** 1928.
Howe, R. J., and Malkin, T. (1951). *J. Chem. Soc., London* p. 2663.
IUPAC-IUB Commission on Biochemical Nomenclature. (1967). *J. Lipid Res.* **8,** 523.
Jackson, J. E., and Lundberg, W. O. (1963). *J. Amer. Oil Chem. Soc.* **40,** 276.
Joo, C. N., and Kates, M. (1968). *Biochim. Biophys. Acta* **152,** 800.
Joo, C. N., and Kates, M. (1969). *Biochim. Biophys. Acta* **176,** 278.
Joo, C. N., Shier, T., and Kates, M. (1968). *J. Lipid Res.* **9,** 782.
Karpova, N. B., Grum-Grzhimailo, M. A., Volkova, L. V., Vernikova, L. M., and Preobrazhenskii, N. A. (1966). *Zh. Org. Khim.* **2,** 789.
Kates, M., Chan, T. H., and Stanacev, N. Z. (1963a). *Biochemistry* **2,** 394.
Kates, M., Sastry, P. S., and Yengoyan, L. S. (1963b). *Biochim. Biophys. Acta* **70,** 705.
Kates, M., Palameta, B., and Yengoyan, L. S. (1965a). *Biochemistry* **4,** 1595.
Kates, M., Yengoyan, L. S., and Sastry, P. S. (1965b). *Biochim. Biophys. Acta* **98,** 252.
Kates, M., Palameta, B., Joo, C. N., Kushner, D. J., and Gibbons, N. E. (1966). *Biochemistry* **5,** 4092.
Kates, M., Park, C. E., Palameta, B., and Joo, C. N. (1971). *Can. J. Biochem.* **49,** 275.
Klenk, E., and Debuch, H. (1954). *Hoppe-Seyler's Z. Physiol. Chem.* **296,** 179.
Kornblum, N., and Holmes, H. N. (1942). *J. Amer. Chem. Soc.* **64,** 3045.
Krabisch, L., and Borgström, B. (1965). *J. Lipid Res.* **6,** 156.
Kramer, J. K. G., Baumann, W. J., and Holman, R. T. (1971a). *Lipids* **6,** 492.
Kramer, J. K. G., Holman, R. T., and Baumann, W. J. (1971b). *Lipids* **6,** 727.
Lands, W. E. M., and Zschocke, A. (1965). *J. Lipid Res.* **6,** 324.
Lawson, D. D., and Getz, H. R. (1961). *Chem. Ind. (London)* p. 1404.
Lundberg, W. O. (1969). *J. Amer. Oil Chem. Soc.* **46,** 145.
Lutton, E. S., and Stewart, C. B. (1970). *Lipids* **5,** 545.
Malkin, T., and Bevan, T. H. (1957). *Progr. Chem. Fats Other Lipids* **4,** 63.
Mangold, H. K. (1968). *J. Label. Compounds* **4,** 3.
Mangold, H. K., and Baumann, W. J. (1967). *Lipid Chromatogr. Anal.* **1,** 339.
Mangold, H. K., and Malins, D. C. (1960). *J. Amer. Oil Chem. Soc.* **37,** 383.
Mangold, H. K., Houle, C. R., and Baumann, W. J. (1965). Unpublished results.
Martin, J. B. (1953). *J. Amer. Chem. Soc.* **75,** 5482.

Mattson, F. H., and Volpenhein, R. A. (1962). *J. Lipid Res.* **3,** 281.
Morgan, R. G. H., and Hofmann, A. F. (1970a). *J. Lipid Res.* **11,** 223.
Morgan, R. G. H., and Hofmann, A. F. (1970b). *J. Lipid Res.* **11,** 231.
Oswald, E. O., Piantadosi, C., Anderson, C. E., and Snyder, F. (1966). *Lipids* **1,** 241.
Palameta, B., and Kates, M. (1966). *Biochemistry* **5,** 618.
Paltauf, F. (1968). *Monatsh. Chem.* **99,** 1277.
Paltauf, F. (1969). *Biochim. Biophys. Acta* **176,** 818.
Paltauf, F., and Spener, F. (1968). *Chem. Phys. Lipids* **2,** 168.
Piantadosi, C., Hirsch, A. F., Yarbro, C. L., and Anderson, C. E. (1963). *J. Org. Chem.* **28,** 2425.
Pietruszko, R., and Gray, G. M. (1962). *Biochim. Biophys. Acta* **56,** 232.
Prey, V., Berbalk, H., and Kausz, M. (1961). *Mikrochim. Acta* p. 968.
Reboul, M. (1860). *Ann. Chim. Phys.* [3] **60,** 5.
Renkonen, O. (1962). *Biochim. Biophys. Acta* **59,** 497.
Renoll, M., and Newman, M. S. (1955). *Org. Syn.* III, 502.
Rose, W. G. (1947). *J. Amer. Chem. Soc.* **69,** 1384.
Rosenthal, A. F. (1966). *J. Lipid Res.* **7,** 779.
Rosenthal, A. F., and Pousada, M. (1965). *Rec. Trav. Chim. Pays-Bas* **84,** 833.
Rosenthal, A. F., and Pousada, M. (1969). *Lipids* **4,** 37.
Rosenthal, A. F., Kosolapoff, G. M., and Geyer, R. P. (1964). *Rec. Trav. Chim. Pays-Bas* **83,** 1273.
Rosenthal, A. F., Chodsky, S. V., and Han, S. C. H. (1969). *Biochim. Biophys. Acta* **187,** 385.
Schmid, H. H. O., and Mangold, H. K. (1966). *Biochem. Z.* **346,** 13.
Schmid, H. H. O., Mangold, H. K., Lundberg, W. O., and Baumann, W. J. (1966). *Microchem. J.* **11,** 306.
Schmid, H. H. O., Baumann, W. J., and Mangold, H. K. (1967). *J. Amer. Chem. Soc.* **89,** 4797.
Sehgal, S. N., Kates, M., and Gibbons, N. E. (1962). *Can. J. Biochem. Physiol.* **40,** 69.
Serdarevich, B., and Carroll, K. K. (1966). *Can. J. Biochem.* **44,** 743.
Shirley, D. A., Zietz, J. R., Jr., and Reedy, W. H. (1953). *J. Org. Chem.* **18,** 378.
Snyder, F. (1968). *Progr. Chem. Fats Other Lipids* **10,** 287.
Sowden, J. C., and Fischer, H. O. L. (1941). *J. Amer. Chem. Soc.* **63,** 3244.
Spener, F., Paltauf, F., and Holasek, A. (1968). *Biochim. Biophys. Acta* **152,** 368.
Stanacev, N. Z., Baer, E., and Kates, M. (1964). *J. Biol. Chem.* **239,** 410.
Stegerhoek, L. J., and Verkade, P. E. (1956a). *Rec. Trav. Chim. Pays-Bas* **75,** 143.
Stegerhoek, L. J., and Verkade, P. E. (1956b). *Rec. Trav. Chim. Pays-Bas* **75,** 467.
Svennerholm, L., and Thorin, H. (1960). *Biochim. Biophys. Acta* **41,** 371.
Teramura, K., and Oda, R. (1951). *J. Chem. Soc. Jap.* **54,** 605.
Thomas, P. J., and Law, J. H. (1966). *J. Lipid Res.* **7,** 453.
Varanasi, U., and Malins, D. C. (1969). *Science* **166,** 1158.
Verkade, P. E., and van Roon, J. D. (1942). *Rec. Trav. Chim. Pays-Bas* **61,** 831.
Verkade, P. E., vander Lee, J., and Meerburg, W. (1937). *Rec. Trav. Chim. Pays-Bas* **56,** 613.
Wood, R., and Snyder, F. (1967). *Lipids* **2,** 161.
Wrigley, A. N., Stirton, A. J., and Howard, E., Jr. (1960). *J. Org. Chem.* **25,** 439.
Zervas, L., and Dilaris, I. (1956). *Chem. Ber.* **89,** 925.

CHAPTER IV

THE CHEMICAL SYNTHESIS OF GLYCEROLIPIDS CONTAINING S-ALKYL, HYDROXY-O-ALKYL, AND METHOXY-O-ALKYL GROUPS

Claude Piantadosi

I. Synthesis of Glycerolipids Containing S-Alkyl Groups

A significant number of the glyceryl derivatives encountered in natural lipids belong to the *O*-ether series; an excellent review describing their occurrence and metabolism, and chemical and physiological properties, has recently appeared (Snyder, 1969). The first chemical synthesis of the sulfur analogs of batyl and chimyl alcohols was accomplished by Lawson *et al.* (1961) as part of their search for potential tuberculostatic drugs. Syntheses of ^{35}S- and ^{14}C-labeled *S*-alkylglycerols have since been described by Piantadosi *et al.* (1969) in connection with studies on the bio-

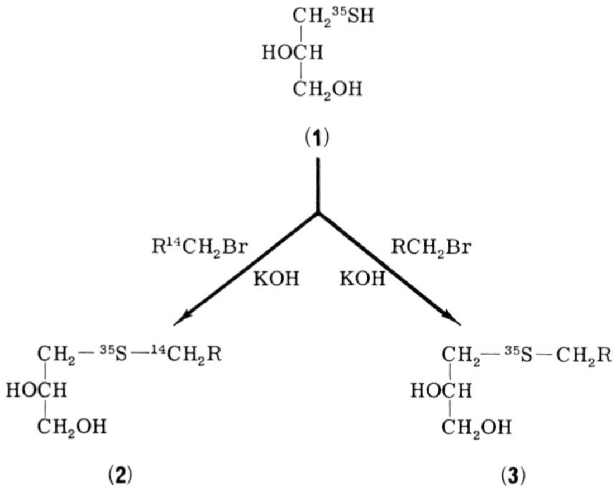

R designates $-CH_2(CH_2)_{16}CH_3$ and $-CH_2(CH_2)_{14}CH_3$

SCHEME I. Synthesis of ^{35}S- and ^{14}C-labeled *S*-alkylglycerols according to Piantadosi *et al.* (1969).

synthesis and mechanism of biocleavage of the *O*-ether linkage in glyceryl ethers. It was felt that the *S*-alkyl glyceryl ethers would be valuable models for these studies, especially when labeled with three different isotopes. The synthetic scheme used by Piantadosi and co-workers (1969) in the preparation of the labeled *S*-alkylglycerol was based on a modification of the Williamson reaction for the synthesis of ethers in general (Lawson *et al.*, 1961; Scheme I). Compound (**1**), ^{35}S-thioglycerol, was reacted wtih octadecyl-l-^{14}C bromide in hexane, followed by treatment with alcoholic KOH. The resulting product, compound (**2**), DL-1-^{35}S-thio-1-^{14}C-octadecyloxy-2,3-propanediol, was purified by silicic acid chromatography. An analogous synthetic procedure was used for the preparation of compound (**3**), DL-1-^{35}S-thiooctadecyloxy-2,3-propanediol. In a biochemical investigation of the 1-^{14}C-labeled synthetic *S*-alkylglycerols, Snyder *et al.* (1969) found that **2** and **3** are not metabolized in the same way as are the *O*-alkylglycerols.

In a separate study, Wood *et al.* (1969) developed procedures that allow one to isolate, characterize, identify, quantify, and distinguish between the *O*-alkyl and *S*-alkyl ethers of glycerol. Based on the principle of these procedures, Ferrell and Radloff (1970) isolated and identified *S*-alkylglycerols as naturally occurring compounds of bovine heart. These workers established the identity of the isolated material by IR,* NMR, mass spectroscopy and synthesis.

II. Synthesis of Glycerolipids Containing Hydroxy-Substituted *O*-Alkyl Groups

Studies by Kasama *et al.* (1971) on the *O*-alkyl glycerolipids in the harderian glands of rabbits describe the characterization of an *O*-alkylglycerol containing a hydroxyl group on the 11- or 12-carbon atom of the *O*-alkyl chain. A 1-*O*-(9- or 10-hydroxy)octadecylglycerol standard was chemically synthesized by Kasama *et al.* (1971) from selachyl alcohol by the method of Rao and Achaya (1970). The route involved sulfation and hydrolysis of the sulfate with an acidified solution of barium chloride. Identification of the 9- and 10-hydroxy-substituted octadecylglycerol was made by IR, TLC, GLC, and mass spectrometry.

1-*O*-(β-Hydroxy)hexadecylglycerol (**4**) has been prepared and investigated as a possible precursor for plasmalogens (Blank *et al.*, 1970).

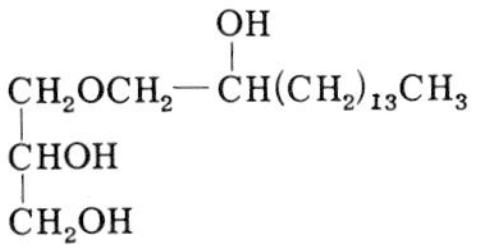

(**4**)

The synthetic scheme involved the reaction of 1,2-epoxyhexadecane with isopropylideneglycerol in the presence of potassium; the isopropylidene group was removed by acid hydrolysis to yield a racemic mixture of the final product. The 1,2-isopropylidene derivative of (β-hydroxy)hexadecylglycerol has also been synthesized as an intermediate in the synthesis of hexadecenylglycerol (Serebrennikova *et al.*, 1966).

III. Synthesis of Glycerolipids Containing Methoxy-Substituted *O*-Alkyl Groups

Three methoxy-substituted alkylglycerols have been isolated from Greenland shark liver oil by Hallgren and Ställberg (1967). These were 1-*O*-(β-methoxy)hexadecylglycerol, 1-*O*-(β-methoxy)-4-hexadecenylglyc-

* Abbreviations used are: IR, infrared; NMR, nuclear magnetic resonance; TLC, thin-layer chromatography; GLC, gas–liquid chromatography.

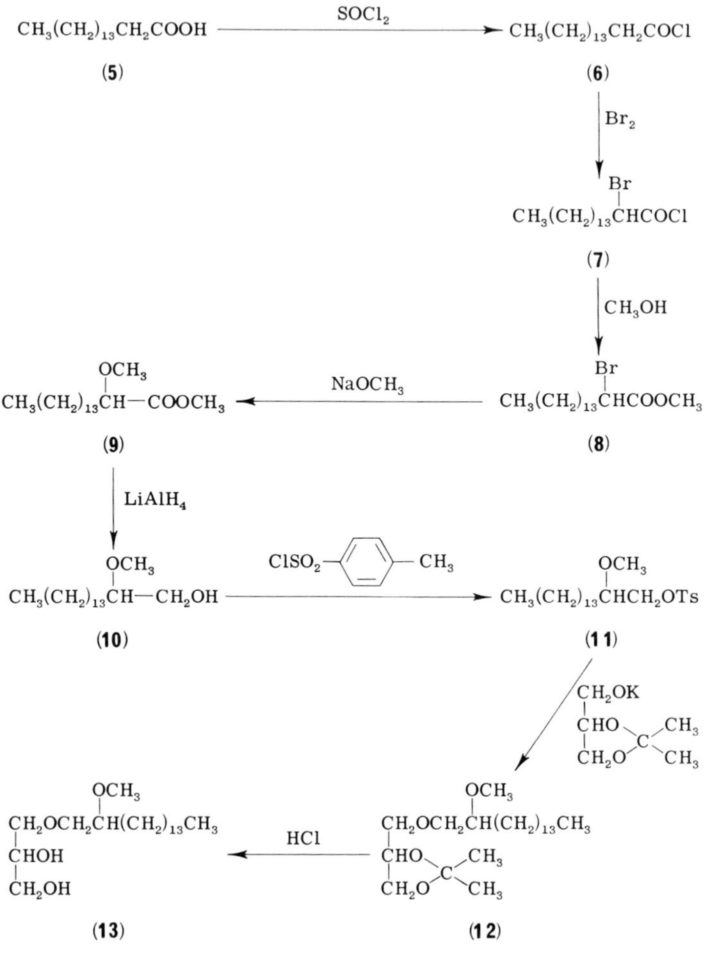

SCHEME II. Synthesis of 1-*O*-(β-methoxy)hexadecylglycerol according to Hallgren and Ställberg (1967).

erol, and 1-*O*-(β-methoxy)-4-octadecenylglycerol. However, only the 1-*O*-(β-methoxy)hexadecylglycerol was synthesized in the laboratory and the sequence of reactions for this synthesis is illustrated in Scheme II (Hallgren and Ställberg, 1967). Compound (**8**), methyl 2-bromohexadeconate, was prepared according to the method of Schwenk and Papa (1948). This was converted to (**10**), β-methoxyhexadecanol, by reduction with lithium aluminum hydride. Treatment of (**10**) with *p*-toluenesulfonyl chloride resulted in (**11**), which was then reacted with 1,2-isopropylidene glycerol as described by Gupta and Kummerow (1959) to give (**12**).

Hydrolysis of this isopropylidene compound with acid resulted in the candidate compound (**13**), which was purified by silicic acid chromatography. No melting point or elemental analysis were given for this compound.

REFERENCES

Blank, M. L., Wykle, R. L., Piantadosi, C., and Snyder, F. (1970). *Biochim. Biophys. Acta* **210,** 442.

Ferrell, W. J., and Radloff, D. M. (1970). *Physiol. Chem. Phys.* **2,** 551.

Gupta, S. C., and Kummerow, F. A. (1959). *J. Org. Chem.* **24,** 409.

Hallgren, B., and Ställberg, G. (1967). *Acta Chem. Scand.* **21,** 1519.

Kasama, K., Rainey, W. T., Jr., and Snyder, F. (1971). Unpublished data.

Lawson, D. D., Getz, H. R., and Miller, D. A. (1961). *J. Org. Chem.* **26,** 615.

Piantadosi, C., Snyder, F., and Wood, R. (1969). *J. Pharm. Sci.* **58,** 1028.

Rao, G. V., and Achaya, K. T. (1970). *J. Amer. Oil Chem. Soc.* **47,** 289.

Schwenk, E., and Papa, D. (1948). *J. Amer. Chem. Soc.* **70,** 3626.

Serebrennikova, G. A., Parferroo, E. A., Perlova, N. A., and Preobrazhenskii, N. A. (1966). *Zh. Org. Khim.* **2,** 1580.

Snyder, F. (1969). *Progr. Chem. Fats Other Lipids* **10,** 287.

Snyder, F., Piantadosi, C., and Wood, R. (1969). *Proc. Soc. Expt. Biol. Med.* **130,** 1170.

Wood, R., Piantadosi, C., and Snyder, F. (1969). *J. Lipid Res.* **10,** 370.

CHAPTER V

THE CHEMICAL SYNTHESES OF PLASMALOGENS

Roy Gigg

I. Introduction

By 1962, when the complete structure of plasmalogens [e.g., (**1**)] had been established with some certainty, it was possible to try to confirm this structure by synthesis, and at the same time to make these interesting compounds more readily available for biological studies.

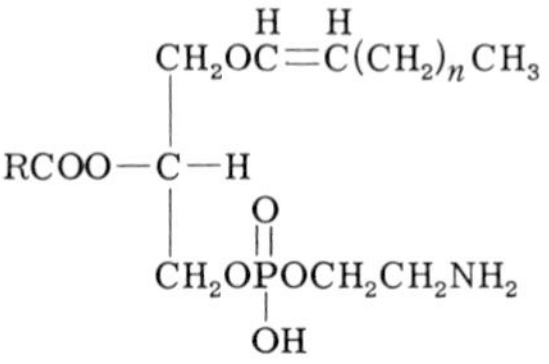

(1)

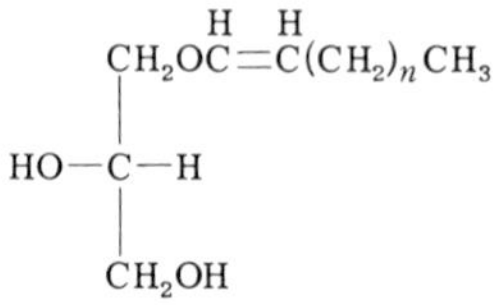

(2)

A logical scheme for the synthesis consisted of the preparation of a 1-*O*-alk-1′-enylglycerol (**2**), then the conversion of this material into a suitable derivative for phosphorylation of the primary hydroxyl group, and finally the phosphorylation itself to give the complete plasmalogen. Each of these stages has posed interesting chemical problems requiring new synthetic approaches. Various methods for the synthesis of alk-1-enyl glycerol ethers have been investigated, with emphasis on the introduction of the *cis* configuration of the double bond. In order to introduce the correct absolute configuration for the glycerol moiety it has been necessary to use protecting groups which later can be removed without affecting the labile alk-1-enyl ether linkage. The preferential esterification of the secondary hydroxyl group of (**2**) and the phosphorylation stage have also required careful choice of conditions because of the reactivity of the alk-1-enyl ether linkage.

II. Preparation of Alk-1-enyl Ethers of Glycerol

A. Reaction of Haloacetals with Alkali Metals

Piantadosi and Hirsch (1961) were the first workers to report the synthesis of long-chain alk-1-enyl ethers of glycerol. They used a method discovered by Wislicenus (1878) for the preparation of simple alk-1-enyl ethers (for a review of methods for alk-1-enyl ether synthesis, see Shostakovskii *et al.*, 1968). Wislicenus prepared ethyl "vinyl" ether (**4**) by treating chloroacetal (**3**) with metallic sodium at 130°C.

$$\underset{(3)}{ClCH_2CH(OEt)_2} \xrightarrow{Na} \underset{(4)}{CH_2{=}CH{-}OEt}$$

Piantadosi and Hirsch (1961) and Piantadosi *et al.* (1963) showed that the reaction could also be carried out if a free hydroxyl group was present in the molecule when they converted the glycerol acetal (**5**) into a mixture of the alk-1-enyl ethers (**6**) and (**7**).

The acetal (**5**) was prepared by a transacetalation reaction (Piantadosi *et al.*, 1958) between glycerol and the dimethyl acetal of a long-chain

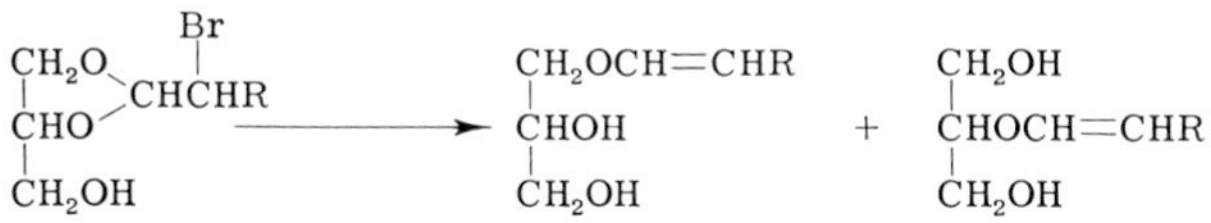

(5) (6) (7)

α-bromoaldehyde. Piantadosi *et al.* (1963) showed that the alk-1-enyl ether formed in this reaction had mainly the *trans* configuration and assumed that the product was entirely the 1-*O*-alk-1′-enyl ether (**6**). However, the preparation of the alk-1-enyl ethers of glycerol by this route was reinvestigated by Cymerman-Craig *et al.* (1965), who showed that both compounds (**6**) and (**7**) were obtained in approximately equal proportions and that each compound was present as a mixture of *cis* and *trans* isomers. As a logical extension of their work, Cymerman-Craig and Hamon (1965a,b) prepared the 1,3-acetals of glycerol [(**8**) and (**9**)] by the route shown and converted these into the 1-*O*-alk-1′-enylglycerol (**10**) by the use of lithium in dimethoxyethane.

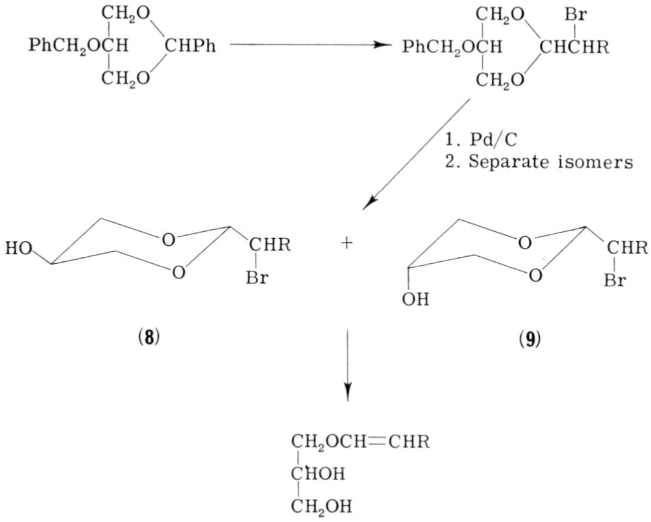

(8) (9)

(10)

R designates $(CH_2)_{13}CH_3$

The alk-1-enyl ether (**10**) obtained from the acetal (**8**) was a mixture of *cis* and *trans* isomers in the ratio 1:2, and the ether obtained from compound (**9**) had a *cis*:*trans* ratio of 1:4. The *cis* and *trans* isomers were separated by gas chromatography of the diacetates of compound (**10**) and pure racemic *cis*- and *trans*-1-*O*-hexadec-1′-enylglycerols were obtained. Unfortunately, this method gives mainly the *trans* isomer and cannot provide a direct synthesis of optically active material, since the starting acetal is a symmetrical compound. This route has also been used (Kramer and Mangold, 1969) for the preparation of 1-*O*-alk-1′-enyl ethers of ethylene glycol.

B. Pyrolysis of Acetals

The pyrolysis of acetals has long been the method of choice for the preparation of simple alk-1-enyl ethers. The method was first used for the preparation of alk-1-enyl glycerol ethers by Zvonkova *et al.* (1964a,b), who prepared a simple long-chain alk-1-enyl ether (**12**) by pyrolysis of the corresponding diethylacetal (**11**) of the long-chain aldehyde. The alk-1-enyl ether (**12**) was condensed with a 1,2-di-*O*-acylglycerol to give the mixed acetal (**13**), which was pyrolyzed in the presence of an acid catalyst to give the alk-1-enyl ether (**14**) and other products. This route presents several problems. The mixed acetals are unstable, and since the first stage in the pyrolysis of compound (**13**) was the rearrangement to the symmetrical acetal (**15**) (Serebryakova *et al.*, 1966), the reaction sequence was simplified (Serebryakova *et al.*, 1966; Serebrennikova *et al.*, 1967a) by preparing the symmetrical acetal (**15**) directly from the diethylacetal (**11**) by trans-acetalation. Moreover, 1,2-di-*O*-acylglycerols are rapidly isomerized under acidic conditions to give an equilibrium mixture containing predominantly 1,3-di-*O*-acylglycerols (**17**) (Mattson and Volpenheim, 1962), so care must also be exercised in the preparation of the acetal (**15**) to avoid this migration during the formation of the alk-1-enyl ether (**14**). Another problem is the possible recombination of the diacylglycerol (**17**) with the alk-1-enyl ether (**14**), which would lead to the formation of the mixed acetal (**18**), and eventually to the formation of the isomeric 2-*O*-alk-1′-enyl ether of glycerol (**19**). The presence of approximately 17% of the isomeric alk-1-enyl ether (**19**) was in fact detected as a contaminant of the alk-1-enyl ether (**14**) prepared by this route by Serebryakova *et al.* (1967). A modification of the method was used by Slotboom *et al.* (1967) for the preparation of 1-*O*-alk-1′-enylglycerols and they reported that the product was entirely *trans*.

In order to avoid some of the difficulties that arise from the ready migration of acyl groups in diacylglycerols, Cunningham and Gigg (1965b) used the cyclic carbonate of glycerol (**21**), which is known to be fairly stable under acidic conditions (Hough *et al.*, 1960), for protection of the hydroxyl groups. The racemic glycerol carbonate (**21**) was prepared (Cunningham and Gigg, 1965a) from racemic 1-*O*-benzylglycerol (**20**) and

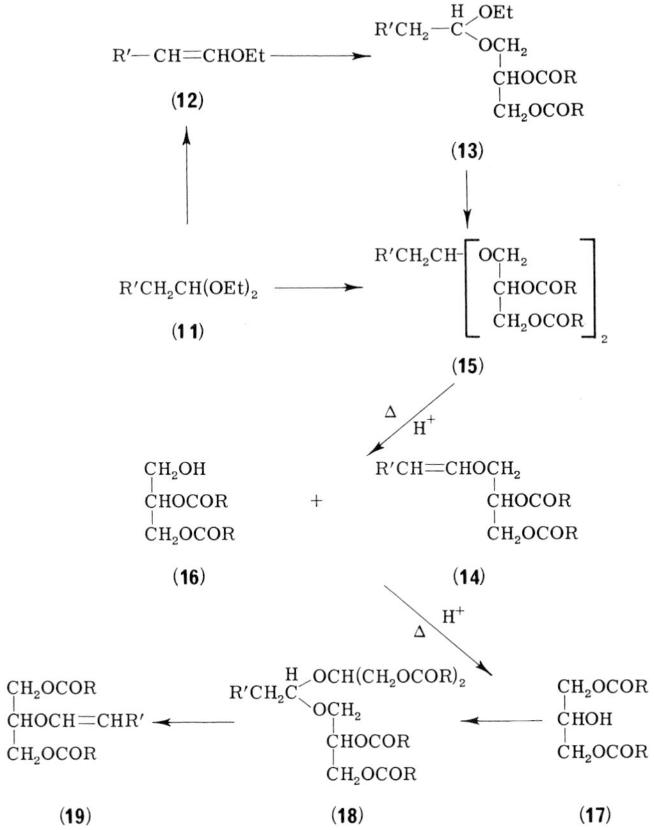

converted into the bis glycerol carbonate acetals (**22**) by transacetalation with the dimethyl acetals of long-chain aldehydes. Pyrolysis of the acetals (**22**) gave the alk-1-enyl ethers (**23**), which were hydrolyzed by base to give the alk-1-enyl glycerol ethers. The alk-1-enyl ether (**23**) was mainly the *trans* form.

This route for the preparation of long-chain alk-1-enyl ethers of glycerol is not very satisfactory in that the *trans* isomer is preferentially formed and the yields are not high. With the diacylglycerol method, acyl migrations can also lead to contamination of the product by the 2-*O*-alk-1'-enylglycerol. Rubtsova *et al.* (1969) have carried out this synthesis using optically active diacylglycerols. The optical rotations reported for the alk-1-enyl ethers obtained are lower than those observed for the same products prepared by different routes (J. Gigg and Gigg, 1968b; Titov *et al.*, 1970b,c), indicating that acyl migrations have led to racemization.

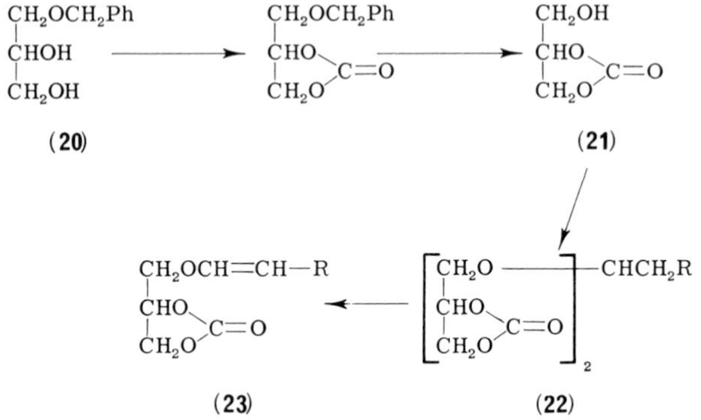

(20) (21) (22) (23)

C. The Acetylenic Ether Route

A method for the preparation of *cis*-alk-1-enyl ethers of glycerol from the corresponding acetylenic ethers was first reported by Berezovskaya *et al.* (1964, 1966). These authors described the condensation of the long-chain 1-bromoacetylene (**24**) with the sodium derivative of 1,2-*O*-isopropylideneglycerol to give the acetylenic ether (**25**), which was reduced with Lindlar catalyst to give the *cis*-alk-1-enyl ether (**26**).

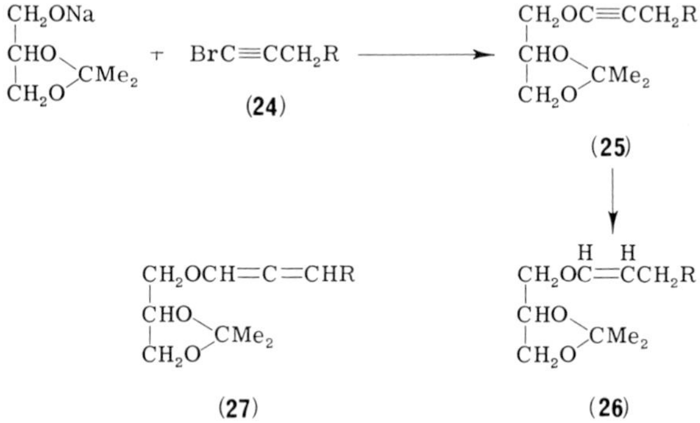

(24) (25) (26) (27)

The authors did not describe a method for the removal of the isopropylidene group, which would probably present some difficulties since the isopropylidene group and the alk-1-enyl ether linkage have the same order of acid lability (Cunningham *et al.*, 1964) and since a free alk-1-enyl ether of glycerol is rapidly cyclized to an acetal under acidic conditions (Piantadosi *et al.*, 1964; Cunningham and Gigg, 1965b). It has been suggested (Chacko, 1966; Slotboom *et al.*, 1967) that boric acid in trimethyl borate would be satisfactory for the removal of the isopropylidene group from (**26**), and a preliminary report has appeared on the hydrolysis of the isopropylidene group from the acetylenic ether (**25**) to form 1-*O*-hexadec-*cis*-1′-enyl-D-glycerol by this method (Cymerman-Craig and Solomon, 1969). However, some controversy exists about this method for the preparation of alk-1-enyl glycerol ethers. Chacko *et al.* (1967) reported that the condensation product of the bromoacetylene (**24**) and the sodium derivative of isopropylideneglycerol was an allenic ether (**27**) rather than the acetylenic ether (**25**). Oswald *et al.* (unpublished, quoted in Piantadosi and Snyder, 1970) were also unable to obtain the alk-1-enyl ether by this method.

D. Isomerization of Allyl Ethers

Prosser (1961) and Price and Snyder (1961) observed that allyl ethers (**28**) were isomerized to *cis*-prop-1-enyl ethers (**29**) by the action of potassium *tert*-butoxide and the isomerization was shown to be particularly rapid in dimethyl sulfoxide. Cunningham *et al.* (1964) and Cunningham and Gigg (1965b) showed that the simple *cis*-prop-1-enyl ethers of glycerol [(**30**) and (**31**)] could readily be prepared by this method. The preparation of compound (**30**) by this route was also described by Cymerman-Craig and Hamon (1965b).

(**28**) (**29**)

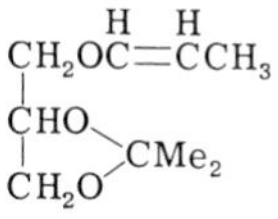

(**30**)

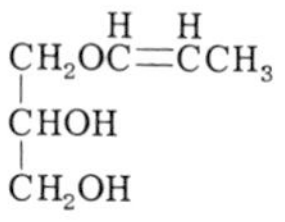

(**31**)

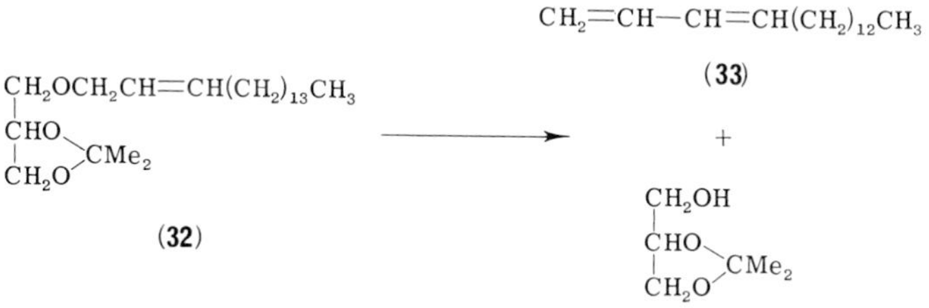

(**32**) (**33**)

In an attempt to prepare long-chain alk-1-enyl ethers of glycerol by the same route, Cunningham and Gigg (1965b) treated the γ-substituted allyl ether (**32**) with potassium *tert*-butoxide in dimethyl sulfoxide and found that the compound rapidly formed heptadecadiene (**33**) and its isomers. This type of elimination reaction has since been observed with related compounds (Kesslin and Orlando, 1966; Mkryan *et al.*, 1967; R. Gigg and Warren, 1968; Vtorov *et al.*, 1970), and is of some significance in connection with the synthesis of alk-1-enyl ethers of glycerol to be described in Section II,E.

E. Eliminations from Derivatives of 1-*O*-[2′-Hydroxyalkyl]glycerols and 1-*O*-[2′-Haloalkyl]glycerols

Parfenov *et al.* (1965) described a synthesis of the isopropylidene derivative (**35**) by elimination of the tosyl group from compound (**34**) with potassium *tert*-butoxide in *tert*-butanol. The possibility that some of the isomer (**36**) was also formed in this reaction was not excluded. In extensions of this type of elimination reaction, Parfenov *et al.* (1966a,b) and Serebrennikova *et al.* (1966a,b) prepared alk-1-enylglycerols (**38**) by the route

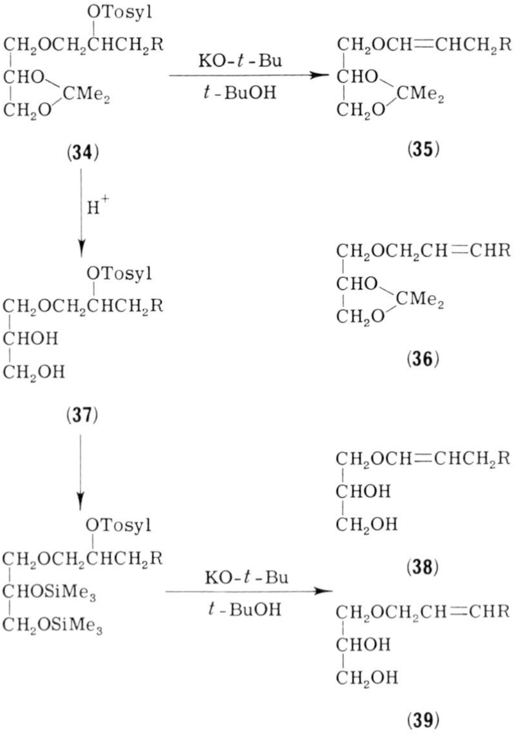

shown, but used the trimethylsilyl group for protection of the intermediate diol (**37**). Again the absence of the alternative elimination product (**39**) was not definitely established.

In a reinvestigation of this reaction, Serebrennikova *et al.* (1969b) observed that the trimethylsilyl groups were not effective as protecting groups, since they were eliminated before the tosyl groups. The dehydrotosylation was therefore carried out on the free diol (**37**). They also observed that the product was completely hydrolyzed by acid to give glycerol and long-chain aldehyde, thus establishing the absence of the isomeric acid-stable ether (**39**) in the product. Although compound (**39**) is probably formed in this reaction, it must be degraded under the conditions of the reaction to give a long-chain diene as described in Section II, D. J. Gigg and Gigg (1968b) observed that γ-substituted allyl ethers such as compound (**39**) were much more rapidly degraded by potassium *tert*-butoxide in dimethyl sulfoxide than the corresponding alk-1-enyl ethers, and this was subsequently confirmed by the Russian workers (Vtorov *et al.*, 1970; Titov *et al.*, 1970c).

In these reactions the tosyl derivatives (**41**) were prepared from the epoxides (**40**) by the route shown. A modification of this route for the preparation of alk-1-enyl ethers was also used by Serebrennikova *et al.* (1967b) and Titov *et al.* (1970c), who eliminated hydrogen iodide from the iodo compound (**42**) that was prepared from the tosyl derivative by the action of sodium iodide.

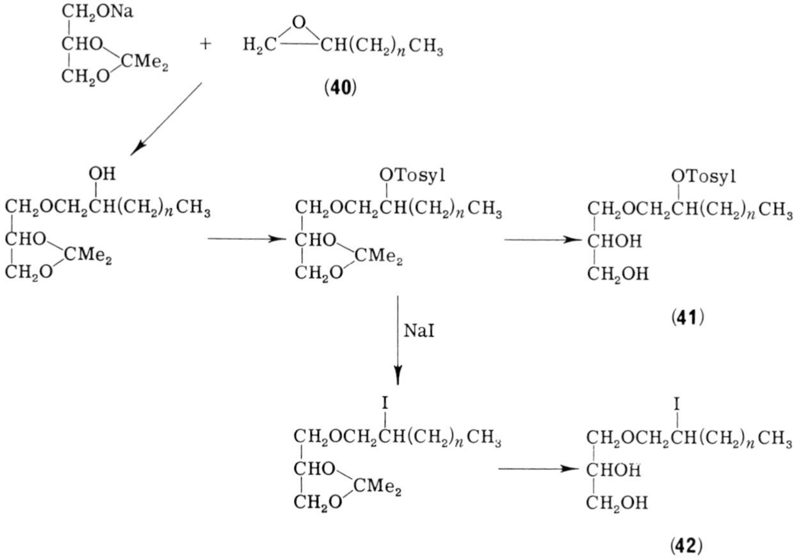

The dehydrotosylation route for the preparation of alk-1-enyl glycerol ethers was extended to the preparation of the corresponding epoxides (**45**) by Parfenov *et al.* (1967a,b). These epoxides are potentially useful intermediates for subsequent phosphorylation reactions. The route chosen initially was via the epoxide (**43**) but because of the instability of such compounds, the tosylate (**44**) route was later preferred. Epoxidation

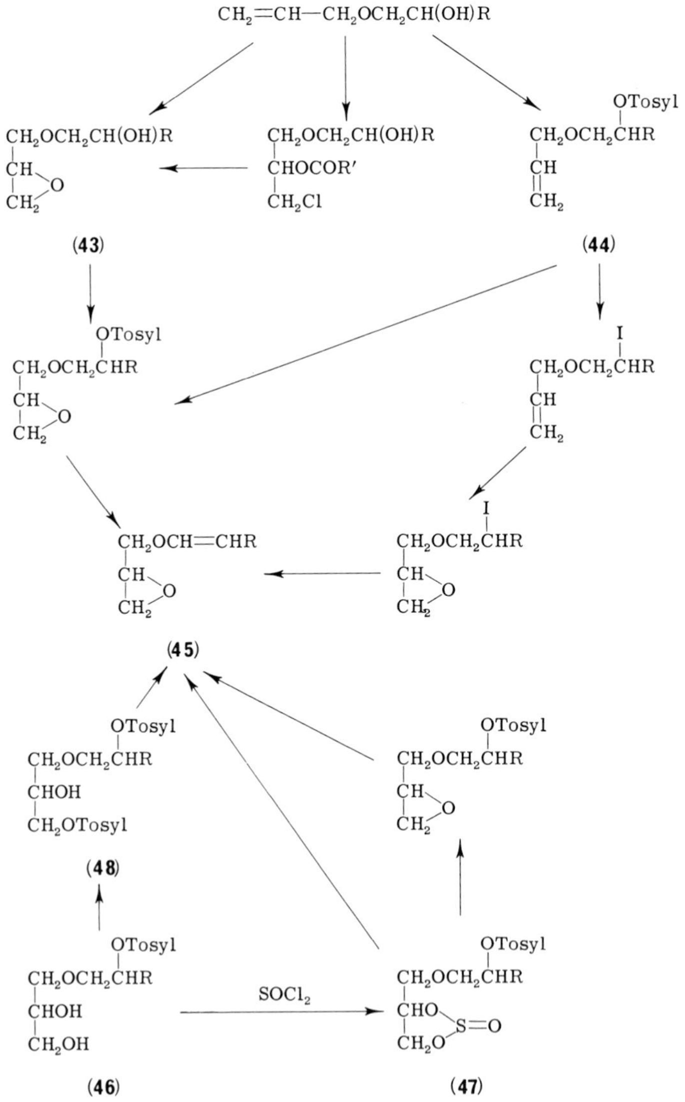

of the double bond in compound (**44**) was achieved by the action of perbenzoic acid or by reaction with *tert*-butyl hypochlorite in acetic acid followed by the action of base. The epoxy alk-1-enyl ethers (**45**) were also prepared (Parfenov *et al.*, 1967d,e) from the corresponding diols (**46**) by the action of potassium *tert*-butoxide on the sulfites (**47**), which were prepared from the diols by the action of thionyl chloride. As a further extension of this method for the preparation of the epoxides of the alk-1-enyl ethers, Serebrennikova *et al.* (1968b) converted the diol (**46**) to the ditosylate (**48**), which was converted directly into the epoxy alk-1-enyl ether (**45**) by the action of potassium tert-butoxide.

The preparation of the alk-1-enyl ethers by these elimination reactions using potassium *tert*-butoxide leads to a mixture of *cis* and *trans* isomers as well as other by-products. Investigations by Oswald *et al.* (unpublished, quoted in Piantadosi and Snyder, 1970) indicate that the required alk-1-

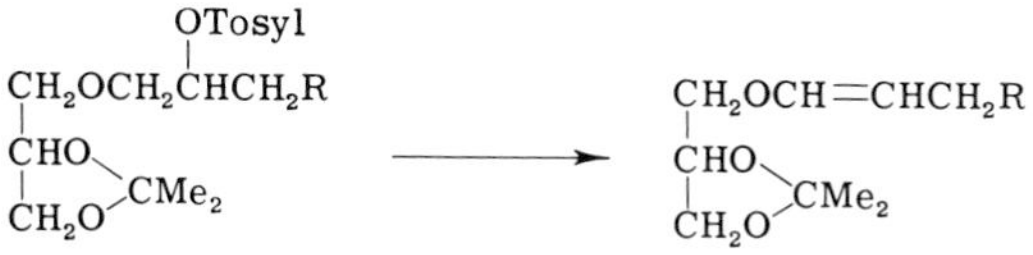

(**49**) (**51**)

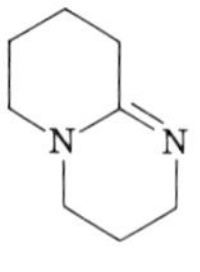

(**50**)

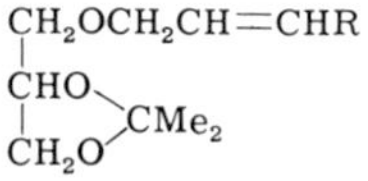

(**52**)

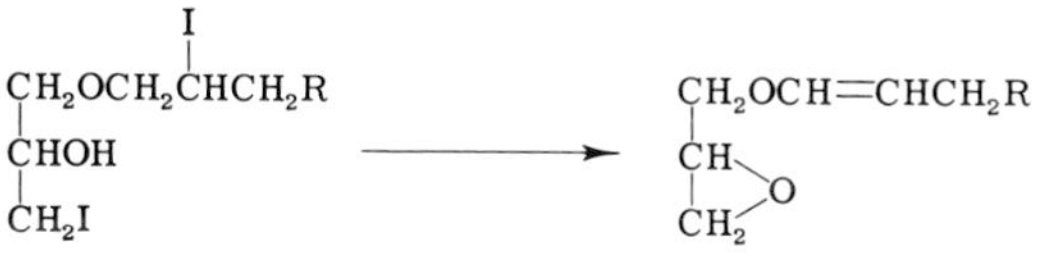

(**53**) (**54**)

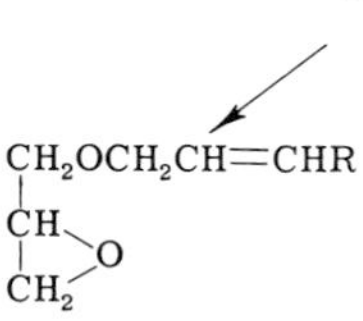

(**55**)

enyl ethers are produced in low yields and are difficult to separate from impurities formed at the same time.

In an effort to improve the yield of *cis*-alk-1-enyl ether formed, Serebrennikova *et al.* (1969c) investigated the use of other bases for the elimination reaction. They found that the elimination of the tosyl group from compound (**49**) with potassium *tert*-butoxide gave an alk-1-enyl ether with a *cis*:*trans* ratio of 3:7, whereas with the organic base (**50**) only the *cis* isomer of the alk-1-enyl ether (**51**) was formed. The fact that the product in the latter

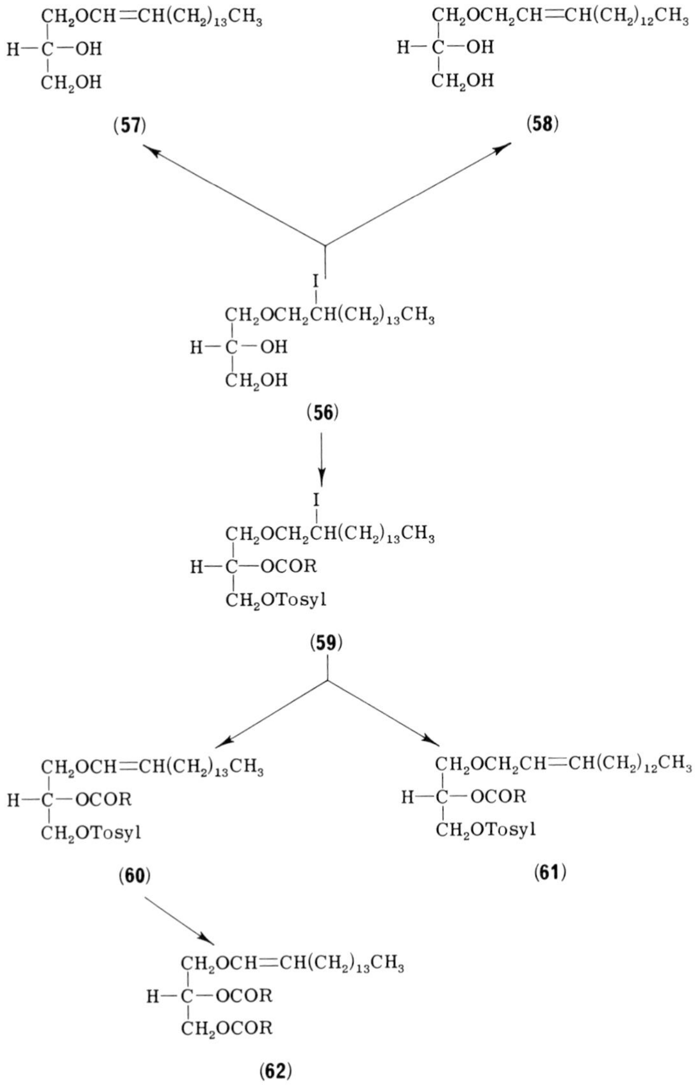

reaction was completely hydrolyzed by acid indicated the absence of the isomer (**52**) in the product.

However, in a further investigation of the elimination of hydrogen iodide from (**53**) with the organic base (**50**), Vtorov *et al.* (1970) found that a mixture of *cis* and *trans* isomers of the alk-1-enyl ether (**54**) was formed as well as some of the isomer (**55**). They also confirmed that compound (**55**) was degraded to a diene by the action of potassium *tert*-butoxide, as suggested by J. Gigg and Gigg (1968b). In a detailed investigation of the action of various bases on the optically active iodo compound (**56**), Titov *et al.* (1970c) found that the bicyclic base (**50**) in dimethyl sulfoxide gave a 16% yield of the alk-1-enyl ether (**57**) (with a *cis*:*trans* ratio of 2:1) and a 68% yield of the isomer (**58**). Potassium *tert*-butoxide in a mixture of dimethyl sulfoxide and *tert*-butanol (1:1) gave a 58% yield of the alk-1-enyl ether (**57**) (*cis*:*trans* ratio 2:3) and potassium *tert*-butoxide in *tert*-butanol gave an 86% yield of the alk-1-enyl ether (**57**) with a *cis*:*trans* ratio of 1:19. Serebrennikova *et al.* (1970) have reported the equilibration of the *trans* form of the alk-1-enyl ether to give a mixture of *cis* and *trans* isomers and this method might prove useful in increasing the yield of the required *cis* isomer. Similar equilibrations of *cis*- and *trans*-alk-1-enyl ethers have been reported previously (Salomaa and Nissi, 1967).

Titov *et al.* (1970b) converted the diol (**56**) into the tosylate (**59**). Elimination of hydrogen iodide from (**59**) with the organic base (**50**) gave the alk-1-enyl ether (**60**) in low yield (8%) and the isomeric compound (**61**) in 40% yield. The tosylate (**60**) was subsequently converted into the alk-1-enyldiacylglycerol ("neutral plasmalogen") (**62**), which had a *cis*:*trans* ratio of 2:1.

F. Dehydrochlorination of α-Chloroethers

In considering other routes for the synthesis of 1-*O*-alk-1′-enylglycerols, Cunningham and Gigg (1965b) and J. Gigg and Gigg (1968a) investigated the elimination of hydrogen chloride from α-chloroethers, a well-established method for the synthesis of alk-1-enyl ethers (Summers, 1955). However, a mild method for the preparation of α-chloroethers that would be applicable in the presence of protecting groups was required, and several investigators reported the reaction of acid chlorides with acetals (Blaise, 1904, 1905; Mylo, 1911; Straus and Heinze, 1932; Post, 1936; Spurlock and Henze, 1939; Grummitt and Stearns, 1955), resulting in the formation of an ester and an α-chloroether.

Cunningham and Gigg (1965b) and J. Gigg and Gigg (1968a) treated racemic glycerol carbonate acetals (**63**) with excess acetyl chloride at 20°C for 4 days, and the α-chloroether (**64**) obtained on evaporation of the acetyl chloride was immediately treated with triethylamine to give the alk-1-enyl ether (**65**). Thin-layer chromatography showed a mixture of *cis* and *trans* isomers of compound (**65**), in which the *cis* isomer predominated.

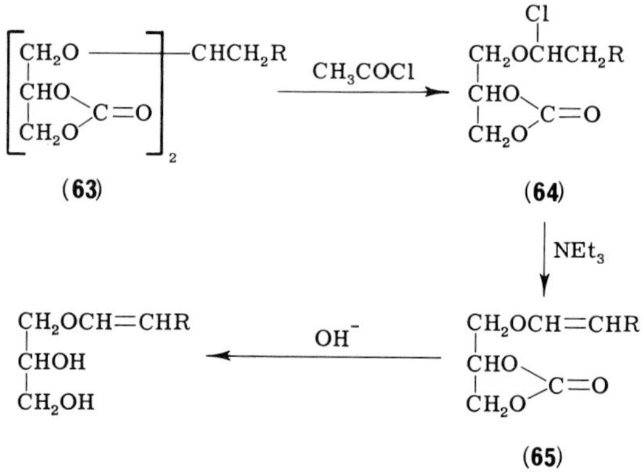

(**63**) (**64**) (**65**)

For a complete synthesis of the required alk-1-enyl ethers of glycerol by this route, the optically active glycerol carbonate acetals (**63**) were required, and a separation of the *cis* and *trans* isomers had to be achieved.

The absolute configuration of the plasmalogens that had been previously established (Thannhauser *et al.*, 1951; Baer and Stancer, 1953; Cymerman-Craig *et al.*, 1966) indicated that the glycerol carbonate required in the synthesis must have the configuration shown in (**68**). This compound was prepared (J. Gigg and Gigg, 1967b) from the D-mannitol derivative (**66**) by way of 1-*O*-benzyl-L-glycerol (**67**) (J. Gigg and Gigg, 1967b, 1968a).

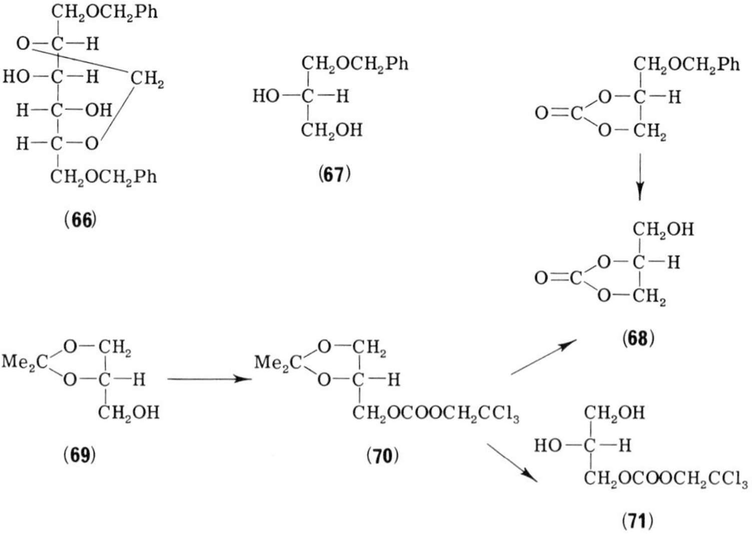

(**66**) (**67**) (**68**) (**69**) (**70**) (**71**)

Recently, Pfeiffer *et al.* (1968, 1970) described another route to the optically active glycerol carbonate (**68**) from 1,2-*O*-isopropylidene-L-glycerol (**69**), which is readily available from D-mannitol. Acid hydrolysis of the isopropylidene derivative (**70**) gave a mixture of the diol (**71**) (74%) and the cyclic carbonate (**68**) (26%). The diol (**71**) was also cyclized to the carbonate (**68**) by heating in pyridine. Pfeiffer *et al.* (1970) did not describe the isolation of the pure carbonate (**68**) by this method, and the procedure may require some investigation before it becomes preparatively useful. The beautifully crystalline triphenylmethyl ether of the carbonate (**68**) (J. Gigg and Gigg, 1967b) is readily prepared in good yield after the cyclization of compound (**71**) (Pfeiffer *et al.*, 1970), and the acid hydrolysis of this triphenylmethyl ether might provide a good method for the purification of the carbonate prepared by this procedure.

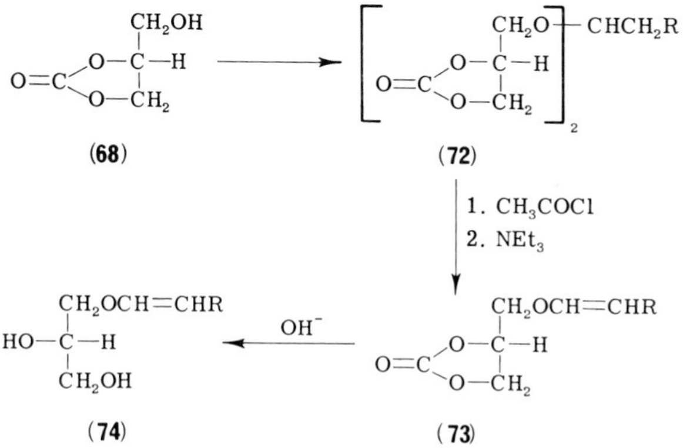

The optically active bis glycerol carbonate acetals (**72**) were prepared (J. Gigg and Gigg, 1968a,b) from the optically active glycerol carbonate and obtained as crystalline solids. These were converted into the alk-1-enyl ethers (**73**), and again a mixture of *cis* and *trans* isomers formed, in which the *cis* isomer predominated. The diacetates of the alk-1-enyl glycerol ether (**74**) were prepared and the *cis* and *trans* isomers separated by thin-layer chromatography on silver nitrate-impregnated silica gel (J. Gigg and Gigg, 1968a). Of considerably more practical value, however, was the finding that recrystallization of the dipalmitate esters of (**74**) gave the pure *cis* isomers (J. Gigg and Gigg, 1968b).

Serebrennikova *et al.* (1970) have described, in a preliminary communication, the conversion of the 1-*O*-alk-1′-enyl-D-glycerol (**75**) (the isomer of the naturally occurring form of the ether) into the diester (**77**). This was achieved by heating the ditosylate (**76**) with the sodium salt of a fatty acid, resulting in inversion of the configuration.

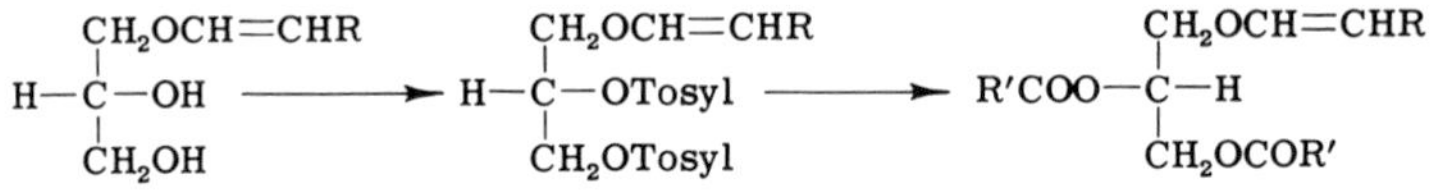

(75) (76) (77)

III. Preparation of Derivatives of Alk-1-enyl Glycerol Ethers Suitable for Phosphorylation

Although much of the effort devoted to the synthesis of the plasmalogens has been directed towards the preparation of the alk-1-enyl ethers of glycerol, the problem of converting these ethers into a suitable form for phosphorylation also had to be solved. J. Gigg and Gigg (1967a) and Slotboom *et al.* (1967) showed that pancreatic lipase (EC 3.1.1.3) did not hydrolyze the alk-1-enyl ether linkage and thus could be used to convert the diacyl derivatives (**78**) into the 1-alk-1-enyl-2-acylglycerols (**79**) ("plasmalogenic diglycerides").

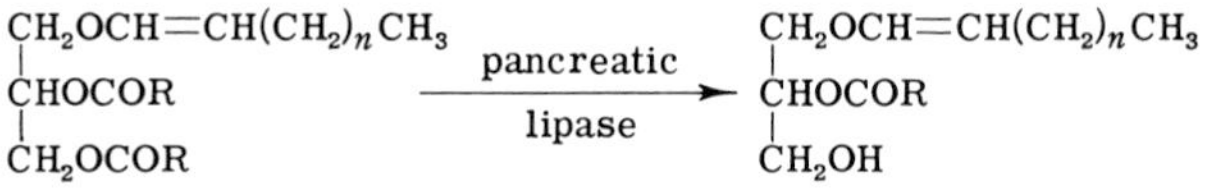

(78) (79)

J. Gigg and Gigg (1967a) used the dipalmitoyl derivative of a short-chain alk-1-enyl ether [(**78**), $n = 4$, R = $CH_3(CH_2)_{14}$–], a solid with a low melting point, easily emulsified and readily cleaved by pancreatic lipase under the conditions normally used for the hydrolysis of triacylglycerols. Slotboom *et al.* (1967) found that their synthetic *trans compound* [(**78**), $n = 13$, R = $CH_3(CH_2)_{14}$–] with a higher melting point was not readily attacked by pancreatic lipase without the presence of hexane and high enzyme concentrations. On the other hand, the dioleoyl compound [(**78**), $n = 13$, R = $CH_3(CH_2)_7CH{=}CH(CH_2)_7$–], an oil, was readily hydrolyzed by pancreatic lipase. This route for the preparation of a suitable intermediate for phosphorylation was also suggested by Chacko (1966).

Another potentially useful intermediate for phosphorylation is the epoxide (**81**). The preparation of a short-chain racemic compound with this structure [(**81**), R = CH_3] by the action of alkali on the tosylate [(**80**), R = CH_3] was described by Cunningham and Gigg (1965b). Optically active material with the configuration shown in (**80**) should give the epoxide with the configuration (**81**). This optically active epoxide (**81**) should also be available from the alk-1-enyl glycerol ether (**82**) of opposite configuration by the route shown, since elimination of the tosyl group from compound (**83**) would lead to inversion of configuration. Serebrennikova *et al.* (1967b, 1968b) have also described the preparation of epoxides (**81**) from the tosylates (**80**).

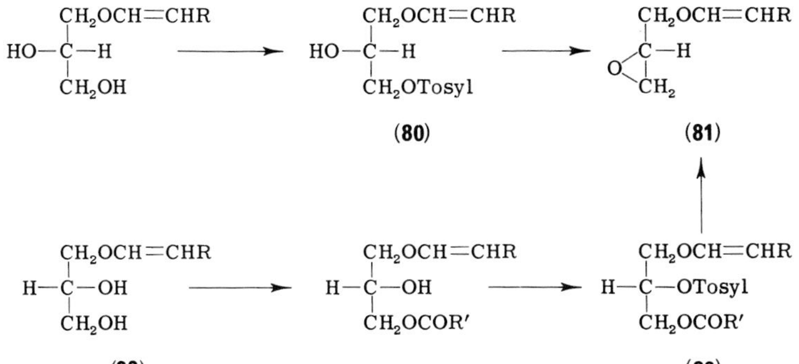

(**80**) (**81**) (**82**) (**83**)

The Russian workers also prepared epoxides (**81**) directly during the synthesis of the alk-1-enyl ether linkage as described in Section II,E. The epoxides are useful intermediates, since they can be opened directly by phosphate anions (Brown, 1963) to give phosphate esters (**84**) and since they also react with acyl halides to give halo esters (**85**) (Parfenov *et al.*, 1967c; Serebrennikova *et al.*, 1968a) that can be phosphorylated by silver phosphates. The halo derivates (**85**) have also been prepared (Serebrennikova *et al.*, 1969b) via the tosyl derivatives (**86**).

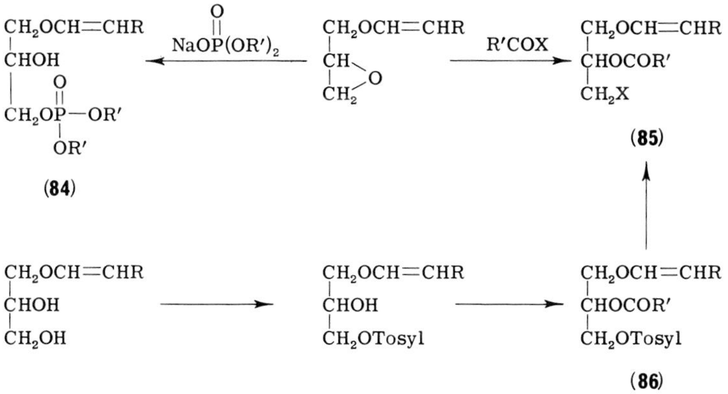

(**84**) (**85**) (**86**)

IV. Synthetic Alk-1-enyl Glycerol Ethers Containing Phosphate Ester Groups

The first reported synthesis of a phosphorylated plasmalogen derivative was by Slotboom *et al.* (1967). They phosphorylated a synthetic, racemic, *trans*-alk-1-enylacylglycerol (**87**) with 2-bromoethyl phosphorodichloridate and subsequently quaternized the product (**88**) with triethylamine to give alk-1-enylacylglycerophosphorylcholine (**89**).

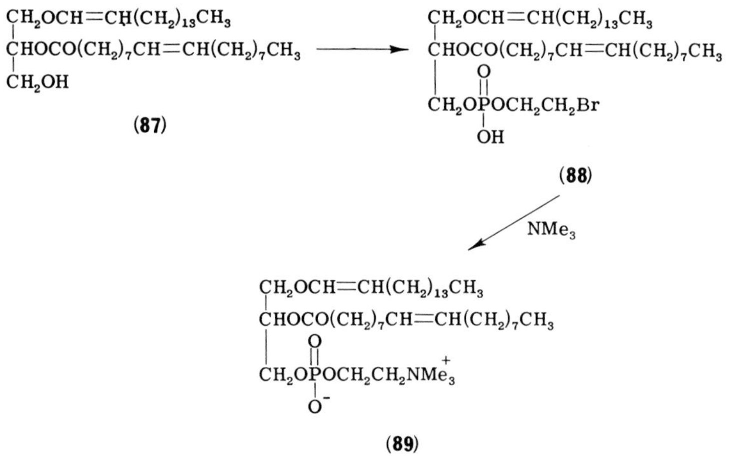

The same reaction sequence was also carried out (Slotboom *et al.*, 1967) on an alk-1-enylacylglycerol obtained from a natural plasmalogen by the action of phospholipase C. Eibl and Lands (1970) have also phosphorylated "natural plasmalogenic diglycerides" with phosphorus oxychloride to give alk-1-enyl analogs of phosphatidic acids.

Parfenov *et al.* (1968) described the condensation of the acyl phosphate (**91**) with the saturated epoxy ether (**90**) to give phosphate esters (**92**). However, with unsaturated epoxy ethers the reaction was complicated by the formation of many side products.

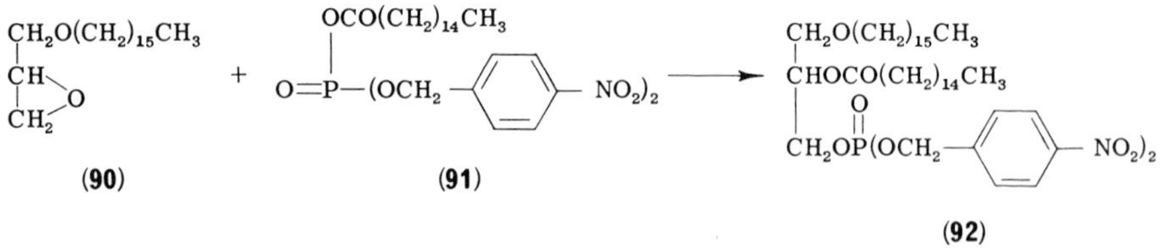

Serebrennikova *et al.* (1969b) and Preobrazhenskii *et al.* (1969) have described the preparation of the phosphate esters [(**94**), $n = 9$] and [(**95**),

$n = 9$] by the action of silver phosphates on the iodo compound (**93**). In extensions of this work, Titov *et al.* (1970a) described the separation of the *cis* and *trans* isomers of compound [(**94**), $n = 13$] prepared by this route, and Serebrennikova *et al.* (1969a) described the conversion of compound (**95**) into alk-1-enylacylglycerophosphoryldimethylethanolamine (**96**) by the route shown.

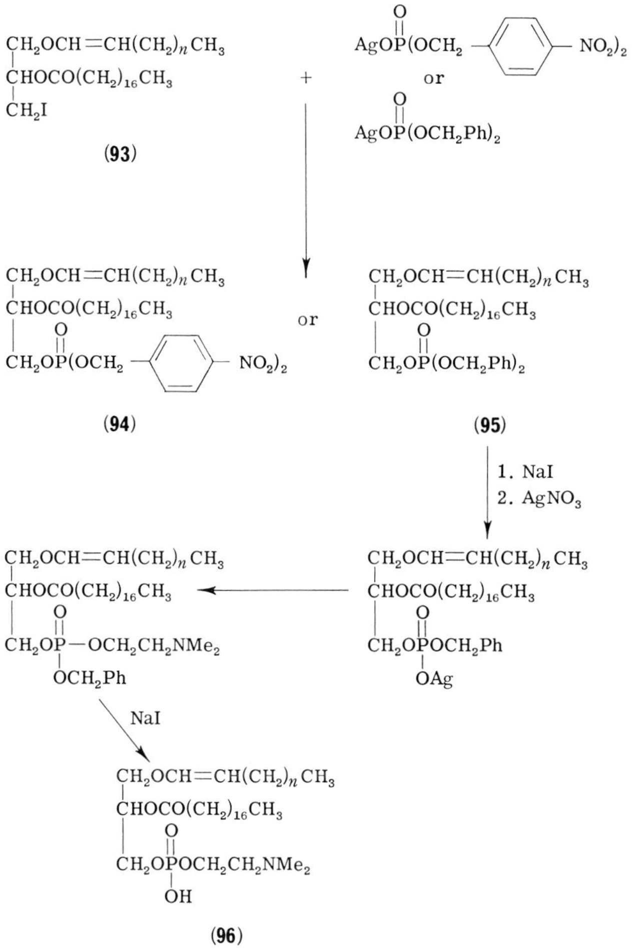

REFERENCES

Baer, E., and Stancer, H. C. (1953). *J. Amer. Chem. Soc.* **75,** 4510.

Berezovskaya, M. V., Sarycheva, I. K., and Preobrazhenskii, N. A. (1964). *Zh. Obshch. Khim.* **34,** 543; *J. Gen. Chem. USSR* **34,** 545 (1964).

Berezovskaya, M. V., Zubkova, T. P., Sarycheva, I. K., and Preobrazhenskii, N. A. (1966). *Zh. Org. Khim.* **2,** 1774; *J. Org. Chem. USSR* **2,** 1745(1966).

Blaise, E. E. (1904). *C. R. Acad. Sci.* **139,** 1211.
Blaise, E. E. (1905). *C. R. Acad. Sci.* **140,** 661.
Brown, D. M. (1963). *Advan. Org. Chem.* **3,** 75.
Chacko, G. K. (1966). Ph.D. Thesis, University of Illinois.
Chacko, G. K., Schilling, K., and Perkins, E. G. (1967). *J. Org. Chem.* **32,** 3698.
Cunningham, J., and Gigg, R. (1965a). *J. Chem. Soc., London* p. 1553.
Cunningham, J., and Gigg, R. (1965b). *J. Chem. Soc., London* p. 2968.
Cunningham, J., Gigg, R., and Warren, C. D. (1964). *Tetrahedron Lett.* p. 1191.
Cymerman-Craig, J., and Hamon, D. P. G. (1965a). *Chem. Ind. (London)* p. 1559.
Cymerman-Craig, J., and Hamon, D. P. G. (1965b). *J. Org. Chem.* **30,** 4168.
Cymerman-Craig, J., and Solomon, M.D. (1969). *Abst. 158th Meet. Amer. Chem. Soc.* ORGN 023.
Cymerman-Craig, J., Hamon, D. P. G., Brewer, H. W., and Harle, H. (1965). *J. Org. Chem.* **30,** 907.
Cymerman-Craig, J., Hamon, D. P. G., Purushothaman, K. K., Roy, S. K., and Lands, W. E. M. (1966). *Tetrahedron* **22,** 175.
Eibl, H., and Lands, W. E. M. (1970). *Biochemistry* **9,** 423.
Gigg, J., and Gigg, R. (1967a). *J. Chem. Soc., C* p. 431.
Gigg, J., and Gigg, R. (1967b). *J. Chem. Soc., C* p. 1865.
Gigg, J., and Gigg, R. (1968a). *J. Chem. Soc., C* p. 16.
Gigg, J., and Gigg, R. (1968b). *J. Chem. Soc., C* p. 2030.
Gigg, R., and Warren, C. D. (1968). *J. Chem. Soc., C* p. 1903.
Grummitt, O., and Stearns, J. A. (1955). *J. Amer. Chem. Soc.* **77,** 3136.
Hough, L., Priddle, J. E., and Theobald, R. S. (1960). *Advan. Carbohyd. Chem.* **15,** 91.
Kesslin, G., and Orlando, C. M. (1966). *J. Org. Chem.* **31,** 2682.
Kramer, J. K. G., and Mangold, H. K. (1969). *Chem. Phys. Lipids* **3,** 176.
Mattson, F. H., and Volpenhein, R. A. (1962). *J. Lipid Res.* **3,** 281.
Mkryan, G. M., Papazyan, N. A., and Pogosyan, A. A. (1967). *Zh. Org. Khim.* **3,** 1160.
Mylo, B. (1911). *Chem. Ber.* **44,** 3212.
Parfenov, E. A., Serebrennikova, G. A., and Preobrazhenskii, N. A. (1965). *Sin. Prir. Soedin., Ikh Analogov Fragmentov, Akad. Nauk SSSR, Otd. Obshch. i Tekhnol. Khim.* p. 12.
Parfenov, E. A., Serebrennikova, G. A., Roitberg, S. Ya., and Preobrazhenskii, N. A. (1966a). *Khim. Prir. Soedin.* **2,** 367; *Chem. Natur. Compounds* **2,** 299(1966).
Parfenov, E. A., Serebrennikova, G. A., and Preobrazhenskii, N. A. (1966b). *Zh. Org. Khim.* **2,** 633; *J. Org. Chem. USSR* **2,** 633 (1966).
Parfenov, E. A., Serebrennikova, G. A., and Preobrazhenskii, N. A. (1967a). *Zh. Org. Khim.* **3,** 1566; *J. Org. Chem. USSR* **3,** 1522 (1967).
Parfenov, E. A., Serebrennikova, G. A., and Preobrazhenskii, N. A. (1967b). *Zh. Org. Khim.* **3,** 1766; *J. Org. Chem. USSR* **3,** 1723 (1967).
Parfenov, E. A., Serebrennikova, G. A., and Preobrazhenskii, N. A. (1967c). *Zh. Org. Khim.* **3,** 1951; *J. Org. Chem. USSR* **3,** 1903 (1967).
Parfenov, E. A., Serebrennikova, G. A., and Preobrazhenskii, N. A. (1967d). *Zh. Org. Khim.* **3,** 1955; *J. Org. Chem. USSR* **3,** 1908 (1967).
Parfenov, E. A., Serebrennikova, G. A., and Preobrazhenskii, N. A. (1967e). USSR Pat. 192,778.
Parfenov, E. A., Serebrennikova, G. A., and Preobrazhenskii, N. A. (1968). *Zh. Org. Khim.* **4,** 2241; *J. Org. Chem. USSR* **4,** 2162 (1968).
Pfeiffer, F. R., Cohen, S. R., Williams, K. R., and Weisbach, J. A. (1968). *Tetrahedron Lett.* p. 3549.

Pfeiffer, F. R., Miao, C. K., and Weisbach, J. A. (1970). *J. Org. Chem.* **35,** 221.
Piantadosi, C., and Hirsch, A. F. (1961). *J. Pharm. Sci.* **50,** 978.
Piantadosi, C., and Snyder, F. (1970). *J. Pharm. Sci.* **59,** 283.
Piantadosi, C., Anderson, C. E., Brecht, E. A., and Yarbro, C. L. (1958). *J. Amer. Chem. Soc.* **80,** 6613.
Piantadosi, C., Hirsch, A. F., Yarbro, C. L., and Anderson, C. E. (1963). *J. Org. Chem.* **28,** 2425.
Piantadosi, C., Frosolono, M. F., Anderson, C. E., and Hirsch, A. F. (1964). *J. Pharm. Sci.* **53,** 1024.
Post, H. W. (1936). *J. Org. Chem.* **1,** 231.
Preobrazhenskii, N. A., Serebrennikova, G. A., Ovechkin, P. L., and Vtorov, I. B. (1969). USSR Pat. 246,496.
Price, C. C., and Snyder, W. H. (1961). *J. Amer. Chem. Soc.* **83,** 1773.
Prosser, T. J. (1961). *J. Amer. Chem. Soc.* **83,** 1701.
Rubtsova, T. I., Serebrennikova, G. A., and Preobrazhenskii, N. A. (1969). *Zh. Org. Khim.* **5,** 2030; *J. Org. Chem. USSR* **5,** 1975 (1969).
Salomaa, P., and Nissi, P. (1967). *Acta Chem. Scand.* **21,** 1386.
Serebrennikova, G. A., Parfenov, E. A., Serebryakova, N. I., and Preobrazhenskii, N. A. (1966a). *Khim. Prir. Soedin.* **2,** 306; *Chem. Natur. Compounds* **2,** 249 (1966).
Serebrennikova, G. A., Parfenov, E. A., Perlova, N. A., and Preobrazhenskii, N. A. (1966b). *Zh. Org. Khim.* **2,** 1580; *J. Org. Chem. USSR* **2,** 1560 (1966).
Serebrennikova, G. A., Rubtsova, T. I., and Preobrazhenskii, N. A. (1967a). *Zh. Org. Khim.* **3,** 1947; *J. Org. Chem. USSR* **3,** 1900 (1967).
Serebrennikova, G. A., Ryaplova, T. I., Parfenov, E. A., and Preobrazhenskii, N. A. (1967b). *Zh. Org. Khim.* **3,** 1958; *J. Org. Chem. USSR* **3,** 1911 (1967).
Serebrennikova, G. A., Vtorov, I. B., Federova. G. N., and Preobrazhenskii, N. A. (1968a). *Zh. Org. Khim.* **4,** 603; *J. Org. Chem. USSR* **4,** 590 (1968).
Serebrennikova, G. A., Trubaichuk, B. A., and Preobrazhenskii, N. A. (1968b). *Zh. Org. Khim.* **4,** 765; *J. Org. Chem. USSR* **4,** 744 (1968).
Serebrennikova, G. A., Ovechkin, P. L., Vtorov, I. B., and Preobrazhenskii, N. A. (1969a). *Zh. Org. Khim.* **5,** 546.
Serebrennikova, G. A., Titov, V. I., and Preobrazhenskii, N. A. (1969b). *Zh. Org. Khim.* **5,** 550; *J. Org. Chem. USSR* **5,** 538 (1969).
Serebrennikova, G. A., Vtorov, I. B., and Preobrazhenskii, N. A. (1969c). *Zh. Org. Khim.* **5,** 676; *J. Org. Chem. USSR* **5,** 663 (1969).
Serebrennikova, G. A., Titov, V. I., and Preobrazhenskii, N. A. (1970). *Abstr. Int. Symp. Chem. Natur. Prod., 7th,* (1970) C6, p. 289.
Serebryakova, T. V., Zvonkova, E. N., Serebrennikova, G. A., and Preobrazhenskii, N. A. (1966). *Zh. Org. Khim.* **2,** 2004; *J. Org. Chem. USSR* **2,** 1966 (1966).
Serebryakova, T. V., Serebrennikova, G. A., and Preobrazhenskii, N. A. (1967). *Zh. Org. Khim.* **3,** 1412; *J. Org. Chem. USSR* **3,** 1374 (1967).
Shostakovskii, M. F., Trofimov, B. A., Atavin, A. S., and Lavrov, V. I. (1968). *Usp. Khim.* **37,** 2070; *Russ. Chem. Rev.* **37,** 907 (1968).
Slotboom, A. J., de Haas, G. H., and Van Deenen, L. L. M. (1967). *Chem. Phys. Lipids* **1,** 192.
Spurlock, J. J., and Henze, H. R. (1939). *J. Org. Chem.* **4,** 234.
Straus, F., and Heinze, H. (1932). *Justus Liebigs Ann. Chem.* **493,** 191.
Summers, L. (1955). *Chem. Rev.* **55,** 301.
Thannhauser, S. J., Boncoddo, N. F., and Schmidt, G. (1951). *J. Biol. Chem.* **188,** 423.

Titov, V. I., Serebrennikova, G. A., and Preobrazhenskii, N. A. (1970a). *Zh. Org. Khim.* **6,** 1147.

Titov, V. I., Serebrennikova, G. A., and Preobrazhenskii, N. A. (1970b). *Zh. Org. Khim.* **6,** 1151.

Titov, V. I., Serebrennikova, G. A., and Preobrazhenskii, N. A. (1970c). *Zh. Org. Khim.* **6,** 1154.

Vtorov, I. B., Serebrennikova, G. A., and Preobrazhenskii, N. A. (1970). *Zh. Org. Khim.* **6,** 669.

Wislicenus, J. (1878). *Justus Liebigs Ann. Chem.* **192,** 106.

Zvonkova, E. N., Sarycheva, I. K., and Preobrazhenskii, N. A. (1964a). *Dokl. Akad. Nauk SSSR, Ser. Khim.* **159,** 1079; *Dokl. Chem.* **159,** 1317 (1964).

Zvonkova, E. N., Sarycheva, I. K., and Preobrazhenskii, N. A. (1964b). USSR Pat. 165,710.

CHAPTER VI

THE CHEMICAL SYNTHESES OF O-ALKYLDIHYDROXYACETONE PHOSPHATE AND RELATED COMPOUNDS

Claude Piantadosi

I. Introduction

Dihydroxyacetone [1,3-dihydroxy-2-propanone] (**1**) was first prepared by Piloty (1897), who observed an indefinite melting point and assumed

$$HOCH_2-\underset{\underset{\displaystyle O}{\|}}{C}-CH_2OH$$

(**1**)

polymerization during the process of melting. Bertrand (1899) and Fischer and Mildbrand (1924) showed that (**1**) will undergo monomer–dimer interconversions, and Wohl and Newberg (1900) proposed the cyclic acetal [2,5-dihydroxy-1,4-dioxane-2,5-dimethanol] (**2**) for the dimeric form.

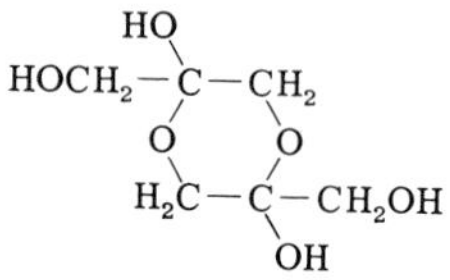

(**2**)

Distillation *in vacuo* converts (**2**) to (**1**) (Fischer and Mildbrand, 1924). Monoacyl derivatives of (**1**) also undergo monomer–dimer interconversion (Schlenk *et al.*, 1952).

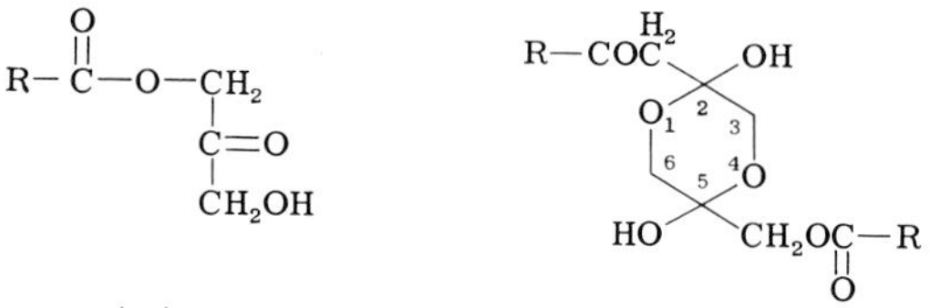

(**1a**) (**1b**)

R designates $C_{15}H_{32}$

When standing at room temperature in the crystalline state, the compound with lower melting point (**1a**) slowly reverts to the dimer (**1b**), which has a higher melting point. Recrystallization of (**1b**) from boiling ethanol produced the monomer (**1a**) (Schlenk *et al.*, 1952). A similar observation was made by Garson *et al.* (1969), who prepared undecanoyl derivatives of (**1**).

Recent studies have shown that 1-*O*-alkyl derivatives of dihydroxyacetone phosphate [1,3-dihydroxy-2-propanone dihydrogen phosphate] (**3**) are important intermediates in the biosynthesis of *O*-alkyl lipids (Snyder *et al.*, 1969; Wykle and Snyder, 1969; Snyder *et al.*, 1970a,b,c; Hajra, 1969). Another new phospholipid, an acyl dihydroxyacetone phosphate, has been detected in the liver of guinea pigs (Hajra and Agranoff, 1967).

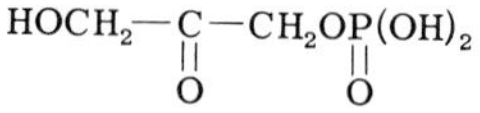

(**3**)

II. Synthesis of Dihydroxyacetone Phosphate

Compound (**3**), which is an intermediate in glycolysis and photosynthesis, is produced by the enzymic dismutation of D-fructose 1,6-diphosphate

(Meyerhof and Lohmann, 1934) and by the phosphorylation of dihydroxyacetone with phosphorus oxychloride (Kiessling, 1934). In 1956, Ballou and Fischer reported the first laboratory preparation of (**3**) from 3-chloro-1,2-propanediol in nine steps, with an overall yield of 15% [Scheme I]. They converted the 3-chloro-1,2-propanediol to isopropylidene-3-chloro-1,2-propanediol (**4**), which they dehydrohalogenated with solid KOH to isopropylidene-2-propane-1,2-diol (**5**). Oxidation of (**5**) with lead tetraacetate, followed by acid hydrolysis of the product (**6**), resulted in 1-acetoxy-3-hydroxy-2-propanone (**7**), which was ketalized to 1-acetoxy-2,2-dimethoxy-3-propanol (**8**). They then phosphorylated (**8**) with diphenyl phosphorochloridate in dry pyridine to give (**9**). Base hydrolysis removed the acetoxy group, while the protecting phenyl groups were removed by catalytic hydrogenolysis with platinum oxide. Compound (**3**) was isolated as a crystalline cyclohexylammonium salt.

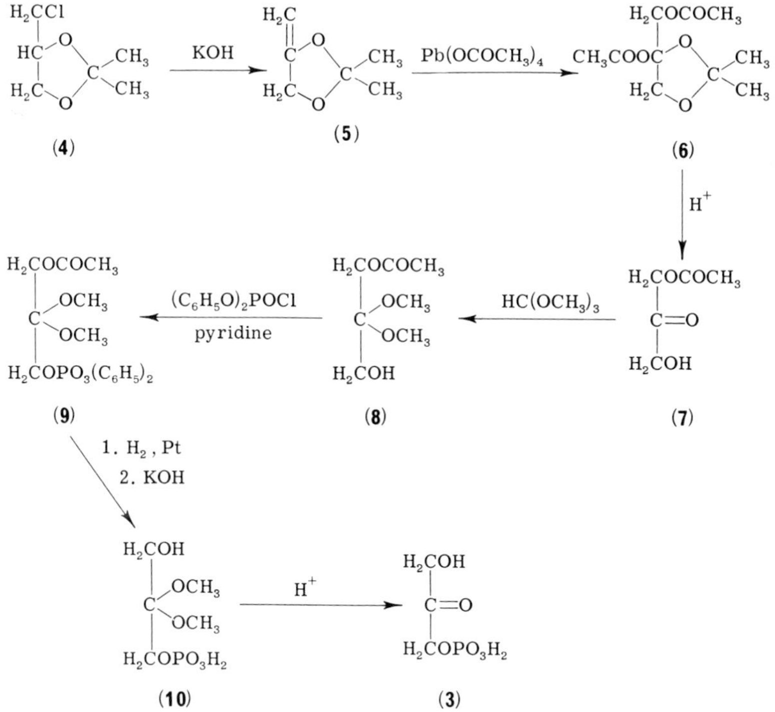

SCHEME I. Synthesis of dihydroxyacetone phosphate according to Ballou and Fischer (1956).

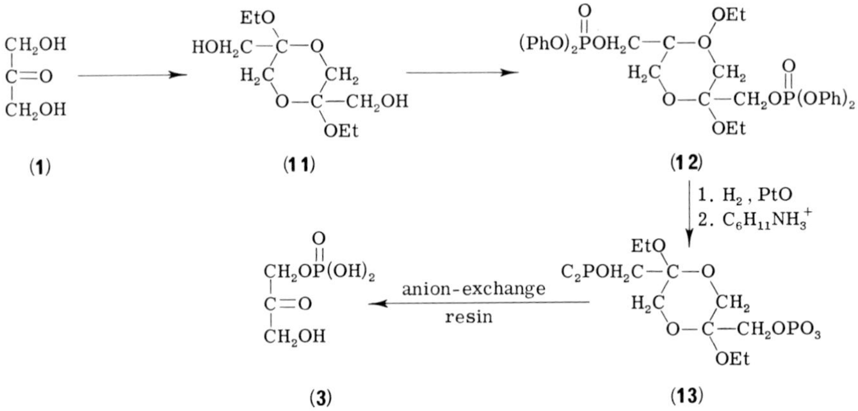

Et represents an ethyl moiety
Ph represents a phenyl moiety

SCHEME II. Synthesis of dihydroxyacetone phosphate according to Colbran *et al.* (1967).

Colbran *et al.* (1967) reported an alternate synthesis from 1,3-dihydroxy-2-propanone, in 27% overall yield (Scheme II). They esterified 2,5-diethoxy-*p*-dioxane-2,5-dimethanol (**11**) (Fischer and Baer, 1924)

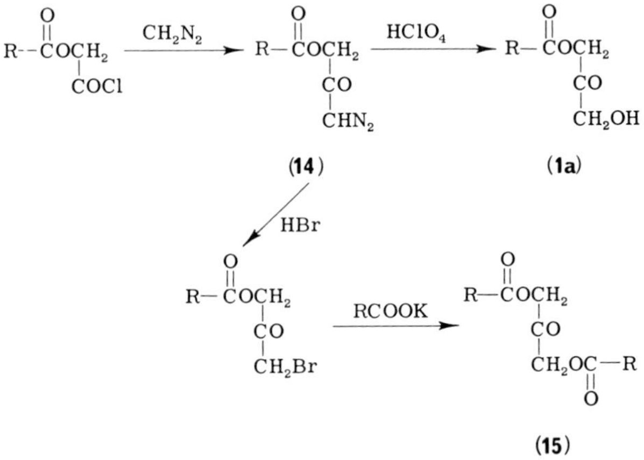

R designates $C_{15}H_{31}$ or $C_{17}H_{35}$

SCHEME III. Synthesis of acyl derivatives of dihydroxyacetone according to Schlenk *et al.* (1952).

with phenyl phosphorochloridate in pyridine to the bis(diphenyl phosphate) (**12**). The cyclic acetal (**11**) was obtained by treating (**1**) with ethyl orthoformate in absolute alchol in the presence of aluminum chloride. Hydrogenolysis of (**12**) with platinum oxide (Adams catalyst), followed by neutralization with cyclohexylamine, produced the bis(dicyclohexylammonium phosphate) (**13**), which was converted to (**3**) by means of anion-exchange resin and isolated as the cyclohexylammonium salt.

III. Synthesis of Acyl Derivatives of Dihydroxyacetone and Dihydroxyacetone Phosphate

In their route to the synthesis of glycerides, Schlenk *et al.* (1952) utilized dihydroxyacetone esterified with fatty acids as an intermediate. They converted the glycolic acid chloride starting materials to the desired diazoketone (**14**) and found that the catalytic decomposition of the diazo group was best accomplished by utilizing perchloric acid in aqueous dioxane. The result was acyl-3-hydroxyacetone in monomer and dimer form. In preparing the 1,3-diacylacetone derivatives (**15**), the diazomethyl ketone was first treated with aqueous hydrobromic acid and then with the appropriate potassium acid salt (Scheme III).

Barry and Craig (1955) subsequently reported a modification of this method that minimized monomer–dimer interconversion. The carbonyl function was protected as a mercaptal, and 1,3-distearoyl and dipalmitoyl

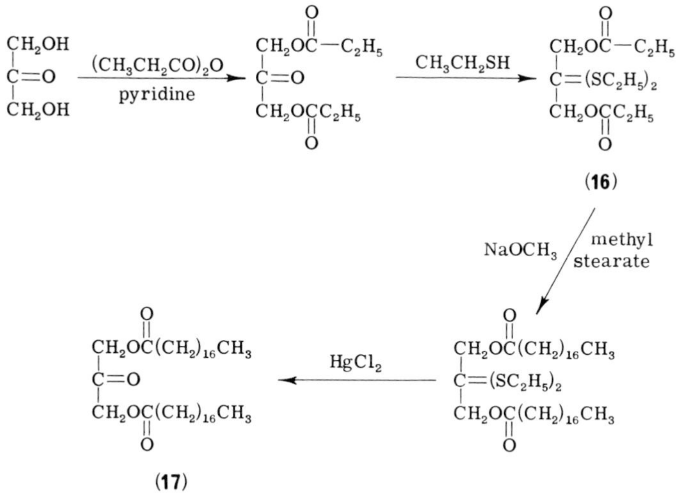

SCHEME IV. Synthesis of acyl derivatives of dihydroxyacetone according to Barry and Craig (1955).

acetone were prepared by interesterification of the methyl esters of the fatty acids with 1,3-dipropionoyl 2-diethylmercaptopropane (**16**). Hydrolysis of the mercaptal resulted in (**17**) (Scheme IV).

Garson *et al.* (1969) employed a direct acylation method in the synthesis of monomer and dimer esters of dihydroxyacetone. They prepared 1-undecanoyl-3-hydroxyacetone (**18**) by the reaction of (**1**) with undecanoyl chloride in dry pyridine. The product with the higher melting point was characterized as the dimer of (**18**). Romo (1956) used lithium aluminum hydride reduction of diethoxy diethyl malonate to produce diethoxy diethyl dihydroxyacetone, which was benzoylated to the dibenzoylated derivatives.

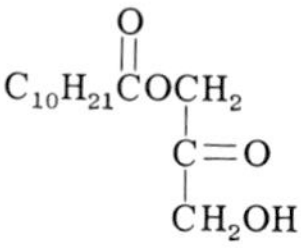

(**18**)

In a preliminary communication, Hajra and Agranoff (1967) reported the preparation of palmitoyl dihydoxyacetone phosphate (**19**) by the reaction of palmitoyl chloride with excess dihydroxyacetone. They phos-

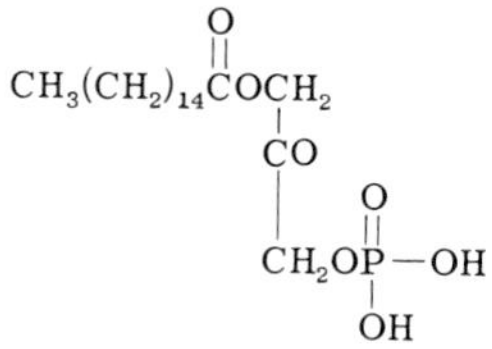

(**19**)

phorylated the resulting compound with $POCl_3$ and purified the product by column chromatography. Both the mono- and dipalmitoyl dihydroxyacetone were formed by direct acylation (Hajra and Agranoff, 1968). The diesters can be removed either by column chromatography using silica gel or by differential crystallization from hot ethanol. Hajra and Agranoff (1968) also reported an alternate synthesis for (**19**) utilizing 1-palmitoxy-3-diazoacetone (**14**), an intermediate reported by Schlenk *et al.* (1952) in their synthetic scheme. Compound (**14**) was treated with 85% H_3PO_4 in dioxane and (**19**) isolated as the cyclohexylamine salt. A major by-product of this reaction was palmitoyl dihydroxyacetone (Hajra and Agranoff, 1968).

IV. Synthesis of *O*-Alkyl Derivatives of Dihydroxyacetone and Dihydroxyacetone Phosphate

A dihydroxyacetone monoether has been produced by reacting 1-hydroxy-3-halo-2-propanone with an alkali salt of a low-molecular-weight aliphatic acid in the presence of alcohol at 100°–150°C (*Chem. Abstr.*, 1944, 1949). Phenoxy-3-hydroxy-2-propanones as well as propenoxy, ethoxy, amylphenoxy, and diethylaminophenoxy derivatives have been synthesized by treating an α-halo-α-hydroxyacetone ether with an alkali metal carboxylic acid salt at 100°–150°C (Grun and Stoll, 1945).

Hajra (1970) reported a synthesis for 1-*O*-hexadecydihydroxyacetone phosphate, but did not give experimental details. He oxidized 1-*O*-hexadecyl-2,3-propanediol to the hexadecyloxy glycolic acid by periodate–permanganate oxidation according to von Rudloff (1956), and then made the diazoketone intermediate, which upon treatment with phosphoric acid presumably gave the desired compound.

The procedures described by Hartman (1970) and Ballou and Fischer (1956) for the synthesis of 3-haloacetol phosphates and dihydroxyacetone phosphate, respectively, have recently been applied with modifications to the complete chemical synthesis of 1-*O*-alkyldihydroxyacetone phosphate (Piantadosi *et al.*, 1971) and *O*-alkyldihydroxyacetone (Piantadosi *et al.*, 1970) (Scheme V). The procedure outlined in Scheme V synthesizes 1-*O*-octadecyl-2-propanone-3-phosphate (**28**) and the 1-*O*-hexadecyl analog in good yields and high purity. Both 1-*O*-octadecylglycerol and 1-*O*-hexadecylglycerol were used as starting materials. Compound (**20**) was benzoylated in the presence of pyridine at −10°C, resulting in a crude mixture of (**21**), in which the monobenzoate predominated in a ratio of approximately 2:1. Compound (**21**) was not isolated but oxidized in high yields to (**22**) with dimethyl sulfoxide and dicyclohexylcarbodiimide in the presence of trifluoroacetic acid (Piantadosi *et al.*, 1970; Pfitzner and Moffatt, 1965; Hartman, 1970). During this oxidation reaction, the dimethyl sulfoxide was converted to a labile intermediate which facilitated the attack on the sulfur atom by the β-hydroxy group of hexadecylglycerol. Trifluoroacetic acid was used to initiate the reaction, while dicyclohexylcarbodiimide acted as a polarizing agent (Fenselau and Moffatt, 1966). The oxidized compound (**22**) was ketalized to (**23**), then hydrolyzed with NaOH to (**24**), and subsequently phosphorylated with diphenyl phosphorochloridate; the phenyl groups were removed with platinum oxide and

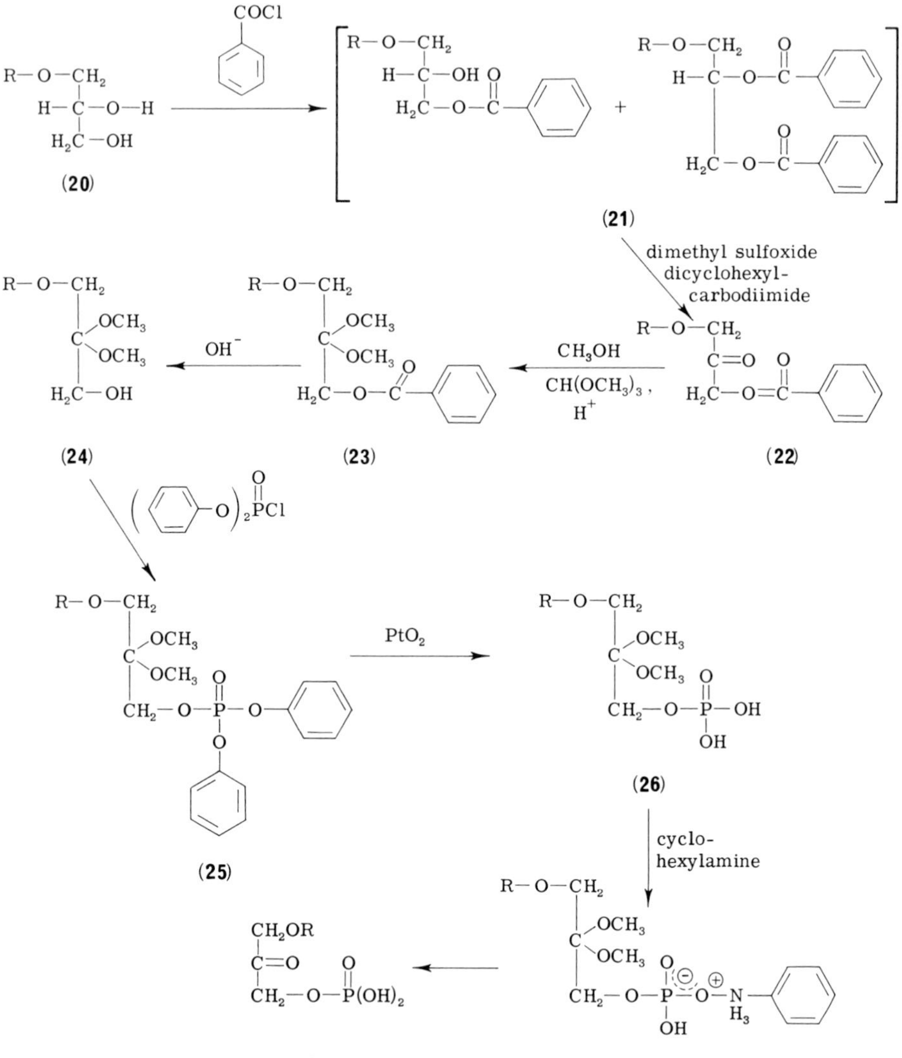

R designates $CH_3(CH_2)_{14}CH_2$– and $CH_3(CH_2)_{16}CH_2$–

SCHEME V. Synthesis of 1-*O*-alkyldihydroxyacetone phosphate according to Piantadosi *et al.* (1970).

compound (**27**) was isolated as the cyclohexylamine salt in 50% yield. Compound (**27**) was shaken with dilute HCl or Dowex-50W ion-exchange resin at room temperature to release (**28**). In this study, *O*-alkyldihydroxyacetone phosphate, whether organically and enzymically synthesized, had identical chemical and chromatographic properties, thereby confirming it as an intermediate in the biosynthesis of alkyl glyceryl ethers (Snyder

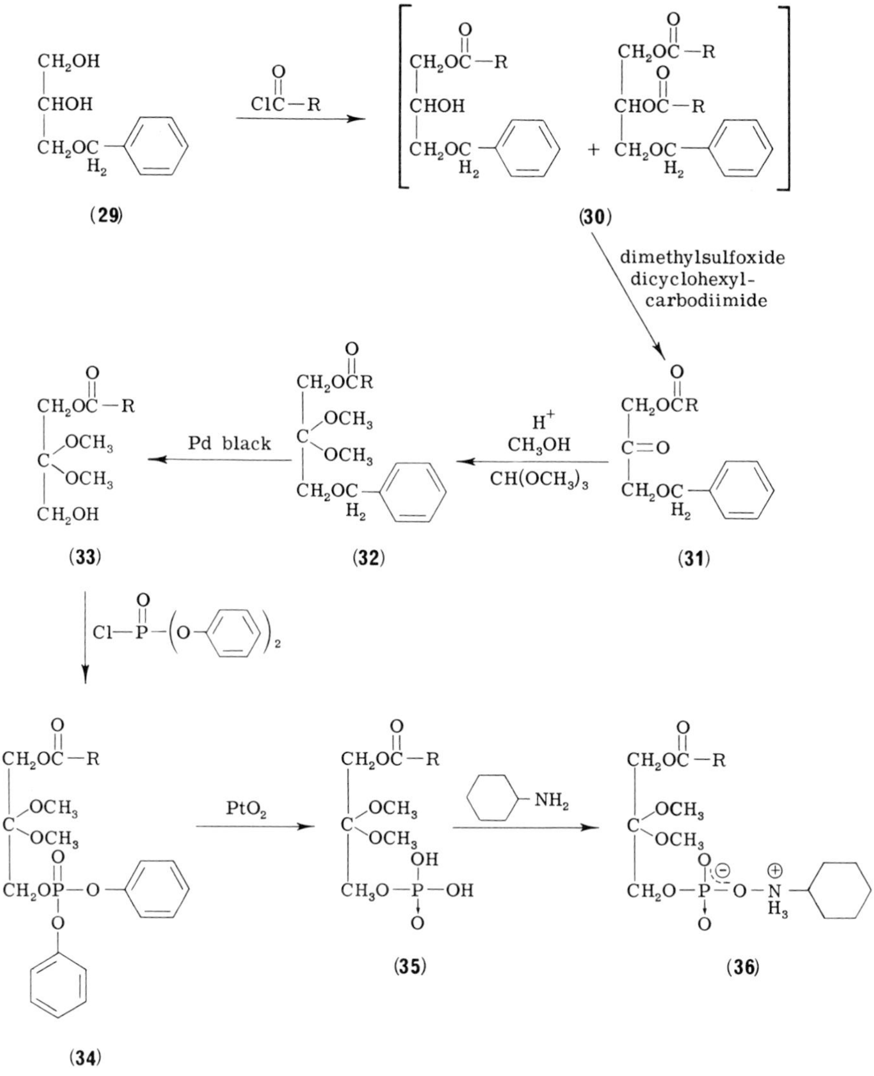

R represents $CH_3(CH_2)_{13}CH_2$–

SCHEME VI. Synthesis of 1-*O*-acyldihydroxyacetone phosphate according to Piantadosi *et al.* (1972).

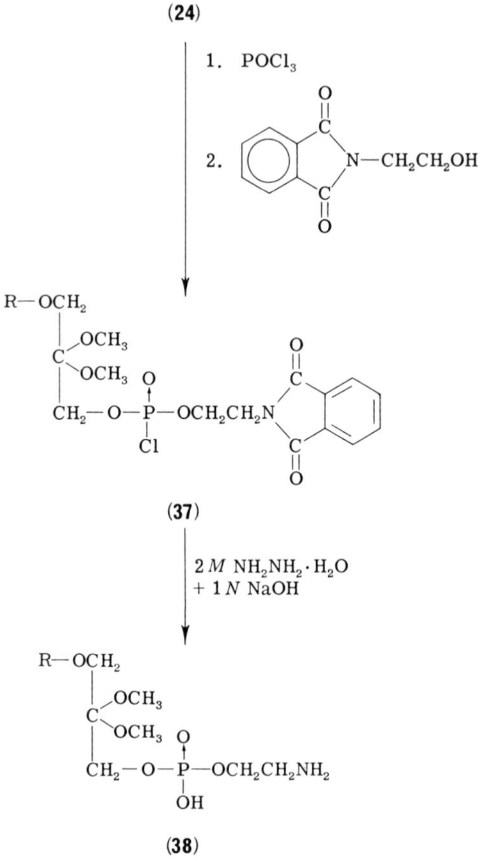

R designates $CH_3(CH_2)_{14}CH_2$– and $CH_3(CH_2)_{16}CH_2$–

SCHEME VII. Synthesis of 1-*O*-alkyl-2,2-dimethoxypropane-3-phosphate ethanolamine according to Piantadosi and Ishaq (1972).

et al., 1970a,b,c; Wykle and Snyder, 1970). The structure of ^{14}C-*O*-alkyldihydroxyacetone isolated from enzyme systems was similarly confirmed in our earlier study (Piantadosi *et al.*, 1970).

In an analogous manner Piantadosi *et al.* (1972) have used 3-*O*-benzylglycerol (**29**) (Howe and Malkin, 1951) for the synthesis of the corresponding 1-*O*-acyldihydroxyacetone phosphate. Compound (**36**) in Scheme VI was obtained in excellent yields and high purity. Compound (**29**) was acylated with palmitoyl chloride in the presence of pyridine at −20°C, resulting in a crude mixture of (**30**) that contained better than 70% of the monopalmitoyl derivative. Compound (**30**) was not isolated, but was oxidized in high yields with dimethyl sulfoxide and dicyclohexylcarbodiimide in the presence of trifluoroacetic acid (Piantadosi *et al.*, 1970; Pfitzner and Moffatt, 1965; Hartman, 1970) in a manner similar to the synthesis of the 1-*O*-alkyl derivative, compound (**28**) in Scheme V. The oxidized (**31**) was ketalized to (**32**), then debenzylated by stirring with palladium black in an atmosphere of hydrogen at 20 psi at room temperature and finally phosphorylated with diphenyl phosphorochloridate. Compound (**34**) was treated with platinum oxide in order to remove the phenyl groups, and the resulting compound (**35**) was subsequently treated with cyclohexylamine and isolated as the salt in 80% overall yield.

The chemical synthesis of keto phosphatidylethanolamine has also been accomplished (Piantadosi and Ishaq, 1972) (see Schemes V and VII). In this sequence, compound (**20**) was benzoylated either with benzoyl chloride or with *p*-nitrobenzoyl chloride, resulting in a crude mixture of compound (**21**) containing better than 70% of the mono derivative. Compound (**21**) was then subjected to the "Pfitzner-Moffatt" oxidation (Pfitzner and Moffatt, 1963, 1965) in a solution of dicyclohexylcarbodiimide–dimethyl sulfoxide containing a proton source such as pyridinium phosphate or pyridinium trifluoroacetate. A similar oxidation procedure has been described by Hartman (1970) for the synthesis of 3-haloacetol phosphates. Compound (**22**) was then ketalized in the usual manner, resulting in (**23**), which was hydrolyzed to (**24**) and phosphorylated with $POCl_3$ and treated with β-hydroxyethylpthalimide, resulting in compound (**37**). This was treated with hydrazine hydrate yielding 1-*O*-hexadecyl-2,2-dimethoxypropane-3-phosphate ethanolamine (**38**). This compound was isolated and purified in 27% overall yield. The availability of these key compounds for use in further studies of the biosynthesis of alkyl glyceryl ethers is of great importance.

Recently 1-*O*-(1-^{14}C-octadecyl)-2,2-dimethoxy-3-hydroxypropane (**27**) has also been synthesized and isolated as the cyclohexylamine salt in good yields with a high specific activity (Piantadosi *et al.*, 1972).

REFERENCES

Ballou, C. E., and Fischer, H. O. L. (1956). *J. Amer. Chem. Soc.* **78**, 1659.
Barry, P. J., and Craig, B. M. (1955). *Can. J. Chem.* **33**, 716.
Bertrand, G. (1899). *C. R. Acad. Sci.* **129**, 341.
Chem. Abstr. (1944). **38**, 5224 (Abstr. 9).
Chem. Abstr. (1949). **43**, 3038 (Abstr.d).
Colbran, R. L., Jones, J. K. N., Matheson, M. K., and Rozema, I. (1967). *Carbohyd. Res.* **4**, 355.

Fenselau, A. H., and Moffatt, J. G. (1966). *J. Amer. Chem. Soc.* **88,** 1762.
Fischer, H. O. L., and Baer, E. (1924). *Chem. Ber.* **57,** 710.
Fischer, H. O. L., and Mildbrand, H. (1924). *Chem. Ber.* **57,** 707.
Garson, L. R., Quintana, R. P., and Lasslo, A. (1969). *Can. J. Chem.* **47,** 1249.
Grun, A., and Stoll, W. (1945). *Chem. Abstr.* **34,** 3535.
Hajra, A. K. (1969). *Biochem. Biophys. Res. Commun.* **37,** 486.
Hajra, A. K. (1970). *Biochem. Biophys. Res. Commun.* **39,** 1037.
Hajra, A. K., and Agranoff, B. W. (1967). *J. Biol. Chem.* **242,** 1074.
Hajra, A. K., and Agranoff, B. W. (1968). *J. Biol. Chem.* **243,** 1617.
Hartman, F. C. (1970). *Biochemistry* **9,** 1776.
Howe, R. J., and Malkin, T. (1951). *J. Chem. Soc., London* p. 2663.
Kiessling, W. (1934). *Chem. Ber.* **67,** 869.
Meyerhof, O., and Lohmann, K. (1934). *Biochem. Z.* **271,** 89.
Pfitzner, K. E., and Moffatt, J. G. (1963). *J. Amer. Chem. Soc.* **85,** 3027.
Pfitzner, K. E., and Moffatt, J. G. (1965). *J. Amer. Chem. Soc.* **87,** 5670.
Piantadosi, C., and Ishaq, K. S. (1972). Unpublished data.
Piantadosi, C., Ishaq, K. S., and Snyder, F. (1970). *J. Pharm. Sci.* **59,** 1201.
Piantadosi, C., Ishaq, K. S., Wykle, R. L., and Snyder, F. (1971). *Biochemistry* **10,** 1417.
Piantadosi, C., Chae, K., Ishaq, K. S., and Snyder , F. (1972). Unpublished data.
Piloty, O. (1897). *Ber. Deut. Chem. Ges.* **30,** 3161.
Romo, J. (1956). *J. Org. Chem.* **21,** 1038.
Schlenk, H., Lamp, B. G., and de Haas, B. W. (1952). *J. Amer. Chem. Soc.* **74,** 2550.
Snyder, F., Wykle, R. L., and Malone, B. (1969). *Biochem. Biophys. Res. Commun.* **34,** 315.
Snyder, F., Blank, M. L., and Malone, B. (1970a). *J. Biol. Chem.* **245,** 4016.
Snyder, F., Blank, M. L., Malone, B., and Wykle, R. L. (1970b). *J. Biol. Chem.* **245,** 1800.
Snyder, F., Malone, B., and Blank, M. L. (1970c). *J. Biol. Chem.* **245,** 1790.
von Rudloff, E. (1956). *Can. J. Chem.* **34,** 1413.
Wohl, A., and Newberg, C. (1900). *Ber. Deut. Chem. Ges.* **33,** 3095.
Wykle, R. L., and Snyder, F. (1969). *Biochem. Biophys. Res. Commun.* **37,** 658.
Wykle, R. L., and Snyder, F. (1970). *J. Biol. Chem.* **245,** 3047.

CHAPTER VII

THE ENZYMIC PATHWAYS OF ETHER-LINKED LIPIDS AND THEIR PRECURSORS

Fred Snyder

I. Enzymic Synthesis of Alkyl Glycerolipids

A. General

The most pertinent data leading to the discovery of an enzyme system that synthesizes alkyl lipids came from two *in vivo* experimental observations: (1) ^{14}C- and ^{3}H-labeled long-chain fatty alcohols could be incorporated into alkyl glycerolipids of cells from various species (Friedberg and Greene, 1967; Ellingboe and Karnovsky, 1967; Snyder and Wood, 1968) and (2) the double bond locations in the alkyl chains of alkyldiacylglycerols and in the fatty alcohols of preputial glands from normal mice were structurally similar (Snyder and Blank, 1969).

A cell-free system was discovered in October of 1968 that made it possible to investigate the biosynthetic reactions involved in the formation of alkyl glycerolipids. This investigation (Snyder *et al.*, 1969a) revealed that 1-^{14}C-hexadecanol was incorporated into alkyl glycerolipids. The alcohol was incubated with CoA,* ATP, Mg^{2+}, and homogenates or microsomal plus soluble supernatant fractions prepared from preputial gland tumors (ESR 586) grown in mice. We found that α-glycerophosphate did not substitute for the soluble fraction when washed microsomes were the sole source of enzymes, but that fructose 1,6-diphosphate stimulated the formation of *O*-alkyl bonds when added along with the microsomes and soluble fraction. Furthermore, the soluble supernatant fraction of normal rat liver could replace the soluble supernatant fraction of the tumor in the biosynthetic system, which suggested that the soluble fraction did not contain enzymes that were directly required in the synthesis of ether bonds. This became apparent when we found that fructose 1,6-diphosphate and aldolase could replace the requirement of the soluble fraction.

At this point, we tried glyceraldehyde-3-P as a "glycerol" substrate

* Abbreviations used in this chapter: CoA, coenzyme A; ATP, adenosine 5′-triphosphate; DHAP, dihydroxyacetone phosphate; DHA, dihydroxyacetone; $NADP^+$, nicotinamide adenine dinucleotide phosphate; NADPH, reduced nicotinamide adenine dinucleotide phosphate; NAD^+, nicotinamide adenine dinucleotide; NADH, reduced nicotinamide adenine dinucleotide; GLC, gas–liquid chromatography; TLC, thin-layer chromatography; FAD, flavin adenine dinucleotide; UDP, uridine diphosphate.

and found that it could react with the fatty alcohol to form the *O*-alkyl bonds in the microsomal system. We immediately proposed a biosynthetic pathway for alkyl glycerolipids (Snyder *et al.*, 1969b). The mechanism for the initial reaction between hexadecanol and glyceraldehyde-3-P was thought to proceed via a hemiacetal structure in which water was removed to yield an ene-diol form that rearranged itself to alkyldihydroxyacetone-P (alkyl-DHAP). The ketone intermediates postulated were later identified and synthesized organically. In the initial experiments (Snyder *et al.*, 1969b) the existence of alkyl-DHAP and alkyldihydroxyacetone (alkyl-DHA) was based primarily on their chromatographic behavior before and after enzymic reduction by NADPH and after chemical reduction with $LiAlH_4$. At the time we discovered the cell-free system from neoplastic cells that synthesized *O*-alkyl lipids, Friedberg and Greene (1968) obtained evidence *in vivo* with 1,3-^{3}H-glycerol and uniformly ^{14}C-labeled glycerol incubated with *Tetrahymena pyriformis* that suggested a triose-P was involved in the biosynthesis of alkyl glyceryl ethers.

Later we found not only that both glyceraldehyde-3-P and DHAP could serve as glycerol acceptors (Snyder *et al.*, 1970a), but also that a triose-P isomerase was present in the microsomes (Wykle and Snyder, 1969). The presence of this enzyme did not appear to be caused by contamination from the soluble fraction (glycerophosphate dehydrogenase could not be detected in the thoroughly washed microsomal preparations), but it made it difficult to ascertain which triose-P was the immediate precursor of *O*-alkyl lipids. However, when a specific and potent inhibitor of triose-P isomerase, 1-hydroxy-3-chloro-2-propanone phosphate (Hartman, 1968, 1970), was made available to us, we conducted experiments that pinpointed DHAP, not glyceraldehyde-3-P, as the obligate triose-P precursor in the biosynthesis of *O*-alkyl lipids (Wykle and Snyder, 1969). Triose-P isomerase is also effectively inhibited by glycidol phosphate (Rose and O'Connell, 1969), but its effect on the biosynthesis of *O*-alkyl lipids is unknown. High levels of inorganic phosphate inhibit both triose-P isomerase (Oesper and Meyerhof, 1950) and the biosynthesis of alkyl-DHAP (Snyder *et al.*, 1970b).

About the same time, Hajra (1969) used liver mitochondria from guinea pigs and brain microsomes from mice to demonstrate that DHAP and glyceraldehyde-3-P could serve as the "glycerol" acceptor in the synthesis of *O*-alkyl lipids. Hajra observed differences in the kinetics of the reaction between fatty alcohols and DHAP or glyceraldehyde-3-P that led him to conclude that DHAP rather than glyceraldehyde-3-P was the true "glycerol" precursor of alkyl-DHAP. Decreased synthesis of *O*-alkyl lipids from both DHAP and glyceraldehyde-3-P in the presence of glycero-

TABLE I

Sources of Alkyl Ether Synthesizing Enzymes

Source[a]	Approximate enzymic activities (nmoles/mg protein/hr)	References
Mammalia		
Preputial gland tumors	8–17	Snyder *et al.* (1969a,b)[b] Snyder *et al.* (1970b)
Ehrlich ascites cells	12	Wykle and Snyder (1970)
Fibroblasts (modified L strain)	9–12	Snyder *et al.* (1970a)
Mouse brain	Not reported	Hajra (1969)
Rat brain	1.0–1.5[c]	Snyder *et al.* (1971a)
Guinea pig liver	Not reported	Hajra (1969)
Rat liver	1.0–1.5[c]	Snyder *et al.* (1971a)
Protozoa		
Tetrahymena pyriformis	Not reported	Kapoulas and Thompson(1969)[b] Friedberg and Greene (1969)[b]
Echinodermata		
Asterias forbesi	10	Snyder *et al.* (1969c)

[a] All preparations were microsomes except those from guinea pig livers (mitochondria) and *Tetrahymena pyriformis* (bulk of activity in 100,000 *g* supernatant fraction).

[b] Glyceraldehyde-3-P was used as the substrate; all other data are based on dihydroxyacetone phosphate as the substrate.

[c] Maximum activities which occurred 5 days after birth.

phosphate dehydrogenase and from glyceraldehyde-3-P in the presence of glyceraldehyde-3-P dehydrogenase supported his conclusion.

Similar enzymic systems for the biosynthesis of alkyl glycerolipids have been found in several tumors, starfish, fibroblasts (L-M cells grown in suspension cultures), brains from mice and rats, livers from guinea pigs and rats, and *Tetrahymena* (Table I). The significance of the *O*-alkyl enzymic system in whole cells has been emphasized by *in vivo* experiments in which long-chain fatty alcohols were used as precursors of the *O*-alkyl moieties. The metabolic pathway for alkyl glycerolipids that exists in the endoplasmic reticulum of tumors and certain healthy cells consists of the following steps (Fig. 1): (1) *alkylation* of DHAP or acyl-DHAP, (2) *reduction* of the ketone by an NADPH-linked reductase, (3) *acylation* to form 1-alkyl-2-acylglycerol-3-P, (4) *dephosphorylation* to form 1-alkyl-2-acylglycerol, and (5) *acylation* to 1-alkyl-2,3-diacylglycerol and/or (6) *transfer of phosphorylcholine* or *phosphorylethanolamine* via cytidine diphosphate to alkylacylglycerol. Once the *O*-alkyl analog of phosphatidic acid is synthesized, the reactions that occur are identical to those characterized for the synthesis of acylated glycerolipids (Kennedy, 1957). In fact, the same enzymes probably utilize the ether analogs as substrates. Ketone intermediates have not been reported in *Tetrahymena*, perhaps because cofactors are present in the fractions from this organism that rapidly catalyze the reduction of the ketone and subsequent acylation to the ether analog of phosphatidic acid.

B. Alkyldihydroxyacetone Phosphate and Alkyldihydroxyacetone

The complete enzymic system for the synthesis of alkyl-DHAP and alkyl-DHA requires the following components: long-chain fatty alcohols, DHAP, CoA, ATP, Mg^{2+}, and microsomes. Table II summarizes the optimal concentrations of cofactors for the reaction catalyzed by microsomal enzymes from preputial gland tumors (Snyder *et al.*, 1970b) and Ehrlich ascites cells (Wykle and Snyder, 1970). The location of the enzyme catalyzing this reaction in neoplasms was clearly documented as being microsomal in a study of the distribution of enzymic activities in organelles that were prepared by conventional and zonal centrifugation techniques from the preputial gland tumor (Snyder *et al.*, 1970b).

^{3}H- and ^{14}C-labeled fatty alcohols and ^{14}C- and ^{32}P-labeled DHAP are incorporated into alkyl-DHAP in a 1:1 molar ratio (Wykle and Snyder,

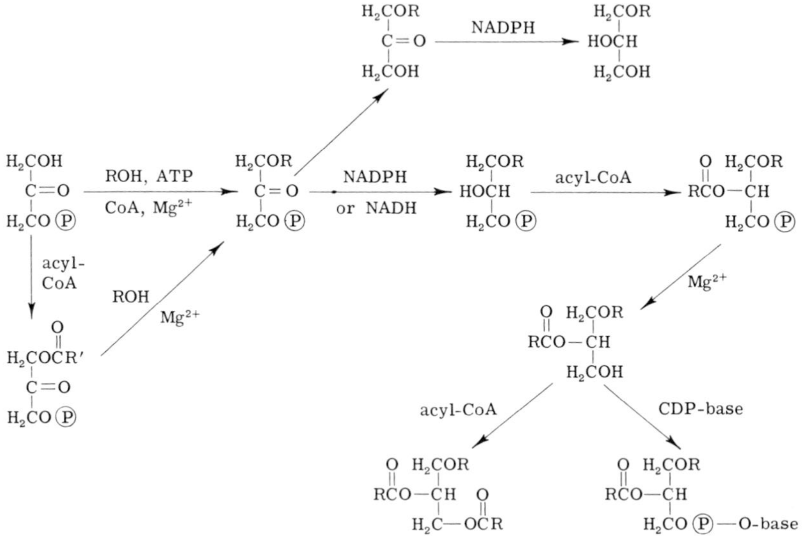

Fig. 1. Enzymic pathways involved in the biosynthesis of alkyl glycerolipids.

TABLE II

Optimal Concentrations for the Complete System Used in Kinetic Studies on the Biosynthesis of Alkyl Ether Bonds by Microsomal Enzymes

Components	Concentration (mM)
ATP	4–10
CoA[a]	0.05–0.1
Mg^{2+}	1–4
DHAP	0.6–1.5
1-^{14}C-Hexadecanol	0.03–0.06
Microsomes (1–2 mg protein)	—
Phosphate buffer (pH 7.1)	100

[a] The optimum concentration of acyldihydroxyacetone phosphate that can replace the CoA requirement is approximately 60 μM.

1970) (Table III). Phosphatases in the microsomal and soluble fractions can dephosphorylate alkyl-DHAP to produce alkyl-DHA. As with other phosphatases, NaF is an inhibitor, but it also inhibits the formation of *O*-alkyl bonds to some extent (Snyder *et al.*, 1969b).

The fact that the cofactors (CoA, ATP, Mg^{2+}) are the same as those necessary for fatty acid activation and that alkyl-DHAP synthesis is completely inhibited when NADPH is present at the beginning of the reaction (Snyder *et al.*, 1969b; Wykle and Snyder, 1970), led Hajra (1970) to suspect that acyl-DHAP might somehow be involved in the mechanism responsible for the biosynthesis of alkyl-DHAP. He has reported data indicating that reaction (1) is catalyzed by mitochondrial enzymes in guinea pig livers and microsomes of mouse brains. Such an exchange reaction also occurs in microsomes from Ehrlich ascites cells and preputial gland tumors (Wykle *et al.*, 1972).

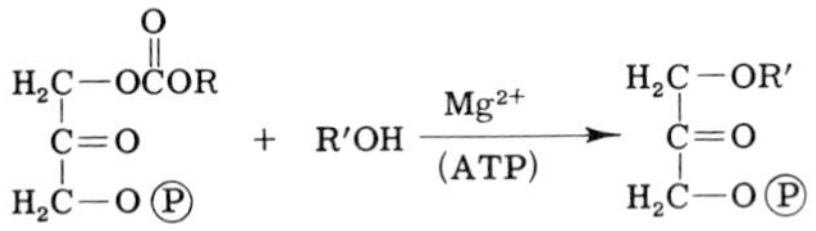

(1)

Murooka and colleagues (1970) have obtained data on the alkylation of homoserine that supports the principle of this mechanism. Recently, Agranoff and Hajra (1971) demonstrated that the biosynthesis of acyl-DHAP is a prominent pathway in Ehrlich ascites cells, which contain a significant quantity of ether-linked glycerolipids.

The possibility of substrate-cofactor complexes being formed in the initial reaction step has also been postulated (Snyder *et al.*, 1970b). The label from ^{18}O-hexadecanol (Snyder *et al.*, 1970e) was retained in the ether linkage of alkyl-DHAP and alkyl-DHA, and any reaction mechanisms depicting the initial step in the synthesis of the *O*-alkyl moiety must take this into account. The hydroxyl moiety of hexadecanol behaves in a manner similar to that observed in reactions between alcohol moieties and acids, e. g., in the formation of acylglycerols (Kornberg and Pricer, 1953; Stewart *et al.*, 1959), phosphodiesters (Kennedy, 1962), and phosphomonoesters (McVicar, 1934).

Identification of the ketone intermediates formed by the *O*-alkyl-synthesizing enzymes was based on chemical and enzymic reactions and the subsequent chromatographic behavior of derivatives (Snyder *et al.*, 1970c; Wykle and Snyder, 1970). Standards were derived by subjecting known

TABLE III

Incorporation of Labeled Precursors into Alkyl Glycerolipids

Alkyl glycerolipid[a]	Molar ratio based on isotope incorporation[b]	
	ROH ($1-^{14}C$-hexadecanol) / PO_4^{3-} (^{32}P-DHAP)	ROH ($9,10-^{3}H$-hexadecanol) / Triose [(U)-^{14}C-DHAP, (U)-^{14}C-GA-3-P]
I. Alkyldihydroxyacetone phosphate	1.1	1.1
II. 1-Alkyl-2-acylglycerophosphate	1.1	1.2
III. Alkyldihydroxyacetone	—	0.9
IV. Alkylglycerol produced by $LiAlH_4$ reduction of alkyldihydroxyacetone	—	0.9

[a] The labeled compounds were prepared in incubations that contained ATP (10 m*M*), CoA (0.1 m*M*), Mg^{2+} (4 m*M*), $1-^{14}C$- or $9,10-^{3}H$-hexadecanol (33 m*M*), microsomes (approximately 2 mg protein), and ^{14}C- or ^{32}P-triose phosphate.

[b] The liquid scintillation spectrometer was set for simultaneous counting of ^{32}P (52% efficiency in the ^{32}P channel and 4% efficiency in the ^{14}C channel) and ^{14}C (0.2% efficiency in the ^{32}P channel and 53% efficiency in the ^{14}C channel) or for simultaneous counting of ^{3}H (20.5% efficiency in the ^{3}H channel and 0.45% efficiency in the ^{14}C channel) and ^{14}C (9.5% efficiency in the ^{3}H channel and 64% efficiency in the ^{14}C channel). The ratios were obtained from samples containing 2,000–10,000 dpm of each isotope.

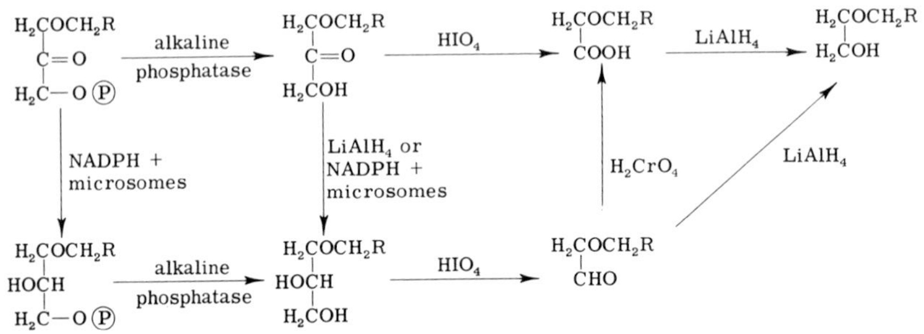

FIG. 2. Chemical and enzymic reactions used to identify alkyl-DHAP and alkyl-DHA synthesized by microsomal enzymes. Reproduced from Snyder *et al.* (1970c) by permission of the *Journal of Biological Chemistry*.

alkylglycerols to oxidation and reduction procedures (Fig. 2). Alkyl-DHA was formed from alkyl-DHAP by incubating the latter with alkaline phosphatase. Periodate oxidation of alkyl-DHA produced alkyl glycolic acid which was subsequently reduced by $LiAlH_4$ to alkyl ethylene glycol. In addition, alkyl-DHA could be reduced to alkylglycerols by $LiAlH_4$, hydrogenation, $NaBH_4$, and NADPH (enzymically by the microsomal reductase). Relative R_f's of alkyl-DHAP, alkyl-DHA, their chemical derivatives and a number of other known lipids separated by thin-layer chromatographic systems are summarized in Table IV. The acetate derivatives of the alkyl ethylene glycols and the isopropylidene derivatives of the alkylglycerols were identified by GLC; these data have demonstrated that the *O*-alkyl moieties were formed at the 1-position on the glycerol moiety. Formation of the 1-isomers was also confirmed by chromatography of the ^{14}C-labeled alkylglycerols obtained from the enzymic system on silica gel layers impregnated with arsenite ions (Snyder *et al.*, 1970b). Essentially no oxidation or chain elongation occurs during the enzymic synthesis of *O*-alkyl lipids since the *O*-alkyl chains are of the same length as the fatty alcohols added to the system. This was verified by data showing that 1-^3H,1-^{14}C-hexadecanol was incorporated into the hexadecylglycerol without any change in the $^3H/^{14}C$ ratio (Wykle and Snyder, 1970).

An important step leading to the identification of alkyl-DHAP was the successful removal of the phosphate group without altering the remaining part of the molecule. This was accomplished with alkaline phosphatase (Snyder *et al.*, 1970b; Wykle and Snyder, 1970; Blank and Snyder, 1970b), which specifically dephosphorylates glycerolipids that have a free phosphate group and a single aliphatic moiety in the 1-position. Unfortunately, the ketone lipids are not stable to acid or prolonged alkaline hydrolysis; in fact, the *O*-alkyl moiety can be quantitatively released as a fatty alcohol by acid hydrolysis (Snyder *et al.*, 1970b; Wykle and Snyder, 1970). These workers suggest that the acid lability of the ketone structure occurs because of the alkenyl linkage that forms according to reaction (2).

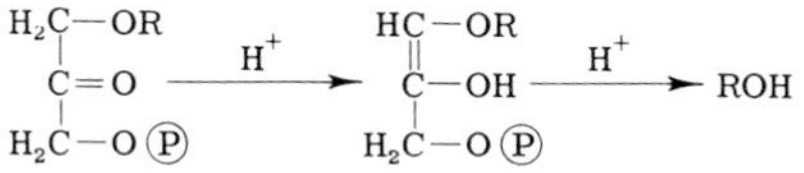

(2)

Alkyl-DHAP and alkyl-DHA have been synthesized chemically (Piantadosi *et al.*, 1970, 1971); a summary of the methods used to prepare and purify the chemically synthesized compounds is given in Chapter VI. These chemically synthesized ketone intermediates have chemical and chro-

TABLE IV

Relative R_f Values of Lipids in Various Chromatographic Systems

Compound	Thin-layer chromatographic system used[a]							
	I	II	III	IV	V	VI	VII	VIII
Long-chain alcohol	0.76	0.68	0.49	0.29	—	—	—	—
Alkylglycerol	0.45	0.29	0.16	0.08	—	—	—	—
Alkyl glycolic acid	0.13	0.41	—	—	—	—	—	—
Alkyl glycolic aldehyde	0.84	—	—	—	—	—	—	—
Alkyl ethylene glycol	0.68	—	—	—	—	—	—	—
Alkyldihydroxyacetone	0.67	—	0.34	0.22	—	—	—	—
Fatty acid	—	—	—	0.45	—	—	—	—
Fatty acid esters of long-chain alcohol	—	—	0.86	—	—	—	—	—
Long-chain aldehyde	—	—	—	0.64	—	—	—	—
Isopropylidene of alkylglycerol	—	—	—	—	0.48	—	—	—
Alkyldihydroxyacetone phosphate	0	0	0	—	—	0.42	—	0.17
1-Alkyl-2-acylglycerophosphate	0	0	0	—	—	0.79	0.82	0.19
Phosphatidylcholine	0	0	0	—	—	0.37	0.41	0.41
Phosphatidylethanolamine	0	0	0	—	—	0.76	0.68	0.58

[a] Solvent systems: I, diethyl ether–water (100:0.5, v/v); II, diethyl ether–glacial acetic acid (100:0.5, v/v); III, hexane–diethyl ether–methanol–glacial acetic acid (70:30:5:1, v/v); IV, hexane–diethyl ether–glacial acetic acid (60:40:1, v/v); V, hexane–diethyl ether (80:20, v/v); VI, chloroform–methanol–glacial acetic acid–water (50:25:6:2, v/v); VII, chloroform–methanol–glacial acetic acid—0.154 *M* (0.9%) NaCl solution (50:25:8:4, v/v); VIII, chloroform–methanol–ammonium hydroxide (65:35:8, v/v). Silica Gel G was used as the adsorbent with solvent systems I–V and Silica Gel HR was used as the adsorbent with solvent systems VI–VIII.

matographic properties identical to those originally isolated and characterized from the enzymic system.

When identifying ether-linked lipids in biochemical studies, it is important to make sure that alkane-1,2-diols (1) and alkane-1,3-diols (2) are not mistaken for alkylglycerols (3), since their structures are similar except for the ether linkage.

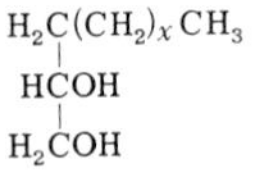

$$
\begin{array}{ccc}
\text{H}_2\text{C(CH}_2)_x\text{CH}_3 & \overset{\displaystyle\text{OH}}{\text{H}_2\text{C}-(\text{CH}_2)_x\text{CH}_3} & \text{H}_2\text{CO(CH}_2)_x\text{CH}_3 \\
\text{HCOH} & \text{CH}_2 & \text{HCOH} \\
\text{H}_2\text{COH} & \text{H}_2\text{COH} & \text{H}_2\text{COH} \\
(1) & (2) & (3)
\end{array}
$$

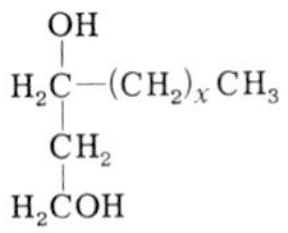

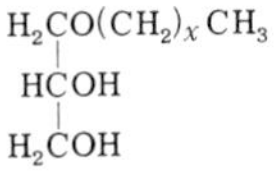

These compounds all have the same R_f in thin-layer chromatographic systems (Blank and Snyder, 1969; Blank *et al.*, 1971), but their isopropylidene derivatives can be resolved to some extent by TLC, and completely by GLC. Alkanediols have been isolated after incubating fatty acids in mitochondrial systems and reducing the extracted lipids with LiAlH$_4$ (R. E. Anderson and Sansone, 1970). Blank *et al.* (1971) that the alkane-1,3-diols produced under these conditions are derived from a β-hydroxy fatty acid (4), an intermediate formed during β-oxidation of fatty acids by

$$
\text{RCHOHCH}_2-\overset{\displaystyle\text{O}}{\overset{\displaystyle\|}{\text{C}}}-\text{SCoA}
$$

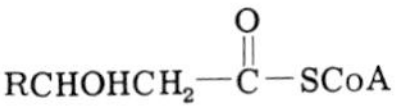

(4)

mitochondrial enzymes. 1-^{14}C-Palmitic acid incubations with beef heart mitochondria and with rat liver mitochondria did not produce hexadecyl-1,3-diol with NAD$^+$ (Blank *et al.*, 1971). The same paper describes the methodology used to distinguish a variety of alkanediols from corresponding alkylglycerols.

C. ALKYLGLYCEROL

Reduction of alkyl-DHA by an NADPH-linked enzyme in microsomes produces alkylglycerols (Snyder *et al.*, 1969b, 1970c; Wykle and Snyder, 1970). The specificity of this reaction appears to depend on NADPH as a source of hydrogens. Only small amounts of alkylglycerols were produced in the presence of NADPH if NaF was added to inhibit the phosphatase and no alkyl lipids were produced in the complete system if NADPH was added at the beginning of the reaction.

In some systems (several mouse and rat tumors and normal rat livers), alkylglycerols (1-isomers) are acylated, although only in the 3-position

(Snyder *et al.*, 1970f); but in microsomes from hamster mucosa, *rac*-1,3-alkylacylglycerols can be acylated to form the triglyceride analog (Paltauf and Johnston, 1971). 2-Alkylglycerols are acylated to form the diacyl products (Snyder *et al.*, 1970f) and have been used extensively as model analogs for 2-acylglycerols in studies of fat absorption. The enantiomers of alkylglycerols (1- and 3-isomers) have been investigated in enzymic preparations from intestinal mucosa of rats and hamsters (Paltauf and Johnston, 1971) and also in rat intestinal mucosa *in vivo* (Paltauf, 1971), demonstrating that the 1-isomer is the preferred substrate in biosynthetic reactions and that the 3-alkylglycerols were, in fact, inert substrates for acyl transferases. Acylation reactions of alkylglycerols are summarized in Fig. 3. Phosphorylation of alkylglycerols by an ATP-Mg^{2+} kinase reaction has not yet been detected in cell-free systems, although it apparently occurs *in vivo*.

D. Alkylacylglycerol Phosphate

Enzymic reduction of alkyl-DHAP can also be accomplished by microsomal enzymes, and unlike the enzymic reduction of alkyl-DHA, which can only use NADPH, either NADPH (Snyder *et al.*, 1970b; Wykle and Snyder, 1970) or NADH (Wykle *et al.*, 1972) can serve as the hydrogen source for reduction of the phosphorylated ketone intermediate. Enzymic reduction of alkyl-DHAP yields a product of decreased polarity that behaves chromatographically like phosphatidic acid. After enzymic reduction of the ketone group, the alkylglycerol phosphate formed is acylated by endogenous fatty acids (in the presence of ATP, CoA, and Mg^{2+}). Acylation at the 2-position does not occur if the phosphate group is absent (Snyder *et al.*, 1970f). The *O*-alkyl analog of phosphatidic acid (1-alkyl-2-acylglycerophosphate) synthesized enzymically was identified by the following methods: mild saponification to 1-alkylglycerophosphate, more drastic saponification to alkylglycerols, treatment by $LiAlH_4$ to yield alkylglycerols, hydrolysis with phospholipase A to yield 1-alkylglycerophosphate, and TLC in acidic and basic solvent systems (Fig. 4).

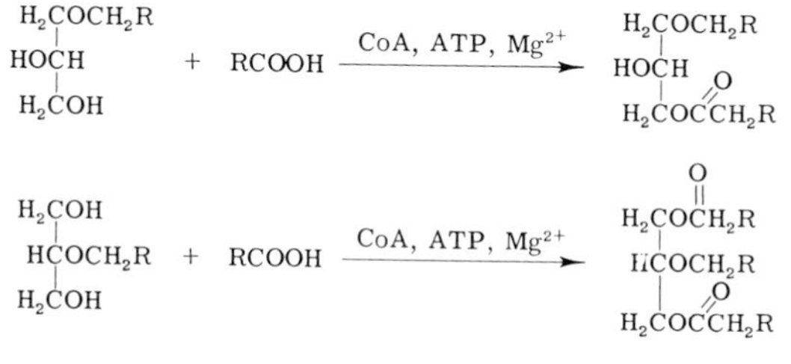

Fig. 3. Acylation of 1- and 2-isomers of alkylglycerols.

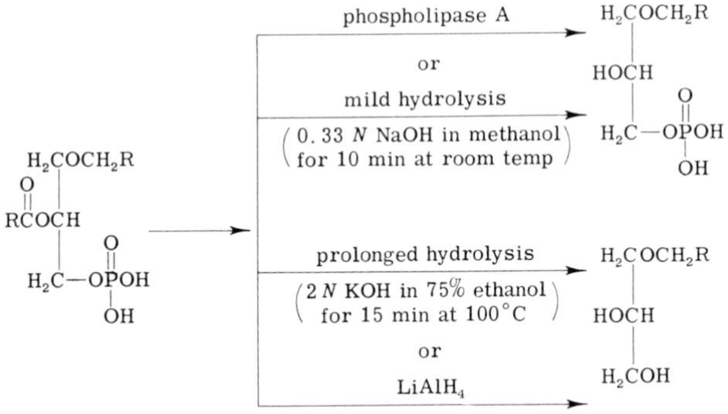

FIG. 4. Methods used to identify the *O*-alkyl analog of phosphatidic acid synthesized by microsomal enzymes. (See Wykle and Snyder, 1970.)

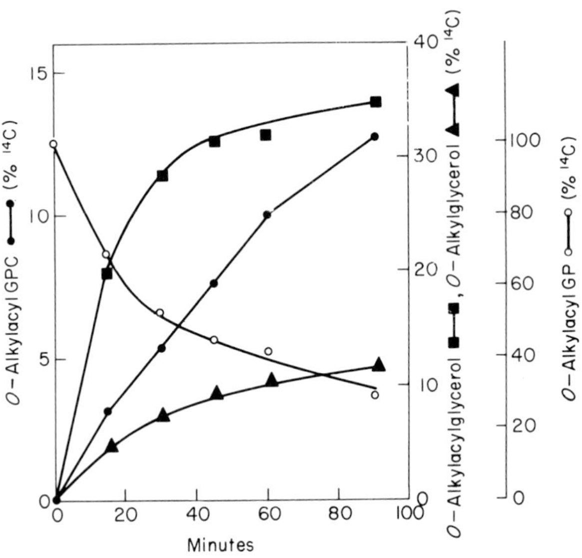

FIG. 5. The enzymic synthesis of alkylacylglycerylphosphorylcholine (●——●) from CDP-choline (3.33 m*M*) and 1-^{14}C-alkylacylglycerol (■——■) (formed from 1-^{14}C-alkylacylglycerophosphate, ○——○); Mg^{2+} (10 m*M*) was also required. The formation of 1-^{14}C-alkylacylglycerol (■——■) and 1-^{14}C-alkylglycerol (▲——▲) is also shown on the graph. Only about 2% of the ^{14}C was found in alkylacylglycerylphosphorylcholine after a 90-min incubation when CDP-choline was omitted from the incubation mixture. Reproduced from Snyder *et al.* (1970d) by permission of the *Journal of Biological Chemistry*.

E. Alkylacylglycerol

Alkylacylglycerol phosphate is dephosphorylated to the *O*-alkyl analog of a diglyceride (alkylacylglycerol) by microsomal enzymes (Snyder *et al.*, 1969b,c, 1970a,c,d; Wykle and Snyder, 1970), magnesium being the only cofactor required (Snyder *et al.*, 1970d). The alkylacylglycerol mimics the central role of diacylglycerols in the biosynthesis of triacylglycerols, phosphatidylcholine, and phosphatidylethanolamine.

F. Alkyldiacylglycerol

Most naturally occurring alkyl ether-linked lipids in the "neutral lipid fraction" of living cells occur as alkyldiacylglycerols, the *O*-alkyl analogs of triacylglycerols. The biosynthesis of these compounds has been demonstrated in homogenates of preputial gland tumors (Snyder *et al.*, 1969a, 1970c) and in microsomal preparations from Ehrlich ascites cells (Wykle and Snyder, 1970). The alkylacylglycerol substrate is apparently enzyme-bound, since acylation did not occur when exogenous alkylacylglycerol was added to homogenates containing CoA, ATP, and Mg^{2+} (Snyder *et al.*, 1970f). In tumor preparations, alkyldiacylglycerols are characteristic components of the neutral lipid fraction (Chapter X).

G. Alkylacylglycerylphosphorylcholine and Alkylacylglycerylphosphorylethanolamine

Cytidine diphosphate choline and cytidine diphosphate ethanolamine are required cofactors in the biosynthesis of *O*-alkyl phospholipids (Snyder *et al.*, 1970d). The reaction between the cytidine nucleotides and alkylacylglycerol is catalyzed by phosphorylcholine- or phosphorylethanolamine-alkylacylglycerol transferases that occur in microsomes; Mg^{2+} is necessary for optimum enzymic activity (Fig. 5). The enzymes that utilize the alkylacylglycerols as substrates may be identical to those that catalyze the analogous reactions in the biosynthesis of diacyl phosphoglycerides.

II. Enzymic Synthesis of Plasmalogens

A. General

Since the elucidation of the *O*-alk-1-enyl structure in plasmalogens by Rapport *et al.* (1957) and Rapport and Franzl (1957), there has been much speculation on possible mechanisms for the biosynthesis of the *O*-alk-1-enyl moiety. A number of review articles summarize these speculations (Thiele and Gowin, 1955; Burton, 1959; Thiele *et al.*, 1960; Cymerman-Craig and Horning, 1960; Hartree, 1964; Goldfine, 1968; Snyder, 1969; Piantadosi and Snyder, 1970). Two short communications have appeared which stated that plasmalogens were synthesized from palmitic-1-^{14}C-acid in homogenates of brain containing added CoA, ATP, and Mg^{2+} (Gambal and Monty, 1959; Carr *et al.*, 1963), but neither group determined whether long-chain fatty alcohols or *O*-alkyl lipids were formed in their systems, and their results were not extended for a subsequent detailed publication. It is possible that more than one pathway for the biosynthesis of plasmalogens exists. However, at present only a cell-free system from tumor preparations has been isolated that forms plasmalogens; data obtained from this system indicate that *O*-alkyl and *O*-alk-1-enyl moieties of glycerolipids are formed from the same precursors (Wykle *et al.*, 1970; Snyder *et al.*, 1971b).

O-Alkyl lipids have been implicated as precursors in plasmalogen biosynthesis in a number of *in vivo* studies (see Chapters VIII, IX, and XI) carried out with ^{3}H/^{14}C-labeled alkylglycerols (Thompson, 1968;Blank *et al.*, 1970); ^{14}C-, ^{3}H-, and ^{18}O-labeled long-chain fatty alcohols (Keenan *et al.*, 1961; Wood and Healy, 1970a,b; Stoffel *et al.*, 1970; Schmid and Takahashi, 1970; Wood *et al.*, 1970; Snyder *et al.*, 1970e; Bell *et al.*, 1971); combinations of labeled hexadecylglycerol, hexadecanol, and hexadecanal (Stoffel and LeKim, 1971); and ^{14}C-alkylacylglycerylphosphorylethanolamine (Debuch *et al.*, 1970). Thompson, using the slug, was the first to obtain fairly strong evidence for the conversion of *O*-alkyl lipids to *O*-alk-1-enyl lipids (see Chapter XII). He was able to show conclusively that 1-^{14}C-hexadecyl-2-^{3}H-glycerol was incorporated into ethanolamine plasmalogens without any change in the ^{3}H/^{14}C ratio (Thompson, 1968). This experiment has been verified in Ehrlich ascites cells (Blank *et al.*, 1970). In the mammalian system, an unknown *O*-alkyl lipid was isolated after $LiAlH_4$ reduction of the total lipids; it had the same ^{3}H/^{14}C ratio as the *O*-alkyl- and *O*-alk-1-enyl-containing ethanolamine plasmalogens and was thought to be derived from an intermediate involved in the biosynthesis of plasmalogens. The unidentified *O*-alkyl lipid, which has also been detected in *in vitro* experiments (Snyder *et al.*, 1970b), is chromatographically similar (but not identical) to synthetic 1-β-hydroxyhexadecylglycerol. The data suggest that the conversion of *O*-alkyl lipids to *O*-alk-1-enyl lipids involves a type of substitution [reaction (3)] on the *O*-alkyl moiety rather than simple dehydrogenation by a hydrogen-accepting nucleotide [reaction (4)].

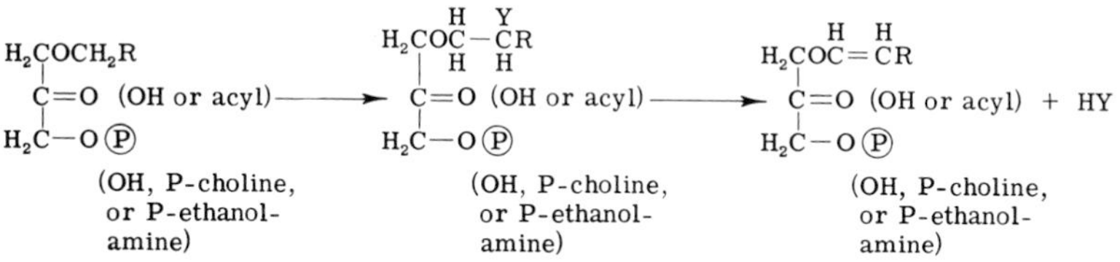

(3)

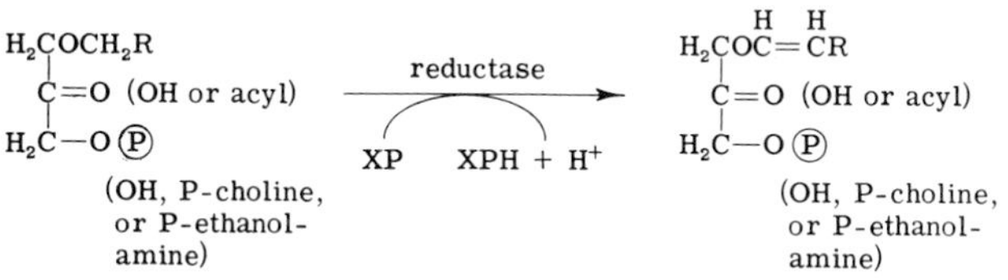

(4)

XP represents a hydrogen acceptor such as NAD^+, $NADP^+$, or FAD*

Stoffel and LeKim (1971) have described some elegant *in vivo* studies with rat brain on the stereospecificity of the dehydrogenation of the *O*-alkyl moiety to the *O*-alk-1-enyl moiety in glycerolipids. They found that 1-^{3}H,1-^{14}C-(erythro-**1***S*,**2***S*)-hexadecanol loses all of its tritium when the alcohol was incorporated into the *O*-alk-1-enyl moiety of plasmalogens, but no loss was encountered when 1-^{3}H,1-^{14}C-(**1***R*)-hexadecanol or 2-^{3}H,1-^{14}C-(**2***R*)-hexadecanol were incorporated, i. e., the specific loss of ^{3}H from the **1***S*,**2***S* configuration of hexadecanol gives the *cis* configuration of the *O*-alk-1-enyl linkage in plasmalogens. Furthermore, Stoffel and LeKim's results on the comparison of the precursor role of fatty aldehydes, fatty alcohols, and alkylglycerols demonstrate that the latter (or a derivative) was the direct precursor of the alk-1-enyl moiety of plasmalogens.

Wykle *et al.* (1970) obtained the first direct enzymic evidence for the role of *O*-alkyl lipids in the biosynthesis of plasmalogens. The enzymic system and the methodology used to document the ethanolamine plasmalogens synthesized are described in the next section (II,B).

B. Alk-1-enylacylglycerylphosphorylethanolamine

Perhaps one of the most difficult problems in the investigation of plasmalogen biosynthesis is to prove unequivocally that the product formed in the biological system is indeed a plasmalogen and not an artifact. To circumvent this problem, one must isolate the ethanolamine phospholipid

* Current results indicate that 1-alkyl-2-acyl-*sn*-glycero-3-phosphorylethanolamine is converted to ethanolamine plasmalogens by microsomal enzymes that require a reduced pyridine nucleotide and molecular oxygen (see "Note Added in Proof," p. 156).

fraction (the main source of plasmalogens in most cells) and identify the plasmalogen structure on the basis of chemical derivatives that take into account all three positions of the glycerol moiety (Fig. 6). Using this approach, Wykle *et al.* (1970) and Snyder *et al.* (1971b) documented the enzymic synthesis of ethanolamine plasmalogens in a cell fraction containing microsomes and the soluble supernatant isolated from Ehrlich ascites cells and preputial gland tumors. Their systems utilized the same substrates (fatty alcohols and DHAP) and cofactors (CoA, ATP, and Mg^{2+}) as required for the biosynthesis of alkyl glycerolipids by microsomes, except for the additional requirement of $NADP^+$ or NAD^+ and a soluble supernatant fraction. Under these conditions, most of the radioactivity found in plasmalogens from labeled fatty alcohols and DHAP is in the ethanolamine-containing fraction. Supplemented CDP-ethanolamine in the system increases the total quantity of plasmalogen synthesized (Wykle *et al.*, 1971). 1-^{14}C-Palmitic acid could not substitute for the 1-^{14}C-hexadecanol in this system, and added unlabeled hexadecanal had little effect on the incorporation of 1-^{14}C-hexadecanol into the ethanolamine plasmalogens. Washed microsomes as the enzyme source will not alone synthesize plasmalogens; they also require the addition of a soluble protein fraction from either the tumor or normal rat liver.

Carbon-14 was recovered in a number of derivatives formed from intact ethanolamine plasmalogens and isolated by TLC and GLC. These included (1) hexadecanal liberated by aqueous HCl, (2) dimethylacetals of hexadecanal formed by HCl-methanol treatment, (3) alk-1-enylglycerols liberated by reduction with $LiAlH_4$ or $NaAlH_2(OCH_2CH_2OCH_3)_2$, (4) alkylglycerols obtained by hydrogenation of the alk-1-enylglycerols, (5) isopropylidene derivatives of the hydrogenated alk-1-enylglycerols, (6) lysoethanolamine plasmalogens after phospholipase A treatment, and (7) dinitrobenzene derivatives of the intact ethanolamine plasmalogens.

The reaction(s) involved in the formation of alk-1-enyl glycerolipids and the roles of $NADP^+$ or NAD^+ and the soluble protein fraction in plasmalogen biosynthesis are not known. However, since no ketone compounds appear to be reduced unless NAD^+ or $NADP^+$ is added, even though the soluble fraction is present, it would seem that the oxidized forms of the nucleotides are related to the reduction–oxidation of alkyl-DHAP; in addition, they could also serve as acceptors of hydrogens from the α,β-carbon atoms of the *O*-alkyl moiety [reaction (4)] or that they could be involved in some other mechanism [e.g., reaction (3)].

The requirements of the cytidine diphosphate derivatives of choline and ethanolamine for the transferase reactions that form plasmalogens are the same as those documented for diacyl phosphatides and *O*-alkylacyl phosphatides. Some years ago, Kiyasu and Kennedy (1960) had already

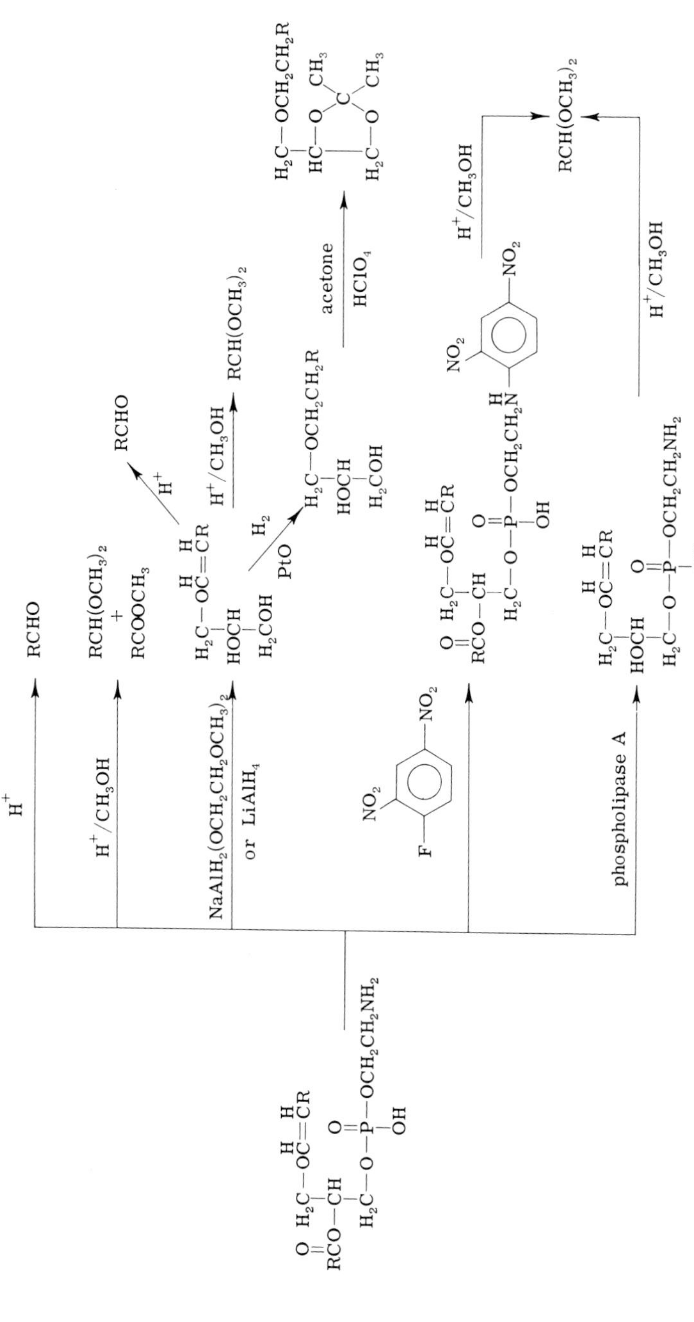

FIG. 6. Procedures used to identify ethanolamine plasmalogens synthesized enzymically. Reproduced from Snyder *et al.* (1971b) by permission of the *Journal of Biological Chemistry*.

found phosphorylcholine- or phosphorylethanolamine-alk-1-enylacylglycerol transferases in the particulate fraction of rat liver that could catalyze the formation of the corresponding plasmalogens. Kiyasu and Kennedy (1960) prepared the alk-1-enylacylglycerol that was used as substrate for the transferases by prolonged hydrolysis of plasmalogens isolated from beef heart with a phospholipase obtained from *Clostridium perfringens.* McMurray (1964) has provided evidence for similar transferase reactions in the biosynthesis of plasmalogens in a brain particulate fraction, but he did not demonstrate a net synthesis of plasmalogens. More recently, Poulos *et al.* (1968) showed that ox heart also contains phosphorylcholine–alk-1-enylacylglycerol transferase activities.

C. Alk-1-enyldiacylglycerol

Enzymic systems that synthesize alk-1-enyldiacylglycerols ("neutral plasmalogens") have not yet been isolated, but presumably these lipids are formed by acyl transferases that utilize alk-1-enylacylglycerols as substrates. Albro and Dittmer (1969) proposed that the alk-1-enyldiacylglycerols might be intermediates in the pathway that forms hydrocarbons in *Sarcina lutea,* but these workers did not study the biosynthesis of *O*-alk-1-enyl lipids. Their data did demonstrate that ^{14}C-labeled alk-1-enyldiacylglycerols (prepared biosynthetically from *Clostridium butyricum*) were much better precursors of hydrocarbons than palmitic acid in *Sarcina lutea.*

D. The Precursor Role of Alkyl Glycerolipids

Although the substrate and cofactor requirements for the biosynthesis of plasmalogens are essentially identical to those for the biosynthesis of *O*-alkyl lipids, this by itself was insufficient evidence to allow the conclusion that an alkyl glycerolipid was converted to an alk-1-enyl glycerolipid. However, thin-layer radiochromatographic data obtained with uniformly ^{14}C-labeled DHAP and 9,10-^{3}H-hexadecanol demonstrated that the $^{3}H/^{14}C$ ratios were the same for the alkylglycerols and alk-1-enylglycerols isolated from the total lipids of the incubation mixture after $LiAlH_4$ reduction. Collection and radioassay of the isopropylidene derivatives of the hexadecylglycerols and hydrogenated hexadec-1-enylglycerols from GLC confirmed the TLC results, providing the first clear-cut evidence of an enzymic system that catalyzes the transformation of an *O*-alkyl moiety to an *O*-alk-1-enyl moiety (Snyder *et al.,* 1971b).

III. Enzymic Synthesis of Ether Bonds in Miscellaneous Lipids and Other Compounds

A. *S*-Alkyl Glycerolipids

The synthesis (Piantadosi *et al.*, 1969), analysis (Wood *et al.*, 1969), and *in vivo* metabolism (Snyder *et al.*, 1969d) of glyceryl thioethers (**5**) were investigated before it was known that they occurred in nature, but Ferrell and Radloff (1970) have now reported that there are glyceryl thioethers in bovine heart.

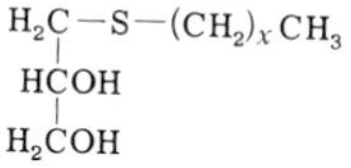

(**5**)

After feeding ^{35}S- and 1-^{14}C-alkyl-labeled glyceryl thioethers to rats (Snyder *et al.*, 1969d), between 45% and 87% of the radioactivity from the *S*-ethers appeared in the urine within 24 hr after stomach tube intubation, as compared to less than 5% of the radioactivity from the *O*-ethers. Essentially none of the 1-^{14}C in the *S*-alkyl moiety appeared in the respiratory $^{14}CO_2$, but urine, intestinal contents, and feces had the highest content of radioactivity. Most of the ^{14}C in these tissues was as the unesterified glyceryl thioether, although detectable quantities were present in the ethanolamine and serine phosphatides. Nothing is yet known about the enzymic synthesis or cleavage of the *S*-ether linkage in glycerolipids.

B. Hydroxy-Substituted Alkyl Glycerolipids

K. Kasama, W. T. Rainey, Jr., and F. Snyder (unpublished data) have identified a new class of alkyl glycerolipids in the triglyceride fraction of lipid extracts from pink harderian glands of rabbits; it yields the following structure after $LiAlH_4$ reduction:

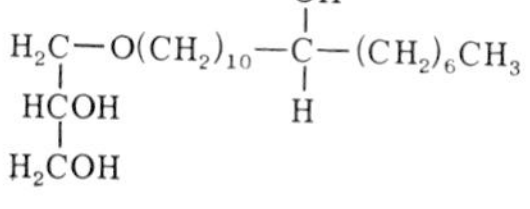

(**6**)

Identification has been based on chromatography, infrared spectroscopy, and mass spectra of various derivatives and on comparison to standards prepared chemically. The hydroxyl group on the *O*-alkyl moiety is attached to carbon atoms 11 or 12. All the hydroxyl groups are thought to be esterified with aliphatic moieties in the natural state; it is possible that one of the acyl moieties contains less than 10 carbon atoms.

Enzymic synthesis of the hydroxy-substituted alkyl glycerolipids occurs when DHAP and octadecenyl-1,12-diol (prepared by $LiAlH_4$ reduction of ricinoleic acid) are incubated with mitochondrial supernatants of pink harderian glands from rabbits, CoA, ATP, and Mg^{2+} (K. Kasama and F. Snyder, unpublished data). Except for the hydroxy-substituted alcohol, the enzymic system is identical to that described for alkyl glycerolipids in Section I.

C. β-Methoxyalkyl Glycerolipids

Hallgren and Ställberg (1967) isolated β-methoxy-substituted alkylglycerols from Greenland shark liver oil. Molecular structures of this class of ether lipid containing 2-methoxyhexadecyl, 2-methoxy-4-hexadecenyl, and 2-methoxy-4-octadecenyl moieties were identified by mass spectrometry, nuclear magnetic resonance spectra, infrared spectroscopy, GLC, and by chemical analysis.

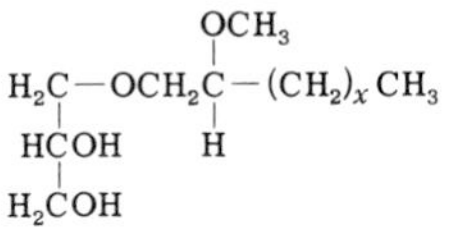

(7)

Since the methoxy-substituted alkyl glycerolipids were isolated from the nonsaponifiable fraction of total lipids, it would appear that these compounds exist naturally with the hydroxyl groups esterified with fatty acids. Biosynthesis of the methoxy-substituted *O*-alkyl lipids has not been investigated.

D. Ether Bonds in Nonlipid Compounds

In bacteria, phosphoenolpyruvate is the precursor of the alkenyl ether bond in uridine diphospho-*N*-acetylmuramic acid (**8**) formed by reduction of uridine diphospho-*N*-acetylenolpyruvylglucosamine (Strominger, 1958; Gunetileke and Anwar, 1968) and in 3-enolpyruvylshikimate 5-phosphate (**9**) (Levin and Sprinson, 1964).

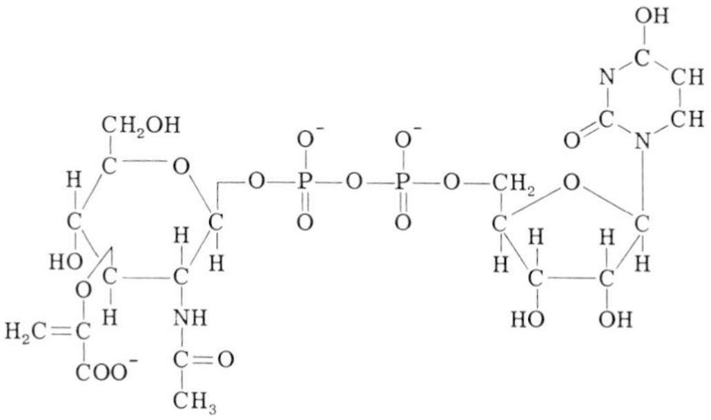

(**8**)

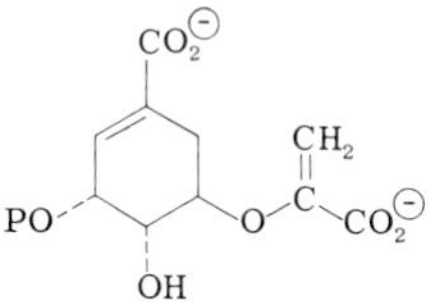

(**9**)

$$\text{UDP-GlcNAc} + \text{P-enolpyruvate} \rightarrow \text{UDP-GlcNAc-enolpyruvate} + P_i$$
$$\text{UDP-GlcNAc-enolpyruvate} + \text{NADPH} \rightarrow \text{UDP-MurNAc} + \text{NADP}^+ \quad (5)$$

$$\text{Phosphoenolpyruvate} + \text{shikimate-5-P} \rightarrow \text{3-enolpyruvylshikimate-5-P} + \text{P} \quad (6)$$

The two reactions [(5) and (6)] are analogous in that both enzymic systems catalyze the condensation of phosphoenolpyruvate with the second substrate (either UDP-N-acetylglucosamine or shikimate-5-P). Enzymes for catalyzing reaction (5) have been obtained from *Staphylococcus aureus*, *Escherichia coli*, *Aerobacter aerogenes*, and *Enterobacter cloacae*, whereas reaction (6) was studied using partially purified extracts of *Escherichia coli*.

Murooka *et al.* (1970) have described a mechanism (7) for the biosynthesis of ether bonds in *O*-alkylhomoserine by an enzyme from *Corynebacterium acetophilum* A51. The reaction appears to proceed via an exchange between an esterified acetyl group and ethanol; it is similar to the mechanism that Hajra (1970) proposed for the biosynthesis of the *O*-alkyl moiety in glycerolipids.

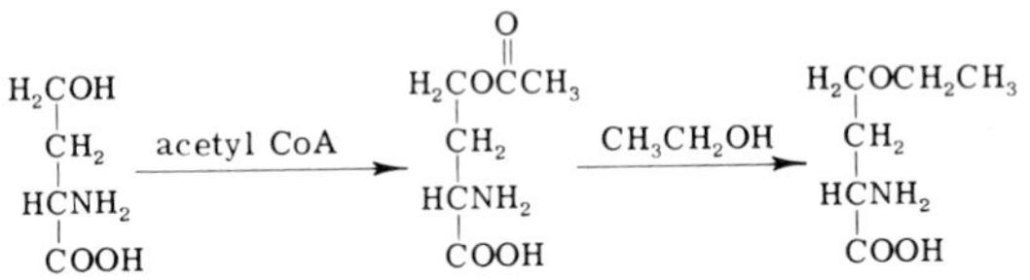

(7)

An important hormone that contains an ether bond is thyroxine (**10**). Taurog (1970) proposed a scheme for the intramolecular formation of thyroxine that is based on the coupling of two diiodotyrosyl radicals (formed by a peroxidase-catalyzed reaction) to form a quinol ether intermediate. A different view of how thyroxine is synthesized has suggested an intermolecular coupling between 4-hydroxy-3,5-diiodophenylpyruvic acid and diiodotyrosine in thyroglobulin (Toi *et al.*, 1965; Blasi *et al.*, (1969).

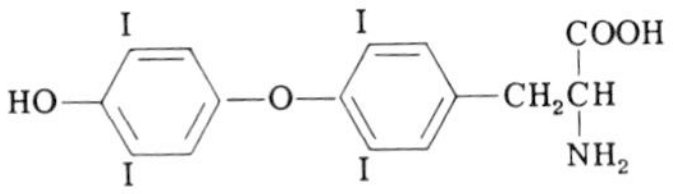

Thyroxine

(**10**)

IV. Enzymic Reactions Responsible for the Biosynthesis of Ether-Linked Precursors

A. Fatty Alcohols

Specific enzymes and mechanisms involved in the complex assembly of fatty acids from acetate by two separate enzymic systems have been known for some time. In contrast, essentially nothing has been known about the biosynthesis of long-chain fatty aldehydes and long-chain fatty alcohols until the past year, except as by-products of several degradative reactions involving complex molecules, e.g., alkylglycerols (Tietz *et al.*, 1964; Pfleger *et al.*, 1967) and sphingosine or dihydrosphingosine (Stoffel and Sticht, 1967; Stoffel and Assmann, 1970; Stoffel *et al.*, 1970). Although the concentration of long-chain fatty alcohols in mammals is low, they have been detected in both free and esterified forms (Blank and Snyder, 1970a; Takahashi and Schmid, 1970). Because fatty alcohols are precursors of ether-linked lipids and waxes, their origin, especially in mammalian tissue, has become an important facet of research in the lipid field.

Oxidation of fatty alcohols to fatty acids (Blomstrand and Rumpf, 1954) and the reduction of fatty acids to fatty alcohols (Sand *et al.*, 1969) occur *in vivo*, but an understanding of the mechanism of such interconversions is not possible from *in vivo* experiments. However, the cell-free systems isolated from *Euglena* (Kolattukudy, 1970), bacteria (Day *et al.*, 1970), and mammals (Tabakoff and Erwin, 1970; Stoffel *et al.*, 1970; Snyder and Malone, 1970) do provide the necessary tools to investigate the specific enzymic reactions that interconvert the acid, aldehyde, and alcohol moieties of long aliphatic carbon chains. Kolattukudy (1970) found that acyl-CoA is converted to fatty alcohols via an aldehyde intermediate by a soluble enzyme fraction prepared from etiolated *Euglena gracilis*. NADH was the sole source of hydrogen in the reduction of fatty acids to fatty alcohols. Although the reduction of the aldehyde intermediate to the alcohol did not show this specificity, NADPH was not as effective as NADH. Kolattukudy (1970) stated that the pH optimum was about 6.5 and the apparent K_m's of myristic acid and NADH were 1.6×10^{-5} *M* and 2.4×10^{-4} *M*, respectively. The reaction (8) observed in *Euglena gracilis* is depicted as:

$$\mathrm{R\overset{O}{\overset{\|}{C}}\cdot\sim SCoA} \xrightarrow[\mathrm{NADH}\ \ \mathrm{NAD^+}]{} \mathrm{RCHO} \xrightarrow[\mathrm{NADH\ (NADPH)}\ \ \mathrm{NAD^+\ (NADP^+)}]{} \mathrm{ROH} \qquad (8)$$

Day *et al.* (1970) found that essentially the same reaction can be catalyzed by a soluble protein fraction isolated from *Clostridium butyricum* (see Chapter XIV). About the only difference between the *Euglena* and the bacterial system is that NADPH had essentially no effect on the reduction of the fatty aldehyde by the bacterial enzyme. Day and co-workers (1970) reported that the reaction step between acyl-CoA and fatty aldehyde was reversible with crude extracts in the absence of added NAD^+ or $NADP^+$.

Mammalian enzyme systems have also provided information on the interrelationships of fatty acids, fatty aldehydes, and fatty alcohols. Stoffel and co-workers (1970) investigated the reduction of fatty aldehydes to fatty alcohols by liver enzymes [reaction (9)] that appear to be most prominent in the soluble supernatant fraction, although a portion of the total activity was also observed in the microsomal and mitochondrial fractions.

$$\mathrm{R\overset{O}{\overset{\|}{C}}H} \xrightarrow[\mathrm{NADPH}]{\mathrm{NADH\ or}} \mathrm{ROH} \qquad (9)$$

This aldehyde reductase from rat liver showed no specificity for NADH or NADPH. Its specificity for the carbon chain length of the fatty aldehyde varied considerably for the wide range of substrates used (C_2–C_{18}), and it appeared to be similar to alcohol dehydrogenase. That alcohol dehydrogenase can indeed utilize long-chain fatty aldehydes as substrates was documented by Stoffel *et al.* (1970), who used a purified commercial preparation of alcohol dehydrogenase from horse liver.

Tabakoff and Erwin (1970) were able to purify 120-fold a nonspecific NADPH-linked aldehyde reductase from the soluble protein fraction of bovine brain. It reduced a number of compounds with aldehyde groups, including decanaldehyde and palmitaldehyde, but not short-chain fatty aldehydes. NADH did not substitute for NADPH in this enzymic system. The reductase, which was also found in rat brain homogenates, had properties different from alcohol dehydrogenase.

Snyder and Malone (1970) detected an enzymic system in unwashed microsomes from preputial gland tumors that interconverts fatty acids and fatty alcohols. Although fatty aldehydes were not found in this system, they were not ruled out as intermediates. It is theoretically possible for a reduction–oxidation reaction of this type to occur by direct transfer of the hydrogens to the nucleotides without an aldehyde being formed. Absolute nucleotide specificity was exhibited for the oxidation of long-chain fatty alcohols to fatty acids, whereas the reverse reaction was dependent primarily on NADPH; a synergistic effect of NADH in the presence of NADPH was observed for the enzymic reduction. The requirement of CoA, ATP, and Mg^{2+} for the reduction of palmitic-1-^{14}C acid suggests that acyl-CoA is the substrate in reaction (10).

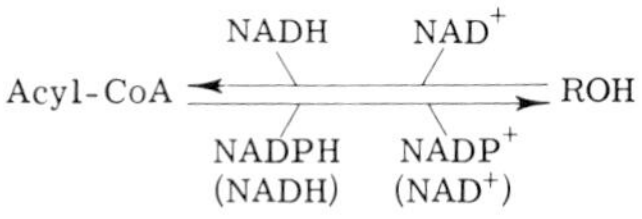

(10)

The high reductase activities in tumors probably account for the accumulation of ether-linked lipids during neoplastic growth. Perhaps the availability of the oxidized and reduced forms of nucleotides plays an important role in the regulatory mechanisms that determine the concentration of ester- and ether-linked glycerolipids in all living cells.

B. Dihydroxyacetone Phosphate

DHAP can be formed by aldolase and by α-glycerophosphate dehydrogenase (Fig. 7). In the glycolytic pathway, fructose 1,6-diphosphate is

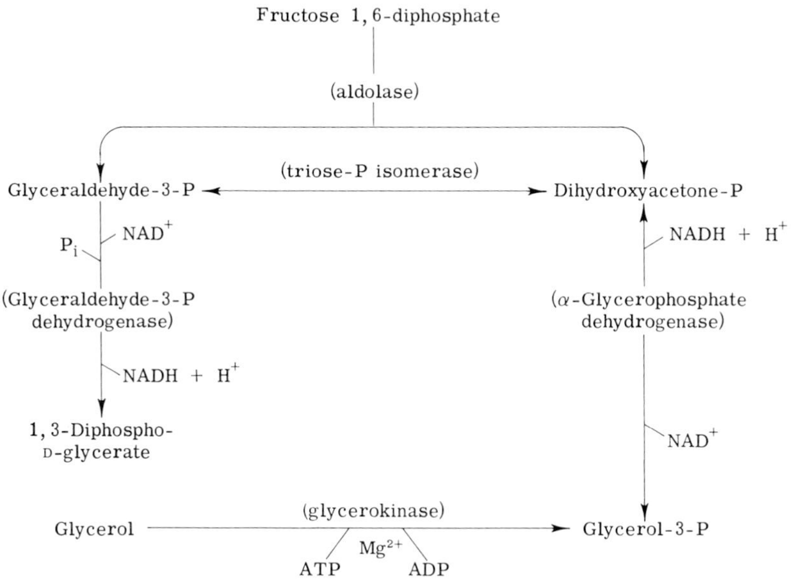

FIG. 7. Biosynthesis of dihydroxyacetone phosphate and related glycolytic intermediates.

converted to DHAP and glyceraldehyde-3-P by aldolase. Equilibrium of these two triose phosphates is under the influence of triose-P isomerase (Meyerhof *et al.*, 1936a,b) and the hexose monophosphate shunt (see Hollmann and Touster, 1964). The isomerase occurs in the soluble fraction of cells; however, traces of activity are also associated with mitochondrial (Boxer and Shonk, 1960) and microsomal fractions (Rao *et al.*, 1968; Wykle and Snyder, 1969, 1970; Snyder *et al.*, 1970b). The chemical equilibrium favors DHAP by about 96 to 4 although more glyceraldehyde-3-P than DHAP is found in the intact tumor cells (Garfinkel and Hess, 1964). Triose-P isomerase is inhibited by high concentrations of phosphate (Oesper and Meyerhof, 1950), 1-hydroxy-3-chloro-2-propanone phosphate (Hartman, 1968, 1970), and by glycidol phosphate (Rose and O'Connell, 1969).

The reaction catalyzed by NAD^+-linked α-glycerophosphate dehydrogenase in the cytoplasm converts α-glycerophosphate to DHAP (Meyerhof and Kiessling, 1933). Mitochondrial α-glycerophosphate dehydrogenase catalyzes the same reaction but the mitochondrial enzyme is flavin-linked (Green, 1936). These reactions are important in the formation of acylated glycerolipids, since in tissues that lack glycerokinase, DHAP can be enzymically reduced by the dehydrogenase to α-glycerophosphate, the key acyl acceptor in lipid metabolism.

V. Enzymic Degradation of Ether Linkages in Glycerolipids

A. Alkyl Glycerolipids

In vivo studies have demonstrated that the *O*-alkyl linkage is cleaved by intestinal enzymes during absorption (Bergström and Blomstrand, 1956; Blomstrand and Ahrens, 1959; Blomstrand, 1959). Similar studies carried out after intravenous (Snyder and Pfleger, 1966) and intraperitoneal (Oswald *et al.*, 1968) injections of ^{14}C-labeled alkylglycerols have shown that other tissues also contain enzymes that can cleave the ether linkage. *In vivo* degradation of alkylglycerols also occurs in *Tetrahymena pyriformis* (Kapoulas *et al.*, 1969a; also see Chapter XIII).

Tietz and co-workers (1964), using rat liver microsomes, discovered a tetrahydropteridine-requiring hydroxylase that could cleave the ether linkage of alkylglycerols. NADPH was required as a cofactor for the reductase that reduced the dihydropteridine produced during the course of the cleavage reaction. On the basis of experiments on the hydroxylation of phenylalanine (Kaufman, 1959), Tietz *et al.* (1964) proposed that hydroxylation of the *O*-alkyl moiety occurred on the α-carbon and yielded an unstable hemiacetal structure that spontaneously decomposed to a long-chain fatty aldehyde and free glycerol. In the presence of NAD^+, the aldehyde was oxidized to the corresponding fatty acid. Later, Pfleger *et al.* (1967) were able to show that in the absence of NAD^+, the aldehyde could be reduced to fatty alcohols, and that a portion of the cleavage enzyme could be solubilized. However, more recent data (Soodsma *et al.*, 1972) indicate that the bulk of the activity is retained in the microsomes even after extensive washing.

The cleavage enzyme appears to be an important regulator of ether lipid levels in cells. The investigation by Pfleger and co-workers (1967) revealed the *O*-alkyl cleavage enzyme system in liver preparations from a variety of species, but found that cells rich in ether-linked lipids exhibited low or no enzymic activities. In fact, activities of the cleavage enzyme are absent or depressed in cancer cells (Soodsma *et al.*, 1970; see Chapter X) and brain tissue (Snyder *et al.*, 1971a), both tissues with a high ether-linked lipid content.

Recently, Soodsma *et al.* (1972) found that ammonium ions and sulfhydryl compounds are required for optimal cleavage of the alkyl ether bonds in rat liver by the tetrahydropteridine-requiring hydroxylase that utilizes alkylglycerols as a substrate. The overall reaction, as we now visualize it, for the cleavage of alkylglycerols that occurs in mammalian liver is summarized in Fig. 8. A reaction which is similar except for the requirement of pteridine has been detected in cell fractions prepared from *Tetrahymena pyriformis* (Kapoulas *et al.*, 1969b; see Chapter XIII); the effect of NH_4^+ and sulfhydryl groups was not investigated in the *Tetrahymena* system.

All enzymic studies published on the cleavage of the alkyl moiety in glycerolipids were carried out with alkylglycerols as substrates. Soodsma *et al.* (unpublished data) demonstrated that at least one or both of the hydroxyl groups of the alkylglycerols must be free to serve as a substrate since the ether linkage was not cleaved when diacylalkylglycerols were used. Data on the cleavage of the *O*-alkyl moiety in phospholipid molecules in mammalian systems have not been published; however, our laboratory has observed that 20–30% of the ether chain was cleaved when 1-^{14}C-1-alkyl-2-acylglycerylphosphorylethanolamine was incubated with homogenates of Ehrlich ascites cells (R. L. Wykle, M. L. Blank, and F. Snyder, unpublished data). Specificity of the cleavage enzyme from *Tetrahymena* for intact *O*-alkyl phospholipids was not conclusive (Kapoulas *et al.*, 1969b) but if cleavage occurred it was at a much lower rate than that obtained for alkylglycerols.

B. Alk-1-enyl Glycerolipids

In vivo investigations designed to determine the fate of administered plasmalogens (Feulgen *et al.*, 1928; Leupold and Büttner, 1953; Robertson and Lands, 1962) have demonstrated that they are removed from the blood fairly rapidly. These data and the *in vitro* experiments described below make it clear that enzymic systems are present in tissues that metabolize

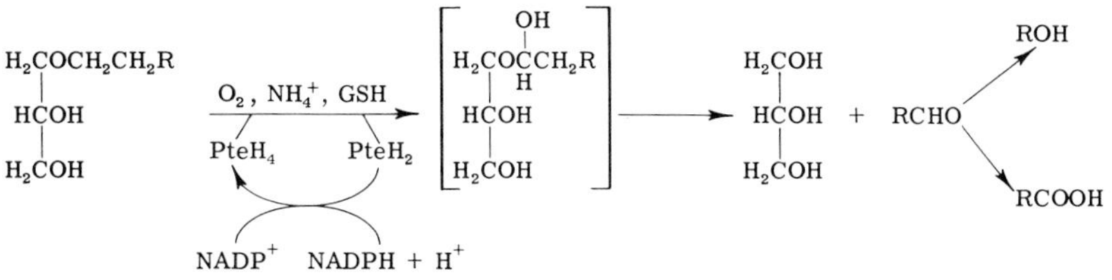

FIG. 8. Enzymic cleavage of alkylglycerols by microsomal enzymes of rat liver. The abbreviations designate glutathione (GSH) and pteridine (Pte). The hemiacetal structure within the brackets has not been isolated. The fatty aldehydes produced during cleavage can be oxidized or reduced by liver enzymes. Reduction of the dihydropteridine tetrahydropteridine is accomplished by pteridine reductase which is also present in liver.

plasmalogens. Bergmann and co-workers (1957) found that bacterial extracts from *Escherichia coli* and *Bacillus subtilis* could cleave the *O*-alk-1-enyl moiety of plasmalogens isolated from horse brain, but extracts from other bacteria or yeasts did not affect the ether linkage. C. E. Anderson and co-workers (1960) noted that rat liver homogenates degraded plasmalogens, but aldehydes could not be detected in the incubation mixture; the authors believed the aldehydes were oxidized and polymerized. Thiele (1959a), however, did report that when acetone extracts of fresh organs were incubated with plasmalogens at 37°C, degradation occurred and aldehydes were produced. Since purified plasmalogens did not serve as substrates and since chelating agents prevented degradation in crude systems, Thiele (1959b) concluded that the cleavage reaction was probably promoted by a stable heavy metal complex.

The first clear description of a specific enzyme system capable of cleaving the *O*-alk-1-enyl moiety of ethanolamine plasmalogens was published by Warner and Lands (1961). These investigators isolated a microsomal preparation from rat liver that cleaved the *O*-alk-1-enyl moiety of 1-alk-1-enylglycerylphosphorylcholine; ethanolamine substituted for choline, or an acyl group substituted for the hydroxyl group, prevented the cleavage of the *O*-alk-1-enyl moiety. The alk-1-enylglycerylphosphorylcholine used as substrate in these studies was isolated from pig and beef hearts by mild alkali treatment of the choline phosphatides. The hydrolytic enzyme was labile to heat, acid, alkali, and chymotrypsin treatments, whereas lyophilization, Mg^{2+}, Ca^{2+}, or chelating agents had no effect on its activity. Products produced by the liver enzymes were long-chain fatty aldehydes and glycerylphosphorylethanolamine [reaction (11)]. The alk-1-enylglycerylphosphorylcholine hydrolase from liver was also

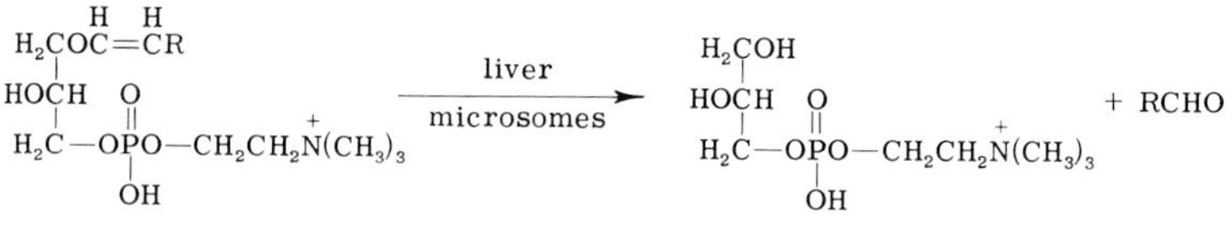

(11)

Alk-1-enylglyceryl-phosphorylcholine   Glycerylphosphoryl-choline

inactivated by treating the microsomes with phospholipase A or C, freezing and thawing, or after the addition of imidazole and some of its derivatives (Ellingson and Lands, 1968). However, enzymic activity was restored by the addition of exogenous phosphoglycerides. Activation of the enzyme by phosphoglycerides appeared to depend on the presence of two hydrocarbon chains in the activator molecule.

Ansell and Spanner (1965) described the preparation of an acetone powder from rat brain that catalyzes the hydrolysis of the ether bond in plasmalogens. To some extent it is similar to the hydrolase in liver, except that it requires Mg^{2+} for optimal activity and that it is directed primarily towards 1-alk-1-enyl-2-acylglycerylphosphorylethanolamine. Lysoethanolamine plasmalogen also serves as a substrate but to a lesser extent than the native plasmalogen. The hydrolytic reaction (12) for plasmalogens that occurs in brain is

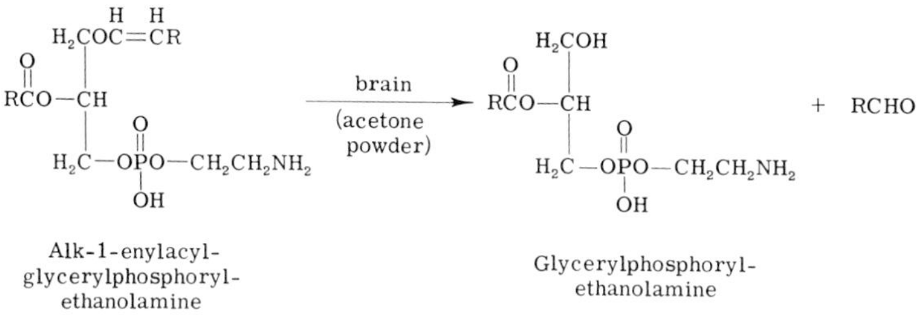

(12)

VI. Ether-Linked Glycerolipids as Substrates for Lipases, Phospholipases, and Phosphatases

Alkyl and alk-1-enyl chains of glycerolipids are not attacked by pancreatic lipase, phospholipases, or phosphatases. In general, the specificities of these lipolytic enzymes, which hydrolyze the acyl, phosphate, or phosphoryl base moieties at the other positions of the glycerol moiety of plasmalogens, are not affected by the presence of the ether bond. However, the lipolytic reactions are somewhat sluggish with the ether-linked lipid substrates when compared to their activity with those glycerolipids that do not contain alkyl or alk-1-enyl moieties. Pancreatic lipase, phospholipases A, C, and D, alkaline phosphatase, and acid phosphatase have been used primarily as tools in the structural analysis of lipids and in the preparation of substrates for metabolic investigations. All of these enzymic activities are expressed in most tissues and they catalyze important pathways for the degradation and redistribution of lipid moieties. Purified pancreatic lipase is specific for catalyzing the hydrolysis of acyl moieties at the 1- and 3-positions of glycerides (Mattson and Volpenhein, 1961; Desnuelle and Savary, 1963). De Haas *et al.* (1965) have also shown this positional specificity for phosphoglycerides; when they used electrophoretically purified pancreatic lipase, the 1-position of the phosphoglyceride was hydrolyzed. This specificity makes it possible to remove contaminating diacyl phospholipids from the native forms of *O*-alkyl and *O*-alk-1-enyl phospholipids, since the monoacyl phospholipids are easily separated from the ether-

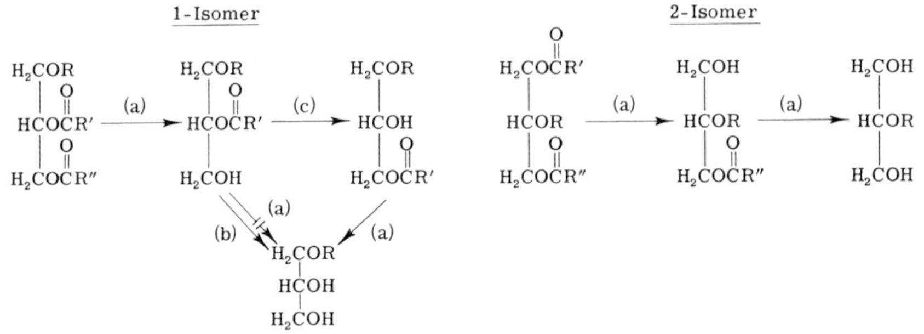

FIG. 9. Deacylation of alkyl- or alk-1-enyldiacylglycerols by lipases. R designates either *O*-alkyl or *O*-alk-1-enyl moieties. The small letters in parentheses above arrows designate: (a) pancreatic lipase, (b) lipoprotein lipase, and (c) acyl migration.

linked phospholipids by adsorption chromatography. An important problem that investigators must take into consideration when using commercially available pancreatic lipase is that enzymic contaminants are often present that can catalyze the removal of the acyl moieties at the 2-position of glycerolipids (Mattson and Volpenhein, 1968).

Pancreatic lipase has been used to deacylate the diacyl forms of alkylglycerols (Snyder and Piantadosi, 1968) and alk-1-enylglycerols (Gigg and Gigg, 1967; Slotboom *et al.*, 1967; also, see Chapter V). The acyl groups are removed from the alkyldiacyl- or alk-1-enyldiacylglycerols (1- and 2-ether isomers) in a manner analogous to that observed when triacylglycerols are substrates (Fig. 9), except that the rate of deacylation is slower for the ether analogs. However, it is possible to overcome this sluggish rate by using higher concentrations of pancreatic lipase and by altering the solvent conditions of the reaction (Slotboom *et al.*, 1967).

Post-heparin plasma from humans and NH_4OH extracts of adipose tissue and hearts from rats contain a triglyceride lipase and a monoglyceride lipase that can hydrolyze the acyl moiety in either 1,2-dioctadecyl-3-oleylglycerol and 1,3-dioctadecyl-2-oleylglycerol (Greten *et al.*, 1970). The lipase isolated from adipose cells exhibits a preferential specificity for the 2-acyl moiety.

Phospholipase specificities (Fig. 10) for substituents on the glycerol moiety of phospholipids are: A (3.1.1.4), hydrolyzes acyl group at position 2; B or lysophospholipase (3.1.1.5), hydrolyzes acyl group at position 1; C (3.1.4.3), hydrolyzes the phosphate ester linkage at position 3; and D (3.1.4.4), hydrolyzes choline, ethanolamine, or other moieties esterified to the phosphate moiety at position 3. The reader is referred to the review by van Deenen and de Haas (1966) and the book by Ansell and Hawthorne (1964) for a good account of work in this field. Variability in the specificities of lipolytic activities for ether-linked lipid substrates among tissues has been reported, but it is not always clear whether such differences are due principally to the techniques used or to the substrate specificities.

Marinetti and collaborators (1959) noted that phospholipase A hydrolyzed the 2-acyl moiety of alk-1-enylacylglycerylphosphorylcholine at a much slower rate than the 2-acyl moiety of diacylglycerylphosphorylethanolamine, and Gottfried and Rapport (1962) purified ethanolamine plasmalogens from phosphatidylcholine by taking advantage of the different reaction rates that these substrates had with phospholipase A. In contrast, Hartree and Mann (1961) observed that phospholipase A activity in ram sperm hydrolyzed the 2-acyl moiety of alk-1-enylacylglycerylphosphorylcholine, but not that of phosphatidylcholine.

The hydrolysis of the phosphoryl base moieties of alk-1-enylacylglycerylphosphorylcholine and alk-1-enylacylglycerylphosphorylethanolamine by phospholipase C (from *Clostridium welchii*) is slower than that of the diacyl counterparts (Warner and Lands, 1963). Ansell and Spanner (1965) carried out similar experiments for the ethanolamine plasmalogens with phospholipase C from *Bacillus cereus*, but a comparison of the enzyme's reactivity with phosphatidylethanolamine was not made. Snyder *et al.* (1970d) successfully used phospholipase C from *Clostridium perfringens* to help identify ^{14}C-labeled alkylacylglycerylphosphorylcholine and alkylacylglycerylphosphorylethanolamine that had been synthesized enzymically from cytidine diphosphate choline and cytidine diphosphate ethanolamine.

The source of the phospholipases represents an important factor in their relative reactivities with lipid substrates, especially for those that contain ether linkages. This point is emphasized in the investigations reported for phospholipase D. Hack and Ferrans (1959) and Slotboom and co-workers (1967) found that freshly prepared phospholipase D from cabbage could slowly hydrolyze the base portion ($\simeq$40% in 20 hr) from *O*-alk-1-enyl phospholipid substrates. On the other hand, van Golde and

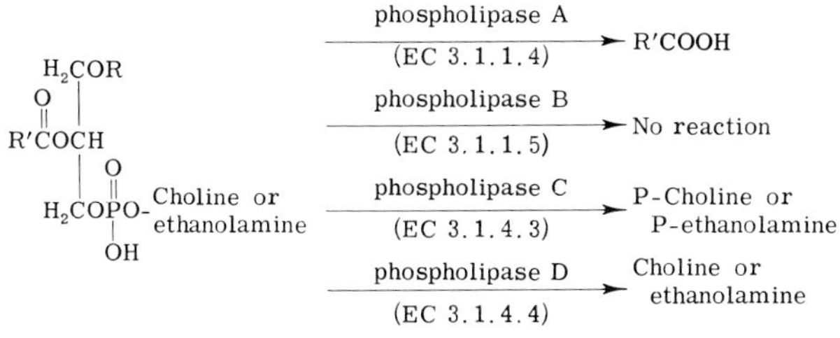

FIG. 10. Phospholipase specificities for ether-containing phospholipids. R designates either *O*-alkyl or *O*-alk-1-enyl moieties.

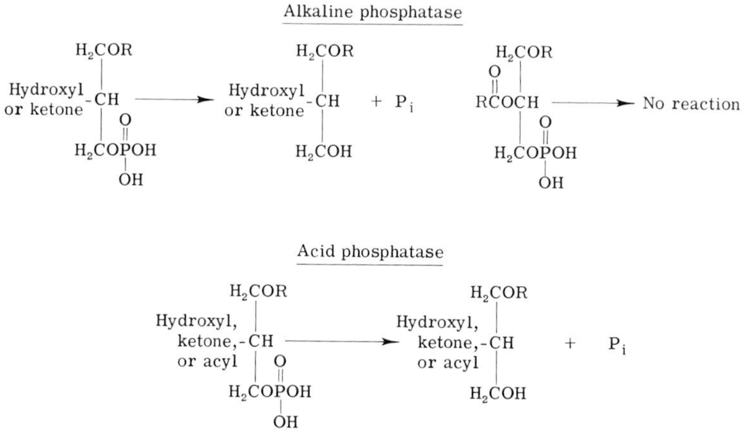

FIG. 11. Dephosphorylation of alkyl glycerolipids by alkaline and acid phosphatases.

van Deenen (1967) reported that phospholipase D isolated from *Bacillus cereus* is extremely active with *O*-alk-1-enyl-containing substrates under conditions of incubation that were similar to those used by the other workers. Lands and Hart (1965) reported that alk-1-enylacylglycerylphosphorylcholine does not serve as a substrate for commercial preparations of phospholipase D obtained from cabbage; therefore, removal of the choline moiety from phosphatidylcholine by phospholipase D from cabbage provides a convenient method for separating native choline plasmalogen from its more common diacyl analog. Lands and Hart (1965) also reported that the ethanolamine could be hydrolyzed from alk-1-enylacylglycerylphosphorylethanolamine by phospholipase D from cabbage, but the extent was much less than that observed for phosphatidylethanolamine.

Enzymic dephosphorylation of ether-linked substrates has not been studied in detail. Phosphatases and phosphatidate phosphohydrolase (microsomal and soluble forms) appear to act on *O*-alkyl and *O*-alk-1-enyl analogs of phosphatidic acid and lysophosphatidic acid. Rosenthal and Pousada (1966) found that dialkylglycerol phosphonic acid inhibits phosphatidate phosphohydrolase (EC 3.1.3.4) when phosphatidic acid was the substrate; the inhibition appears to reflect both the physical state and the structural characteristic of the added ether derivative. Blank and Snyder (1970b) investigated the specificities of acid and alkaline phosphatases for key substrates involved in the biosynthesis of ester- and ether-linked phospholipids. Alkaline phosphatase exhibits a high substrate specificity since it only removes the free phosphate group from glycerolipids that contain a hydroxyl or ketone moiety in the 2-position. In contrast, acid phosphatase hydrolyzes free phosphate groups from the 3-position of glycerolipids regardless of the groupings at the other two positions of the glycerol moiety (Fig. 11). The relationship of lipid metabolism and phosphatase specificities, particularly in diseases exhibiting abnormal phosphatase levels, remains to be investigated.

REFERENCES

Agranoff, B. W., and Hajra, A. K. (1971). *Proc. Nat. Acad. Sci. U.S.* **68,** 411.

Albro, P. W., and Dittmer, J. C. (1969). *Lipids* **5,** 320.

Anderson, C. E., Williams, J. N., Jr., and Yarbro, C. L. (1960). *Biochem. Lipids Proc. Int. Conf. Biochem. Probl. Lipids, 5th,* **1958** pp. 105–113.

Anderson, R. E., and Sansone, G. L. (1970). *Lipids* **5,** 577.

Ansell, G. B., and Hawthorne, J. N. (1964). "Phospholipids—Chemistry, Metabolism and Function." Elsevier, Amsterdam.

Ansell, G. B., and Spanner, S. (1965). *Biochem. J.* **97,** 375.

Bell, O. E., Jr., Blank, M. L., and Snyder, F. (1971). *Biochim. Biophys. Acta* **231,** 579.

Bergmann, H., Schneweis, K-E., and Thiele, O. W. (1957). *Naturwissenschaften* **44,** 380.

Bergström, S., and Blomstrand, R. (1956). *Acta Physiol. Scand.* **38,** 166.

Blank, M. L., and Snyder, F. (1969). *Biochim. Biophys. Acta* **187,** 154.

Blank, M. L., and Snyder, F. (1970a). *Lipids* **5,** 337.

Blank, M. L., and Snyder, F. (1970b). *Biochemistry* **9,** 5034.

Blank, M. L., Wykle, R. L., Piantadosi, C., and Snyder, F. (1970). *Biochim. Biophys. Acta* **210,** 442.

Blank, M. L., Cress, E. A., Stephens, N., and Snyder, F. (1971). *J. Lipid Res.* **12,** 638.

Blasi, F., Fragomele, F., and Covelli, I. (1969). *Endocrinology* **85,** 542.

Blomstrand, R. (1959). *Proc. Soc. Exp. Biol. Med.* **102,** 662.

Blomstrand, R., and Ahrens, E. H., Jr. (1959). *Proc. Soc. Exp. Biol. Med.* **100,** 802.

Blomstrand, R., and Rumpf, J. A. (1954). *Acta Physiol. Scand.* **32,** 374.

Boxer, G. E., and Shonk, C. E. (1960). *Biochim. Biophys. Acta* **37,** 194.

Burton, R. M. (1959). *Progr. Neurobiol.* **4,** 301.

Carr, H. G., Haerle, H., and Eiler, J. J. (1963). *Biochim. Biophys. Acta* **70,** 205.

Cymerman-Craig, J., and Horning, E. C. (1960). *J. Org. Chem.* **25,** 2098.

Day, J. I. E., Goldfine, H., and Hagen, P-O. (1970). *Biochim. Biophys. Acta* **218,** 179.

Debuch, H., Friedemann, H., and Müller, J. (1970). *Hoppe-Seyler's Z. Physiol. Chem.* **351,** 613.

De Haas, D. H., Sarda, L., and Roger, J. (1965). *Biochim. Biophys. Acta* **106,** 638.

Desnuelle, P., and Savary, P. (1963). *J. Lipid Res.* **4,** 369.

Ellingboe, J., and Karnovsky, M. L. (1967). *J. Biol. Chem.* **242,** 5693.

Ellingson, J. S., and Lands, W. E. M. (1968). *Lipids* **3,** 111.

Ferrell, W. J., and Radloff, D. M. (1970). *Physiol. Chem. Phys.* **2,** 551.

Feulgen, R., Imhäuser, K., and Westhues, M. (1928). *Biochem. Z.* **193,** 251.

Friedberg, S. J., and Greene, R. C. (1967). *J. Biol. Chem.* **242,** 5709.

Friedberg, S. J., and Greene, R. C. (1968). *Biochim. Biophys. Acta* **164,** 602.

Friedberg, S. J., and Greene, R. C. (1969). *J. Amer. Oil Chem. Soc.* **46,** No. 8, Abstr. 82.

Gambal, D., and Monty, K. J. (1959). *Fed. Proc., Fed. Amer. Soc. Exp. Biol.* **18,** 232.

Garfinkel, D., and Hess, B. (1964). *J. Biol. Chem.* **239,** 971.

Gigg, J., and Gigg, R. (1967). *J. Chem. Soc., C,* p. 431.

Goldfine, H. (1968). *Annu. Rev. Biochem.* **37,** 303.

Gottfried, E. L., and Rapport, M. M. (1962). *J. Biol. Chem.* **237,** 329.

Green, D. E. (1936). *Biochem. J.* **30,** 629.

Greten, H., Levy, R. I., Fales, H., and Fredrickson, D. S. (1970). *Biochim. Biophys. Acta* **210,** 39.

Gunetileke, K. G., and Anwar, R. A. (1968). *J. Biol. Chem.* **243,** 5770.

Hack, M. H., and Ferrans, V. J. (1959). *Hoppe-Seyler's Z. Physiol. Chem.* **315,** 157.

Hajra, A. K. (1969). *Biochem. Biophys. Res. Commun.* **37, 486.**

Hajra, A. K. (1970). *Biochem. Biophys. Res. Commun.* **39,** 1037.

Hallgren, B., and Ställberg, G. (1967). *Acta Chem. Scand.* **21,** 1519.

Hartman, F. C. (1968). *Biochem. Biophys. Res. Commun.* **33,** 888.

Hartman, F. C. (1970). *Biochemistry* **9, 1776**

Hartree, E. F. (1964). *In* "Conference on Metabolism and Physiological Significance of Lipids" (R. M. C. Dawson and D. N. Rhodes, eds.), pp. 207–218. Wiley, New York.

Hartree, E. F., and Mann, T. (1961). *Biochem. J.* **80,** 464.

Hollmann, S., and Touster, O. (1964). "Non-Glycolytic Pathways of Metabolism of Glucose." Academic Press, New York.

Kapoulas, V. M., and Thompson, G. A., Jr. (1969). *Biochim. Biophys. Acta* **187,** 594.

Kapoulas, V. M., Thompson, G. A., Jr., and Hanahan, D. J. (1969a). *Biochim. Biophys. Acta* **176,** 237.

Kapoulas, V. M., Thompson, G. A., Jr., and Hanahan, D. J. (1969b). *Biochim. Biophys. Acta* **176,** 250.

Kaufman, S. (1959). *J. Biol. Chem.* **234,** 2677.

Keenan, R. W., Brown, J. B., and Marks, B. H. (1961). *Biochim. Biophys. Acta* **51,** 226.

Kennedy, E. P. (1957). *Fed. Proc., Fed Amer. Soc. Exp. Biol.* **16,** 847.

Kennedy, E. P. (1962). *Harvey Lect.* **57,** 143.

Kiyasu, J. Y., and Kennedy, E. P. (1960). *J. Biol. Chem.* **235,** 2590.

Kolattukudy, P. E. (1970). *Biochemistry* **9,** 1095.

Kornberg, A., and Pricer, W. E., Jr. (1953). *J. Biol. Chem.* **204,** 329.

Lands, W. E. M., and Hart, P. (1965). *Biochim. Biophys. Acta* **98,** 532.

Leupold, F., and Büttner, H. (1953). *Verh. Deut. Ges. Inn. Med.* **59,** 210.

Levin, J. G., and Sprinson, D. B. (1964). *J. Biol. Chem.* **239,** 1142.

McMurray, W. C. (1964). *J. Neurochem.* **11,** 315.

McVicar, G. A. (1934). Ph.D. Thesis, University of Toronto.

Marinetti, G. V., Erbland, J., and Stotz, E. (1959). *Biochim. Biophys. Acta* **33,** 403.

Mattson, F. H., and Volpenhein, R. A. (1961). *J. Lipid Res.* **2,** 58.

Mattson, F. H., and Volpenhein, R. A. (1968). *J. Lipid Res.* **9,** 79.

Meyerhof, O., and Kiessling, W. (1933). *Biochem. Z.* **264,** 40.

Meyerhof, O., Lohmann, K., and Schuster, Ph. (1936a). *Biochem. Z.* **286,** 319.

Meyerhof, O., Lohmann, K., and Schuster, Ph. (1936b) *Biochem. Z.* **286,** 301.

Murooka, Y., Seto, K., and Harada, T. (1970). *Biochem. Biophys. Res. Commun.* **41,** 407.

Oesper, P., and Meyerhof, O. (1950). *Arch. Biochem. Biophys.* **27,** 223.

Oswald, E. O., Anderson, C. E., Piantadosi, C., and Lim, J. (1968). *Lipids* **3,** 51.

Paltauf, F. (1971). *Biochim. Biophys. Acta* **239,** 38.

Paltauf, F., and Johnston, J. M. (1971). *Biochim. Biophys. Acta* **239,** 47.

Pfleger, R. C., Piantadosi, C., and Snyder, F. (1967). *Biochim. Biophys. Acta* **144,** 633.

Piantadosi, C., and Snyder, F. (1970). *J. Pharm. Sci.* **59,** 283.

Piantadosi, C., Snyder, F., and Wood, R. (1969). *J. Pharm. Sci.* **58,** 1028.

Piantadosi, C., Ishaq, K. S., and Snyder, F. (1970). *J. Pharm. Sci.* **59,** 1201.

Piantadosi, C., Ishaq, K. S., Wykle, R. L., and Snyder, F. (1971). *Biochemistry* **10,** 1417.
Poulos, A., Hughes, B. P., and Cumings, J. N. (1968). *Biochim. Biophys. Acta* **152,** 629.
Rao, G. A., Sorrells, M. F., and Reiser, R. (1968). *Biochem. Biophys. Res. Commun.* **31,** 252.
Rapport, M. M., and Franzl, R. E. (1957). *J. Neurochem.* **1,** 303.
Rapport, M. M., Lerner, B., Alonzo, N., and Franzl, R. E. (1957). *J. Biol. Chem.* **225,** 859.
Robertson, A. F., and Lands, W. E. M. (1962). *J. Clin. Invest.* **41,** 2160.
Rose, I. A., and O'Connell, E. L. (1969). *J. Biol. Chem.* **244,** 6548.
Rosenthal, A. F., and Pousada, M. (1966). *Biochim. Biophys. Acta* **125,** 265.
Sand, D. M., Hehl, J. L., and Schlenk, H. (1969). *Biochemistry* **8,** 4851.
Schmid, H. H. O., and Takahashi, T. (1970). *J. Lipid Res.* **11,** 412.
Slotboom, A. J., de Haas, G. H., and van Deenen, L. L. M. (1967). *Chem. Phys. Lipids* **1,** 192.
Snyder, F. (1969) *Progr. Chem. Fats Other Lipids* **10,** 287–335.
Snyder, F., and Blank, M. L. (1969). *Arch. Biochem. Biophys.* **130,** 101.
Snyder, F., and Malone, B. (1970). *Biochem. Biophys. Res. Commun.* **41,** 1382.
Snyder, F., and Pfleger, R. C. (1966). *Lipids* **1,** 328.
Snyder, F., and Piantadosi, C. (1968). *Biochim. Biophys. Acta* **152,** 794.
Snyder, F., and Wood, R. (1968). *Cancer Res.* **28,** 972.
Snyder, F., Malone, B., and Wykle, R. L. (1969a). *Biochem. Biophys. Res. Commun.* **34,** 40.
Snyder, F., Wykle, R. L., and Malone, B. (1969b). *Biochem. Biophys. Res. Commun.* **34,** 315.
Snyder, F., Malone, B., and Blank, M. L. (1969c). *Biochim. Biophys. Acta* **187,** 302.
Snyder, F., Piantadosi, C., and Wood, R. (1969d). *Proc. Soc. Exp. Biol. Med.* **130,** 1170.
Snyder, F., Malone, B., and Cumming, R. B. (1970a). *Can. J. Biochem.* **48,** 212.
Snyder, F., Malone, B., and Blank, M. L. (1970b). *J. Biol. Chem.* **245,** 1790.
Snyder, F., Blank, M. L., Malone, B., and Wykle, R. L. (1970c). *J. Biol. Chem.* **245,** 1800.
Snyder, F., Blank, M. L., and Malone, B. (1970d). *J. Biol. Chem.* **245,** 4016.
Snyder, F., Rainey, W. T., Jr., Blank, M. L., and Christie, W. H. (1970e). *J. Biol. Chem.* **245,** 5853.
Snyder, F., Piantadosi, C., and Malone, B. (1970f). *Biochim. Biophys. Acta* **202,** 244.
Snyder, F., Hibbs, M., and Malone, B. (1971a). *Biochim. Biophys. Acta* **231,** 409.
Snyder, F., Blank, M. L., and Wykle, R. L. (1971b). *J. Biol. Chem.* **246,** 3639.
Soodsma, J. F., Piantadosi, C., and Snyder, F. (1970). *Cancer Res.* **30,** 309.
Soodsma, J. F., Piantadosi, C., and Snyder, F. (1972). *J. Biol. Chem.*, in press.
Stewart, J. E., Kallio, R. E., Stevenson, D. P., Jones, A. C., and Schissler, D. O. (1959). *J. Bacteriol.* **78,** 441.
Stoffel, W., and Assmann, G. (1970). *Hoppe-Seyler's Z. Physiol. Chem.* **351,** 1041.
Stoffel, W., and LeKim, D. (1971). *Hoppe-Seyler's Z. Physiol. Chem.* **352,** 501.
Stoffel, W., and Sticht, G. (1967). *Hoppe-Seyler's Z. Physiol. Chem.* **348,** 1345.
Stoffel, W., LeKim, D., and Heyn, G. (1970). *Hoppe-Seyler's Z. Physiol. Chem.* **351,** 875.
Strominger, J. L. (1958). *Biochim. Biophys. Acta* **30,** 645.
Tabakoff, B., and Erwin, V. G. (1970). *J. Biol. Chem.* **245,** 3263.
Takahashi, T., and Schmid, H. H. O. (1970). *Chem. Phys. Lipids* **4,** 243.
Taurog, A. (1970). *Recent Progr. Horm. Res.* **26,** 189–241.
Thiele, O. W. (1959a). *Hoppe-Seyler's Z. Physiol. Chem.* **315,** 117.
Thiele, O. W. (1959b). *Hoppe-Seyler's Z. Physiol. Chem.* **316,** 137.

Thiele, O. W., and Gowin, E. (1955). *Hoppe-Seyler's Z. Physiol. Chem.* **299,** 151.
Thiele, O. W., Schröder, H., and von Berg, W. (1960). *Hoppe-Seyler's Z. Physiol. Chem.* **322,** 147.
Thompson, G. A., Jr. (1968). *Biochim. Biophys. Acta* **152,** 409.
Tietz, A., Lindberg, M., and Kennedy, E. P. (1964). *J. Biol. Chem.* **239,** 4081.
Toi, K., Salvatore, G., and Cahnmann, H. J. (1965). *Biochim. Biophys. Acta* **97,** 523.
van Deenen, L. L. M., and de Haas, G. H. (1966). *Annu. Rev. Biochem.* **35,** 157.
van Golde, L. M. G., and van Deenen, L. L. M. (1967). *Chem. Phys. Lipids* **1,** 157.
Warner, H. R., and Lands, W. E. M. (1961). *J. Biol. Chem.* **236,** 2404.
Warner, H. R., and Lands, W. E. M. (1963). *J. Amer. Chem. Soc.* **85, 60.**
Wood, R., and Healy, K. (1970a). *Biochem. Biophys. Res. Commun.* **38,** 205.
Wood, R., and Healy, K. (1970b). *J. Biol. Chem.* **245,** 2640.
Wood, R., Piantadosi, C., and Snyder, F. (1969). *J. Lipid Res.* **10,** 370.
Wood, R., Walton, M., Healy, K., and Cumming, R. B. (1970). *J. Biol. Chem.* **245,** 4276.
Wykle, R. L., and Snyder, F. (1969). *Biochem. Biophys. Res. Commun.* **37,** 658.
Wykle, R. L., and Snyder, F. (1970). *J. Biol. Chem.* **245,** 3047.
Wykle, R. L., Blank, M. L., and Snyder, F. (1970). *FEBS Lett.* **12,** 57.
Wykle, R. L., Blank, M. L., and Snyder, F. (1971). *Fed. Proc., Fed. Amer. Soc. Exp. Biol.* **30,** 1243.
Wykle, R. L., Piantadosi, C., and Snyder, F. (1972). *J. Biol. Chem.* **247,** 2944.

Note Added in Proof

Recent publications by Snyder *et al.* (1972), Paltauf (1972), Blank *et al.* (1972), and Wykle *et al.* (1972) indicate that ethanolamine plasmalogens are formed from 1-alkyl-2-acyl-*sn*-glycero-3-phosphorylethanolamine; the microsomal enzyme system catalyzing this reaction has the properties of a mixed-function oxidase in that a reduced pyridine nucleotide (NADH or NADPH) and molecular oxygen are required. Thus the synthesis of the *O*-alk-1-enyl moiety from an *O*-alkyl moiety is similar to the desaturation process involved in the biosynthesis of monoenoic acids.

Blank, M. L., Wykle, R. L., and Snyder, F. (1972). *Biochem. Biophys. Res. Commun.*, in press.
Paltauf, F. (1972). *FEBS Lett.* **20,** 79.
Snyder, F., Wykle, R. L., Blank, M. L., Lumb, R. H., Malone, B., and Piantadosi, C. (1972). *Fed. Proc., Fed. Amer. Soc. Exp. Biol.* **31,** 454.
Wykle, R. L., Blank, M. L., Malone, B., and Snyder, F. (1972). *J. Biol. Chem.*, in press.

CHAPTER VIII

BIOLOGICAL EFFECTS AND BIOMEDICAL APPLICATIONS OF ALKOXYLIPIDS

Helmut K. Mangold

I. Introduction

It is generally conceded that many questions concerning the biological significance of the naturally occurring alkoxylipids are unanswered. Why, for instance, is there such a striking difference between the contents of alkyldiacylglycerols in normal and neoplastic tissues? What are the functions of the high proportions of alk-1-enylacyl phosphoglycerides in the central and peripheral nervous system? Can the composition of the various a.koxylipids in a tissue be changed and how do such changes affect the organism?

In the last 25 years, numerous biological activities have been attributed to "chimyl alcohol," "batyl alcohol," and "selachyl alcohol," the 1-alkylglycerols that are the most common constituents of naturally occurring alkoxylipids. Investigators have claimed that these compounds have therapeutic value in a number of disease states. However, reports of the biological and therapeutic effects of the 1-alkylglycerols are at variance, and in the opinion of this author, none of these effects has been unequivocally proven. Obviously it is difficult to repeat the experiments reported from one laboratory in another, and it is particularly difficult to substantiate claims of biological effects, since the preparations used in some biological studies were rather ill-defined. In most cases, the alkoxylipids were isolated from animal tissues, usually shark livers. The alkoxylipid content and the composition of the alkoxylipid fraction were not determined, and it is more than likely that the preparations contained a large variety of compounds other than alkoxylipids. Thus, the biological effects of some of the preparations might have been due to the presence of fat-soluble vitamins.

In recent years, synthetic lipids containing alkyl groups or alk-1-enyl groups have become available, and methods for the characterization and analysis of these compounds have been developed. Experiments that have led to reports of biological activities and even therapeutic effects of various alkoxylipids should now be repeated using pure synthetic compounds. Such studies should also consider compounds that, as a rule, occur naturally in trace amounts only, such as alkoxylipids derived from polyhydric alcohols other than glycerol, and alkoxylipids containing polyunsaturated alkyl or alk-1-enyl moieties.

Glycerol-derived lipids containing one, two, or three alkyl groups instead of acyl groups of comparable chain lengths resemble the glycerides closely in molecular size and shape, and also in their physical properties. However, the alkyl ethers are more stable, or even totally resistant to hydrolytic enzymes. Therefore, individual synthetic alkoxylipids are finding wide application as model substances in systems where lipids containing ester bonds cannot be used because of their relative instability. Alkoxylipids that contain more than one alkyl moiety are of particular interest, especially the trialkylglycerols. These compounds resemble the triacylglycerols ("triglycerides") in many respects, but are extraordinarily stable against enzymic action, and remarkably resistant to chemical attack.

In this chapter the various biological activities and therapeutic effects that have been ascribed to lipids containing alkyl or alk-1-enyl groups will be discussed, and an attempt will be made to present a balanced view of the many conflicting reports. However, since these aspects have been reviewed quite recently (Snyder, 1969), emphasis will be placed on the description of studies in which pure individual alkoxylipids were put to good use because of their relative stability.

II. Toxicity of Alkoxylipids

A. 1-Alkylglycerols and 1-Alkyl-2,3-diacylglycerols

1. *Toxicity in Animals*

Alkylglycerols and their esters have been administered orally to mice, rats, and dogs. Agduhr *et al.* (1934) reported that mice given small oral doses of batyl alcohol for several months developed lesions in their hearts, kidneys, livers, and adrenal glands. These observations could not be confirmed by later investigators. Thus, Alexander *et al.* (1959) reported that mice which had received a diet containing 18% alkyldiacylglycerols showed no ill effects, even after 2 years. And Brohult (1963) mentioned unpublished work by B. Melander, who fed large doses of alkylglycerols as well as alkyldiacylglycerols to mice. Melander reportedly found that mice could tolerate as much as about 4.4 g of alkyldiacylglycerols per kg of body weight per day, for 18 days. The weight gain of the mice was found to be normal, and pathological changes could not be observed. Thus, Brohult (1963) concluded that the toxicity of purified alkylglycerols and their esters to mice is low. But she also noted that one preparation tested, crude liver oil of Greenland shark (*Somniosus microcephalus*), must have contained some components that had an irritating effect on the gastrointestinal tract of the mice and caused the death of some of the animals. Work by Kaneda and Ishii (1952), Kaneda *et al.* (1955), Brohult (1963), Peifer *et al.* (1965), and Carlson (1966), as well as more recent experiments (Snyder *et al.*, 1971; Bandi *et al.*, 1971a), show that alkylglycerols and their esters, when given orally, are also relatively nontoxic to rats. Carlson (1966) found that dogs, as well as rats, fed chimyl, batyl, or selachyl alcohols at a level of 2.4 g per kg of body weight per day, did not manifest any ill effects.

When batyl alcohol was administered subcutaneously to mice (Berger, 1948; Penny *et al.*, 1964), the LD_{50} dose was found to be greater than 3 g per kg of animal weight (Berger, 1948). Arturson and Lindbäck (1951) found that intraperitoneal injections of batyl alcohol led to an increase of reticulocytes in mice. Evenstein *et al.* (1958) subcutaneously injected increasing amounts of batyl alcohol in rats over a period of 41 days; they did not observe any pathological changes in the animals. Hietbrink *et al.* (1962), also working with rats, found that intraperitoneal injections of batyl alcohol, at a level of 5 to 10 mg per kg of body weight per day, caused an increase in the weight of the spleen, but did not affect the activity of adenosine triphosphatase in this tissue or in the thymus gland of the animals. I. A. Evans *et al.* (1953; W. C. Evans *et al.*, 1957) administered batyl alcohol to cattle, both by intramuscular and intraperitoneal injections, and found that 1 g of this compound could be given intramuscularly over a period of 4 to 5 days without ill effects. Many other investigators administered rather small doses of alkylglycerols to experimental animals exposed to X-rays.

2. *Toxicity in Humans*

Sandler (1949) reported that he had given healthy adult men 45 mg of batyl alcohol per day for 10 days with no sign of ill effects. Several investigators administered rather small oral or intramuscular doses of alkylglycerols to patients suffering from radiation leukopenia or cancer. Brohult (1963) mentioned that amounts of alkoxylipids corresponding to 10–100 mg batyl alcohol per day are consumed in food by the average human being. There is no doubt that occasionally much larger amounts of alkyldiacylglycerols are ingested, at least by people eating shark products. Not only is the liver of Atlantic and Pacific dogfish (*Squalus acanthias* and *Squalus suckleyi*) a rich source of alkyldiacylglycerols (Mangold, 1961), but the rather oily flesh of these and other sharks also contains substantial amounts. For many decades, shark meat has been used for human consumption in Australia and New Zealand (*Callorhynchus milii*, *Galeorhinus australis*, and *Mustelus antarcticus*), in Germany and Great Britain (*Squalus acanthias*), in Japan, especially the northern islands (several species), and in the United States (*Lamna nasus*). In the West Indies the demand for shark meat is said to exceed the supply landed by local fishermen. To the best of this author's knowledge, there are no reports of any ill effects from eating meat of the species mentioned, but meat of Greenland shark (*Somniosus microcephalus*) is allegedly unfit for human consumption because of its purgative effect.

A few isolated cases are known where the consumption of shark products has been harmful. Ill effects caused by eating shark livers and shark liver oils in New Zealand could be related to an overdose of fat-soluble vitamins and when, during the war, a restaurant in Japan used shark liver oil for frying sea food, thereby giving its patrons diarrhea, this was probably related to the oil's high squalene content.

B. 2-Alkylglycerols and 2-Alkyl-1,3-diacylglycerols

Symmetrical alkylglycerols and their acyl derivatives have not been found in nature. The behavior of synthetic 2-alkylglycerols in the gastrointestinal tract has been studied (see Section III, B), but attempts have not been made to assess the toxicity of these compounds and their acyl derivatives.

C. 1,2- and 1,3-Dialkylglycerols and Their Acyl Derivatives

It has been reported that phospholipids derived from 1,2-dialkylglycerols occur in the human heart (Popović, 1965). The behavior of synthetic 1,2- and 1,3-dialkylglycerols and of various 1,2-dialkyl phospholipids in the gastrointestinal tract has been studied (see Section III, C), and acyl derivatives of the dialkylglycerols have been used in biochemical investigations (see Section IV), but whether these compounds exhibit toxic effects has not been determined.

D. Trialkylglycerols

These compounds have not been found in nature. However, synthetic trialkylglycerols (Baumann and Mangold, 1966) have recently attracted considerable interest. Because they are not hydrolyzed or otherwise changed in the gastrointestinal tract, they may become useful as nonabsorbable substitutes for dietary fats, and in fat absorption studies the trialkylglycerols may become useful as reference markers for dietary lipids.

Studies on rats eating a labeled trialkylglycerol, (9,10-^{3}H)-hexadecyldidodecylglycerol, in cumulative amounts of more than 200 mg over 28 days, showed no evidence of toxicity at autopsy (Hofmann, 1969). On histological examination all of the following tissues appeared normal: brain, heart, liver, spleen, kidney, adrenal gland, pancreas, lung, intestine, bone marrow, and adipose tissue (Karlson, 1969).

E. Steryl Ethers

Very small amounts of cholesteryl alkyl ethers (Funasaki and Gilbertson, 1968) and cholesteryl alk-1-enyl ethers (Gilbertson *et al.*, 1970) were recently detected in bovine cardiac muscle; toxicological studies were not reported. Kaufmann *et al.* (1970) found disteryl ethers such as dicholesteryl, disitosteryl, distigmasteryl, and dibrassicasteryl ethers in refined vegetable oils. They showed that the disteryl ethers are formed during bleaching of the crude oils with mineral clay. Dicholesteryl ether was found to be noncarcinogenic, at least in rats.

III. Absorption of Alkoxylipids and Use of Alkoxylipids in "Fat" Absorption Studies

A. 1-Alkylglycerols and 1-Alkyl-2,3-diacylglycerols

With the aid of ^{14}C-labeled compounds it was established that dietary 1-alkylglycerols are almost completely absorbed by rats (Bergström and Blomstrand, 1956; Blomstrand, 1959; Swell *et al.*, 1965) as well as humans (Blomstrand and Ahrens, 1959). Some of the dietary 1-alkylglycerol is absorbed intact and acylated in the intestinal mucosa (Blomstrand, 1959; Bandi *et al.*, 1971b), but the major portion is cleaved at the ether linkage, and the long-chain alkyl moiety is converted to a fatty acid (Blomstrand and Ahrens, 1959). The same reactions occurred when a 1-alkyl-2,3-diacylglycerol was fed (Blomstrand, 1959). The alkyl moieties in naturally occurring alkoxylipids are almost exclusively saturated and monounsaturated. Bandi *et al.* (1971a) fed rats a synthetic polyunsaturated 1-alkylglycerol, *cis,cis*-9,12-octadecadienylglycerol (α-linoleylglycerol) and demonstrated that it underwent the same reactions as the labeled saturated 1-alkylglycerols.

The results of *in vitro* studies by Sherr *et al.* (1963), Sherr and Treadwell (1965), and Gallo *et al.* (1968) confirmed that, in the intestinal mucosa of the rat, 1-alkylglycerols can be acylated to form 1-alkyl-2,3-diacylglycerols, even though to a very limited extend. These authors showed, too, that 2-alkylglycerols can also be acylated. Kern and Borgström (1965) demonstrated the acylation of 1-alkylglycerols and 2-alkylglycerols in hamster intestinal rings. Snyder *et al.* (1970) found that cellfree homogenates from fat liver contain enzymes that catalyze the acylation of alkylglycerols. Accordinging to these authors, acylation of racemic 1-alkylglycerols produces only 1-alkyl-3-acylglycerols, whereas the acylation of 2-alkylglycerols leads to 2-alkyl-1,3-diacylglycerols. Of great interest, in this context, is a publication by Lewis (1966), who reported the occurrence of large proportions of 1-alkyl-2,3-diacylglycerols in the stomach oil of a marine bird, the Leach's petrel (*Oceanodroma leucorhoa*). The origin and function of these acylated alkylglycerols is unknown.

Whereas both enantiomeric 1-acylglycerols are acylated in the intestinal mucosa, acylation of the 1-alkylglycerols proceeds in a strictly stereospecific manner in that only the naturally occurring enantiomers serve as substrates *in vivo* and *in vitro* (Paltauf, 1971).

B. 2-Alkylglycerols and 2-Alkyl-1,3-diacylglycerols

Because of their stability against lipases, and because of the fact that they do not undergo isomerization, the 1-alkyl- and 2-alkylglycerols were used as model substances in studies aimed at elucidating the reactions that lead to the resynthesis of triacylglycerols in the intestinal mucosa. As mentioned in the preceding section, several authors found that both the 1-alkylglycerols and 2-alkylglycerols are acylated to the respective alkylacylglycerols and alkyldiacylglycerols almost as readily as the acylglycerols are further acylated. Using the isomeric octadecylglycerols as model substances, Gallo *et al.* (1968) tried to establish which of the isomeric acylglycerols, and which diacylglycerol, are preferred for the resynthesis of triacylglycerols. The authors concluded that the mucosa of the rat intestine utilizes 2-acylglycerols for the formation of triacylglycerols, which means that the intermediates are 1,2-diacylglycerols and that utilization of 1,3-diacylglycerols probably requires their isomerization to the 1,2-isomer.

C. 1,2- and 1,3-Dialkylglycerols and Their Acyl Derivatives

Only one investigator, Paltauf (1969) has studied the intestinal absorption of the isomeric dialkylglycerols. He fed various ^{14}C-labeled 1,2-dialkylglycerols and 1,3-dialkylglycerols to rats and found that isomeric compounds having the same chain lengths were absorbed at equal rates, but that the chain lengths of the alkyl moieties had a pronounced effect on the extent these compounds were absorbed. Thus, less than 10% of the dioctadecylglycerols were absorbed, but over 50% of the dioctylglycerols. The rate of acylation during passage through the gut was found to be quite different for isomeric compounds, the 1,2-dialkylglycerol being the favored substrate. This finding certainly supports the conclusions that Gallo *et al.* (1968) had drawn when interpreting their results (see Section III, B).

Paltauf (1969) also studied the absorption of 1,2-dialkyl analogs of phosphatidic acid as well as those of choline phospholipids and ethanolamine phospholipids. He found that these compounds were rather poorly absorbed. After feeding the ionic compounds to rats, he could detect only traces of alkoxylipids in their lymph.

D. Trialkylglycerols

Long-chain trialkylglycerols are isosteric to triacylglycerols, the "fats," but they are neither hydrolyzed nor absorbed in the alimentary canal. These facts were demonstrated first by Spener *et al.* (1968), and their findings have been confirmed since (Carlson and Bayley, 1970; Morgan and Hofmann, 1970b).

Spener *et al.* (1968) used a ^{14}C-labeled trialkylglycerol to assess the amount of "fat" a rat can absorb, possibly through pinocytosis, without prior hydrolysis of the triacylglycerols. They administered tri-(1-^{14}C)-*cis*-9-octadecenylglycerol (trioleylglycerol) to normal rats and to rats provided with a thoracic duct fistula. Spener *et al.* measured the radioactivity of the lipids in the lymph during the first 24 hours after feeding. In addition they determined the radioactivity in various organs, and in the remaining carcass. Spener *et al.* found that only 0.14% of the trialkylglycerol fed had been absorbed (most of the adsorbed material was detected in the liver) and they concluded that intact long-chain triacylglycerols are probably absorbed at about the same rate. Simple pinocytosis of lipid droplets can therefore be ruled out as a quantitatively significant process of fat absorption.

In acute feeding experiments, Morgan and Hofmann (1970a) found that rats absorbed less than 0.2% of an ^{3}H-labeled trialkylglycerol, (9,10-^{3}H)-hexadecyldidodecylglycerol. The trialkylglycerol was found unchanged in the feces, indicating that it had not been degraded by digestive or bacterial enzymes. Chronic feeding experiments in rats confirmed that hexadecyldidodecylglycerol is absorbed no more than a fraction of 1%, that it is nontoxic, and that it does not influence the absorption of dietary triacylglycerols. Almost all of the trialkylglycerol absorbed was found in the liver, spleen, and adipose tissue of the animals.

Carlson and Bayley (1970) showed that tridodecylglycerol passes through the rat's digestive tract in association with other lipids, but remains unabsorbed; these authors were able to recover about 97% of the fed trialkylglycerol from the feces. This agrees substantially with the findings of Spener *et al.* (1968) and Morgan and Hofmann (1970a).

In a recent publication, Morgan and Hofmann (1970b) proposed (9,10-^{3}H)-hexadecyldidodecylglycerol as a marker for the quantitative evaluation of fat malabsorption. They fed this compound, together with a ^{14}C-labeled triacylglycerol, to normal rats and to rats with cholestyramine-induced steatorrhea. Fat absorption was estimated both from the ratio of $^{3}H/^{14}C$ in the meal and in the feces, and from the total fecal excretion of ^{14}C-labeled material. The values for fat absorption obtained by the two methods agreed within a range of 50–100%. Reference is made to a recent and more extensive review on the chemistry and biochemistry of trialkylglycerols (Mangold *et al.*, 1972).

E. Steryl Ethers

It is not known if, and to what extent, cholesteryl esters are absorbed intact by the intestinal mucosa, or whether only cholesterol and fatty acids can be absorbed after hydrolytic cleavage of the esters. In order to get some idea of the magnitude of the absorption of intact cholesteryl esters, Borgström (1968) studied the behavior of the corresponding ethers, as these can be expected to be more resistant to the enzymes present in the digestive tract. He found that ingested cholesteryl alkyl ethers with alkyl chains up to ten carbon atoms long are absorbed by the intestinal musosa and transported in the thoracic lymph duct of rats. The recovery of the cholesteryl alkyl ethers from the lymph decreased with increasing chain length of the alkyl moiety. About 4% of the cholesteryl decyl ether could be recovered. It can be assumed that ethers having chain lengths similar to those of the naturally occurring cholesteryl esters, namely, C_{16}, C_{18}, and C_{20}, are absorbed to an extent considerably less than 1%. The synthesis of such cholesteryl alkyl ethers was reported (Paltauf, 1968); however, the absorption of these substances has not been studied. The fate of disteryl ethers (Kaufmann *et al.*, 1970) in the alimentary tract is unknown.

IV. Alkoxylipids as Substrates for Acyl Hydrolases

Pure synthetic alkoxylipids have been successfully used in studies of the physicochemical state of lipids in intestinal content during digestion and absorption. Hofmann (1963) determined the solubilization of monosubstituted glycerols, including 1-alkylglycerols, under *in vitro* conditions simulating the conditions prevailing in the human small intestine. As an extension of this work, Hofmann and Borgström (1963) presented observations on the action of pancreatic lipase on mixed bile salt/lipid micelles. More recently, Paltauf (1969) showed a good correlation between micellar solubility and rate of absorption of alkylglycerols, dialkylglycerols, and trialkylglycerols. Borgström (1965) used gel filtration to determine the dimensions of the micelles formed between bile salts and a series of lipids that included 1-alkylglycerols.

Several authors used alkoxylipids as substrates in studies of the specificities of enzyme preparations. Thus, Greten *et al.* (1970) studied the positional specificity of lipases with two isomeric dialkylacylglycerols as substrates. And Slotboom *et al.* (1970) investigated the action of purified lipase preparations from porcine pancreas and from the mold *Rhizopus arrhizus* on synthetic 1,2-diacyl-, 1-alkyl-2-acyl-, and 1-alk-1′-enyl-2-acyl-choline phosphoglycerides. They found that the enzyme(s) from the two sources hydrolyze only the 1-acyl ester bond of choline phosphoglycerides, which means that phospholipase A_1 may actually be identical to lipase.

Anatol *et al.* (1964) mentioned that they were preparing a peralkylated analog of cardiolipin for use in the Wassermann test.

V. Applications of Alkoxylipids in Therapy

A. Bacteriostatic Properties

In 1952, Emmerie *et al.* used the nonsaponifiable matter of cod liver oil to isolate a fraction of 1-alkylglycerols that had strong inhibitory action *in vitro* upon the growth of human tubercle bacillus (*Mycobacterium tuberculosis*). The material was toxic to guinea pigs, and therefore its effectiveness in experimental tuberculosis could not be studied in these animals. Emmerie *et al.* ascribed the bacteriostatic properties of the fraction they had isolated to the presence of unsaturated 1-alkylglycerols, although pure natural selachyl alcohol was less active than their mixture of compounds. Ten years after the above-mentioned publication, Emmerie and Engel (1962) provided evidence for the occurrence of polyunsaturated 1-alkylglycerols in the nonsaponifiable matter of cod liver oil. They claimed that these compounds were in fact responsible for the bacteriostatic properties they had found in 1952. The interesting experiments of Emmerie and his associates certainly ought to be repeated with pure synthetic compounds, keeping in mind that tuberculostatic properties have also been attributed to diaryl ethers (Barry *et al.*, 1947).

B. Hemopoietic Effects

More than 40 years ago, bone marrow was successfully used for the treatment of secondary anemia (Giffin and Watkins, 1930). A few years later, Marberg and Wiles (1938) demonstrated that the material stimulating the formation of erythrocytes was present in the nonsaponifiable matter of yellow bone marrow. After Holmes *et al.* (1941) had found batyl alcohol in the nonsaponifiable fraction of bovine yellow bone marrow, several investigators studied the hemopoietic effects of the common 1-alkylglycerols. Sandler (1949) reported that not only the "nonsaponifiables" of bone marrow, but also batyl alcohol had a beneficial effect on the erythrocyte count of both normal rats and those poisoned with benzene. Moreover, Sandler found that subcutaneous injections of pure, synthetic, optically inactive, 1-octadecylglycerol produced an increase in the circulating reticulocytes in human subjects. The erythropoietic, thrombopoietic, and granulopoietic stimulatory activities of both optically active and racemic 1-octadecylglycerol were later confirmed by several investigators (Arturson and Lindbäck, 1951; Brohult and Holmberg, 1954; Brohult, 1957; Hasegawa *et al.*, 1961; Linman, 1958, 1960; Linman *et al.*, 1958, 1959a; De Gaetani and Baiotti, 1959; Suki and Grollman, 1960; Osmond *et al.*, 1963). Brohult (1957) noted that there is an optimum dose which should not be exceeded, and Linman and Bethell (1956, 1961) pointed out that 1-octadecylglycerol possesses the same properties as the plasma fraction that stimulates erythropoiesis. However, batyl alcohol had no effect on the uptake of ^{59}Fe by hemoglobin *in vitro* and *in vivo* (Linman *et al.*, 1959a,b).

Suki and Grollman (1960) found that chimyl alcohol is also able to stimulate hemopoiesis, whereas selachyl alcohol is rather inactive. Linman (1960) and Osmond *et al.* (1963) found that both natural and synthetic *cis*-9-octadecenylglycerol are devoid of hemopoietic stimulatory activity in normal rats. Evenstein *et al.* (1958) questioned the erythropoietic effect of 1-octadecylglycerol in rats. And other investigators found this compound of little or no value in the treatment of leukopenia caused by irradiation (see Section V, C) or bracken poisoning in cattle (see Section V, D), conditions which also produce damage to the bone marrow.

In several laboratories the effects of "radiomimetic compounds," substances causing damage similar to that observed after irradiation, could be counteracted by intravenous injections of bone marrow (Talbot and Elson, 1958; Tran Ba Loc and Bernard, 1958). However, the aplastic anemia induced in cattle or calves by feeding trichloroethylene-extracted soybean meal could not be prevented by injecting synthetic 1-octadecylglycerol by various routes (Schultze *et al.*, 1958). A publication on the distribution and biosynthesis of alkylacyl phosphoglycerides in hematopoietic bone marrow certainly is of interest in this connection (Thompson and Hanahan, 1963).

Reference is made to a paper by Heller *et al.* (1963) on the presence of a lipophilic agent in shark liver that stimulates the reticuloendothelial system. This material was said to be rather unstable. Although it was not characterized further, it may well be an alkoxylipid, possibly identical or similar to the 1-alk-1′-enyl-2,3-diacylglycerols ("neutral plasmalogens") that are found in the liver of a species related to the sharks, the ratfish (*Hydrolagus colliei*), a chimaera (Schmid *et al.*, 1967).

A few years ago, the Astra Nutrition AB (1965) claimed a patent on the use of shark liver oil plus several vitamins in a "vitamin capsule." This preparation is meant to be used "to raise the percentage of glycerol ethers in natural fats and foods." Supposedly, "the glycerol ethers stimulate the formation of white and red blood corpuscles and promote the growth of intestinal Lactobacilli." The latter claim is based on a publication by Brohult (1960).

C. Protection against Radiation Damages

Marberg and Wiles' (1938) report on the treatment of leukopenia with extracts of bone marrow, and Sandler's finding (1949) of the hemopoietic activity of "batyl alcohol," stimulated research on the effects of 1-alkylglycerols and their acyl derivatives on radiation injuries to erythropoietic tissues. Simultaneously, but independently, Brohult and Holmberg (1954) and Edlund (1954) published that they had been successful in protecting human subjects and mice, respectively, against radiation damage by giving them various alkoxylipids. Brohult and Holmberg (1954) administered concentrates of 1-alkylglycerols and their esters orally to patients suffering from radiation leukopenia, whereas Edlund (1954) gave subcutaneous injections of synthetic 1-octadecylglycerol dissolved in peanut oil to mice that had been given total body irradiation. Brohult (1957, 1958, 1962; Brohult *et al.*, 1970) extended these studies and, in 1963, presented a comprehensive review of her results.

In the course of the last 10 years the effects of 1-alkylglycerols in the treatment of radiation-induced leukopenia have been studied extensively. Some investigators confirmed the beneficial action of these compounds, especially of selachyl alcohol and its esters (Alexander *et al.*, 1959; Dudin, 1961; Mozharova *et al.*, 1961; Rusanov *et al.*, 1962; Sviridov *et al.*, 1964; Chebotarev, 1965). Others did not find the desired effects (Mizuno *et al.*, 1960; Ghys, 1962; Hietbrink *et al.*, 1962; Bassi and Dunjic, 1962; Prokhonchukov and Panikarovskii, 1963; Snyder *et al.*, 1971). Snyder *et al.* (1971) studied not only the 1-alkylglycerols and their acyl derivatives, but also 1-alk-1′-enyl-2,3-diacylglycerols, various diol lipids, long-chain alcohols, and other compounds. These substances were administered orally, intraperitoneally, and intramuscularly to rats that had been exposed to total body radiation. The number of circulating leukocytes was chosen as the criterion for effectiveness, and it was found that none of the various lipids investigated was able to alter the course of radiation-induced leukopenia.

It is interesting that 1-alkylglycerols occur in the nonsaponifiable matter from the bone marrow of rats even after exposure to total body irradiation (Snyder and Cress, 1963).

According to Maqsood and Ashikawa (1961; Ashikawa, 1961), intraperitoneal injection of olive oil and other vegetable oils increases survival of mice after whole-body radiation. These oils are known to contain little alkoxylipids, if any. (See Chapter XVI.)

D. Treatment of Bracken Poisoning

After eating fern, cattle and calves develop severe leukopenia that is almost invariably fatal. W. C. Evans *et al.* (1957; I. A. Evans *et al.*, 1953) claimed that this "bracken poisoning" can be alleviated by subcutaneous injections of a solution of batyl alcohol in olive oil. However, neither Dalton (1964) nor Penny *et al.* (1964), who tested this therapy, found it effective.

E. Healing of Wounds

For generations various fats and oils, including fish oils, have been used as household medicine in the treatment of burns. Bodman and Maisin (1958) and Maisin *et al.* (1959) claimed that the alleged beneficial effects of certain fish oils are due to their alkoxylipid content. They reported striking results in the treatment of wounds by topical application of 1-alkylglycerols. Such therapy was reportedly beneficial in the treatment of nonhealing septic wounds in elderly patients, of nonhealing wounds associated with surgical treatment of fractures and malignant tumors, and of wounds and burns resulting from X-ray treatment. However, local application of 1-alkylglycerols was ineffective in the treatment of accidental burns in normal healthy human beings.

In a patent issued in 1966, Chalmers *et al.* claimed that 1-alkylglycerols dissolved in cottonseed oil are of value in the treatment of inflammatory diseases, chimyl alcohol supposedly being as efficacious as hydrocortisone.

Recently, the beneficial effects of the 1-alkylglycerols in wound healing have been questioned. Stansby *et al.* (1967) investigated the effects of various fats and oils in the treatment of wounds and burns of hairless mice. The authors found that the topical application of any oil had a slightly accelerating effect on the healing process when compared to no treatment at all. A fish liver oil having a high vitamin A content and a fish body oil to which 1-alkylglycerols had been added had the same effect as plain mineral oil. In the opinion of this author the alleged beneficial effects of 1-alkylglycerols in wound healing are not yet proven.

F. Inhibition of Neoplastic Growth

Abaturova and Shubina (1964) fed about 1 mg of batyl alcohol or selachyl alcohol per day for 3 months to rats bearing subcutaneous transplantable tumors. They claimed that both of these compounds slightly retarded growth of the tumors, selachyl alcohol being the more effective of the two. In 1970, Brohult *et al.* reported the use of 1-alkylglycerols in cancer therapy in connection with radiation treatment of a large number of patients. 1-Alkylglycerols from the liver of Greenland shark (*Somniosus microcephalus*) were fed for several weeks, at a level of 0.6 g/day, to patients with cancer of the uterine cervix. One group of patients received these substances prophylactically 8 days before treatment with X-rays; another group received them only during the treatment. According to Brohult *et al.*, the 1-alkylglycerols given to the patients affected their tumors before as well as after radiation treatment, and reduced the mortality rates. The authors recommended that therapy should be started with 1-alkylglycerols before X-ray treatment. In experiments with rats, Delmon and Biraben (1966) did not observe any inhibitory effect of 1-octadecylglycerol on the development of carcinoma T 8; however, treatment with this compound appeared to lengthen the survival time of the animals.

It should be noted that most healthy human and animal tissues contain 1-alk-1′-enyl-2,3-diacylglycerols and 1-alkyl-2,3-diacylglycerols in small and roughly equal amounts (Tuna and Mangold, 1963; Schmid and Mangold, 1966). Recently, neutral and ionic lipids derived from 1-alkylglycerols were found to be present at a much higher level in transplantable rat and mouse tumors (Snyder and Wood, 1968) as well as in neoplastic human tissues (Snyder and Wood, 1969). The corresponding derivatives of 1-alk-1′-enylglycerols occur in rather small proportions in the neutral lipid fraction, but are at relatively high levels in the ionic lipids. (See Chapter X.) The role of alkoxylipids in tumors and the reported inhibition of neoplastic growth by 1-alkylglycerols remains a field wide open to further exploration.

G. Neuromuscular Activities

Several 1-alkylglycerols and 1-arylglycerols, when given subcutaneously or intraperitoneally, were found to produce transient muscular relaxation and paralysis in mice, probably due to a depressant action on their central nervous system (Berger and Bradley, 1946, 1948). Many of the compounds studied reduced blood pressure and heart rate and also caused vasodilatation (Berger, 1948). One aryl ether of glycerol, 1-guajacylglycerol, has become a widely used medicinal remedy and has been incorporated into several pharmacopoeas.

VI. Possible Functions of Naturally Occurring Alkoxylipids

In surveying the literature it becomes very evident that, although much work has been done on the alkylglycerols, the biological activities and functions of their naturally occurring derivatives, namely the alkyldiacylglycerols and the various alkylacyl phospholipids, have hardly been studied. Malins and Barone (1970) as well as Lewis (1970) showed that 1-alkyl-2,3-diacylglycerols and triacylglycerols have different specific gravities. Malins and Barone (1970) suggested that a regulatory mechanism involving the selective metabolism of these two types of lipids is used by sharks in the maintenance of neutral buoyancy during vertical migration. This regulatory mechanism may serve as a substitute for the commonly found gas-filled swim bladder. (See Chapter XI.)

Little is known about the biological effects of neutral and ionic lipids containing alk-1-enyl moieties. Shah and Schulman (1965) demonstrated that the surface potential of alk-1-enylacylcholine phosphoglycerides is lower than that of diacylcholine phosphoglycerides. The authors attributed this to the presence of an additional induced dipole in the double bond of the alk-1-enyl moiety. Gottfried and Rapport (1963) compared the hemolytic activities of choline lysophosphoglycerides having an alkyl, alk-1-enyl, or acyl moiety. They did not find appreciable differences in the activities of these substances, but it should be noted that according to Safanda and Holocek (1965), chimyl alcohol markedly inhibits hemolysis by choline lysophosphoglycerides. Robertson and Lands (1962) injected emulsions of alk-1-enylacylcholine phosphoglycerides intravenously into rabbits and observed that 90% of this material disappeared from the bloodstream within 5 hours. Of great interest is a recent publication by Roots and Johnston (1968), who determined the plasmalogen content in the brain of goldfish maintained at different temperatures. These authors found a significant increase in the content of alk-1-enylacyl phosphoglycerides in goldfish brain with increasing environmental temperature. According to Roots and Johnston, this change is the most striking modification of brain lipids in response to changes in temperature so far recorded. Thiele *et al.* (1960) found that, in muscle, the concentration of plasmalogens decreased during exercise, despite an increase in the total phosphoglyceride content. Vogt (1949, 1957) reported the isolation of an acidic lipid fraction, from equine intestine, that acted as smooth muscle stimulant. Quite recently, Wiley *et al.* (1970) pointed out that this fraction was most likely a mixture of compounds derived from alk-1-enylacyl phospholipids by treatment with alkali. They thought that the major constituents of this mixture were acetals of long-chain aldehydes with glycerol phosphate.

Stowe (1960) reported that selachyl alcohol promotes the growth of pea stem sections. This finding is of particular interest as very little is known about the occurrence of alkoxylipids in plant tissues. (See Chapter XVI.)

In addition to the rather well known derivatives of the alkyl and alk-1-enylglycerols occurring in nature, several "new" alkoxylipids have been found in recent years. The following paragraph lists some of these substances and the sources from which they were isolated. In a few instances the structures of the new compounds have not been fully elucidated.

Methoxy-substituted 1-alkylglycerols were isolated from shark livers (Hallgren and Ställberg, 1967), dialkylpentanediols from the jaw oil of a porpoise (*Phocoena phocoena*) (Varanasi and Malins, 1969), alkylacylglycerol galactosides from bovine brain (Norton and Brotz, 1963), and an alkoxy inositol phosphoglyceride (Klenk and Hendricks, 1961) from human brain. "New" lipids containing alk-1-enyl moieties include 1-alk-1′-enyl-2,3-diacylglycerols from human perinephric fat (Schmid and Mangold, 1966) and from the liver of the ratfish (*Hydrolagus colliei*) (Schmid *et al.*, 1967), alk-1-enylacyl ethanediols from rat liver, cod liver, mutton fat and other sources (Bergelson *et al.*, 1966), and "sphingoplasmalogens" from bovine brain (Kochetkov *et al.*, 1964). The function of these substances in living organisms and their effects on biological systems are totally unknown.

The alkyl and alk-1-enyl moieties in neutral and ionic alkoxylipids are, as a rule, exclusively saturated and monounsaturated, but polyunsaturated alk-1-enyl moieties were found in choline phospholipids of a human brain meningioma (Bell *et al.*, 1967). It is of interest, in this connection, that saturated and monounsaturated acids and alcohols are interconverted in the gastrointestinal tract (Stetten and Schoenheimer, 1940; Bandi and Mangold, 1971), whereas polyunsaturated acids are not readily reduced (Bandi and Mangold, 1971). Thus, it appears that polyunsaturated alkyl and alk-1-enyl moieties are not normally found in the alkoxylipids of human and animal organs because precursors, i.e., polyunsaturated alcohols, are not produced in these tissues (Bandi *et al.*, 1971b). Recently, it was shown that alkoxylipids containing polyunsaturated alkyl and alk-1-enyl moieties can be synthesized, at least in the rat, from dietary polyunsaturated alcohols (Bandi *et al.*, 1971a,b).

Recently, alkoxylipids have been identified in membranes, such as the synaptic plasma membranes from rat brain (Cotman *et al.*, 1969). It would be most interesting to learn how the changes in composition reported by Bandi *et al.* (1971a,b) affect the functions of membranes in various tissues.

VII. Conclusions

It can be seen from the foregoing discussion that knowledge of the biological effects of the various alkoxylipids is slim. This is in large measure due to the fact that such compounds had not been available in pure form. There is little doubt that in the future the functions and the biological effects of derivatives of the alkyl- and alk-1-enylglycerols will be studied extensively, as pure substances are now available. Obviously, particular attention will be given to the exploration of the role the alkoxylipids may play in central and peripheral nerve tissues.

REFERENCES

Abaturova, E. A., and Shubina, A. V. (1964). *Byull. Eksp. Biol. Med.* **57,** 81.

Agduhr, E., Blix, G., and Vahlquist, B. (1934). *Upsala Laekarefoeren. Foerh.* **40,** 183.

Alexander, P., Connell, D. I., Brohult, A., and Brohult, S. (1959). *Gerontologia* **3,** 147.

Anatol, J., Berecoechea, J., and Giraud, D. (1964). *C. R. Acad. Sci.* **258,** 6466.

Arturson, G., and Lindbäck, M. (1951). *Acta Soc. Med. Upsal.* **56,** 19.

Ashikawa, J. K. (1961). *U.S. At. Energy Comm., Rep.* UCRL-9592, 1–147.

Astra Nutrition AB (1965). Neth. Pat. Appl. 6,409,944.

Bandi, Z. L., and Mangold, H. K. (1971). *FEBS Lett.* **13,** 198.

Bandi, Z. L., Mangold, H. K., Hølmer, G., and Aaes-Jørgensen, E. (1971a). *FEBS Lett.* **12,** 217.

Bandi, Z. L., Aaes-Jørgensen, E., and Mangold, H. K. (1971b). *Biochim. Biophys. Acta* **239,** 357.

Barry, V. C., O'Rourke, L., and Twomey, D. (1947). *Nature (London)* **160,** 800.

Bassi, P., and Dunjic, A. (1962). *Rev. Fr. Etud. Clin. Biol.* **7,** 187.

Baumann, W. J., and Mangold, H. K. (1966). *J. Org. Chem.* **31,** 498.

Bell, O. E., Jr., Cain, C. E., Sulya, L. L., and White, H. B., Jr. (1967). *Biochim. Biophys. Acta* **144,** 481.

Bergelson, L. D., Vaver, V. A., Prokazova, N. V., Ushakov, A. N., and Popkova, G. A. (1966). *Biochim. Biophys. Acta* **116,** 511.

Berger, F. M. (1948). *J. Pharmacol. Exp. Ther.* **93,** 470.

Berger, F. M., and Bradley, W. (1946). *Brit. J. Pharmacol. Chem. Ther.* **1,** 265.

Berger, F. M., and Bradley, W. (1948). *Lancet* **1,** 367.

Bergström, S., and Blomstrand, R. (1956). *Acta Physiol. Scand.* **38,** 166.

Blomstrand, R. (1959). *Proc. Soc. Exp. Biol. Med.* **102,** 662.

Blomstrand, R., and Ahrens, E. H., Jr. (1959). *Proc. Soc. Exp. Biol. Med.* **100,** 802.

Bodman, J., and Maisin, J. H. (1958). *Clin. Chim. Acta* **3,** 253.

Borgström, B. (1965). *Biochim. Biophys. Acta* **106,** 171.

Borgström, B. (1968). *Proc. Soc. Exp. Biol. Med.* **127,** 1120.

Brohult, A. (1957). *In* "Advances in Radiobiology" (G. C. de Hevesy, A. G. Forssberg, and J. D. Abbatt, eds.), pp. 241–247. Oliver & Boyd, Edinburgh.

Brohult, A. (1958). *Nature (London)* **181,** 1484.

Brohult, A. (1960). *Nature (London)* **188,** 591.

Brohult, A. (1962). *Nature (London)* **193,** 1304.

Brohult, A. (1963). *Acta Radiol., Suppl.* **223,** 7.

Brohult, A., and Holmberg, J. (1954). *Nature (London)* **174,** 1102.

Brohult, A., Brohult, J., and Brohult, S. (1970). *Acta Chem. Scand.* **24,** 730.

Carlson, W. E. (1966). M.S. Thesis, University of British Columbia, Vancouver, Canada.

Carlson, W. E., and Bayley, H. S. (1970). *Fed. Proc., Fed. Amer. Soc. Exp. Biol.* **29,** 300.

Chalmers, W., Wood, A. C., Shaw, A. J., and Majnarich, J. J. (1966). U. S. Pat. 3,294,639.

Chebotarev, E. E. (1965). *Fiziol. Zh. Akad. Nauk. Ukr. RSR* **11,** 385.

Cotman, C., Blank, M. L., Moehl, A., and Snyder, F. (1969). *Biochemistry* **8,** 4606.

Dalton, R. G. (1964). *Vet. Rec.* **76,** 411.

De Gaetani, G. F., and Baiotti, G. (1959). *Boll. Soc. Ital. Biol. Sper.* **35,** 1156.

Delmon, G., and Biraben, J. (1966). *C. R. Soc. Biol.* **160,** 76.

Dudin, V. N. (1961). *Med. Radiol.* **6,** 82.

Edlund, T. (1954). *Nature* (*London*) **174,** 1102.

Emmerie, A., and Engel, C. (1962). *Fette, Seifen, Anstrichm.* **64,** 813.

Emmerie, A., Engel, C., and Klip, W. (1952). *J. Sci. Food Agr.* **3,** 264.

Evans, I. A., Thomas, A. J., Evans, W. C., and Edwards, C. M. (1953). *Brit. Vet. J.* **114,** 253.

Evans, W. C., Evans, I. A., Edwards, C. M., and Thomas, A. J. (1957). *Biochem. J.* **65,** 6P.

Evenstein, D., Gordon, A. S., and Eisler, M. (1958). *Anat. Rec.* **132,** 435.

Funasaki, H., and Gilbertson, J. R. (1968). *J. Lipid Res.* **9,** 766.

Gallo, L., Vahouny, G. V., and Treadwell, C. R. (1968). *Proc. Soc. Exp. Biol. Med.* **127,** 156.

Ghys, R. (1962). *Laval Med.* **30,** 331.

Giffin, H. Z., and Watkins, C. H. (1930). *J. Amer. Med. Ass.* **95,** 587.

Gilbertson, J. R., Garlich, H. H., and Gelman, R. A. (1970). *J. Lipid Res.* **11,** 201.

Gottfried, E. L., and Rapport, M. M. (1963). *J. Lipid Res.* **4,** 57.

Greten, H., Levy, R. I., Fales, H., and Fredrickson, D. S. (1970). *Biochim. Biophys. Acta* **210,** 39.

Hallgren, B., and Ställberg, G. (1967). *Acta Chem. Scand.* **21,** 1519.

Hasegawa, Y., Wakui, N., Shimada, A., Nakamura, S., and Kunii, K. (1961). *Nippon Rinsho* **19,** 1793.

Heller, J. H., Pasternak, V. Z., Ransom, J. P., and Heller, M. S. (1963). *Nature* (*London*) **199,** 904.

Hietbrink, B. A., Raymund, A. B., and Ryan, B. A. (1962). U.S. Air Force Radiat. Lab. Quart. Progr. Rep. AEC Rep. NP–11660, pp. 96–103. University of Chicago, Chicago.

Hofmann, A. F. (1963). *Biochim. Biophys. Acta* **70,** 306.

Hofmann, A. F. (1969). Private communication.

Hofmann, A. F., and Borgström, B. (1963). *Biochim. Biophys. Acta* **70,** 317.

Holmes, H. N., Corbet, R. E., Geiger, W. B., Kornblum, N., and Alexander, W. (1941). *J. Amer. Chem. Soc.* **63,** 2607.

Kaneda, T., and Ishii, S. (1952). *Bull. Jap. Soc. Sci. Fish.* **18,** 39.

Kaneda, T., Sakai, H., Ishii, S., and Arai, K. (1955). *Bull. Tokai Reg. Fish. Res. Lab.* **30,** 41.

Karlson, A. G. (1969). Private communication.

Kaufmann, H. P., Vennekel, E., and Hamza, Y. (1970). *Fette, Seifen, Anstrichm.* **72,** 242.

Kern, F., Jr., and Borgström, B. (1965). *Biochim. Biophys. Acta* **98,** 520.

Klenk, E., and Hendricks, U. W. (1961). *Biochim. Biophys. Acta* **50,** 602.

Kochetkov, N. K., Zhukova, I. G., and Glukhoded, I. S. (1964). *Biokhimiya* **29,** 570.

Lewis, R. W. (1966). *Comp. Biochem. Physiol.* **19,** 363.

Lewis, R. W. (1970). *Lipids* **5,** 151.

Linman, J. W. (1958). *J. Clin. Invest.* **37,** 913.
Linman, J. W. (1960). *Proc. Soc. Exp. Biol. Med.* **104,** 703.
Linman, J. W., and Bethell, F. H. (1956). *Blood* **11,** 310.
Linman, J. W., and Bethell, F. H. (1961). *Haemopoiesis: Cell Prod. Regul., Ciba Found. Symp., 1960* pp. 369–396.
Linman, J. W., Bethell, F. H., and Long, M. J. (1958). *J. Lab. Clin. Med.* **52,** 596.
Linman, J. W., Korst, D. R., and Bethell, F. H. (1959a). *Ann. N. Y. Acad. Sci.* **77,** 638.
Linman, J. W., Long, M. J., Korst, D. R., and Bethell, F. H. (1959b). *J. Lab. Clin. Med.* **54,** 335.
Maisin, J., Keusters, J., Guidetto, H., and Lambert, G. (1959). *J. Radiol., Electrol., Med. Nucl.* **40,** 454.
Malins, D. C., and Barone, A. (1970). *Science* **167,** 79.
Mangold, H. K. (1961). *J. Amer. Oil Chem. Soc.* **38,** 708.
Mangold, H. K., Spener, F., and Baumann, W. J. (1972). *Chem. Phys. Lipids* (in press).
Maqsood, M., and Ashikawa, J. K. (1961). *Int. J. Radiat. Biol.* **4,** 521.
Marberg, C. M., and Wiles, H. O. (1938). *AMA Arch. Intern. Med.* **61,** 408.
Mizuno, N. S., Perman, V., Joel, D. D., Bates, F. W., Sautter, J. H., and Schultze, M. O. (1960). *Proc. Soc. Exp. Biol. Med.* **105,** 317.
Morgan, R. G. H., and Hofmann, A. F. (1970a). *J. Lipid Res.* **11,** 223.
Morgan, R. G. H., and Hofmann, A. F. (1970b). *J. Lipid Res.* **11,** 231.
Mozharova, E. N., Rusanov, A. M., and Komarova, R. S. (1961). *Med. Radiol.* **6,** 13.
Norton, W. T., and Brotz, M. (1963). *Biochem. Biophys. Res. Commun.* **12,** 198.
Osmond, D. G., Roylance, P. J., Webb, A. J., and Joffey, J. M. (1963). *Acta Haematol.* **29,** 180.
Paltauf, F. (1968). *Monatsh. Chem.* **99,** 1277.
Paltauf, F. (1969). *Biochim. Biophys. Acta* **176,** 818.
Paltauf, F. (1971). *Biochim. Biophys. Acta* **239,** 38.
Peifer, J. J., Lundberg, W. O., Ishio, S., and Warmanen, E. (1965). *Arch. Biochem. Biophys.* **110,** 270.
Penny, R. H. C., Wright, A. J., and Stoker, J. W. (1964). *Brit. Vet. J.* **120,** 286.
Popović, M. (1965). *Hoppe-Seyler's Z. Physiol. Chem.* **340,** 18.
Prokhonchukov, A. A., and Panikarovskii, V. V. (1963). *Teor. Prakt. Stomatol.* **6,** 61.
Robertson, A. F., and Lands, W. E. M. (1962). *J. Clin. Invest.* **41,** 2160.
Roots, B. I., and Johnston, P. V. (1968). *Comp. Biochem. Physiol.* **26,** 553.
Rusanov, A. M., Mozharova, E. N., and Komarova, R. S. (1962). *Med. Radiol.* **7,** 42.
Safanda, J., and Holocek, V. (1965). *Folia Haematol. (Leipzig)* **83,** 171.
Sandler, O. E. (1949). *Acta Med. Scand.* **133,** Suppl. 225, 72.
Schmid, H. H. O., and Mangold, H. K. (1966). *Biochem. Z.* **346,** 13.
Schmid, H. H. O., Baumann, W. J., and Mangold, H. K. (1967). *Biochim. Biophys. Acta* **144,** 344.
Schultze, M. O., Perman, V., Bates, F. W., and Sautter, J. H. (1958). *Proc. Soc. Exp. Biol. Med.* **98,** 470.
Shah, D. O., and Schulman, J. H. (1965). *J. Lipid Res.* **6,** 341.
Sherr, S. I., and Treadwell, C. R. (1965). *Biochim. Biophys. Acta* **98,** 539.
Sherr, S. I., Swell, L., and Treadwell, C. R. (1963). *Biochem. Biophys. Res. Commun.* **13,** 131.
Slotboom, A. J., de Haas, G. H., Bonsen, P. P. M., Burbach-Westerhuis, G. J., and van Deenen, L. L. M. (1970). *Chem. Phys. Lipids* **4,** 15.
Snyder, F. (1969). *Progr. Chem. Fats Other Lipids* **10,** 287–335.

Snyder, F., and Cress, E. A. (1963). *Radiat. Res.* **19,** 129.
Snyder, F., and Wood, R. (1968). *Cancer Res.* **28,** 972.
Snyder, F., and Wood, R. (1969). *Cancer Res.* **29,** 251.
Snyder, F., Piantadosi, C., and Malone, B. (1970). *Biochim. Biophys. Acta* **202,** 244.
Snyder, F., Cress, E. A., Arrington, J. H., Schmid, H. H. O., and Mangold, H. K. (1971). Unpublished data.
Spener, F., Paltauf, F., and Holasek, A. (1968). *Biochim. Biophys. Acta* **152,** 368.
Stansby, M. E., Zollman, P. E., and Winkelmann, R. K. (1967). *Fish. Ind. Res.* **3,** 25.
Stetten, D., Jr., and Schoenheimer, R. (1940). *J. Biol. Chem.* **133,** 347.
Stowe, B. B. (1960). *Plant Physiol.* **35,** 262.
Suki, W. N., and Grollman, A. (1960). *Tex. Rep. Biol. Med.* **18,** 662.
Sviridov, N. K., Abaturova, A. V., Shubina, A. V., and Elpatevskaya, G. N. (1964). *Moscow Med. Sb.* p. 254.
Swell, L., Law, M. D., and Treadwell, C. R. (1965). *Arch. Biochem. Biophys.* **110,** 231.
Talbot, T. R., and Elson, L. A. (1958). *Nature* (*London*) **181,** 684.
Thiele, O. W., Schröder, H., and von Berg, W. (1960). *Hoppe-Seyler's Z. Physiol. Chem.* **322,** 147.
Thompson, G. A., Jr., and Hanahan, D. J. (1963). *Biochemistry* **2,** 641.
Tran Ba Loc, G. M., and Bernard, J. (1958). *Rev. Fr. Clin. Biol.* **3,** 461.
Tuna, N., and Mangold, H. K. (1963). *In* "Evolution of the Atherosclerotic Plaque" (R. D. Jones, ed.), pp. 85–108. Univ. of Chicago Press, Chicago.
Varanasi, U., and Malins, D. C. (1969). *Science* **166,** 1158.
Vogt, W. (1949). *Naunyn-Schmiedebergs Arch. Exp. Pathol. Parmakol.* **206,** 1.
Vogt, W. (1957). *J. Physiol.* (*London*) **137,** 154.
Wiley, R. A., Sumner, D. D., and Walaszek, E. J. (1970). *Lipids* **5,** 803.

CHAPTER IX

CONTENT, COMPOSITION, AND METABOLISM OF MAMMALIAN AND AVIAN LIPIDS THAT CONTAIN ETHER GROUPS

Lloyd A. Horrocks

I. Introduction

Recent improvements in methodology (see Chapter II) have led to many advances in our knowledge of the metabolism, composition, and content of alk-1-enyl and alkyl groups in various organs and tissues. I will attempt to evaluate the information available on the ether-linked lipids in birds and mammals. For reviews of the historical development of this area, the reader is directed to Debuch (Chapter I), Hanahan and Thompson (1963), Rapport and Norton (1962), Klenk and Debuch (1963), Thiele (1964), Ansell and Hawthorne (1964), and Hartree (1964). These reviews are concerned primarily with plasmalogens, except for the review of alkyl groups by Hanahan and Thompson. Ansell and Hawthorne (1964) and Rossiter (1966, 1967) have described the metabolism and Eichberg *et al.* (1969) have reviewed the composition of brain phospholipids. The metabolism of alk-1-enyl and alkyl groups has been reviewed by Goldfine (1968), Snyder (1969), Hill and Lands (1970), and Piantadosi and Snyder (1970).

The naturally occurring lipids with alk-1-enyl and alkyl groups are found primarily in membrane phospholipids. Apart from the general structural function of all membrane phospholipids, no specific function has been assigned to the plasmalogens. Terner and Hayes (1961) speculated that plasmalogens may have a role in ion transport systems, since vinyl radicals are known to have an affinity for some naturally occurring trace metal ions, and because of the relatively high concentrations of plasmalogens in brain, muscle, and kidney. They also related the reactivity of mercuric ions toward alk-1-enyl groups to the injury these ions cause in heart muscle and kidney.

The presence of alk-1-enyl or alkyl groups in membrane phospholipids should logically affect the properties of the membrane because these lipids have a dipole moment different from that of the corresponding diacyl compound. Shah and Schulman (1965) found that the surface potential of a monolayer of 1-alk-1′-enyl-2-acyl-GPC was 25–30% lower than the surface potential of 1,2-diacyl-GPC. Similar results were obtained by Colacicco and Rapport (1966) but the surface potential of the 1-alk-1′-enyl-2-acyl-GPC was not reported. The surface area of the 1-alk-1′-enyl-2-acyl-GPC was 80 Å^2 per molecule at 10 dynes/cm as compared with values of 90 and 103 Å^2 for egg and yeast diacyl-GPC at the same pressure (Shah and Schulman, 1965). These differences in monolayer properties suggest that 1,2-diacyl, 1-alk-1′-enyl-2-acyl, and 1-alkyl-2-acyl phosphoglycerides differ from one another in their interactions with membrane proteins (Cornwell and Horrocks, 1964).

Abbreviations

br	Branched
CDPCh	Cytidine diphosphate choline

CDPEa	Cytidine diphosphate ethanolamine
CPG	Choline phosphoglycerides
CTP	Cytidine triphosphate
DMA	Dimethyl acetal
EPG	Ethanolamine phosphoglycerides
GLC	Gas–liquid chromatography
GP	*sn*-Glycero-3-phosphate(s)
GPC	*sn*-Glycero-3-phosphorylcholine(s)
GPE	*sn*-Glycero-3-phosphorylethanolamine(s)
GPI	*sn*-Glycero-3-phosphorylinositol(s)
GPS	*sn*-Glycero-3-phosphorylserine(s)
IPG	Inositol phosphoglycerides
ND	Not detected
PL	Phospholipids
SPG	Serine phosphoglycerides
SRA	Specific radioactivity (or radioactivities)
TLC	Thin-layer chromatography
tr	Trace

II. Content

A. Histochemistry

The distribution of alk-1-enyl groups in various organs and tissues can be determined from unfixed cryostat sections of tissues after staining with Schiff's reagent, which reacts with the aldehydogenic moiety (Hayes, 1947; Terner and Hayes, 1961). Free long-chain aldehydes have not been detected histochemically in cardiac tissue or in any other tissue that has been examined (Ferrans *et al.*, 1962; Flygare *et al.*, 1970). Hayes (1947) and Broderson (1967) have made histochemical surveys of most tissues of the rat. The most intense staining was found in myelinated tracts and nerves of the central and peripheral nervous systems. The adrenal medulla gave a strong reaction. In the male rat adrenal cortex, only the zona glomerulosa was well stained, but in the guinea pig adrenal cortex the staining was quite uniform. In the kidney, the strongest reaction was seen in the convoluted tubules and in the corticomedullary zone (Hayes, 1947; Helmy and Hack, 1970). Helmy and Hack (1963) described an intense staining of epithelial cells in the human choroid plexus. Amniotic epithelial cells of the human placenta (Helmy and Hack, 1964) and striated ducts of guinea pig submandibular glands (Hohenwald and Braedel, 1967) also stain intensely. Ferrans *et al.* (1962) examined the heart muscle from a number of vertebrates and found plasmalogens in the sarcoplasm, sarcosomes, and intercalated disks, but not in the myofibrils. The staining was strongest in cardiac tissue from higher mammals, somewhat less in that from the rat, and lowest in tissue from lower vertebrates. In the latter, cardiac muscle and skeletal muscle were quite similar.

B. Less Polar Lipids

Since free long-chain alcohols and aldehydes have been reported to be involved in reactions leading to the biosynthesis of alkyl and alk-1-enyl groups, it became important to establish the content of alcohols and aldehydes in tissues. Free aldehyde concentrations (Table I) are difficult to measure because the alk-1-enyl ether bond is labile under acidic conditions and a plasmalogenase (Ansell and Spanner, 1965a) is known to exist in brain. Schmid *et al.* (1967a) found that the amount of free aldehyde in lipid extracts increases during storage at −40°C. Schmid and Takahashi (1970) found no evidence for any free long-chain aldehydes in rat brain although a sensitive method was used. Free aldehydes were undetectable in mouse brain lipid extracts examined by thin-layer chromatography (TLC) and tested with Schiff's reagent (Horrocks, 1972), a method which detects free aldehydes at a concentration of 0.02 μmole/g tissue. Gilbertson *et al.* (1967) stated that the free aldehydes from the rat had a specific radioactivity that was more than 30 times that of the alk-1-enyl groups after four daily intraperitoneal injections of ^{14}C-acetic acid. This finding suggests that the small amounts of free aldehydes did not originate from a plasmalogen pool during processing. The compositions of the aldehydes reported by Gilbertson *et al.* (1967) are rather unusual—no 18:1 aldehydes in canine and bovine heart and more than 20% of an unidentified aldehyde in canine heart. Until further studies have been done we must conclude that free long-chain aldehydes are not present in detectable concentrations in the brain but are present in small amounts in cardiac and skeletal muscle.

The presence of small amounts of free long-chain alcohols in brains, hearts, and preputial glands has been definitely established. Takahashi and Schmid (1970) found 2 μg of long-chain alcohol per gram of pig brain and similar amounts have been found in ox hearts and brains and mouse preputial glands (Blank and Snyder, 1970), ox hearts and brains and pig hearts (Takahashi and Schmid, 1970), and rat brains (Schmid and Takahashi, 1970). Spener *et al.* (1969) found large amounts of alkyl acetates in male mouse preputial glands. Schmid and Takahashi (1970) found that the free long-chain alcohols in the rat brain had a very high specific radioactivity a short time after an intracerebral injection of 1-^{14}C-palmitic

TABLE I

The Content of Free Long-Chain Aldehydes in Mammalian Tissues

Species, tissue	Free long-chain aldehyde µmoles/g tissue	Free long-chain aldehyde µmoles/g lipid	Total long-chain aldehyde % free	References
Man, heart, age 46	—	19.1	11.9	Radloff and Ferrell, 1970
Man, heart, age 57	—	33.7	12.0	Radloff and Ferrell, 1970
Man, heart, age 72	—	91.4	32.8	Radloff and Ferrell, 1970
Dog, heart	0.078	—	1.4	Gilbertson *et al.*, 1967
Dog, heart	0.31	—	4.3	Gilbertson *et al.*, 1967
Ox, heart	0.057	—	2.6	Gilbertson *et al.*, 1967
Rat, heart	0.11	—	3.5	Gilbertson *et al.*, 1967
Mouse, heart	—	0.49	3.7	Ferrell *et al.*, 1969
Mouse, liver	—	0.98	1.3	Ferrell *et al.*, 1969
Mouse, kidney	—	0.52	1.5	Ferrell *et al.*, 1969
Mouse, brain	—	53.8	38.1	Ferrell *et al.*, 1969
Mouse, skeletal muscle	—	0.39	2.7	Ferrell *et al.*, 1969

acid. The alkyl composition of the free alcohols from ox and pig heart and brain is quite similar to that of the alkylacyl-GPE, except for the presence of up to 19% of 20:0 in the free alcohols (Takahashi and Schmid, 1970).

In terms of gross tissue weight, adipose tissue is a rich source of alk-1-enyldiacylglycerols and alkyldiacylglycerols, although comparisons of the concentrations are difficult. These compounds were isolated and characterized by Schmid and Mangold (1966a) from human perinephric adipose tissue and by Schmid *et al.* (1967b) from human subcutaneous adipose tissue (3 mg of each compound per gram lipid), but these workers did not find them in the adipose tissue from a newborn. Todd and Rizzi (1964) measured the diol content of the nonsaponifiable fraction of tissue neutral lipids by a technique which measures a mixture of alkyl- and alk-1-enylglycerols with an unknown degree of esterification. For rabbit adipose tissue, they reported a value of 0.6 mg/g lipid (about 1.7 µmole/g tissue). Schogt *et al.* (1960) found aldehydes (0.06 mg/g lipid) in beef tallow. According to Gilbertson and Karnovsky (1963), rat adipose tissue contains 2.1 and 9.5 µmoles/g tissue of alkylglycerols and alk-1-enylglycerols, respectively. However, later Gilbertson (1969) found only 0.41 µmole/g tissue of alkyl- and alk-1-enylglycerols, mostly in the form of the diacyl derivatives. Wood and Snyder (1968) found about 4 mg/g neutral lipid of alkylglycerols but only traces of alk-1-enylglycerols in rat perirenal adipose tissue. These results for neutral lipids should be multiplied by 2.3 if concentrations of the diacyl derivatives are desired, nevertheless the overall agreement of the quantitative values for adipose tissue is quite poor. Brown adipose tissue, with 0.7 and 0.9 μmole/g tissue of alkylglycerols and alk-1-enylglycerols (Gilbertson and Karnovsky, 1963), seems to have a much lower content of these compounds than does white adipose tissue.

In mammals, bone marrow is a richer source of ether-linked lipids than adipose tissue. Bovine yellow bone marrow contains 9.7 and 15.4 μmoles/g tissue of alkylglycerols and alk-1-enylglycerols according to Gilbertson and Karnovsky (1963), 2–10 μmoles/g tissue according to Todd and Rizzi (1964), but only 0.1 mg/g lipid according to Hallgren and Larsson (1962). The presence of alkyldiacylglycerols in this tissue was detected by Thompson and Hanahan (1963). The yellow bone marrow of the pig, sheep, and rabbit (Todd and Rizzi, 1964) contains similar amounts of alkylglycerols. Thompson (cited by Thompson and Hanahan, 1963) detected alkyldiacylglycerols in rat femoral marrow. Wood and Snyder (1968) found 1 mg alk-1-enylglycerols and 4 mg alkylglycerols/g neutral lipid in rat femoral marrow. These compounds were not detected in rat marrow by Todd and Rizzi (1964). Human red bone marrow contains about 2 mg of alkylglycerols/g lipid according to Hallgren and Larsson (1962) and Todd and Rizzi (1964).

Heart neutral lipids have been used as a source of small amounts of alkyl- and alk-1-enylglycerols (Schmid and Takahashi, 1968; Gilbertson, 1969; Wood and Snyder, 1968; Schmid *et al.*, 1967a; Gilbertson and Karnovsky, 1963; Todd and Rizzi, 1964). Part or all of these compounds could have been present in the pericardial adipose tissue. Wood and Snyder (1968) found small amounts of ether lipids in skeletal muscle. *S*-Alkylglycerols were recently detected in bovine heart neutral lipids by Ferrell and Radloff (1970).

The intestines of the rabbit contain moderate amounts of alkylglycerols (Todd and Rizzi, 1964), and its liver smaller amounts. In rat liver, Snyder *et al.* (1969a) found no alkyldiacylglycerols or alk-1-enyldiacylglycerols, but small amounts of alkylglycerols and alk-1-enylglycerols were found after hydrogenolysis (Wood and Snyder, 1968). Both lipid classes have been detected in human β-lipoproteins and chylomicrons (Gilbertson and Karnovsky, 1963) and in rat plasma (Wood and Snyder, 1968). Human aorta contains 0.8 μmole of alkylglycerols/g tissue (Miller *et al.*, 1964) and Schmid *et al.* (1967b) detected alkyldiacylglycerols but not alk-1-enyldiacylglycerols, in aortic plaques.

Rabbit kidneys contain small amounts of alkylglycerols (Todd and Rizzi,

1964) and after hydrogenolysis, alk-1-enylglycerols and alkylglycerols can be found in rat kidney, brain, and spleen (Wood and Snyder, 1968). Alkylglycerols were detected in human spleen by Hallgren and Larsson (1962). Both alk-1-enylglycerols and alkylglycerols were measured in ox brain and guinea pig polymorphonuclear leukocytes (Gilbertson and Karnovsky, 1963). Alk-1-enyldiacylglycerols were detected in rat and mouse placenta by Helmy and Hack (1967) and some evidence for the presence of free alkylglycerols in human amniotic fluid has been presented (Helmy and Hack, 1962).

Human and bovine milk contain alkylglycerols according to Hallgren and Larsson (1962) and Schogt *et al.* (1960). Schogt *et al.* (1960) also found alk-1-enyl groups in bovine milk. Yolks of chicken eggs probably contain some alk-1-enyldiacylglycerols (Prostenik and Popovic, 1963) but alkylglycerols were not detected in the neutral lipids (Hallgren and Larsson, 1962).

The preputial glands of the mouse and rat are rudimentary sebaceous glands that contain a large amount of alk-1-enyldiacylglycerols (Helmy and Hack, 1967). Snyder and Blank (1969) have studied the lipid composition of the gland in young adult male mice and found about 80 mg alk-1-enyldiacylglycerols, 60 mg alkyldiacylglycerols, 7 mg alk-1-enylacylglycerols, 3 mg alkylacylglycerols, and 480 mg of acylalkanols (waxes) in each gram of total lipid.

Alkylcholesterols in ox heart (Funasaki and Gilbertson, 1968) and alk-1-enylcholesterols in ox and pig heart (Gilbertson *et al.*, 1970) exist in trace quantities (2–6 μmoles/g tissue). Bergelson *et al.* (1966) have found traces of diols with two to five carbon atoms in the neutral plasmalogens from rat liver, mutton fat, and egg yolks.

C. Polar Lipids

1. *General Comments*

The differential hydrolysis method of Dawson *et al.* (1962) has been widely used to determine the content of alk-1-enyl groups in phospholipids, but it has been found that the acyl groups are not completely removed from the plasmalogens during mild alkaline hydrolysis (Ansell and Spanner, 1963b; Dawson, 1967). As a result, the unmodified method gives low values. To complicate matters, Wells and Dittmer (1966) have stated that during the subsequent mild acid hydrolysis used in Dawson's method part of the lysoplasmalogens are converted to cyclic acetals and are found in the acid-stable fraction. The effect of this artifact production is to increase the apparent content of the alkylacyl-GPE. Only a trace of cyclic acetals could be found in the acid-stable fraction from rat brain EPG by Horrocks and Ansell (1967a), who pointed out that the exact conditions for the mild acid hydrolysis are quite important.

The isolation of *p*-nitrophenylhydrazone derivatives (Wittenberg *et al.*, 1956) has also been widely used for the measurement of alk-1-enyl group contents and concentrations. Rapport and Norton (1962) concluded that both this method and the more widely used iodine addition method gave results that were only 92% of the true value, primarily because "pure" preparations of alk-1-enylacyl-GPC from bovine heart gave an alk-1-enyl group-to-lipid phosphorus molar ratio of 0.92. In my opinion, this value would be expected because of the small amounts of alkylacyl-GPC in this tissue. In this review the original uncorrected values for alk-1-enyl group contents and concentrations are reported.

Phospholipid compositions are often determined after separation of the lipid classes by two-dimensional TLC. If the alk-1-enyl groups are cleaved from the plasmalogens before development in the second dimension (Owens, 1966; Schmid and Mangold, 1966b; Horrocks, 1968b), the amount of plasmalogen in each phospholipid class can be determined at the same time as the phospholipid composition. The plasmalogen contents are in excellent agreement with those determined by iodine addition (Horrocks, 1968b). The assay of the alkylacyl types of phosphoglycerides is more difficult. The preferred method in our laboratory is to saponify the acid-stable phosphoglycerides that are obtained from the separation-reaction-separation two-dimensional TLC. The phosphorus contents of the water-soluble and chloroform-soluble products are measures of the diacyl and alkylacyl forms of that class of phosphoglyceride (Horrocks, 1967b; Sun and Horrocks, 1969b; Porcellati *et al.*, 1970b; Horrocks and Sun, 1972).

Wood and Snyder (1968) described another approach for the isolation and measurement of alkyl and alk-1-enyl groups from phosphoglycerides. After hydrogenolysis of the phospholipids with lithium aluminum hydride, the resultant alkylglycerols, alk-1-enylglycerols, and fatty alcohols were separated by thin-layer chromatography and quantitated by densitometry after charring of the spots. Based on the starting amount of phospholipid (Snyder, personal communication), the recoveries of alkyl and alk-1-enyl groups from rat heart and brain were much lower than expected. Cotman *et al.* (1969) reported a higher proportion of alk-1-enyl and alkyl groups in the reaction products from hydrogenolysis of rat brain phospholipids. Quantitative results can be obtained with chloroform-methanol extractions of the lithium aluminum hydride reaction mixture (Horrocks, 1968c) or by the use of $NaAlH_2(OCH_2CH_2OCH_3)_2$, Vitride, as the reducing agent (Snyder *et al.*, 1971b).

Some evidence for mammalian phosphoglycerides with two alkyl groups has been reported recently by Popovic (1965), Viswanathan *et al.* (1968c), and Horrocks and Ansell (1967a), but such compounds or their derivatives have not been purified and characterized. Since I have been unable to detect any material with the same chromatographic properties as authentic 1,2-dialkylglycerols in the products from complete phospholipase C (*Bacillus cereus*) hydrolyses of brain EPG preparations, I have concluded that the reported presence of dialkyl phosphoglycerides in mammals is probably due to the resistance of the 2-acyl group of the alkylacyl-GPE to hydrolysis.

Column chromatography is often used to separate larger amounts of phosphoglyceride classes before the alk-1-enyl and alkyl group contents are measured. Quantitative separations of the EPG and SPG can be accomplished with the use of diethylaminoethyl cellulose columns. Hydrolysis of the plasmalogens can be a problem under some conditions (Gluck *et al.*, 1966; Horrocks and Ansell, 1967a; Sanders, 1967), but high plasmalogen values, for SPG in particular, have been reported by O'Brien and Sampson (1965a,b) for human brain white matter and myelin and by O'Brien *et al.* (1967) for bovine spinal root myelin. The same laboratory has reported much lower SPG plasmalogen values for human brain gray matter (O'Brien and Sampson, 1965a,b), bovine brain gray matter (Yabuuchi and O'Brien, 1968), and bovine optic nerve myelin (MacBrinn and O'Brien, 1969). The abnormally high values for SPG plasmalogens may have some relationship to the "altered EPG" after drying in glass vessels that was described by Rouser and Kritchevsky (cited by Fleischer and Rouser, 1965) or to the two forms of EPG described by Das and Rouser (1967). The results from diethylaminoethyl cellulose column fractions cannot be evaluated properly unless the recovery of plasmalogens has been proven quantitative.

Debuch and Wendt (1967) isolated and characterized a new plasmalogen, alk-1-enylacylglycerophosphoryl-*N*-acetylethanolamine, from bovine and human placenta. The *N*-acetyl ethanolamine phosphoglycerides from bovine brain seem to have the same alk-1-enyl group content and the same alk-1-enyl and acyl group compositions as the EPG. A compound with a similar chromatographic mobility and composition as that described by Debuch and Wendt (1967) accounts for about 10% of the plasmalogens from human brain (McMartin and Horrocks, 1969). Helmy and Hack (1966) had previously reported a similar plasmalogen from cat placenta that they believed to be the acetone imine of the alk-1-enylacyl-GPE.

Most of the available information about the alk-1-enyl and alkyl groups of polar lipids from various tissues in the higher animals is tabulated and discussed on the following pages. In many cases, the tabulated values have been recalculated, and this reviewer would appreciate correspondence on errors and omissions, and apologizes for them. Several kinds of information on the content of alk-1-enyl and alkyl groups are of value. The tables include the total content of alk-1-enyl groups in tissues and their subcellular fractions, and when the nitrogen base is known, the contents of specific phosphoglyceride plasmalogens are given. The rather small amount of corresponding data on the content of phosphoglycerides with alkyl groups appear in separate tables. Another group of tables describes the proportions of the alk-1-enylacyl and alkylacyl forms in specific phosphoglyceride classes. For the best evaluation of the data, an inspection of each of the tables pertaining to a particular tissue is necessary. A critical commentary on the tabulated data appears in the accompanying text.

2. *Nervous System*

In the central nervous system, the most concentrated source of plasmalogens is the myelin sheath (Table V). Nearly all the reliable values lie in the range of 0.31–0.37 for the mole ratio of alk-1-enyl groups to lipid phosphorus. Lower values may be due to impurities in the myelin or to autolysis. High values reflect methodological differences. Values for white matter, which includes a large amount of myelin, are generally close to 0.30. In contrast, the median value for gray matter is about 0.17. In whole brain or in any morphological part of the brain, the mole ratio for alk-1-enyl groups to lipid-P can be considered a reflection of the relative proportions of white and gray matter. This is evident in the limited data on anatomical areas of the human brain in Table IV, and from the increase in plasmalogen content during myelination of the rat brain, as illustrated in Fig. 1. The relatively undeveloped tissue which is present before myelination has a mole ratio of alk-1-enyl groups to lipid phosphorus of about 0.08.

The plasmalogen content of brain tissue and myelin during development has been studied in considerable detail. The rat brain (Fig. 1) and mouse brain (Hogan and Joseph, 1970; Hogan *et al.*, 1970) accumulate a large amount of alk-1-enylacyl-GPE during the period of the greatest rate of deposition of myelin. In addition, the amounts of alk-1-enylacyl-GPC, alk-1-enylacyl-GPI, and alkylacyl-GPE in the rat brain increase by more than twofold between 12 and 24 days of age (Wells and Dittmer, 1967). Ansell and Spanner (1961) had also reported an increase in concentration of the alkylacyl-GPE with development. The concentration of the alk-1-enylacyl-GPE in the purified myelin of rat brain also increases from the immature level of 21–28% of the total phospholipid to the adult level of 32–37% (Norton and Poduslo, 1967, cited by Norton, 1972; Eng and

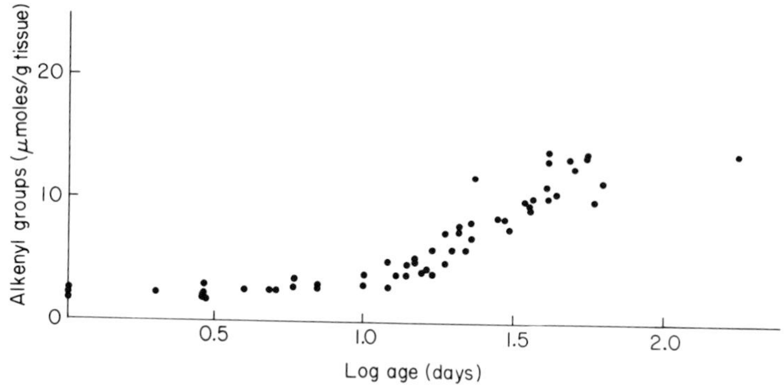

FIG. 1. A graph of the content of alk-1-enyl groups in rat brain tissue as a function of age. The points are taken from Cuzner and Davison (1968), Erickson and Lands (1959), Freysz *et al.* (1963), Geison and Waisman (1970a,b), Korey and Orchen (1959), and Wells and Dittmer (1967).

Noble, 1968; Horrocks, 1968a). This is in agreement with the finding by Erickson and Lands (1959) that the highest rate of plasmalogen deposition was several days later than the highest rate of myelin deposition in the rat brain.

The wide range of values for the plasmalogen content of the adult rat brain (Table II) is primarily due to the use of rats of different ages; the age or body weight should be specified for experiments on the lipids of rat brain. In some cases, the investigative method or omission of the brain stem from the total extract may account for the variance. The values reported by Webster (1960) seem to be about 50% too high for the rat cerebrum and spinal cord and chicken spinal cord (Table III), but the other values from Webster (1960) in Tables III and IV are in good agreement with the consensus.

In the human brain, the plasmalogen concentration increases from about 11% of the total phospholipid in the fetal brain to 19% at 1 year of age (Altrock and Debuch, 1968), but this rise is retarded by malnourishment (Fishman *et al.*, 1969). Rouser and Yamamoto (1968) derived formulas for calculating the alk-1-enylacyl-GPE content of the human brain from 1 day to 98 years of age. According to their calculations, the amount of alk-1-enylacyl-GPE increases most rapidly at the earliest age and then continues to increase at a decreasing rate until 30 years of age. After 30 years of age, the amount of alk-1-enylacyl-GPE decreases. At 80 years of age, the mole ratio of alk-1-enylacyl-GPE to total lipid phosphorus is less than it was at 1 year.

TABLE II

The Content of Alk-1-enyl Groups in the Phospholipids from the Central Nervous System of the Rat

Alk-1-enyl groups (μmoles/g tissue)	Mole ratio, alk-1-enyl groups:lipid P	Age	Tissue, reference
—	0.16	—	Brain, Norton *et al.*, 1962
10.0	0.17	60 days	Brain, Korey and Orchen, 1959
10.3	0.19	42 days	Brain, Freysz *et al.*, 1963
10.8	—	—	Brain, Norton, 1960
11.9	0.20	—	Brain, Keenan *et al.*, 1968
12.2	0.19	—	Brain, Wittenberg *et al.*, 1956
—	0.21	—	Brain, Horrocks and Ansell, 1967a
13.3	—	—	Brain, J. N. Williams *et al.*, 1962
13.3	0.22	—	Brain, Cuzner *et al.*, 1965a
13.6	—	—	Brain, Rapport and Lerner, 1959
13.7	0.20	6 months	Brain, Wells and Dittmer, 1967
13.9	0.21	56 days	Brain, Geison and Waisman, 1970a
14.7	0.24	—	Brain, Ansell and Spanner, 1963b
15.7	—	—	Brain, Ansell and Spanner, cited by Ansell and Hawthorne, 1964
19.1	0.26	—	Brain, Gasteva and Dvorkin, 1967
19.5	0.25	—	Brain, Freysz *et al.*, 1963
—	0.26	—	Brain, Bieth *et al.*, 1961
19.6	—	—	Brain, Broderson, 1967
9.1	0.14	—	Cerebrum, Seminario *et al.*, 1964
10.5	0.18	—	Cerebrum, Dvorkin and Gasteva, 1969
12.5	0.22	—	Cerebrum, Lapetina *et al.*, 1968
20.6	0.26	—	Cerebrum, Webster, 1960
26.5	0.31	—	Medulla, Dvorkin and Gasteva, 1969
28.0	0.27	—	Spinal cord, Keenan *et al.*, 1968
64.3	0.48	—	Spinal cord, Webster, 1960

Plasmalogens account for about 18% of the total phospholipids in microsomes from whole brain in a variety of animals (Table V). The corresponding value for mitochondria is only 8–10%. Higher values in mitochondria may be due to contamination by nerve-endings. Eichberg *et al.* (1964) obtained mole ratios of alk-1-enyl groups to lipid phosphorus of 0.09 for mitochondria and 0.16 for nerve endings. The values obtained from rat brain have been nearly the same for these two fractions, although different from investigation to investigation (Seminario *et al.*, 1964; Lapetina *et al.*, 1968). These authors and Mandel and Nussbaum (1966) have given the only values for the cytosol fraction, but the range is very wide.

The differences in plasmalogen content between different brain structures could be due to differences in the relative content of neurons and glial cells. Quantitative separations of cell types from brain tissue are not yet possible, but some analyses have been done on enriched fractions. Freysz *et al.* (1968) found no differences between neurons and glia in their relative contents of alk-1-enylacyl-GPE, alkylacyl-GPE, alk-1-enylacyl-GPS, and alk-1-enylacyl-GPC. Fewster and Mead (1968a,b) found that the plasmalogens from bovine oligodendroglial cells were quite similar to those from bovine white matter. Quite pure beef neuron cell bodies contained alk-1-enylacyl-GPE (Roots and Johnston, 1965). Plasmalogens account for about 10% of the phospholipids from rat brain neurons and astrocytes according to Norton and Poduslo (1971).

Almost all measurements of peripheral nerve plasmalogens have been

TABLE III

The Content of Alk-1-enyl Groups in the Phospholipids from the Central Nervous System

Alk-1-enyl groups (μmoles/g tissue)	Mole ratio, alk-1-enyl groups:lipid P	Animal, tissue, reference
—	0.24	Chicken, brain, Sheltawy and Dawson, 1969
15.4	0.25	Chicken, cerebrum, Webster, 1960
65.3	0.49	Chicken, spinal cord, Webster, 1960
11.8	0.20	Mouse, brain, Sun and Horrocks, 1968
11.0	0.18	Guinea pig, brain, Eichberg *et al.*, 1964
15.2	0.24	Rabbit, brain, Rapport and Lerner, 1959
18.6	0.25	Rabbit, brain, Owens, 1966
—	0.21	Sheep, brain, Getz *et al.*, 1968
12.0	0.17	Sheep, brain, Scott *et al.*, 1967b
8.1	0.11	Sheep, brain, Dawson, 1960
—	0.24	Ox, brain, De Rooij and Hooghwinkel, 1967
—	0.21	Ox, brain, Dawson *et al.*, 1962
—	0.17	Ox, gray matter, Gerstl *et al.*, 1969
10.3	0.17	Ox, cortex, Webster, 1960
37.0	0.30	Ox, internal capsule, Webster, 1960
52.6	0.31	Ox, medulla, Webster, 1960
—	0.26	Ox, corpus callosum, De Rooij and Hooghwinkel, 1967
—	0.28	Ox, white matter, Norton and Autilio, 1966
—	0.32	Ox, white matter, Gerstl *et al.*, 1969
43.4	0.36	Monkey, spinal cord, Webster, 1960

TABLE IV

The Content of Alk-1-enyl Groups in the Phospholipids from the Human Central Nervous System

Alk-1-enyl groups (μmoles/g tissue)	Mole ratio, alk-1-enyl groups:lipid P	Tissue, reference
16.3	0.22	Brain (20 years), Rouser and Yamamoto, 1968
15.4	0.22	Brain (40 years), Rouser and Yamamoto, 1968
—	0.23	Brain, Gueuning and Graff, 1967
—	0.23	Brain, Panganamala *et al.*, 1971
32.6	0.33	White matter, Gerstl *et al.*, 1963, 1965, 1967, 1969
26.5	0.26	White matter, Yanagihara and Cumings, 1968
—	0.29	White matter, Yanagihara and Cumings, 1969
34.7	0.35	White matter, Rapport and Lerner, 1959
34.7	0.33	Internal capsule, Webster, 1960
33.7	0.32	Corpus callosum, Webster, 1960
32.8	0.34	Spinal cord, Webster, 1960
—	0.18	Gray matter, Gerstl *et al.*, 1969
7.6	0.14	Gray matter, Yanagihara and Cumings, 1968, 1969
9.5	0.19	Gray matter, Rapport and Lerner, 1959
—	0.16	Cortex, Cumings *et al.*, 1968
8.2	0.18	Cortex, Webster, 1960
11.4	0.19	Putamen, Webster, 1960
14.6	0.22	Thalamus, Gerstl *et al.*, 1963

made on the sciatic nerve. Since most of the lipids of this nerve are present in the myelin sheath, the relatively high median value of 0.28 for the mole ratio of alk-1-enyl groups to phosphorus is as expected (Table VI). In fact, a similar ratio of alk-1-enylacyl-GPE to phosphorus is found in the two reports on peripheral nerve myelin (Table VII), which is formed by Schwann cells as opposed to the central nervous system myelin, which is made by oligodendroglia. Horrocks (1967a) found a significant difference in the plasmalogen content of the two kinds of myelin isolated from the same animals. The splenic nerve, which is not myelinated and contains small amounts of lipid, has a much smaller proportion of plasmalogens among the phospholipids (Sheltawy and Dawson, 1966).

Most of the values for the content of alkyl groups in phospholipids from the nervous system are based on the amount of phosphoglycerides that are resistant to mild alkaline and acid hydrolyses. These values, which range from 2 to 6% of the phospholipids, seem to be rather high. Svennerholm and Thorin (1960) isolated alkyl-GPE (>1% of total lipids) from the nonsaponifiable fraction of calf and human brain lipids. The alkyl-GPE from rat brain was characterized by Ansell and Spanner (1961, 1963a).

The alkylacyl-GPE accounts for 1–2% of the phospholipids from mammalian brains. Mitochondrial phospholipids contain less than 1% of alkylacyl-GPE. Alkylacyl-GPE has been detected qualitatively in porcine brain and spinal cord and in bovine spinal cord (Viswanathan *et al.*, 1969).

TABLE V

The Content of Alk-1-enyl Groups in the Phospholipids from Subcellular Fractions from the Central Nervous System

Mole ratio, alk-1-enyl groups:lipid P			
Myelin	Microsomes	Mitochondria	Animal, reference
—	—	0.19	Chicken, Patrikeeva, 1965
0.50	—	—	Pigeon, Cuzner *et al.*, 1965b
0.37	0.21	—	Mouse, Horrocks, 1968a
0.33	0.16	—	Mouse, Horrocks, 1968b
0.33	0.18	0.10	Mouse (2 years old), Horrocks, 1970
0.32	0.19	0.12	Mouse, Sun and Horrocks, 1970
0.32	—	—	Rat, Norton and Poduslo, 1967, cited by Norton, 1972
0.33	—	—	Rat, Eng and Noble, 1968
0.35	0.20	0.20	Rat, Mandel and Nussbaum, 1966
0.29	0.22	0.19	Rat, Lapetina *et al.*, 1968
0.20	0.23	0.08	Rat, Seminario *et al.*, 1964
0.39	—	—	Rat, Cuzner *et al.*, 1965a
0.49	0.14	—	Rat, Cuzner *et al.*, 1965b
0.25	0.15	0.09	Guinea pig, Eichberg *et al.*, 1964
0.34	0.20	—	Rabbit, Cuzner *et al.*, 1965b
—	0.08	—	Sheep, Getz *et al.*, 1968
—	—	0.09	Ox, Parsons and Basford, 1967
—	—	0.17	Ox, Gerstl *et al.*, 1969
0.29	—	—	Ox, Cuzner *et al.*, 1965b
0.35	—	—	Ox, Norton and Autilio, 1966
0.33	—	—	Ox, spinal cord, Horrocks, 1968b
0.37	—	—	Ox, optic nerve, Leitch *et al.*, 1969
0.28	—	—	Ox, optic nerve, MacBrinn and O'Brien, 1969
0.35	—	—	Squirrel monkey, spinal cord, Horrocks, 1967a
0.34	—	—	Rhesus monkey, Horrocks, 1968b
—	—	0.17	Man, gray matter, Gerstl *et al.*, 1969
—	—	0.20	Man, white matter, Gerstl *et al.*, 1969
0.32	—	—	Man, Yanagihara and Cumings, 1968
0.33	—	—	Man, Yanagihara and Cumings, 1969
0.28	—	—	Man, Norton *et al.*, 1966
0.31	—	—	Man, Gerstl *et al.*, 1967
0.31	0.16	—	Man, Cuzner *et al.*, 1965b
0.42	—	—	Man, Cuzner *et al.*, 1965b
0.52	—	—	Man, O'Brien and Sampson, 1965a,b

TABLE VI

The Content of Alk-1-enyl Groups in the Phospholipids from Peripheral Nerves and Myelin

Alk-1-enyl group content (μmoles/g tissue)	Mole ratio, alk-1-enyl groups:lipid P	Species, tissue, reference
16.9	0.34	Chicken, sciatic nerve, Sheltawy and Dawson, 1966
—	0.28	Chicken, sciatic nerve, Berry *et al.*, 1965
18.0	0.24	Chicken, sciatic nerve, Porcellati and Mastrantonio, 1964
29.2	0.39	Chicken, sciatic nerve, Webster, 1960
15.6	0.30	Rabbit, sciatic nerve, Sheltawy and Dawson, 1966
47	0.50	Rabbit, sciatic nerve, Domonkos and Heiner, 1968
20.0	0.23	Rabbit, sciatic nerve, Gueuning and Graff, 1967
18.3	0.28	Sheep, sciatic nerve, Sheltawy and Dawson, 1966
—	0.21	Cat, sciatic nerve, Berry *et al.*, 1965
15.3	0.28	Rhesus monkey, sciatic nerve, Sheltawy and Dawson, 1966
15.4	0.34	Man, sciatic nerve, Webster, 1960
1.6	0.15	Ox, splenic nerve, Sheltawy and Dawson, 1966
—	0.34	Ox, spinal root, myelin, O'Brien *et al.*, 1967
—	0.27	Squirrel monkey, brachial plexus, myelin, Horrocks, 1967a

Evidence for presence of alkylacyl-GPC in brain tissue from the mouse, rat, ox, and man has been obtained (Tables VIII and IX). Cotman *et al.* (1969) found more alkylacyl-GPC than alk-1-enylacyl-GPC in rat brain. Horrocks and Ansell (1967a) found a small amount of phospholipid that was resistant to hydrolysis in the SPG plus IPG fraction from rat brain.

The choline phosphoglycerides from mammalian brains (Table IX) include 1–4% alkenylacyl-GPC and 1–2% alkylacyl-GPC. Small proportions of alk-1-enylacyl-GPC have been found in nearly every analysis of nervous system phospholipids (Table VII) when the CPG have been examined carefully, and although the concentration of choline plasmalogens is usually less than 1% of the phospholipid, the content in micromoles per gram of tissue is comparable to the total plasmalogen content in many other tissues. Very high contents have been reported for the alk-1-enylacyl-GPC from sciatic nerves (Table IX) but Sheltawy and Dawson (1966) did not detect alk-1-enylacyl-GPC in the sciatic nerve of the rabbit and other mammals. Further studies are required to reconcile these results.

The high content of plasmalogens in the nervous system is primarily in the form of alk-1-enylacyl-GPE (Table VII), accounting for one-half to two-thirds of the EPG from whole brain (Table X). This proportion is lower in gray matter and higher in white matter, myelin, and sciatic nerves. About 4% of the EPG from whole brain is alkylacyl-GPE.

The serine phosphoglycerides of nervous tissue (Table X) may include small amounts of alk-1-enylacyl- and alkylacyl-GPS. Wells and Dittmer (1966) found no alk-1-enylacyl-GPS but did find 0.2% each of alk-1-enylacyl-GP and alk-1-enylacyl-GPI in the total phospholipids of rat brain: these workers also reported serine plasmalogens in sheep brain. With the same methods, Parsons and Basford (1967) reported 0.1% of the inositol plasmalogen and no serine plasmalogen in bovine brain mitochondria, and Slagel *et al.* (1967) detected small amounts of alk-1-enyl groups in the phosphatidic acids, IPG, and CPG but not in the SPG from human brain tissue adjacent to glial tumors. Some of the other values reported for the SPG plasmalogen are from mixtures of the SPG and IPG. Another group of values is based on a TLC separation of the SPG and IPG, which is often difficult. The existence of serine plasmalogens themselves in rodent brains has not been firmly established. In higher animals, plasmalogens are probably present in small amounts in the CPG, IPG, SPG, and phosphatidic acids. Relatively high values for serine plasmalogens were obtained for human white matter and myelin and bovine spinal root myelin after separation of the SPG with diethylaminoethyl cellulose columns, but similar columns were used for the separation of SPG from human and bovine gray matter and from bovine optic nerve myelin in which a maximum content of 3% plasmalogen was reported (Table X). The actual content of alk-1-enylacyl-GPS in the SPG from nervous tissues is probably 4% or less. For rat brain, the alkylacyl-GPS content was 2% of the SPG (Horrocks and Ansell, 1967a). The first report on inositol plasmalogens was by Hack (1961).

From the results of mild acid and base hydrolyses of *O*-radyl cerebrosides, Kochetkov *et al.* (1963) concluded that *O*-alk-1-enyl and *O*-alkyl cerebrosides might be present in brain tissue, but subsequently, Klenk and Doss (1966), Klenk and Löhr (1967), and Tamai (1968) have stated that *O*-alk-1-enyl cerebrosides could not be found. Norton and Brotz (1963) isolated and characterized a mixture of alkylacylglycerogalactoses that accounted for 0.19% of the total lipid from bovine brain. This lipid class was not detected by Wells and Dittmer (1966) in rat brains but has been found by Rumsby (1967), Rumsby and Rossiter (1968), and Kishimoto *et al.* (1968) in ox, pig, and sheep brains. The concentration of the alkylacylglycerogalactoses is 0.01 − 0.05 μmole/g of tissue (Rumsby and Rossiter, 1968).

TABLE VII

The Alk-1-enyl Group Content of Phosphoglycerides from the Nervous System

Alk-1-enyl-acyl-GPC (μmole/g tissue)	Alk-1-enyl-acyl-GPE (μmole/g tissue)	Alk-1-enyl-acyl-GPS (μmole/g tissue)	Mole ratio, component:lipid P			Animal, tissue, reference
			Alk-1-enyl-acyl-GPC	Alk-1-enyl-acyl-GPE	Alk-1-enyl-acyl-GPS	
—	—	—	—	0.235	—	Chicken, brain, Sheltawy and Dawson, 1969
0.22	—	—	0.004	—	—	Chicken, cerebrum, Webster, 1960
1.57	—	—	0.012	—	—	Chicken, spinal cord, Webster, 1960
—	—	—	—	0.20	—	Chicken, microsomes, Porcellati *et al.*, 1970a
—	—	—	—	0.17	—	Chicken, mitochondria, Porcellati *et al.*, 1970a
—	—	—	0.014	0.195	0.005	Rat, brain, Horrocks and Ansell, 1967a
0.35	13.0	ND	0.005	0.193	ND	Rat, brain (6 months), Wells and Dittmer, 1967
0.62	11.7	—	—	—	—	Rat, brain, Ansell and Spanner, 1968a,b
0.47	—	—	0.006	—	—	Rat, cerebrum, Webster, 1960
—	—	—	0.06	0.28	—	Rat, myelin, Cuzner *et al.*, 1965a
—	—	—	0.08	0.35	—	Rat, myelin, Cuzner *et al.*, 1965b
2.73	—	—	0.020	—	—	Rat, spinal cord, Webster, 1960
—	—	—	0.009	0.186	0.007	Mouse, brain, Sun and Horrocks, 1968
0.8	17.1	0.8	0.010	0.226	0.010	Rabbit, brain, Owens, 1966
—	—	—	0.024	0.177	—	Sheep, brain, Getz *et al.*, 1968
0.62	11.4	—	0.009	0.165	—	Sheep, brain, Scott *et al.*, 1967b
ND	7.8	0.28	ND	0.110	0.004	Sheep, brain, Dawson, 1960
—	—	—	0.020	0.236	—	Ox, brain, De Rooij and Hooghwinkel, 1967
—	—	—	ND	0.211	tr	Ox, brain, Dawson *et al.*, 1962
1.32	—	—	0.008	—	—	Ox, medulla, Webster, 1960
1.30	—	—	0.010	—	—	Ox, internal capsule, Webster, 1960
0.25	—	—	0.004	—	—	Ox, cortex, Webster, 1960
—	—	—	0.006	0.26	0.017	Ox, white matter, Norton and Autilio, 1965

—	—	—	0.002	0.087	ND	Ox, mitochondria, Parsons and Basford, 1967
—	—	—	0.01	0.34	0.01	Ox, myelin, Norton and Autilio, 1966
—	—	—	—	0.28	0.001	Ox, optic nerve myelin, MacBrinn and O'Brien, 1969
—	—	—	—	0.333	ND	Ox, optic nerve myelin, Horrocks, 1968b
—	—	—	—	0.334	ND	Ox, spinal cord myelin, Horrocks, 1968b
1.04	—	—	0.009	—	—	Monkey, spinal cord, Webster, 1960
—	—	—	—	0.356	ND	Squirrel monkey, spinal cord myelin, Horrocks, 1968b
—	—	—	—	0.354	ND	Rhesus monkey, medulla myelin, Horrocks, 1968b
—	—	—	—	0.336	ND	Rhesus monkey, corpus callosum myelin, Horrocks, 1968b
—	—	—	0.032	0.189	0.009	Man, brain, Gueuning and Graff, 1967
—	—	—	0.009	0.224	—	Man, brain, Panganamala *et al.*, 1971
—	—	—	0.007	0.242	0.004	Man, white matter, Cumings *et al.*, 1968
—	—	—	0.009	0.248	0.005	Man, white matter, Yanagihara and Cumings, 1968
—	—	—	—	0.291	—	Man, white matter, De Rooij and Hooghwinkel, 1967
—	—	—	—	0.122	—	Man, gray matter, De Rooij and Hooghwinkel, 1967
—	—	—	ND	0.157	ND	Man, gray matter, Cumings *et al.*, 1968
—	—	—	—	0.136	—	Man, gray matter, Yanagihara and Cumings, 1969
0.13	—	—	0.003	—	—	Man, cortex, Webster, 1960
0.30	—	—	0.005	—	—	Man, putamen, Webster, 1960
1.14	—	—	0.011	—	—	Man, internal capsule, Webster, 1960
1.08	—	—	0.010	—	—	Man, corpus cellosum, Webster, 1960
1.03	—	—	0.012	—	—	Man, spinal cord, Webster, 1960
—	—	—	0.010	0.299	0.006	Man, myelin, Yanagihara and Cumings, 1968
—	—	—	0.009	0.313	0.004	Man, myelin, Yanagihara and Cumings, 1969

TABLE VII (*Continued*)

Alk-1-enyl-acyl-GPC (μmole/g tissue)	Alk-1-enyl-acyl-GPE (μmole/g tissue)	Alk-1-enyl-acyl-GPS (μmole/g tissue)	Mole ratio, component:lipid P			Animal, tissue, reference
			Alk-1-enyl-acyl-GPC	Alk-1-enyl-acyl-GPE	Alk-1-enyl-acyl-GPS	
—	—	—	—	0.327	—	Man, myelin, Sun and Horrocks, 1971
—	—	—	0.006	0.384	0.133	Man, myelin, O'Brien and Sampson, 1965a,b
0.68	16.2	—	0.014	0.329	ND	Chicken, sciatic nerve, Sheltawy and Dawson, 1966
1.3	7.5	9.2	0.017	0.10	0.12	Chicken, sciatic nerve, Porcellati and Mastroantanio, 1964
0.35	—	—	0.005	—	—	Chicken, sciatic nerve, Webster, 1960
—	—	—	ND	0.301	ND	Rabbit, sciatic nerve, Sheltawy and Dawson, 1966
2.6	17.4	—	0.03	0.20	—	Rabbit, sciatic nerve, Gueuning and Graff, 1967
—	11.9	1.4	ND	0.248	0.029	Sheep, sciatic nerve, Sheltawy and Dawson, 1966
—	14.5	0.8	ND	0.270	0.015	Rhesus monkey, sciatic nerve, Sheltawy and Dawson, 1966
0.53	—	—	0.012	—	—	Man, sciatic nerve, Webster, 1960
—	—	—	ND	0.148	ND	Ox, splenic nerve, Sheltawy and Dawson, 1966
—	—	—	—	0.286	0.056	Ox, spinal root myelin, O'Brien *et al.*, 1967
—	—	—	—	0.27	ND	Squirrel monkey, brachial plexus myelin, Horrocks, 1967a, 1968c

Alkyl groups (μmoles/g tissue)	Alkylacyl-GPE (μmoles/g tissue)	Mole ratio, component:lipid P			Animal, tissue, reference
		Alkyl groups	Alkylacyl-GPC	Alkylacyl-GPE	
—	—	0.023	—	—	Chicken, brain, Sheltawy and Dawson, 1969
—	—	—	—	0.022	Chicken, mitochondria, Porcellati *et al.*, 1970a
—	—	—	—	0.030	Chicken, microsomes, Porcellati *et al.*, 1970a
—	—	—	—	0.016	Mouse, brain, Sun and Horrocks, 1969b
—	—	—	0.008	0.014	Mouse, brain, Horrocks, 1970
—	—	—	—	0.014	Mouse, brain, myelin, Horrocks, 1970
—	—	—	—	0.017	Mouse, brain, microsomes, Horrocks, 1970
—	—	—	—	0.008	Mouse, brain, mitochondria, Horrocks, 1970
—	—	0.039	0.004	0.031	Rat, brain, Horrocks and Ansell, 1967a
—	—	—	—	0.031	Rat, brain, Ansell and Spanner, 1961
—	—	—	—	0.023	Rat, brain, Ansell and Spanner, 1963a
—	1.04	—	—	0.015	Rat, brain (6 months), Wells and Dittmer, 1967
1.96	—	0.032	—	—	Guinea pig, brain, Eichberg *et al.*, 1964
—	—	0.032	—	—	Guinea pig, brain, myelin, Eichberg *et al.*, 1964
—	—	0.019	—	—	Guinea pig, brain, microsomes, Eichberg *et al.*, 1964
—	—	0.022	—	—	Guinea pig, brain, mitochondria, Eichberg *et al.*, 1964
—	—	0.021	—	—	Guinea pig, brain, nerve endings, Eichberg *et al.*, 1964
2.07	—	0.030	—	—	Sheep, brain, Scott *et al.*, 1967b
3.97	—	0.056	—	—	Sheep, brain, Dawson, 1960
—	—	0.021	—	—	Ox, brain, Dawson *et al.*, 1962
—	—	—	—	0.004	Ox, brain, mitochondria, Parsons and Basford, 1967
—	—	—	0.009	0.011	Man, brain, Panganamala *et al.*, 1971
2.35	—	0.048	—	—	Chicken, sciatic nerve, Sheltawy and Dawson, 1966
1.64	—	0.032	—	—	Rabbit, sciatic nerve, Sheltawy and Dawson, 1966
1.03	—	0.021	—	—	Sheep, sciatic nerve, Sheltawy and Dawson, 1966
1.94	—	0.036	—	—	Rhesus monkey, sciatic nerve, Sheltawy and Dawson, 1966
0.19	—	0.018	—	—	Ox, splenic nerve, Sheltawy and Dawson, 1966

TABLE IX

The Relative Amounts of Alk-1-enylacyl- and Alkylacylglycerophosphorylcholines in the Nervous System

% of CPG		
Alk-1-enylacyl-GPC	Alkylacyl-GPC	Animal, tissue, reference
4	—	Rat, brain, Ansell and Spanner, 1968b
4	1	Rat, brain, Horrocks and Ansell, 1967a
1.4	ND	Rat, brain, Wells and Dittmer, 1967
2.2	—	Mouse, brain, Sun and Horrocks, 1968
4.4	2.0	Mouse, brain, Horrocks, 1970
3	—	Rabbit, brain, Owens, 1966
7.2	—	Sheep, brain, Getz *et al.*, 1968
2.4	—	Sheep, brain, Scott *et al.*, 1967b
2.2	—	Sheep, brain, microsomes, Getz *et al.*, 1968
5.1	—	Ox, brain, De Rooij and Hooghwinkel, 1967
1.9	2.2	Ox, brain, Renkonen, 1966
1	—	Ox, brain, white matter, Norton and Autilio, 1966
0.6	—	Ox, brain, mitochondria, Parsons and Basford, 1967
4	—	Ox, brain, light myelin, Norton and Autilio, 1966
8	—	Hog, brain, Rhee *et al.*, 1967
8	—	Man, brain, Gueuning and Graff, 1967
2.9	—	Man, brain, Panganamala *et al.*, 1971
3.1	—	Man, brain, white matter, Cumings *et al.*, 1968
3.3	—	Man, brain, white matter, Yanagihara and Cumings, 1968, 1969
3.9	—	Man, brain, myelin, Yanagihara and Cumings, 1968, 1969
5.5	—	Man, brain, meninges, Bell *et al.*, 1967
8.0	—	Chicken, sciatic nerve, Sheltawy and Dawson, 1966
6	—	Chicken, sciatic nerve, Porcellati and Mastrantonio, 1964
19	—	Rabbit, sciatic nerve, Gueuning and Graff, 1967

3. *Ocular Tissues*

The retina of the eye is a nervous tissue that can be compared to gray matter. R. E. Anderson (1970) has reported the content of 16:0 and 18:0 alk-1-enyl groups as a proportion of the total alk-1-enyl and acyl groups of EPG. According to this report, the plasmalogen content of the retina from man, dog, and sheep may be higher than that from the ox, rabbit, or pig. The values for the plasmalogen content of ox retina reported by

TABLE X

The Relative Amounts of Alk-1-enylacyl- and Alkylacylglycerophosphorylethanolamines and Alk-1-enylacylglycerophosphorylserines in the Nervous System

Percent of lipid class			
Alk-1-enyl acyl-GPE	Alkyl- acyl-GPE	Alk-1-enyl acyl-GPS	Animal, tissue, reference
59	—	—	Chicken, brain, Sheltawy and Dawson, 1969
41	7	—	Chicken, brain, mitochondria, Porcellati *et al.*, 1970a
52	6	—	Chicken, brain, microsomes, Porcellati *et al.*, 1970a
60	8	—	Rat, brain, Ansell and Spanner, 1963b
66	—	—	Rat, brain, Ansell and Spanner, 1968a
56	3.3	—	Rat, brain, Brown and Dittmer, 1968
51	—	—	Rat, brain, Keenan *et al.*, 1968
50	7.9	3	Rat, brain, Horrocks and Ansell, 1967a
53	4.2	—	Rat, brain, Wells and Dittmer, 1967
79	—	—	Rat, brain, myelin, Cuzner *et al.*, 1965b
74	—	—	Rat, brain, myelin, Ansell and Spanner, 1968a
59	—	—	Rat, brain, microsomes, Ansell and Spanner, 1968a
36	—	—	Rat, brain, mitochondria, Ansell and Spanner, 1968a
59	—	—	Rat, brain, cytosol, Ansell and Spanner, 1968a
47	—	4.6	Mouse, brain, Sun and Horrocks, 1968
49	3.6	—	Mouse, brain, Horrocks, 1970
70	2.9	—	Mouse, brain, myelin, Horrocks, 1970
70	—	—	Mouse, brain, myelin, Sun and Horrocks, 1970
55	5.0	—	Mouse, brain, microsomes, Horrocks, 1970
51	—	—	Mouse, brain, microsomes, Sun and Horrocks, 1970
26	2.5	—	Mouse, brain, mitochondria, Horrocks, 1970
36	—	—	Mouse, brain, mitochondria, Sun and Horrocks, 1970
64	—	6	Rabbit, brain, Owens, 1966
39	—	2.4	Sheep, brain, Dawson, 1960
56	—	—	Sheep, brain, Getz *et al.*, 1968
68	—	—	Sheep, brain, Scott, *et al.*, 1967b
26	—	—	Sheep, brain, microsomes, Getz *et al.*, 1968
58	—	—	Ox, brain, De Rooij and Hooghwinkel, 1967
48	6	—	Ox, brain, Albro and Dittmer, 1968
48	—	—	Ox, brain, Rouser, 1968
56	4	—	Ox, brain, Renkonen, 1963a
66	—	—	Ox, brain, Dawson *et al.*, 1962
68	—	—	Ox, brain, corpus callosum, De Rooij and Hooghwinkel, 1967
84	—	1	Ox, brain, white matter, Norton and Autilio, 1966
—	3	—	Ox, brain, white matter, Ansell and Spanner, 1963a

TABLE X (*Continued*)

Percent of lipid class			
Alk-1-enyl-acyl-GPE	Alkyl-acyl-GPE	Alk-1-enyl-acyl-GPS	Animal, tissue, reference
48	—	tr	Ox, brain, gray matter, Yabuuchi and O'Brien, 1968
32	1.4	—	Ox, brain, mitochondria, Parsons and Basford, 1967
70	—	1	Ox, optic nerve, myelin, MacBrinn and O'Brien, 1969
81	—	—	Ox, optic nerve, myelin, Horrocks, 1968b
72	—	—	Ox, spinal cord, myelin, Horrocks, 1968b
77	—	4	Ox, brain, light myelin, Norton and Autilio, 1966
56	—	—	Hog, brain, Rhee *et al.*, 1967
90	—	—	Hog, spinal cord, Viswanathan *et al.*, 1968a
80	—	—	Rhesus monkey, brain, myelin, Horrocks, 1968b
80	—	—	Squirrel monkey, spinal cord, myelin, Horrocks, 1968b
63	—	—	Man, brain, Rouser and Yamamoto, 1968
67	—	6	Man, brain, Gueuning and Graff, 1967
63	3.1	—	Man, brain, Panganamala *et al.*, 1971
77	—	—	Man, white matter, De Rooij and Hooghwinkel, 1967
66	—	2.7	Man, white matter, Cumings *et al.*, 1968
94	—	25	Man, white matter, O'Brien and Sampson, 1965a,b
68	—	—	Man, white matter, Yanagihara and Cumings, 1968
76	—	2.2	Man, white matter, Yanagihara and Cumings, 1969
36	—	—	Man, gray matter, De Rooij and Hooghwinkel, 1967
43	—	—	Man, gray matter, Cumings *et al.*, 1968
39	—	—	Man, gray matter, Yanagihara and Cumings, 1968
38	—	—	Man, gray matter, Yanagihara and Cumings, 1969
42	—	0.6	Man, gray matter, O'Brien and Sampson, 1965a,b
74	—	4.2	Man, myelin, Yanagihara and Cumings, 1968
77	—	2.8	Man, myelin, Yanagihara and Cumings, 1969
78	—	—	Man, myelin, Sun and Horrocks, 1971
100	—	73	Man, myelin (55 yrs), O'Brien and Sampson, 1965a,b
88	—	—	Chicken, sciatic nerve, Sheltawy and Dawson, 1966
50	6.1	41	Chicken, sciatic nerve, Porcellati and Mastrantonio, 1964
85	—	—	Chicken, sciatic nerve, Berry *et al.*, 1965

TABLE X (*Continued*)

Percent of lipid class			
Alk-1-enyl-acyl-GPE	Alkyl-acyl-GPE	Alk-1-enyl-acyl-GPS	Animal, tissue, reference
82	—	—	Rabbit, sciatic nerve, Sheltawy and Dawson, 1966
88	—	—	Rabbit, sciatic nerve, Gueuning and Graff, 1967
85	—	15	Sheep, sciatic nerve, Sheltawy and Dawson, 1966
78	2.8	—	Ox, sciatic nerve, J. F. Berry, B. Kaye, and K. Chang, cited by Berry *et al.*, 1965
67	—	—	Cat, sciatic nerve, Berry *et al.*, 1965
87	—	8.5	Rhesus monkey, sciatic nerve, Sheltawy and Dawson, 1966
70	—	—	Ox, splenic nerve, Sheltawy and Dawson, 1966
79	—	29	Ox, spinal root, myelin, O'Brien *et al.*, 1967

various laboratories (Table XI) do not agree. A lower value for ox retinal rod outer segments than for calf retina is consistent with data obtained using the frog retina which has 3% plasmalogen as compared to 0.6% in the frog retinal rod outer segments (Eichberg and Hess, 1967). Adams (1969) found no plasmalogens in bovine rhodopsin isolated from rod outer segments, but acidic conditions had been used and 9% lyso-EPG was found. The human lens has a high sphingomyelin content and a low alk-1-enylacyl-GPE content in comparison with the bovine lens. Broekhuyse and Veerkamp (1968) have reported that the alk-1-enylacyl-GPE accounts for 54% to 44% of the EPG in whole calf lens and in four parts (nucleus, cortex inner layer, cortex outer layer, and epithelium). No information is available on the alkyl groups, and serine plasmalogens have not as yet been detected in ocular tissues.

4. *Glandular Tissues*

Glandular tissues (Table XII) contain an appreciable amount of alk-1-enylacyl-GPE, small amounts of alk-1-enylacyl-GPC, and only traces, if any, of alk-1-enylacyl-GPS (Debuch and Winterfeld, 1966). Canine adrenal glands may be an exception, as only traces of plasmalogens were found by Chang and Sweeley (1963). Small amounts of phospholipids with alkyl groups have been reported in canine thyroid glands (Scott *et al.*, 1966) and human pituitary and adrenal glands (Debuch and Winterfeld, 1966).

5. *Muscle Tissues*

There is a fairly high content of alk-1-enyl groups present in cardiac and skeletal muscles (Tables XIII and XIV). These tissues differ from

TABLE XI

The Content of Alk-1-enyl Groups in the Phospholipids from Ocular Tissues

Mole ratio, component:lipid P			Percent of lipid class		
Alk-1-enyl groups	Alk-1-enyl-acyl-GPC	Alk-1-enyl-acyl-GPE	Alk-1-enyl-acyl-GPC	Alk-1-enyl-acyl-GPE	Animal, tissue, reference
0.10	—	—	—	28	Rabbit, retina, R. E. Anderson *et al.*, 1970a
0.04	—	—	—	10	Ox, retina, R. E. Anderson *et al.*, 1970a
0.12	0.005	0.110	1.2	33	Ox (calf), retina, Broekhuyse, 1968
0.04	—	—	—	10	Ox, retinal rod outer segments, Borggreven *et al.*, 1970
0.04	—	—	—	10	Ox, retinal rod outer segments, Horrocks and McConnell, 1969
0.18	0.008	0.167	2.5	52	Ox (calf), iris, Broekhuyse, 1968
0.15	—	—	—	51	Ox, iris, R. E. Anderson *et al.*, 1970b
0.14	0.007	0.138	2.1	50	Ox (calf), choroid, Broekhuyse, 1968
0.12	0.008	0.108	2.3	43	Ox (calf), cornea, Broekhuyse, 1968
0.06	—	—	—	—	Ox, cornea, Thiele and Denden, 1967
0.12	—	—	—	—	Ox, cornea, Thiele and Denden, 1967
0.11	—	—	—	—	Ox, cornea, epithelium, Thiele and Denden, 1967
0.18	0.012	0.166	3.0	67	Ox (calf), sclera, Broekhuyse, 1968
0.05	0.005	0.048	1.5	31	Ox (calf), vitreous body, Broekhuyse, 1968
0.17	0.003	0.166	1.1	50	Ox (calf), lens, Broekhuyse, 1968
0.12	—	—	ND	35	Ox, lens, R. E. Anderson *et al.*, 1969c
0.08	—	—	ND	29	Rabbit, lens, R. E. Anderson *et al.*, 1969c
0.02	0.004	0.014	7	9	Man, lens (20 yr), Broekhuyse, 1969
0.02	0.005	0.017	10	12	Man, lens (32 yr), Broekhuyse, 1969
0.02	0.003	0.011	14	16	Man, lens (66 yr), Broekhuyse, 1969

Alk-1-enyl groups (μmoles/g tissue)	Mole ratio, component:lipid P				Percent of lipid class		
	Alk-1-enyl groups	Alk-1-enyl-acyl-GPC	Alk-1-enyl-acyl-GPE	Alkyl groups	Alk-1-enyl-acyl-GPC	Alk-1-enyl-acyl-GPE	Animal, gland, reference
2.7	0.17	—	—	—	—	—	Man, pituitary, Winterfeld and Debuch, 1966
—	—	0.01	0.08	—	2	28	Man, pituitary, Debuch and Winterfeld, 1966
2.3	0.11	—	—	—	—	—	Man, pineal, Basinska *et al.*, 1969
2.6	0.08	—	—	—	—	—	Ox, pineal, Basinska *et al.*, 1969
3.0	0.10	—	—	—	—	—	Sheep, pineal, Basinska *et al.*, 1969
1.3	0.12	0.015	0.107	0.016	3	46	Dog, thyroid, Scott *et al.*, 1966
3.5	0.14	—	—	—	—	—	Man, adrenal, Winterfeld and Debuch, 1966
—	—	0.01	0.08	—	2	28	Man, adrenal, Debuch and Winterfeld, 1966
5.5	0.13	—	—	—	—	—	Ox, adrenal, Norton, 1960
—	0.14	—	—	—	—	—	Ox, adrenal, Norton *et al.*, 1962
4.9	0.13	—	—	—	—	—	Ox, adrenal cortex, Norton, 1960
—	0.11	—	—	—	—	—	Ox, adrenal cortex, Norton *et al.*, 1962
5.3	0.18	—	—	—	—	—	Ox, adrenal medulla, Norton, 1960
—	0.19	—	—	—	—	—	Ox, adrenal medulla, Norton *et al.*, 1962
—	0.17	0.014	0.155	0.096	3	60	Sheep, adrenal cortex, Getz *et al.*, 1968
—	0.12	0.007	0.112	0.067	2	51	Sheep, adrenal cortex, mitochondria, Getz *et al.*, 1968
—	0.12	0.026	0.098	0.058	4	64	Sheep, adrenal cortex, microsomes, Getz *et al.*, 1968
—	0.17	0.018	0.150	0.070	4	62	Sheep, adrenal medulla, Getz *et al.*, 1968
3.1	—	—	—	—	—	—	Rat, adrenal, Broderson, 1967

TABLE XIII

The Content of Alk-1-enyl and Alkyl Groups in the Phospholipids from the Heart

Alk-1-enyl groups (μmoles/g tissue)	Mole ratio, component:lipid P				Percent of lipid class				Animal, reference
	Alk-1-enyl groups	Alk-1-enyl-acyl-GPC	Alk-1-enyl-acyl-GPE	Alkyl groups	Alk-1-enyl-acyl-GPC	Alkyl-acyl-GPC	Alk-1-enyl-acyl-GPE	Alkyl-acyl-GPE	
—	0.32	0.16	0.15	0.033	36	5.2	59	3.6	Man, Panganamala *et al.*, 1971
—	0.28	0.14	0.13	—	36	—	49	—	Man, Hughes and Frais, 1967 (38 years old)
—	0.35	0.19	0.16	—	48	—	59	—	Man, Hughes and Frais, 1967 (65 years old)
—	0.22	0.14	0.08	—	24	—	35	—	Man, Debuch and Winterfeld, 1966
—	0.29	—	—	—	—	—	—	—	Man, Winterfeld and Debuch, 1966
—	0.45	0.27	0.17	—	49	—	51	—	Ox, De Rooij and Hooghwinkel, 1967
—	—	—	—	—	33	4	—	—	Ox, Renkonen, 1963a
2.12	—	—	—	—	—	—	—	—	Ox, Gilbertson *et al.*, 1967
7.59	0.42	0.24	0.18	—	57	—	47	—	Ox, Gray and Macfarlane, 1958
—	—	—	—	—	58	—	—	—	Ox, Poulos, 1970
—	—	—	—	—	44	1.6	—	—	Ox, Ansell and Spanner, 1965b
—	—	—	—	—	—	—	50	—	Ox, Viswanathan *et al.*, 1968b
—	—	—	—	—	45	—	33	—	Ox, Warner and Lands, 1961
4.8	0.26	—	—	—	—	—	—	—	Ox, Spanner, 1966
—	—	—	—	—	52	4.2	—	—	Ox, Viswanathan *et al.*, 1968c

9.6	0.45	0.28	0.17	—	56	—	52	—	Ox, Wheeldon *et al.*, 1965
—	—	—	—	—	50	3.2	42	2	Ox, Pietruszko and Gray, 1962
—	0.42	0.24	0.18	—	57	—	42	—	Ox, Gray, 1960b
—	0.29	0.18	0.11	—	42	—	40	—	Ox, Dawson *et al.*, 1962
—	0.40	—	—	—	—	—	—	—	Ox, mitochondria, Fleischer *et al.*, 1961
—	0.39	0.20	0.15	—	50	—	41	—	Ox, mitochondria, Fleischer *et al.*, 1967
—	0.44	0.29	0.15	—	64	—	48	—	Ox, mitochondria, Wheeldon *et al.*, 1965
—	0.48	0.26	0.22	—	58	—	79	—	Ox, microsomes, Wheeldon *et al.*, 1965
—	0.42	0.27	0.15	—	61	—	59	—	Ox, myofibrils, Wheeldon *et al.*, 1965
—	0.43	0.28	0.13	0.012	58	—	68	—	Sheep, Getz *et al.*, 1968
8.31	0.31	0.20	0.11	0.032	44	—	47	—	Sheep, Scott *et al.*, 1967b
5.31	0.18	0.11	0.07	—	29	—	26	—	Sheep, Dawson, 1960
—	0.39	0.26	0.10	0.017	61	—	61	—	Sheep, mitochondria, Getz *et al.*, 1968
—	—	—	—	—	40	—	46	—	Pig, Waku and Lands, 1968
—	—	—	—	—	30	—	40	—	Pig, Matsumoto *et al.*, 1967
—	—	—	—	—	34	—	—	—	Pig, Warner and Lands, 1963
—	—	0.23	—	—	46	—	—	—	Pig, Gray, 1960b
—	—	—	—	0.007	—	—	—	—	Doǵ, Nakagawa and McKibbin, 1962
5.34	—	—	—	—	—	—	—	—	Dog, Gilbertson *et al.*, 1967
6.1	0.27	—	—	—	—	—	—	—	Cat, Spanner, 1966
7.1	0.25	—	—	—	—	—	—	—	Ferret, Spanner, 1966
9.9	0.34	—	—	—	—	—	—	—	Guinea pig, Spanner, 1966
—	0.12	—	—	—	—	—	—	—	Mouse, Hughes and Frais, 1967

TABLE XIII (*Continued*)

Alk-1-enyl groups (μmoles/g tissue)	Mole ratio, component:lipid P				Percent of lipid class				Animal, reference
	Alk-1-enyl groups	Alk-1-enyl-acyl-GPC	Alk-1-enyl-acyl-GPE	Alkyl groups	Alk-1-enyl-acyl-GPC	Alkyl-acyl-GPC	Alk-1-enyl-acyl-GPE	Alkyl-acyl-GPE	
4.3	0.13	—	—	—	—	—	—	—	Mouse, Spanner, 1966
3.4	0.13	—	—	—	—	—	—	—	Rat, Spanner, 1966
3.12	0.09	—	—	—	—	—	—	—	Rat, Gilbertson *et al.*, 1967; Gilbertson, 1969
4.5	0.12	0.02	—	—	—	—	—	—	Rat, Webster, 1960
2.6	0.12	—	—	—	—	—	—	—	Rat, Rapport and Lerner, 1959
—	—	—	—	—	8	—	23	—	Rat, Marinetti *et al.*, 1961
—	—	—	0.05	—	—	—	—	—	Rat, Keenan *et al.*, 1968
3.16	0.09	—	—	—	—	—	—	—	Rat, Wittenberg *et al.*, 1956
—	0.09	—	—	—	—	—	—	—	Rat, Gottfried and Rapport, 1963
2.6	—	—	—	—	—	—	—	—	Rat, Norton, 1960
3.7	—	—	—	—	—	—	—	—	Rat, Broderson, 1967
8.3	0.31	0.15	0.15	—	36	—	50	—	Rabbit, Owens, 1966
6.9	0.29	—	—	—	—	—	—	—	Rabbit, Rapport and Lerner, 1959
7.7	0.21	0.09	—	—	—	—	—	—	Chicken, Webster, 1960
—	0.05	0.03	0.02	—	8	—	9	—	Pigeon, Belsare and Chowdhuri, 1968
—	—	0.06	—	—	12	—	—	—	Pigeon, Gray and Macfarlane, 1961

most other tissues in having a relatively high proportion of both the CPG and the EPG in the plasmalogen form, but serine plasmalogens have not been detected by most investigators. Owens (1966) found 0.7% of the adult rabbit heart lipid phosphorus in the alk-1-enylacyl-GPS and Kunze and Olthoff (1968) found the same amount in human skeletal muscle. A trace of alk-1-enylacyl-GPS was found in human uterine muscle by Debuch and Winterfeld (1966). According to Fleischer *et al.* (1967), the mole ratio of alk-1-enyl groups to lipid phosphorus in the cardiolipin isolated from bovine heart mitochondria is 0.38. Hack and Ferrans (1960) and Hack and Helmy (1967a) have reported that more than 5% of the plasmalogen from an infarcted heart had the same chromatographic mobility as cardiolipin. Further studies are required on the concentration and characterization of cardiolipin plasmalogens in normal and infarcted cardiac muscle.

During development, Hughes and Frais (1967) found an increasing content of alk-1-enylacyl-GPC in human heart and skeletal muscle. In the skeletal muscle, the choline plasmalogen increased from 1.0 to 9.7% of the total phospholipids from 10 weeks of fetal life to 27 years of age. During the same period, the ethanolamine plasmalogen increased from 10.4 to 15.1% with a maximum of 18.9% in a 13-month-old infant. Similar changes were found in heart muscle. A substantial increase in alk-1-enylacyl-GPC and a small decrease in alk-1-enylacyl-GPE were found during development in bovine heart (De Rooij and Hooghwinkel, 1967) and ovine heart (Scott *et al.*, 1967b). A gradual increase in the proportion of choline plasmalogens in the total phospholipids from hearts during development of the cat, dog, and man was detected qualitatively by Hack and Helmy (1965). In contrast, small decreases in the proportion of total alk-1-enyl phospholipids in the total heart and skeletal muscle were found in rats from 1 to 8 weeks of age (Gottfried and Rapport, 1963).

There seem to be genuine species differences in the total plasmalogen concentration of mature animal muscles. In cardiac muscles, plasmalogens account for more than 40% of the phospholipids in bovine hearts, about 32% in human hearts, but only 12 to 13% in hearts from rats and mice. Human skeletal muscle apparently has a lower plasmalogen concentration than does the human heart, and the mouse and rat also have relatively low concentrations of plasmalogens in the phospholipids from skeletal muscle when compared to their heart tissue.

Skeletal muscles from the pigtail monkey are quite similar to human skeletal muscle in their plasmalogen concentration. Masoro *et al.* (1966) have commented on differences in the relative amounts of choline and ethanolamine plasmalogens in fast and slow muscles. The content of choline plasmalogens is higher in the soleus muscle whereas the content of

TABLE XIV

The Content of Alk-1-enyl and Alkyl Groups in the Phospholipids from Skeletal Muscle

Alk-1-enyl groups (μmoles/g tissue)	Mole ratio, component:lipid P				Percent of lipid class		Animal, reference
	Alk-1-enyl groups	Alk-1-enyl-acyl-GPC	Alk-1-enyl-acyl-GPE	Alkyl groups	Alk-1-enyl-acyl-GPC	Alk-1-enyl-acyl-GPE	
—	0.25	0.10	0.15	—	19	63	Man, Hughes and Frais, 1967
2.9	0.29	—	—	—	—	—	Man, Rapport and Lerner, 1959
3.2	0.26	—	—	—	—	—	Man, Winterfeld and Debuch, 1966
—	—	0.11	0.10	—	19	45	Man, Debuch and Winterfeld, 1966
1.8	0.20	—	—	—	—	—	Man, uterus, Winterfeld and Debuch, 1966
—	—	0.02	0.10	—	4	40	Man, uterus, Debuch and Winterfeld, 1966
2.0	0.18	—	—	—	—	—	Man, uterus, Rapport and Lerner, 1959
2.5	0.23	0.08	0.13	—	—	—	Pigtail monkey, gastrocnemius, Masoro *et al.*, 1966
2.5	0.24	—	—	—	17	53	Pigtail monkey, gastrocnemius, Masoro *et al.*, 1964
2.6	0.22	—	—	—	—	—	Pigtail monkey, soleus, Masoro *et al.*, 1966
2.8	0.24	—	—	—	28	28	Pigtail mokney, soleus, Masoro *et al.*, 1964
—	—	—	—	0.016	—	—	Dog, Nakagawa and McKibbin, 1962
—	0.27	0.14	0.13	—	28	—	Ox, longissimus dorsi, Davenport, 1964
1.6	0.09	0.02	0.07	0.034	5	31	Sheep, Dawson, 1960
—	0.09	—	—	—	—	—	Rat, Gottfried and Rapport, 1963
1.6	0.12	—	—	—	—	—	Rat, Rapport and Lerner, 1959
2.4	0.13	0.02	—	—	—	—	Rat, Webster, 1960

1.8	—	—	—	—	—	—	Rat, Broderson, 1967
1.3	—	—	—	—	—	—	Rat, Norton, 1960
1.4	0.11	—	—	—	—	—	Rat, Wittenberg *et al.*, 1956
—	—	—	0.10	—	—	—	Rat, Keenan *et al.*, 1968
—	—	—	0.10	—	—	—	Rat, uterus, Keenan *et al.*, 1968
1.2	—	—	—	—	—	—	Rat, J. N. Williams *et al.*, 1962
—	0.10	—	—	—	—	—	Rat, gastrocnemius, Gueuning and Graff, 1967
—	0.14	—	—	—	—	—	Rat, Masoro, 1967
—	0.09	—	—	—	3	29	Mouse, Owens and Hughes, 1970; Owens, 1966
—	—	—	0.07	—	—	26	Mouse, myofibrils, Owens and Hughes, 1970
—	—	—	0.06	—	—	19	Mouse, mitochondria, Owens and Hughes, 1970
—	—	—	0.06	—	—	28	Mouse, microsomes, Owens and Hughes, 1970
—	—	0.11	—	—	17	—	Rabbit, Gray and Macfarlane, 1961
—	—	0.14	—	—	22	—	Rabbit, myofibrils, Gray and Macfarlane, 1961
—	—	—	—	—	10	—	Rabbit, sarcoplasmic reticulum, Waku and Lands, 1968
1.7	0.20	—	—	—	—	—	Rabbit, Rapport and Lerner, 1959
—	0.12	0.08	0.04	—	19	—	Pigeon, breast, Davenport, 1964
—	—	0.16	—	—	33	12	Pigeon, breast, Gray and Macfarlane, 1961, 1964
—	0.30	0.11	—	—	—	—	Chicken, Webster, 1960
7.0	0.15	0.06	0.09	—	11	42	Turkey, breast, Neudoerffer and Lea, 1968
10.4	0.14	0.05	0.10	—	10	36	Turkey, leg, Neudoerffer and Lea, 1968

TABLE XV

The Content of Alk-1-enyl and Alkyl Groups in the Phospholipids from the Kidney

Alk-1-enyl groups (μmoles/g tissue)	Mole ratio, component:lipid P					Percent of lipid class			Animal, reference
	Alk-1-enyl groups	Alk-1-enyl-acyl-GPC	Alk-1-enyl-acyl-GPE	Alk-1-enyl-acyl-GPS	Alkyl groups	Alk-1-enyl-acyl-GPC	Alk-1-enyl-acyl-GPE	Alk-1-enyl-acyl-GPS	
6.0	0.20	0.03	0.14	0.03	—	8	45	30	Pig, Gray and Macfarlane, 1961
—	—	—	—	—	0.005	—	—	—	Dog, Nakagawa and McKibbin, 1962
2.5	0.06	0.01	0.05	tr	0.03	3	21	—	Sheep, Dawson, 1960
2.2	0.06	0.01	0.05	—	0.024	3	30	—	Sheep, Scott *et al.*, 1967b
—	0.14	0.03	0.07	0.02	—	8	41	32	Sheep, Getz *et al.*, 1968
—	0.10	0.04	0.05	0.01	—	9	24	38	Sheep, mitochondria, Getz *et al.*, 1968
—	0.10	0.04	0.05	0.01	—	12	27	19	Sheep, microsomes, Getz *et al.*, 1968
—	—	—	—	—	—	—	42	—	Rabbit, Gray, 1967b
3.3	0.12	—	—	—	—	—	—	—	Rabbit, Rapport and Lerner, 1959
5.6	0.18	—	—	ND	—	2	25	ND	Rabbit, cortex, Morgan *et al.*, 1963
3.2	0.18	—	—	—	—	5	18	1	Rabbit, medulla, Morgan *et al.*, 1963
3.5	0.09	—	—	—	—	—	—	—	Rat, Rapport and Lerner, 1959
4.4	0.08	0.006	—	—	—	—	—	—	Rat, Webster, 1960
3.2	—	—	—	—	—	—	—	—	Rat, Broderson, 1967
—	—	—	0.08	—	—	—	—	—	Rat, Keenan *et al.*, 1968
2.8	0.07	—	—	—	—	—	—	—	Rat, Wittenberg *et al.*, 1956
2.7	—	—	—	—	—	—	—	—	Rat, Norton, 1960
—	0.04	0.01	0.03	ND	—	3	10	ND	Pigeon, Belsare and Chowdhuri, 1968

ethanolamine plasmalogens is higher in the gastrocnemius muscle. A very low content of alk-1-enylacyl-GPC was found in human uterine muscle (Debuch and Winterfeld, 1966). In most other cases, the muscles used for plasmalogen assays were not identified.

The data on alkyl group concentrations in muscle tissue are rather sparse, but about 3 to 5% of the CPG and 2 to 4% of the EPG exist in the alkylacyl form in cardiac muscle. Similar data are not available for any skeletal muscle. In the human heart, 2.4% of the lipid phosphorus was in the alkylacyl-GPC and 0.9% in the alkylacyl-GPE (Panganamala *et al.*, 1971). Alkyl groups have also been reported in phospholipids from ox heart (Schmid and Takahashi, 1968; Takahashi and Schmid, 1968; Viswanathan *et al.*, 1969), hog heart (Viswanathan *et al.*, 1969), and rat heart and skeletal muscle (Wood and Snyder, 1968).

6. *Renal Tissues*

The ethanolamine plasmalogens account for most of the alk-1-enyl groups in the kidney (Table XV; Helmy and Longley, 1966; Hack and Helmy, 1967b). In pig, rabbit, and sheep kidneys, from 14 to 20% of the phospholipids are plasmalogens, including more than 40% of the EPG. The ethanolamine plasmalogen concentration is lower in rat and pigeon kidney phospholipids and only a small amount of alk-1-enylacyl-GPC is present. Most investigators found a small amount of alk-1-enylacyl-GPS, but Morgan *et al.* (1963) found more alk-1-enyl groups in the IPG than in the SPG.

Scott *et al.* (1967b) found that the concentration of alk-1-enyl groups in the phospholipids was greater in fetal sheep kidneys than in adult sheep kidneys and Getz *et al.* (1968) and Broderson (1967) determined that renal mitochondria and microsomes have similar plasmalogen concentrations. Alkyl groups have been detected in the kidney phospholipids from man (Svennerholm and Thorin, 1960) and rat (Wood and Snyder, 1968).

An interesting compound that causes guinea pig ileum to contract has been isolated from the rabbit kidney medulla by Wiley *et al.* (1970). This compound has been identified as the cyclic acetal of octadec-9-enal with glycerol 3-phosphate. The pharmacologic action mimics that of acetylcholine and "Darmstoff."

7. *Liver*

The low content of alk-1-enyl groups and the presence of considerable amounts of retinal have led to some difficulties in the measurement of alk-1-enyl groups in the liver (Rapport and Norton, 1962; Camejo *et al.*, 1964), but it seems that the alk-1-enyl group content of rat and rabbit liver is 0.5–0.7 μmole/g, a value that is considerably lower than in most

TABLE XVI

The Content of Alk-1-enyl and Alkyl Groups in the Phospholipids from the Liver

Alk-1-enyl groups (μmoles/g tissue)	Mole ratio, component:lipid P				Percent of lipid class		
	Alk-1-enyl groups	Alk-1-enyl-acyl-GPC	Alk-1-enyl-acyl-GPE	Alkyl groups	Alk-1-enyl-acyl-GPC	Alk-1-enyl-acyl-GPE	Animal, reference
—	0.008	0.002	0.006	0.013	0.4	2	Man, Pries *et al.*, 1966
—	0.023	0.15	0.008	0.015	3	4	Man, Blomstrand *et al.*, 1962
—	—	—	—	0.006	—	—	Dog, Nakagawa and McKibbin, 1962
3.2	—	—	—	—	—	—	Ox, J. N. Williams *et al.*, 1962
—	0.051	0.015	0.036	0.005	3	28	Ox, Dawson *et al.*, 1962
0.12	0.002	ND	0.002	0.005	—	0.7	Sheep, Dawson, 1966
2.1	0.044	0.008	0.036	0.021	2	11	Sheep, Scott *et al.*, 1967b
—	0.035	0.014	0.014	0.017	2	9	Sheep, Getz *et al.*, 1968
—	0.029	0.005	0.015	0.013	1	7	Sheep, mitochondria, Getz *et al.*, 1968
—	0.044	0.021	0.008	0.017	3	6	Sheep, microsomes, Getz *et al.*, 1968
—	0.010	0.004	0.003	0.016	0.6	4	Sheep, cytosol, Getz *et al.*, 1968
0.7	0.02	—	—	—	—	—	Rabbit, Rapport and Lerner, 1959
0.1	—	—	—	—	—	—	Rabbit, Robertson and Lands, 1962
0.6	—	—	—	—	—	—	Rabbit, Camejo *et al.*, 1964
1.0	—	—	—	—	—	—	Guinea pig, Camejo *et al.*, 1964
0.4	—	—	—	—	—	—	Mouse, Camejo *et al.*, 1964
0.5	—	—	—	—	—	—	Rat, Camejo *et al.*, 1964
1.0	0.03	—	—	—	—	—	Rat, Rapport and Lerner, 1959
—	0.034	—	0.014	—	—	8	Rat, Getz *et al.*, 1968

—	—	—	0.019	—	—	—	Rat, Keenan *et al.*, 1968
0.7	—	—	—	—	—	—	Rat, Broderson, 1967
0.7	—	—	—	—	—	—	Rat, Norton, 1960
1.1	0.024	—	—	—	—	—	Rat, Wittenberg *et al.*, 1956
—	0.02	—	—	—	—	—	Rat, Gottfried and Rapport, 1963
0.7	—	—	—	—	—	—	Rat, J. N. Williams *et al.*, 1962
—	0.005	—	—	—	—	—	Rat, Gurr *et al.*, 1963
—	—	—	—	—	0.1	2	Rat, Snyder *et al.*, 1969a
0.8	0.02	0.003	—	—	—	—	Rat, Webster, 1960
—	0.024	0.008	0.008	—	2	3	Rat, mitochondria, Getz *et al.*, 1968
—	0.018	—	—	—	—	—	Rat, mitochondria, Wittenberg *et al.*, 1956
—	0.023	—	0.014	—	—	11	Rat, microsomes, Getz *et al.*, 1968
—	0.045	—	—	—	—	—	Rat, microsomes, Wittenberg *et al.*, 1956
—	0.032	—	0.013	—	—	17	Rat, cytosol, Getz *et al.*, 1968
—	0.079	—	—	—	—	—	Rat, cytosol, Wittenberg *et al.*, 1956
—	0.034	—	—	—	—	—	Rat, nuclei, Wittenberg *et al.*, 1956
—	0.011	—	—	—	—	—	Rat, nuclei, Gurr *et al.*, 1963
—	0.025	0.011	0.014	—	—	—	Rat, nuclei, Song *et al.*, 1970
—	0.033	0.018	0.014	—	4	4	Rat, plasma membrane, Takeuchi and Terayama, 1965
—	0.01	—	—	0.05	—	—	Rat, plasma membrane, Pfleger *et al.*, 1968
1.0	0.03	0.003	—	—	—	—	Chicken, Webster, 1960
—	0.004	—	0.004	—	—	2	Pigeon, Belsare and Chowdhuri, 1968

other tissues (Table XVI). The alk-1-enyl group content may be severalfold higher in ruminant livers than in rodent livers. From 1 to 5% of the phospholipids in mammalian livers contain alk-1-enyl groups which are found in both the EPG and the CPG. Almost 10% of the EPG contain alk-1-enyl groups. Takeuchi and Terayama (1965) reported that serine plasmalogens accounted for 0.1% of the phospholipids from rat liver plasma membranes. Hübscher and Clark (1960) found small amounts of alk-1-enyl groups in a phospholipid fraction from rat, ox, and pig liver that was thought to be phosphatidic acids. The high content of 18:2 acyl groups in this fraction suggests that cardiolipins were also present.

During development, the plasmalogen content of the liver increases more than twofold in sheep (Scott *et al.*, 1967b) and in rats (Vennart, 1963). Gottfried and Rapport (1963) found a small decrease in the concentration of alk-1-enyl groups in rat liver phospholipids during development.

From 1 to 3% of liver phospholipids contain alkyl groups. The alkyl groups are present in both the CPG and the EPG from rat liver (Snyder *et al.*, 1969a) and in the EPG from human liver (Svennerholm and Thorin, 1960). Wood and Snyder (1968) found a higher concentration of alkyl groups in rat liver than in rat kidney or heart.

8. *Blood Plasma*

The amount of alkyl groups is greater than the amount of alk-1-enyl groups in human serum. The concentration of plasmalogens in human plasma phospholipids is about 2% (Table XVII), a value similar to that for liver phospholipids, and about 4% of the phospholipids from human bile are plasmalogens, primarily alk-1-enylacyl-GPC (Nakayama and Blomstrand, 1961). The low content of plasmalogens in the phospholipids from these body fluids may be due to their hepatic origin. The small proportion of EPG in plasma from man and horse includes some alk-1-enylacyl-GPE. Serine plasmalogens have not been found in the plasma from any mammal. Renkonen (1962a, 1963b) reported that human serum contains 40 μmoles of alkylacyl-GPC, 20 μmoles of alk-1-enylacyl-GPC, and 5 μmoles each of alkylacyl-GPE and alk-1-enylacyl-GPE per liter. The presence of acyl groups in the serum phosphoglycerides that contain alkyl groups was established by Renkonen (1962a,b, 1963a,b), and alk-1-enyl and alkyl groups were detected in rat serum phosphoglycerides by Wood and Snyder (1968).

9. *Male Gonads and Spermatozoa*

In plasmalogen content and concentration, the testis (Table XVIII) is similar to the kidney and spleen. A large portion of the EPG contain alk-1-

TABLE XVII

The Alk-1-enyl and Alkyl Group Contents in the Phospholipids from Blood Plasma

Mole ratio, component:lipid P				Percent of lipid class		
Alk-1-enyl groups	Alk-1-enyl-acyl-GPC	Alk-1-enyl-acyl-GPE	Alkyl groups	Alk-1-enyl-acyl-GPC	Alk-1-enyl-acyl-GPE	Animal, reference
0.019	0.019	—	—	3.6	—	Man, De Rooij and Hooghwinkel, 1967
0.022	0.008	0.014	—	1.1	58	Man, J. H. Williams *et al.*, 1966a
0.006	0.003	0.003	0.026	0.4	12	Man, Pries *et al.*, 1966
—	—	—	—	1.3	9	Man (serum), Renkonen, 1962a, 1963b
0.010	0.007	—	0.15	1.1	—	Man, Blomstrand *et al.*, 1962
0.038	0.038	—	—	4.8	—	Man, Dawson *et al.*, 1960
0.027	0.027	—	—	3.3	—	Pig, Dawson *et al.*, 1960
0.023	—	—	0.028	—	—	Horse (serum), R. E. Anderson *et al.*, 1969a
0.025	0.015	0.010	—	1.8	67	Horse, Dawson *et al.*, 1960
0.064	0.064	—	—	7.2	—	Ox, Dawson *et al.*, 1960
0.021	0.021	—	—	—	—	Sheep, Dawson *et al.*, 1960
0.020	0.020	—	—	—	—	Goat, Dawson *et al.*, 1960
0.022	0.022	—	—	2.7	—	Pigeon, Belsare and Chowdhuri, 1968

TABLE XVIII

The Alk-1-enyl and Alkyl Group Content in the Phospholipids from Male Gonads and Spermatozoa

Alk-1-enyl groups (μmoles/g tissue)	Mole ratio, component:lipid P				Percent of lipid class		Tissue	Animal, reference
	Alk-1-enyl groups	Alk-1-enyl-acyl-GPC	Alk-1-enyl-acyl-GPE	Alkyl groups	Alk-1-enyl-acyl-GPC	Alk-1-enyl-acyl-GPE		
1.8	0.11	0.02	0.07	—	4	40	Testis	Man, Winterfeld and Debuch, 1966; Debuch and Winterfeld, 1966
—	0.16	0.06	0.10	0.059	12	57	Testis	Ram, Scott and Setchell, 1968
1.6	0.08	0.01	0.06	—	2	23	Testis	Rat, Oshima and Carpenter, 1968
1.8	—	—	—	—	—	—	Testis	Rat, Broderson, 1967
1.6	0.08	—	—	—	—	—	Testis	Rat, Wittenberg *et al.*, 1956
2.5	0.14	0.05	0.09	0.072	9	37	Testis	Rat, Scott *et al.* (1963)
4.1	—	—	—	—	—	—	Prostate	Rat, Broderson, 1967
1.6	—	—	—	—	—	—	Seminal vesicle	Rat, Broderson, 1967
3.0	0.14	0.02	0.12	0.021	5	48	Epididymis (head)	Rat, Scott *et al.* (1963)
3.1	0.17	0.06	0.10	0.034	12	46	Epididymis (tail)	Rat, Scott *et al.* (1963)
—	0.20	0.10	0.10	—	21	30	Spermatozoa (ejaculated)	Boar, Grogan *et al.*, 1966
—	0.20	0.06	0.13	—	12	41	Spermatozoa (tail of epididymis)	Boar, Grogan *et al.*, 1966

—	0.20	0.05	0.15	—	9	46	Spermatozoa (mid-epididymis)	Boar, Grogan *et al.*, 1966
—	0.19	0.05	0.14	—	9	40	Spermatozoa (head of epididymis)	Boar, Grogan *et al.*, 1966
—	0.35	0.28	0.07	—	44	26	Spermatozoa	Bull, Pursel and Graham, 1967
—	0.50	0.42	0.06	0.126	79	61	Spermatozoa (ejaculated)	Ram, Scott *et al.*, 1967a
—	0.47	0.33	0.12	0.081	71	70	Spermatozoa (testicular)	Ram, Scott *et al.*, 1967a
—	0.40	—	—	—	50	—	Semen	Ram, Gray, 1960a
—	—	—	—	—	33	17	Spermatozoa (tail of epididymis)	Rat, Dawson and Scott, 1964
—	—	—	—	—	12	14	Spermatozoa (head of epididymis)	Rat, Dawson and Scott, 1964
—	0.40	0.24	0.16	—	44	61	Seminal plasma	Bull, Pursel and Graham, 1967

enyl groups, but the prostate gland and the epididymis probably contain more alk-1-enyl groups than the testis. Serine plasmalogens were detected by Debuch and Winterfeld (1966) in human testicular tissue at a concentration of 4% of the phospholipids. An appreciable content of alkyl groups has also been found in testicular and epididymal tissue.

Marked species differences are apparent in the plasmalogen concentrations of spermatozoa. These range from 20% of the phospholipids from boar spermatozoa to 35% from bull spermatozoa to 50% from ram spermatozoa; in the latter, 80% of the CPG contained alk-1-enyl groups. Hartree and Mann (1961) reported the same value after freezing and thawing but found that only 50% of the CPG from fresh ram spermatozoa contained alk-1-enyl groups. Large differences in plasmalogen concentrations are also found during maturation of the spermatozoa. Most of these metabolic changes seem to be related to changes in the content of the other phospholipids. A large content of plasmalogens was found in seminal plasma from the bull, but not in the seminal plasma from man, boar, or stallion (Hartree and Mann, 1959).

10. *Blood Cells*

About 15% of the phospholipids from human red blood cells and platelets are alk-1-enylacyl-GPE, which account for half of the EPG (Table XIX). Small amounts of alk-1-enylacyl-GPC and possibly alk-1-enylacyl-GPS (Pries *et al.*, 1966; Farquhar, 1962) are also present in the erythrocytes. Marcus *et al.* (1969) found small amounts of alk-1-enylacyl-GPS and alk-1-enylacyl-GPI in human platelets, and according to R. D. Zilversmit *et al.* (1961) the content of alk-1-enyl groups is 3.0 μmoles/g of platelets. The concentration of alk-1-enylacyl-GPE is higher in white cells than in red cells or platelets. Sequential hydrolysis methods have given low results for the alk-1-enyl group concentration of erythrocytes from a number of species (Dawson *et al.*, 1960; Pries *et al.*, 1966; Belsare and Chowdhuri, 1968).

Less than 2% of the phospholipids from human red blood cells but almost 10% of platelet phospholipids contain alkyl groups according to Pries *et al.* (1966). Dawson *et al.* (1962) gave a value of 1.5% for human red cells but Hanahan (cited by Thompson and Hanahan, 1963) did not detect alkyl groups in the phospholipids. In contrast, the alkylacyl-GPE accounts for over 75% of the EPG from bovine erythrocytes (Hanahan *et al.*, 1963; Hanahan and Watts, 1961). These reports were the first to establish the presence of 2-acyl groups in phosphoglycerides containing 1-alkyl groups. Alkyl groups have also been detected in the phospholipids from sheep red blood cells by Nelson (cited by Rouser *et al.*, 1968).

TABLE XIX

The Content of Alk-1-enyl Groups in the Phospholipids from Human Blood Cells

Mole ratio, component:lipid P			Percent of lipid class		
Alk-1-enyl groups	Alk-1-enyl-acyl-GPC	Alk-1-enyl-acyl-GPE	Alk-1-enyl-acyl-GPC	Alk-1-enyl-acyl-GPE	Blood cell type, reference
0.14	0.01	0.13	4	46	Erythrocytes, Cohen and Derksen, 1969
0.10	0.01	0.09	2	34	Erythrocytes, De Rooij and Hooghwinkel, 1967
0.16	ND	0.16	—	54	Erythrocytes, Gottfried, 1967
0.14	—	—	—	—	Erythrocytes, Dodge and Phillips, 1967
0.16	0.01	0.15	4	52	Erythrocytes, J. H. Williams *et al.*, 1966b
0.12	—	—	—	35	Erythrocytes, Ways and Hanahan, 1964
0.12	0.01	0.10	3	37	Erythrocytes, Dawson *et al.*, 1960
0.19	—	—	10	67	Erythrocytes, Farquhar, 1962
0.23	ND	0.23	—	75	Leukocytes, Gottfried, 1967
0.13	ND	0.13	—	45	Lymphocytes, Gottfried, 1967
0.22	ND	0.22	—	66	Polymorphonuclear leukocytes, Gottfried, 1967
0.14	tr	0.14	—	51	Platelets, Cohen and Derksen, 1969
0.16	—	0.16	2	60	Platelets, Marcus *et al.*, 1969
—	—	0.10	—	30	Platelets, Nordöy and Lund, 1968
0.13	0.01	0.10	3	45	Platelets, Pries *et al.*, 1966

11. *Spleen*

The ethanolamine plasmalogens also account for most of the alk-1-enyl groups in the spleen, and the EPG from the spleen contain a high proportion of alk-1-enyl groups (Table XX). Small amounts of choline plasmalogens, but no serine plasmalogens, were found by Getz *et al.* (1968). They also reported a higher proportion of alkyl groups in spleen phospholipids than in the phospholipids from most of the other organs. A similar observation was reported for rat spleen (Wood and Snyder, 1968). Alkyl groups were also found in the EPG from human spleen (Svennerholm and Thorin, 1960).

12. *Bone Marrow*

Plasmalogen concentrations in bone marrow are similar to those in spleen (Pietruszko, 1962; Thompson and Hanahan, 1963), but a direct comparison of the concentration of alkyl groups cannot be made. In bovine bone marrow, alkylacyl phosphoglycerides account for 28% of the EPG and 12% of the CPG (Thompson and Hanahan, 1963). In the red bone marrow from pig epiphyses, 20% of the CPG is alkylacyl-GPC (Pietruszko, 1962), which accounts for 2.7 μmoles/g of red bone marrow. The phospholipids from rat femoral bone marrow also have a relatively high concentration of alkyl groups (Wood and Snyder, 1968).

13. *Lung*

Very little detailed information is available for lung tissue (Table XX), but the concentration and content of alk-1-enyl groups are similar to those found for the kidney and spleen.

14. *Gastrointestinal Tract*

Intestinal tissue is another interesting tissue for which very little information is available on the content of alk-1-enyl and alkyl groups (Table XX). Forstner *et al.* (1968) did not detect alk-1-enyl groups in the microvillus plasma membrane or whole brush border fractions from rat intestine, but alkyl groups were found in the intestinal mucosa from man, pig, rabbit, and ox by Paltauf and Polheim (1967). In the pig, from 2 to 6% of the EPG, SPG, and CPG contained alkyl groups. Very high contents of choline plasmalogens have been reported in rat intestinal mucosa by Di Costanzo and Clement (1963, 1965). The earlier of two CPG fractions eluted from a silicic acid column gave a positive reaction with Schiff's reagent but the amount of alk-1-enyl groups in the fraction was not assayed (Di Costanzo and Clement, 1963, 1965). The positive reaction for aldehydes was probably due to oxidation products from the polyunsaturated acyl groups.

TABLE XX

The Content of Alk-1-enyl and Alkyl Groups in the Phospholipids from the Spleen, Lung, and Other Tissues

Alk-1-enyl groups (μmoles/g tissue)	Mole ratio, component:lipid P				Percent of lipid class		Tissue	Animal, reference
	Alk-1-enyl groups	Alk-1-enyl-acyl-GPC	Alk-1-enyl-acyl-GPE	Alkyl groups	Alk-1-enyl-acyl-GPC	Alk-1-enyl-acyl-GPE		
3.2	0.16	—	—	—	—	—	Spleen	Man, Rapport and Lerner, 1959
—	—	—	—	0.005	—	—	Spleen	Dog, Nakagawa and McKibbin, 1962
—	0.14	0.03	0.10	0.08	7	66	Spleen	Sheep, Getz *et al.*, 1968
—	0.15	0.03	0.10	0.10	9	60	Spleen, mitochondria	Sheep, Getz *et al.*, 1968
2.9	0.14	—	—	—	—	—	Spleen	Rabbit, Rapport and Lerner, 1959
3.2	0.15	—	—	—	—	—	Spleen	Rat, Rapport and Lerner, 1959
—	0.12	—	—	—	—	—	Spleen	Rat, Gottfried and Rapport, 1963
4.0	—	—	—	—	—	—	Spleen	Rat, Broderson, 1967
—	—	—	0.13	—	—	—	Spleen	Rat, Keenan *et al.*, 1968
2.0	0.07	—	—	—	—	—	Spleen	Rat, Wittenberg *et al.*, 1956
—	—	—	—	0.004	—	—	Lung	Dog, Nakagawa and McKibbin, 1962
2.3	0.07	0.008	0.06	0.02	2	31	Lung	Sheep, Dawson, 1960
1.5	—	—	—	—	—	—	Lung	Rabbit, Robertson and Lands, 1962
3.5	0.13	—	—	—	—	—	Lung	Rabbit, Rapport and Lerner, 1959

TABLE XX (*Continued*)

Alk-1-enyl groups (μmoles/g tissue)	Mole ratio, component:lipid P				Percent of lipid class		Tissue	Animal, reference
	Alk-1-enyl groups	Alk-1-enyl-acyl-GPC	Alk-1-enyl-acyl-GPE	Alkyl groups	Alk-1-enyl-acyl-GPC	Alk-1-enyl-acyl-GPE		
3.6	0.14	—	—	—	—	—	Lung	Rat, Rapport and Lerner, 1959
3.0	0.09	—	—	—	—	—	Lung	Rat, Wittenberg *et al.*, 1956
3.1	—	—	—	—	—	—	Lung	Rat, Norton, 1960
3.4	—	—	—	—	—	—	Lung	Rat, J. N. Williams *et al.*, 1962
—	0.13	—	—	—	—	—	Lung	Rat, Gottfried and Rapport, 1963
4.3	—	—	—	—	—	—	Lung	Rat, Broderson, 1967
—	—	—	—	—	—	42	Lung	Rat, Gray, 1967b
—	—	—	0.14	—	—	—	Lung	Rat, Keenan *et al.*, 1968
2.1	0.16	—	—	—	—	—	Stomach	Man, Rapport and Lerner, 1959
2.0	0.14	—	—	—	—	—	Stomach, fundus	Man, Rapport and Lerner, 1959
1.6	0.19	—	—	—	—	—	Colon	Man, Rapport and Lerner, 1959
—	—	—	—	0.010	—	—	Intestine	Dog, Nakagawa and McKibbin, 1962
2.6	0.14	—	—	—	—	—	Stomach	Rabbit, Rapport and Lerner, 1959
2.0	0.12	—	—	—	—	—	Stomach	Rat, Rapport and Lerner, 1959
1.0	—	—	—	—	—	—	Upper stomach	Rat, Broderson, 1967
1.8	—	—	—	—	—	—	Lower stomach	Rat, Broderson, 1967

2.9	0.06	—	—	—	—	—	Pancreas	Ox, Prottey and Hawthorne, 1966
2.4	0.08	—	—	—	—	—	Pancreas	Guinea pig, Prottey and Hawthorne, 1966
—	—	—	0.08	—	—	—	Pancreas	Rat, Keenan *et al.*, 1968
1.6	0.17	—	—	—	—	—	Ovary	Man, Winterfeld and Debuch, 1966
—	0.10	0.016	0.09	0.012	3	39	Corpus luteum	Ox, Scott *et al.*, 1968
—	0.15	—	—	—	—	—	Placenta	Man, Winterfeld and Debuch, 1968
1.1	0.14	0.024	0.10	—	7	36	Skin	Man, Gerstein, 1963
—	—	—	—	0.03	10	75	Preputial gland	Mouse, Snyder and Blank, 1969
—	0.12	0.08	0.04	—	19	18	Fibroblast L cells	Mouse (tissue culture), Weinstein *et al.*, 1969
—	—	—	—	0.004	—	—	Aorta	Dog, Nakagawa and McKibbin, 1962
1.4	0.08	0.016	0.06	0.036	3	21	Fetal brown adipose tissue	Sheep, Scott *et al.*, 1967b
0.7	0.29	—	—	—	—	—	Adipose tissue	Rat, Gilbertson, 1969

The gastric and intestinal tissues include a large proportion of muscle cells that could account for the plasmalogens present. Further studies on the epithelial cells are required, since a large volume of water is transported through these cells. A complete absence of plasmalogens would be surprising, because other tissues that transport water contain appreciable concentrations of alk-1-enyl or alkyl groups.

15. *Miscellaneous Tissues*

Levels of alkyl and alk-1-enyl lipids for a variety of tissues are summarized in Table XX. The epithelial cells from human skin contain an appreciable concentration of plasmalogens. About 1 μmole of alk-1-enyl groups per gram of tissue have been found in human aorta (Miller *et al.*, 1964; Buddecke and Andresen, 1959). Pries *et al.* (1966) reported that the plasmalogens accounted for less than 6% of the phospholipids from the intima and media, but Miller *et al.* (1964) found a value of 17%. Alk-1-enyl groups exist in human placenta mostly in the EPG (Helmy and Hack, 1964), and alk-1-enyl groups are present in both EPG and CPG from human amniotic fluid (Helmy and Hack, 1962). Serine plasmalogens were found in fibroblast L cells by Weinstein *et al.* (1969). According to R. E. Anderson *et al.* (1969a), fibroblast L–M cells contain considerable amounts of alk-1-enylacyl-GPE, alkylacyl-GPE, and alkylacyl-GPC.

The phospholipids from rat adipose tissue have a high concentration of plasmalogens (Gilbertson, 1969), approaching that of white matter from the brain. Morrill and Rapport (1964) found that the alk-1-enyl group content of rat adipose tissue was from 2.1 to 2.7 μmoles/g of tissue between 1 and 8 days of age. If all of the alk-1-enyl groups were in phospholipids, the mole ratio of alk-1-enyl groups to lipid phosphorus was 0.21 at 8 days of age. For the same age a value of 0.04 was found by Yarbro and Anderson (1956) who found no plasmalogens in adipose tissue from the adult rat (Yarbro and Anderson, 1957). Although further studies are required, I believe that the content of plasmalogens per adipose cell may be quite high in comparison to that of many other tissues.

16. *Chicken Eggs*

Carter *et al.* (1958) found that about 1% of the phospholipids of egg yolks contained alkyl groups. Values of 3% (Holub and Kuksis, 1969) and 8% (Renkonen, 1967b) of the total lipids have been reported for alkylacyl-GPE in the EPG, whereas only traces of plasmalogens are present (Renkonen, 1968).

III. Composition

A. Alk-1-enyl and Alkyl Groups

Alk-1-enyl group compositions have not been studied nearly as much as acyl group compositions, although most phospholipid mixtures contain an appreciable proportion of alk-1-enyl groups. Unfortunately, phospholipids are commonly subjected to an acid-catalyzed methanolysis followed by an analysis of the fatty acid methyl esters by gas–liquid chromatography (GLC) on a polyester column. Two of the resulting peaks may be identified as 16:0 and 18:0 dimethyl acetal (DMA) derivatives of aldehydes, but the 18:1 DMA peak coincides with the 18:0 methyl ester peak. As a result, neither the alk-1-enyl groups nor the acyl groups are analyzed correctly. A better procedure is to separate the DMA from the esters before GLC. The recommended procedure is to form the alk-1-enyl group derivatives prior to the formation of acyl group derivatives (Horrocks and Sun, 1972).

The dimethyl acetal derivatives of the aldehydogenic moiety are used most often for determining the chain length and degree of unsaturation of the alk-1-enyl groups (Mahadevan, 1970). Rao *et al.* (1967) and Sun and Horrocks (1969b) have reported the same compositions for the alk-1-enyl groups from ox heart CPG and from mouse brain EPG with dimethyl acetal (dimethoxyalkane) and cyclic acetal (alkyl dioxolane) derivatives. Panganamala *et al.* (1971) found the same compositions for the alk-1-enyl groups from human heart CPG and EPG by chromatographing the free aldehydes and the alkyl dioxolanes. However, some comparable analyses of alk-1-enyl groups as the DMA derivatives have included much higher contents of branched-chain alk-1-enyl groups.

Stein and Slawson (1966) reported that methoxyalkenes could be formed from dimethoxyalkanes (dimethyl acetals of aldehydes) under some conditions. Neudoerffer (1967) found that this could occur during a sulfuric acid-catalyzed methanolysis and that the methoxyalkenes had retention volumes very much like those of branched-chain dimethoxyalkanes. After injection into a gas chromatograph, this decomposition can produce *cis*- and *trans*-methoxyalkenes (Mahadevan *et al.*, 1968) which could both be reported as branched-chain alk-1-enyl groups. These observations may explain a discrepancy in the analyses of alk-1-enyl groups from chicken erthrocytes (Kates and James, 1961). The alk-1-enyl groups from the total lipids contained less than 1% of br 15:0 plus br 17:0, but the calculated value was about 12% for the same components from the individual phospholipids. Results from the use of the dimethyl acetal derivatives of

TABLE XXI

Composition of the Alk-1-enyl Groups from Mammalian and Avian Ethanolamine and Serine Phosphoglycerides

Percent of alk-1-enyl groups					
16:0	17:0	18:0	18:1	Other	Lipid, source, reference
22	2	33	42	—	EPG, brain, man, Panganamala *et al.*, 1971
14	1	57	23	—	EPG, gray matter, man, O'Brien and Sampson, 1965b
32	3	20	42	—	EPG, white matter, man, MacBrinn and O'Brien, 1969
35	3	20	42	—	SPG, white matter, man, MacBrinn and O'Brien, 1969
29	2	21	42	—	EPG, myelin, man, O'Brien and Sampson, 1965b
26	1	16	56	—	EPG, myelin, man, Horrocks, 1970
28	1	18	52	—	EPG, white matter microsomes, man, Horrocks, 1970
31	—	24	44	—	EPG, spinal cord, pig, Viswanathan *et al.*, 1968a
33	—	28	37	—	EPG, brain, pig, Viswanathan *et al.*, 1968a
25	2	33	40	—	EPG, brain, ox, Sun and Horrocks, 1969b
32	—	30	38	—	EPG, brain, ox, Albro and Dittmer, 1968
20	—	22	52	—	*N*-Acetyl EPG, brain, ox, Debuch and Wendt, 1967
27	4	24	44	—	EPG, optic nerve myelin, ox, MacBrinn and O'Brien, 1969
35	3	24	36	—	SPG, optic nerve myelin, ox, MacBrinn and O'Brien, 1969
24	2	22	51	—	EPG, brain myelin, ox, Sun and Horrocks, 1970
19	1	40	39	—	EPG, brain, mouse (3 months), Sun and Horrocks, 1969b
26	tr	28	46	—	EPG, brain, mouse (2 years), Sun and Horrocks, 1968
17	—	32	51	—	EPG, brain myelin, mouse, Sun and Horrocks, 1970
22	1	50	27	—	EPG, brain microsomes, mouse, Sun and Horrocks, 1970
22	—	49	30	—	EPG, brain mitochondria, mouse, Sun and Horrocks, 1970
24	—	40	36	—	EPG, brain, rat, Paltauf and Polheim, 1970
23	1	41	35	—	EPG, brain, rat, Cotman *et al.*, 1969
24	—	39	36	—	EPG, brain, rat, Albro and Dittmer, 1968
23	tr	5	68	—	EPG, lens (eye), ox, R. E. Anderson *et al.*, 1969c
26	tr	12	58	—	EPG, lens (eye), rabbit, R. E. Anderson *et al.*, 1969c
34	3	38	33	2	EPG, heart, man, Panganamala *et al.*, 1971
34	4	41	19	2	EPG, heart, man, Spener and Mangold, 1969
29	3	43	18	7	EPG, heart, ox, Schmid and Takahashi, 1968

TABLE XXI (*Continued*)

Percent of alk-1-enyl groups					
16:0	17:0	18:0	18:1	Other	Lipid, source, reference
30	4	43	11	12	EPG, heart, ox, Viswanathan *et al.*, 1968b
37	2	47	8	6	EPG, heart, ox, Rao *et al.*, 1967
33	4	43	7	13	EPG, heart, ox, Gray, 1960b
44	3	34	11	8	EPG, heart, pig, Gray, 1960b
43	3	37	10	7	EPG, breast muscle, pigeon, Gray and Macfarlane, 1961
50	6	12	5	27[a]	EPG, kidney, pig, Gray and Macfarlane, 1961
33	7	19	12	29[a]	EPG, kidney cortex, rabbit, Morgan *et al.*, 1963
36	4	39	19	—	EPG, liver, rat, Snyder *et al.*, 1969a
31	2	46	18	—	EPG, blood plasma, man, Panganamala *et al.*, 1969
17	3	26	14	40[a]	EPG, blood serum, man, J. H. Williams *et al.*, 1966a
24	2	53	18	—	EPG, erythrocyte, man, Panganamala *et al.*, 1969
16	2	40	26	16[a]	EPG, erythrocyte, man, Farquhar, 1962
31	—	60	9	—	EPG, platelet, man, Marcus *et al.*, 1969
42	—	51	7	—	EPG, platelet, man, Nordöy and Lund, 1968
52	tr	27	8	13[a]	EPG, erythrocyte, chicken, Kates and James, 1961
41	3	40	18	—	EPG, spleen, pig, Gray and Macfarlane, 1961
80	2	6	6	—	EPG, lung, pig, Gray and Macfarlane, 1961
28	—	47	11	14[b]	EPG, intestinal mucosa, rat, Paltauf and Polheim, 1970
39	2	42	16	—	EPG, aortic intima, man, Panganamala *et al.*, 1969
62	—	22	8	—	EPG, fibroblast L cells, mouse, Weinstein *et al.*, 1969

[a] The high proportion of "other" alk-1-enyl groups indicates the possible presence of artifacts.

[b] Includes 14% of 20:0 alk-1-enyl groups.

alk-1-enyl groups should be evaluated carefully because of the possible complications encountered in GLC identification.

A br 15:0 component has been reported in all analyses of alk-1-enyl groups from ruminant tissue lipids with the exception of the brain. Schmid and Takahashi (1968) found br 15:0 alk-1-enyl and alkyl groups in the radyl diacylglycerols from bovine heart. The br 15:0 components in ruminants may have a bacterial origin (Gray, 1967a).

Only three major components, 16:0, 18:0, and 18:1, are found in the alk-1-enyl and alkyl groups of most lipids. In Tables XXI–XXIV, the proportions of the 16:0, 17:0, 18:0, and 18:1 groups are given. If the proportion of these four components is less than 91%, the sum of the remaining components is listed under "others." Six percent is the highest proportion of "other" alk-1-enyl groups in any analysis of alk-1-enyl groups from the central nervous system.

In the EPG (Table XXI), the proportion of 16:0 alk-1-enyl groups is generally rather low; the range being 17–33% in the central nervous system. The marked differences in the ratio of 18:0 to 18:1 alk-1-enyl groups in

TABLE XXII

Composition of the Alk-1-enyl Groups from Mammalian and Avian Choline Phosphoglycerides

Percent of alk-1-enyl groups					
16:0	17:0	18:0	18:1	Other	Source, reference
22	2	44	26	—	Gray matter, man, O'Brien *et al.*, 1964
25	tr	19	50	—	White matter, man, O'Brien *et al.*, 1964
63	—	12	21	—	Brain, ox, Renkonen, 1966
41	tr	30	30	—	Brain, mouse (2 years) Sun and Horrocks, 1968
33	—	32	35	—	Brain, rat, Paltauf and Polheim, 1970
53	2	26	10	9	Brain, meninges, man, Bell *et al.*, 1967
62	2	14	18	4	Heart, man, Panganamala *et al.*, 1971
61	tr	17	14	8	Heart, man, Spener and Mangold, 1969
62	2	11	7	18	Heart, ox, Schmid and Takahashi, 1968
67	3	17	4	9	Heart, ox, Sun and Horrocks, 1969b
69	1	12	2	16	Heart, ox, Hoevet *et al.*, 1968
65	3	13	4	15	Heart, ox, Viswanathan *et al.*, 1968b,c
62	2	16	3	17	Heart, ox, Rao *et al.*, 1967
59	—	21	3	17	Heart, ox, Gottfried and Rapport, 1962
70	1	14	3	12	Heart, ox, Warner and Lands, 1963
68	2	9	tr	21	Heart, ox, Gray, 1960b
71	3	9	5	12	Heart, pig, Warner and Lands, 1963
83	3	6	5	3	Heart, pig, Gray, 1960b
82	1	10	6	1	Breast muscle, pigeon, Gray and Macfarlane, 1961
82	—	3	2	13[a]	Kidney, pig, Gray and Macfarlane, 1961
50	2	26	15	—	Blood plasma, man, Panganamala *et al.*, 1969
51	2	14	12	21[a]	Blood serum, man, J. H. Williams *et al.*, 1966a
56	4	26	12	—	Erythrocyte, man, Panganamala *et al.*, 1969
52	2	24	8	14[a]	Erythrocyte, chicken, Kates and James, 1961
67	2	12	9	10[a]	Spleen, pig, Gray and Macfarlane, 1961
82	2	5	6	—	Lung, pig, Gray and Macfarlane, 1961
32	—	25	16	27[b]	Intestinal mucosa, rat, Paltauf and Polheim, 1970
74	0	13	12	—	Aortic intima, man, Panganamala *et al.*, 1969
57	—	28	12	—	Fibroblast L cells, mouse, Weinstein *et al.*, 1969

[a] The high proportion of "other" alk-1-enyl groups indicates the possible presence of artifacts.

[b] Includes 18% of 20:0 alk-1-enyl groups.

TABLE XXIII

Composition of the Alk-1-enyl Groups from Mammalian and Avian Lipids

Percent of alk-1-enyl groups					
16:0	17:0	18:0	18:1	Other	Source (age), reference
					Total phospholipids
41	0	48	11	—	Brain, man (newborn), Altrock and Debuch, 1968
28	1	56	15	—	Brain, man (3 months), Altrock and Debuch, 1968
35	1	45	19	—	Brain, man (7 months), Altrock and Debuch, 1968
38	1	38	22	—	Brain, man (13 months), Altrock and Debuch, 1968
—	—	20	44	—	White matter, man, Gerstl *et al.*, 1965
23	1	36	40	—	Brain, pig, Debuch, 1962
27	1	24	47	—	Brain, horse, Debuch, 1962
29	2	34	35	—	Brain, ox, Debuch, 1962
21	2	36	40	—	Brain, mouse (2 years), Sun and Horrocks, 1968
17	3	42	38	—	Brain, rabbit, Debuch, 1962
22	1	48	29	—	Brain, rat, Debuch, 1962
23	1	43	33	—	Brain, rat, Wood *et al.*, 1969
22	1	43	34	—	Brain, rat, R. E. Anderson *et al.*, 1969b
44	—	47	8	—	Brain, rat (14 days), Debuch, 1964
42	—	40	18	—	Brain, rat (14 days), Etzrodt and Debuch, 1970
40	—	41	15	—	Brain, rat (17 days), Joffe, 1969
36	—	46	17	—	Brain, rat (17 days), Debuch, 1966
36	—	42	16	—	Brain, rat (19 days), Joffe, 1969
42	—	42	16	—	Brain, rat (19 days), Bickerstaffe and Mead, 1967
31	—	45	25	—	Brain, rat (21 days), Debuch, 1966
36	—	43	18	—	Brain, rat (22 days), Joffe, 1969
33	—	43	22	—	Brain, rat (28 days), Debuch, 1966
17	—	46	36	—	Brain, rat (90 days), Debuch, 1966
62	tr	21	16	—	Heart, man, Ferrell *et al.*, 1970
59	—	14	11	16[a]	Blood serum, horse, R. E. Anderson *et al.*, 1969a
55	—	30	11	—	Blood serum, ox (fetal), Weinstein *et al.*, 1969
27	—	27	27	19[a]	Erythrocyte, rat, Walker and Kummerow, 1964
68	1	24	6	—	Erythrocyte, chicken, Kates and James, 1961
45	2	32	13	—	Placenta, man, Winterfeld and Debuch, 1970
					Alk-1-enyldiacylglycerols
63	—	20	13	—	Heart, man, Spener and Mangold, 1969
38	2	24	7	29[a]	Heart, ox, Schmid and Takahashi, 1968
55	—	27	13	—	Adipose tissue, man, Schmid and Mangold, 1966a
60	—	25	15	—	Adipose tissue, man, Schmid *et al.*, 1967b
56	—	21	22	—	Aorta, man, Schmid *et al.*, 1967b
43	tr	1	1	55	Male preputial gland, mouse, Snyder and Blank, 1969
50	—	1	—	49	Male preputial gland, mouse, Sansone and Hamilton, 1969
					Alk-1-enylcholesterols
43	—	28	0	29	Heart, ox, Gilbertson *et al.*, 1970

[a] The high proportion of "other" alk-1-enyl groups indicates the possible presence of artifacts.

TABLE XXIV

Composition of the Alkyl Groups from Mammalian Lipids

Percent of alkyl groups					
16:0	17:0	18:0	18:1	Other	Lipid, source, reference
41	—	20	38	—	EPG, brain, ox, Albro and Dittmer, 1968
36	—	24	40	—	EPG, brain, rat, Paltauf and Polheim, 1970
37	1	30	32	—	EPG, brain, rat, Cotman *et al.*, 1969
38	—	27	35	—	EPG, brain, rat, Albro and Dittmer, 1968
27	—	40	33	—	EPG, brain, rat, Wells and Dittmer, 1966
40	—	41	15	—	EPG, brain, rat (19 days), Joffe, 1969
41	1	26	15	17	EPG, heart, man, Spener and Mangold, 1969
30	2	42	19	7	EPG, heart, ox, Schmid and Takahashi, 1968
43	3	25	25	—	EPG, erythrocyte, ox, Hanahan *et al.*, 1963
34	—	29	37	—	EPG, bone marrow, ox, Thompson and Hanahan, 1963
10	—	68	21	—	EPG, intestinal mucosa, pig, Paltauf and Polheim, 1967
9	—	39	8	44[a]	EPG, intestinal mucosa, rat, Paltauf and Polheim, 1970
23	tr	18	59	—	EPG, fibroblast L-M cells, mouse, R. E. Anderson *et al.*, 1969a
54	—	10	32	—	CPG, brain, ox, Albro and Dittmer, 1968
56	—	12	32	—	CPG, brain, rat, Paltauf and Polheim, 1970
52	1	11	33	—	CPG, brain, rat, Cotman *et al.*, 1969
50	tr	22	21	7	CPG, heart, man, Spener and Mangold, 1969
52	1	10	18	19	CPG, heart, ox, Schmid and Takahashi, 1968
40	—	15	42	—	CPG, bone marrow, ox, Thompson and Hanahan, 1963
29	—	32	39	—	CPG, intestinal mucosa, pig, Paltauf and Polheim, 1967
36	—	19	23	22[a]	CPG, intestinal mucosa, rat, Paltauf and Polheim, 1970
42	tr	6	44	—	CPG, fibroblast L-M cells, mouse, R. E. Anderson *et al.*, 1969a
19	—	58	22	—	SPG, intestinal mucosa, pig, Paltauf and Polheim, 1967
41	tr	26	32	—	PL, brain, rat, Wood *et al.*, 1969
59	—	3	23	15[b]	PL, blood serum, horse, R. E. Anderson *et al.*, 1969a
37	—	47	15	—	PL, intestinal mucosa, man, Paltauf and Polheim, 1967
24	—	43	34	—	PL, intestinal mucosa, pig, Paltauf and Polheim, 1967
22	—	69	9	—	PL, intestinal mucosa, rabbit, Paltauf and Polheim, 1967

TABLE XXIV (*Continued*)

Percent of alkyl groups					
16:0	17:0	18:0	18:1	Other	Lipid, source, reference
47	—	38	15	—	PL, intestinal mucosa, ox, Paltauf and Polheim, 1967
33	1	26	28	12[b]	TL[c], spleen, man, Hallgren and Larsson, 1962
29	4	25	17	25[b]	TL[c], bone marrow (red), man, Hallgren and Larsson, 1962
24	4	23	34	15[b]	TL[c], milk, man, Hallgren and Larsson, 1962
83	—	2	2	13	Galactosylglycerides, brain, ox, Norton and Brotz, 1967
70	3	5	6	16	Galactosylglycerides, brain, ox, Rumsby and Rossiter, 1968
72	2	3	5	18	Galactosylglycerides, brain, pig, Rumsby and Rossiter, 1968
70	2	8	7	13	Galactosylglycerides, brain, sheep, Rumsby and Rossiter, 1968
39	—	36	25	—	Alkyldiacylglycerols, heart, man, Spener and Mangold, 1969
32	2	34	21	11	Alkyldiacylglycerols, heart, ox, Schmid and Takahashi, 1968
35	—	39	22	—	Alkyldiacylglycerols, adipose tissue, man, Schmid and Mangold, 1966a
35	—	35	29	—	Alkyldiacylglycerols, adipose tissue, man, Schmid *et al.*, 1967b
34	—	32	34	—	Alkyldiacylglycerols, aorta, man, Spener and Mangold, 1969
50	—	2	—	48	Alkyldiacylglycerols, male preputial gland, mouse, Sansone and Hamilton, 1969
48	tr	1	tr	51	Alkyldiacylglycerols, male preputial gland, mouse, Snyder and Blank, 1969
69	tr	4	tr	27	Acylalkanols, male preputial gland, mouse, Snyder and Blank, 1969
75	3	5	2	15	Alkyl acetates, male preputial gland, mouse, Spener *et al.*, 1969
90	—	—	—	—	Alkylcholesterols, heart, ox, Funasaki and Gilbertson, 1968

[a] Includes 44% and 22% of 20:0 alkyl groups in the EPG and CPG, respectively.

[b] The high proportion of "other" alkyl groups indicates the possible presence of artifacts.

[c] TL, total lipids.

brain seem to depend on their relative amounts in neurones and oligodendroglia. For example, the 18:0 to 18:1 alk-1-enyl group ratio in mouse brain microsomes is 50:27, but in human myelin the same ratio is 16:56. Similar differences are seen in the alk-1-enyl groups from the CPG (Table XXII) and during development (Table XXIII). In human brain, the proportion of 18:0 alk-1-enyl groups decreases during development from 50 to 33%. At the same time, the proportion of 18:1 alk-1-enyl groups increases from 11 to 40% of the total brain alk-1-enyl groups. The developmental changes are also quite marked for the alk-1-enyl groups from rat brain. In this case, the proportion of 16:0 alk-1-enyl groups decreases but the 18:0 alk-1-enyl groups do not. The alk-1-enyl groups from human, porcine, and bovine brain EPG have similar compositions. The lens has an alk-1-enyl group composition similar to that of myelin, with a ratio of 18:0 to 18:1 alk-1-enyl groups of 5:68.

The alk-1-enyl group composition of the bovine heart CPG has been studied extensively. Almost two-thirds of the alk-1-enyl groups are 16:0, but small amounts of 18:0 and 18:1 groups in addition to 9–21% "other" alk-1-enyl groups have been reported. Human heart CPG have a slightly lower proportion of 16:0 alk-1-enyl groups and a higher proportion of 18:1 alk-1-enyl groups. The proportion of 16:0 alk-1-enyl groups in CPG from heart is nearly twice that in the EPG, with the EPG having much higher levels of 18:0 alk-1-enyl groups.

With the exception of a value of 70% for the proportion of 16:0 alk-1-enyl groups in mouse skeletal muscle (Owens and Hughes, 1970), the only alk-1-enyl group compositions reported for skeletal muscle are for the EPG and CPG from pigeon breast muscle. These compositions are similar to those of mammalian heart. Other tissues with alk-1-enyl group compositions similar to those of the corresponding heart lipids are rat liver EPG, human plasma EPG and CPG, human erythrocyte EPG and CPG, human aortic intima EPG and CPG, porcine spleen EPG and CPG, and porcine lung and kidney CPG. Human erythrocytes have a high ratio (53:18) of 18:0 to 18:1 alk-1-enyl groups (Table XXI; Dodge and Phillips, 1967) and this ratio is even higher (56:8) in platelets. The alk-1-enyl group compositions of the subcellular fractions and whole platelets are very similar (Marcus *et al.*, 1969). The reported proportions of 16:0 alk-1-enyl groups in the EPG and CPG from pig tissues are unusually high except in the central nervous system. Other lipids with high proportions of 16:0 in the alk-1-enyl groups are the EPG from chicken erythrocytes and mouse fibroblasts (Table XXI), the CPG from pigeon breast muscle (Table XXII), and the phospholipids from mouse adipose tissue (72%, Owens and Hughes, 1970).

Small proportions of 16:1, 17:1, and 18:2 alk-1-enyl groups are found quite often. Proportions of 18:2 alk-1-enyl groups in the range of 3–7% have been found in rat erythrocytes (Walker and Kummerow, 1964), bovine heart EPG (Viswanathan *et al.*, 1968b), and mouse fibroblasts (Weinstein *et al.*, 1969). The composition of the alk-1-enyl groups from erythrocytes (Farquhar and Ahrens, 1963; Walker and Kummerow, 1964) and other tissues (Bandi *et al.*, 1971) is influenced by the dietary levels of 18:2 acyl groups and alcohols. Saturated and monounsaturated alk-1-enyl groups with 20, 22, and 24 carbon atoms have been found in human placenta (Winterfeld and Debuch, 1970). Small amounts of 20:0 and 20:1 alk-1-enyl groups have also been found in human erythrocytes (Dodge and Phillips, 1967) and human adrenal glands (Winterfeld and Debuch, 1966), and 20:0 alk-1-enyl groups were reported in the lens (R. E. Anderson *et al.*, 1969c). According to Paltauf and Polheim (1970), large proportions of 20:0 alk-1-enyl and alkyl groups are present in rat intestinal mucosa.

Alk-1-enyl group compositions have also been determined for the alk-1-enyldiacylglycerols from several sources (Table XXIII). All of the compositions from human tissues are quite similar. This is also true for the alkyl group compositions of the alkyldiacylglycerols from heart, adipose tissue, and aorta (Table XXIV), but the alk-1-enyl and alkyl group compositions are quite dissimilar. The same compounds from the preputial gland of the male mouse have rather similar alk-1-enyl and alkyl group compositions. These compositions are unusual because of the high proportions of 14:0 and 16:1 groups and very low proportions of 18:0 and 18:1 groups. In the preputial gland, the alkyl groups from the acylalkanols have more 16:0 and less 14:0 than do the alkyl or alk-1-enyl groups from the glycerides.

The alkyl groups from mammalian phosphoglycerides are generally similar in composition to the corresponding alk-1-enyl groups. The proportion of 16:0 alkyl groups is higher in the CPG than in the EPG. Brain and bone marrow are the source of the alkyl groups with the highest proportion of 18:1. A very high proportion of 16:0 alkyl groups was found in the galactosyl glycerides from brain.

B. Acyl Groups

When the phospholipid classes have been separated, the most useful information can be obtained by first separating the alk-1-enyl groups and then separating the resulting lysophosphoglycerides from the unchanged phosphoglycerides. If the latter separation is not done, the acyl groups will contain a larger proportion of acyl groups from the 2-position than from the 1-position. Detailed directions for the analysis of ethanolamine phosphoglycerides have been given (Horrocks and Sun, 1972). Renkonen

TABLE XXV

Composition of the 2-Acyl Groups from Phosphoglycerides that Contain Alk-1-enyl or Alkyl Groups

Percent of acyl groups											
16:0	18:0	18:1	18:2	20:1	20:3	20:4	20:5	22:4	22:5	22:6	Source, reference
								1-Alk-1'-enyl-2-acyl-GPE			
2	2	36		10		11		15		12	Brain, ox, Sun and Horrocks, 1969b
3	5	28		13		14		10		24	Brain, mouse, Sun and Horrocks, 1969b
8	17	15		3		20		10[a]		25	Brain, rat (14 days old), Etzrodt and Debuch, 1970
3	1	15				23		11[a]		35	Brain, rat (19 days old), Joffe, 1969
3	4	48		13		7		17		1	Myelin, man, Horrocks, 1970
2	1	48		16		8		16		3	Myelin, ox, Sun and Horrocks, 1970
2	1	43		21		11		12		6	Myelin, mouse, Sun and Horrocks, 1970
4	5	47		10		8		18		1	White matter microsomes, man, Horrocks, 1970
1	1	22		9		14		11		40	Microsomes, mouse, Sun and Horrocks, 1970
2	4	10		3		18		10		52	Mitochondria, mouse, Sun and Horrocks, 1970
								1-Alk-1'-enyl-2-acyl-GPC			
15	8	63		6		4		4		—	Brain, ox, Renkonen, 1966
16	tr	66		7		4		—		—	Brain, ox, Renkonen *et al.*, 1965
								1-Alkyl-2-acyl-GPC			
38	2	41		12		2		3		—	Brain, ox, Renkonen, 1966
41	2	38		8		2		—		—	Brain, ox, Renkonen *et al.*, 1965

						1-Alk-1'-enyl-2-acyl-GPE					
2	4	4	34	—	7	42	2	3	3	—	Heart, ox, Schmid and Takahashi, 1968
0	1	2	61	—	4	31	—	—	—	—	Heart, ox, Viswanathan *et al.*, 1968b
4	4	10	3	—	1	37	7	10	12	10	Erythrocyte, man, Cohen and Derksen, 1969
2	1	9	4	—	1	44		15	8	11	Erythrocyte, man, J. H. Williams *et al.*, 1966b
1	1	2	1	—	—	66	5	10	9	5	Platelet, man, Cohen and Derksen, 1969
						1-Alkyl-2-acyl-GPE					
6	6	9	18	—	6	27	6	4	17	—	Heart, ox, Schmid and Takahashi, 1968
						1-Alk-1'-enyl-2-acyl-GPC					
2	1	12	53	—	6	21	tr	2	1	—	Heart, ox, Schmid and Takahashi, 1968
2	1	12	59	—	9	16	—	—	—	—	Heart, ox, Viswanathan *et al.*, 1968c
0	0	13	52	—	9	16	—	—	—	—	Heart, ox, Gottfried and Rapport, 1962
2	2	43	19	—	—	33	—	—	—	—	Skeletal muscle sarcosomes, rabbit, Waku and Lands, 1968
14	6	18	20	—	4	24	—	4	1	6	Blood serum, man, J. H. Williams *et al.*, 1966a
14	8	6	5	—	3	15	10	—	17	7	Erythrocyte, man, Cohen and Derksen, 1969
17	3	22	26	—	3	20	—	1	1	4	Erythrocyte, man, J. H. Williams *et al.*, 1966b
8	13	9	3	—	4	13	10	—	22	4	Platelet, man, Cohen and Derksen, 1969
						1-Alkyl-2-acyl-GPC					
11	2	13	51	—	5	14	tr	2	1	—	Heart, ox, Schmid and Takahashi, 1968
5	4	14	50	—	5	10	6	—	—	—	Heart, ox, Viswanathan *et al.*, 1968b

[a] Reported as 22:3.

and Varo (1967) have reviewed an alternate procedure, based on a phospholipase C hydrolysis of the phosphoglycerides followed by separation of the diradylglycerols by TLC. Another approach has been to use mild alkaline hydrolysis for the isolation of intact plasmalogens or successive mild acid and alkaline hydrolyses for the isolation of alkylacyl phosphoglycerides.

Relatively few analyses of the 2-acyl groups have been reported (Table XXV). Only 19 of these analyses include less than 10% saturated acyl groups. Since saturated acyl groups are generally found primarily in the 1-position in phosphoglycerides, it is likely that the other analyses were done on preparations that had not been separated completely from the diacyl phosphoglycerides. In some cases, a possible explanation is transesterification of 2-acyl groups with 1-acyl groups during isolation. For these reasons, the 2-acyl group analyses reported by Gray (1960b) and Gray and Macfarlane (1961) have not been tabulated.

The acyl group compositions, like the alk-1-enyl group compositions, of the ethanolamine plasmalogens from brain are dependent on the cell type. The 2-acyl groups from myelin contain up to 64% monounsaturated acyl groups including from 13 to 21% 20:1 acyl groups. In contrast, 52% of the 2-acyl groups from mouse brain mitochondria are 22:6 acyl groups. The major acyl groups from the alk-1-enylacyl-GPE from bovine heart are the 18:2 and 20:4 acyl groups. The latter is the major acyl group from human erythrocytes and platelets.

The choline plasmalogens from brain have a high proportion of 18:1 acyl groups. In bovine heart, the 18:2 acyl groups account for more than half of the acyl groups. Three analyses of alk-1-enylacyl-GPC and two analyses of alkylacyl-GPC from bovine hearts have given quite similar results for their acyl group compositions. Both 18:1 and 20:4 acyl groups are major components of the choline plasmalogens from rabbit skeletal muscle. In human platelets and erythrocytes, the 20-carbon and 22-carbon polyunsaturated acyl groups each account for at least 25% of the 2-acyl groups in the alk-1-enylacyl-GPC. Earlier data were reviewed by Klenk and Debuch (1963).

Very little information is available on the molecular species that are present in the alk-1-enylacyl or alkylacyl phosphoglycerides. Renkonen (1965) separated the alk-1-enylacylglycerols from bovine brain EPG on $AgNO_3$-impregnated thin-layer plates, but alk-1-enyl and acyl group analyses were not reported. Subsequently, the same technique was used for alk-1-enylacylglycerols and alkylacylglycerols from bovine brain CPG (Renkonen, 1966, 1967a). The three most common species in order of prevalence were 1–16:0, 2–18:1; 1–18:1, 2–18:1; and 1–16:0, 2–16:0. The first species accounted for about 30%. Many more analyses of molecular species distributions must be done so that membrane structures can be better understood.

C. Double-Bond Position

Very little attention has been given to the position of the double bond in the monounsaturated alk-1-enyl and alkyl groups even though these lipid classes may represent an appreciable proportion of the total phospholipids. Probably the highest proportion of phosphoglycerides with 18:1 alk-1-enyl groups (15–20% of the total phospholipids) is found in the myelin sheath. Farquhar (1962) reported that two isomers were present in the 18:1 alk-1-enyl groups from human erythrocytes and the presence of positional isomers in the 18:1 acyl groups from animal fats has been known for many years. A hemolytic fatty acid that was isolated from several tissues was identified as *cis*-octadec-11-enoic acid by Laser (1950). The amount of this compound in equine brain was 0.25 mg/g tissue (Morton and Todd, 1950). The *cis* $(n - 7)$ isomer accounts for 2–6% of the 18:1 acyl groups from body fats of the camel, hippopotamus, hog, and ox (Kuemmel and Chapman, 1968), 10–45% of the 18:1 acyl groups from rat and mink liver EPG and CPG (Brockerhoff and Ackman, 1967), 17% of the 18:1 acyl groups from pig brain phospholipids (Kishimoto and Radin, 1964), and 8–32% of the 18:1 acyl groups from rat and beef brain EPG, CPG, and SPG (Spence, 1970). After an intraperitoneal injection of 1-^{14}C-palmitic acid into rats, Holloway and Wakil (1964) isolated the 18:1 acyl groups from the lipids that were extracted from the pooled heart, liver, brain, kidney, and spleen. The $(n - 7)$ isomer contained 36% of the radioactivity. Presumably, the 18:1 $(n - 7)$ acyl groups were elongated from 16:1 $(n - 7)$ acyl groups, which in turn were formed by desaturation of 16:0 acyl groups.

The only two quantitative studies on monounsaturated alk-1-enyl groups from mammalian tissues are summarized in Table XXVI. The monounsaturated alk-1-enyl groups from human heart and placenta have an average composition of about two-thirds from the $(n - 9)$ series and one-third from the $(n - 7)$ series. In contrast, the 18:1 alk-1-enyl groups from human brain contain only 30% of the $(n - 9)$ isomer. The $(n - 7)$ isomer is present in a much higher proportion in the 18:1 alk-1-enyl groups than in the 18:1 acyl groups. The 18:1 alkyl groups from brain and heart have not been studied.

The monounsaturated alkyl groups from the alkyldiacylglycerols and the acylalkanols isolated from male mouse preputial glands were studied by Snyder and Blank (1969). These lipids are unusual because of their low

TABLE XXVI

Position of the Double Bond in Monounsaturated Alk-1-enyl Groups from Human Tissues

$(n-9)$ (%)	$(n-7)$ (%)	$(n-5)$ (%)	Group	Tissue, lipid, reference
45	41	10	18:1 alk-1-enyl	Placenta, PL, Debuch and Winterfeld, 1970
67	23	10	20:1 alk-1-enyl	Placenta, PL, Debuch and Winterfeld, 1970
67	33	—	18:1 alk-1-enyl	Heart, CPG, Panganamala *et al.*, 1971
62	38	—	18:1 alk-1-enyl	Heart, EPG, Panganamala *et al.*, 1971
30	70	—	18:1 alk-1-enyl	Brain, EPG, Panganamala *et al.*, 1971

content of 18:1 alkyl groups and their high content of 14:1 and 16:1 alkyl groups. The double-bond positions in these alkyl groups are also unusual. The major components were $(n - 10)$ 16:1 alkyl groups and $(n - 9)$ and $(n - 8)$ 14:1 alkyl groups. The ether-linked lipids from preputial glands are atypical when compared to the ether-linked lipids from a number of other mammalian tissues.

In earlier studies of the position of the double bond in alk-1-enyl and alkyl groups, the oxidation fragments were subjected to crystallization as part of the isolation procedures. Because the isolation procedures were not quantitative, the results can only be used qualitatively. In the brain monounsaturated alk-1-enyl groups, Leupold (1950) found that the major component was 18:1 $(n - 7)$, with a smaller amount of 18:1 $(n - 9)$ alk-1-enyl groups. Klenk *et al.* (1952) detected only 18:1 $(n - 9)$ alk-1-enyl groups in the monounsaturated alk-1-enyl groups from bovine cardiac muscle and equine skeletal muscle. The major component of the monounsaturated alkyl groups of the EPG from bovine erythrocytes is 18:1 $(n - 9)$ (Hanahan *et al.*, 1963).

The availability of reference compounds for identification and calibration purposes has been a problem for investigators of double-bond positions. For unsaturated alkylglycerols, the problem has been solved by the formation and analysis of α,ω-diiodoalkanes from the carbon chain between the ether and the double bond (Ramachandran *et al.*, 1968). These diiodoalkanes may be further identified by mass spectrometry (Foltz *et al.*, 1969). For alk-1-enyl groups, the alkyl dioxolane derivatives are recommended for double-bond position studies (Panganamala *et al.*, 1971) because of the chemical stability of the dioxolanes.

IV. Metabolism

A. Introduction

Intact tissues have been used to establish the overall pathways including some precursor–product relationships. The advantage of using intact tissues is that the conditions are much more physiological. For example, the correct cofactors are present and the correct spatial relationships within the cell are maintained, but the number of substrates that can be used is limited by considerations of pool sizes and entry into cells, and the interpretation of results can be complicated by the reutilization of the substrate or metabolites.

Cell-free systems are necessary for the study of individual reactions and details of metabolic pathways. Such systems are also used for assays of enzyme activities and measurements of enzyme properties. For many lipid substrates, the physical characteristics of the substrate and the choice of surfactant can markedly influence the extent of reaction. Many physiologically important substrates can be used only with cell-free systems. However, the presence of a proposed intermediate in intact tissues must be shown in order to establish that an *in vitro* reaction takes place *in vivo*.

The criteria for precursor–product relationships were discussed by D. B. Zilversmit *et al.* (1943). If A is the precursor and B is the product, the SRA of A must be greater than the SRA of B before the peak SRA is reached. Conversely, after the peak SRA of B is reached, the SRA of B must be greater than the SRA of A. Therefore, when the maximum SRA of B is found, the SRA of A and B should be equal. In other words, the SRA of the precursor and product should cross over.

In many *in vivo* experiments, the reproducibility of results between animals is not as good as the reproducibility of SRA ratios. For this reason, precursor–product relationships are often best detected by comparing the ratios of the SRA of A and B at various times after injection. If A and B have a precursor–product relationship, the ratio of the SRA of A to the SRA of B should decrease with time from a ratio of more than 1 to a ratio of less than 1. If the SRA ratios of two products, B and C, from the same precursor are compared, the ratio of the SRA of B to the SRA of C should be constant with time. A ratio of 1 would indicate an equal turnover rate and ratios different from 1 would indicate that one product was more active metabolically than the other.

These relationships are valid if pulse labeling is achieved, if A and B are present in steady state concentrations, if there is a constant rate of appearance and disappearance of A and B, and if there is a random appearance and disappearance of A and B (D. B. Zilversmit *et al.*, 1943). The last assumption is often violated by the presence of more than one cell type in an organ or tissue, or by the presence of subcellular pools. Since lipid classes are mixtures of many different compounds, differences in metabolic rates of different molecular species should be measured.

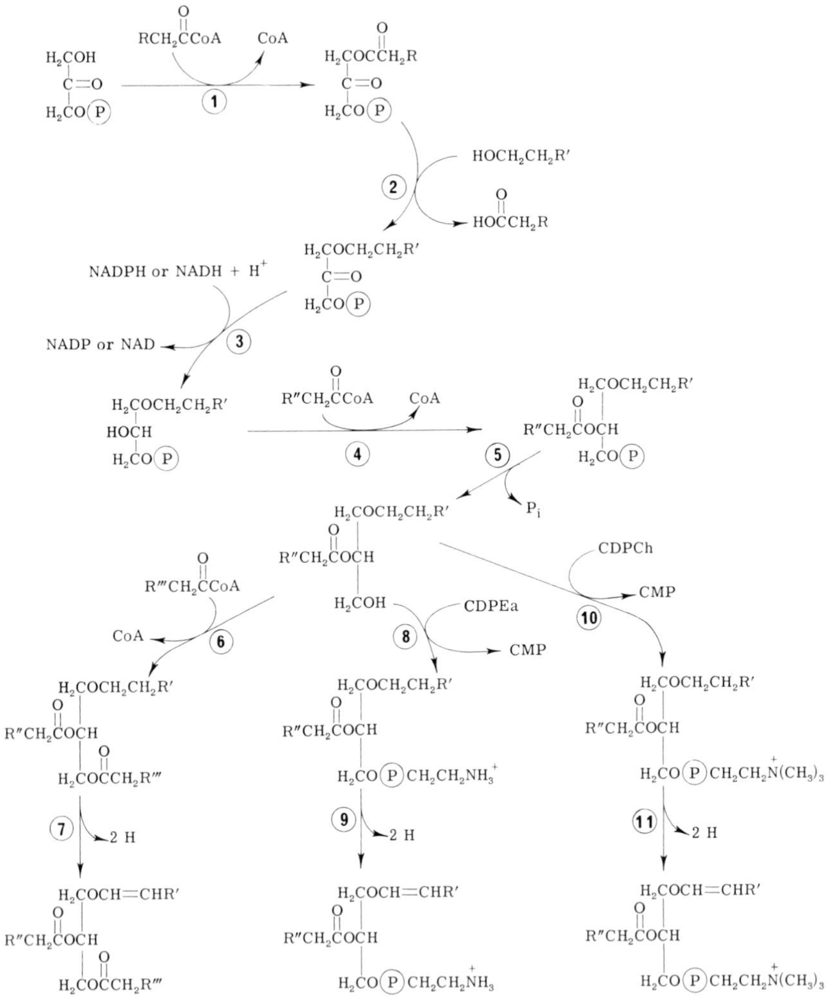

FIG. 2. The most probable pathway for the biosynthesis of alkenyl glycerides. The reactions are designated by numbers which are used in the text.

B. Cell-Free Systems

1. *Synthetic Reactions*

A breakthrough leading to the discovery of a pathway for the biosynthesis of alkyl glycerolipids (see Chapter VII) was reported by Snyder *et al.* (1969b). Further contributions to understanding of the mammalian pathway (Fig. 2) were made by Snyder *et al.* (1969c,e), Hajra (1969, 1970a,b, 1971), Wykle and Snyder (1969, 1970), and Snyder *et al.* (1970a,b,c,e,f, 1971a,b). The reader should consult Chapter VII for a detailed discussion of this pathway and the enzymes involved.

The enzymes CTP: ethanolamine phosphate cytidyl transferase (EC 2.7.7.14) and CTP: choline phosphate cytidyl transferase (EC 2.7.7.15) catalyze the formation of CDPEa and CDPCh, respectively. Diacyl-GPE and diacyl-GPC are formed by reactions catalyzed by CDP-ethanolamine: 1,2-diglyceride ethanolaminephosphotransferase (EC 2.7.8.1) and CDP-choline: 1,2-diglyceride cholinephosphotransferase (EC 2.7.8.2). The specificity of phosphotransferases with regard to the type of diradylglycerol has been studied a number of times (reactions 8, 10, and 11, Fig. 3).

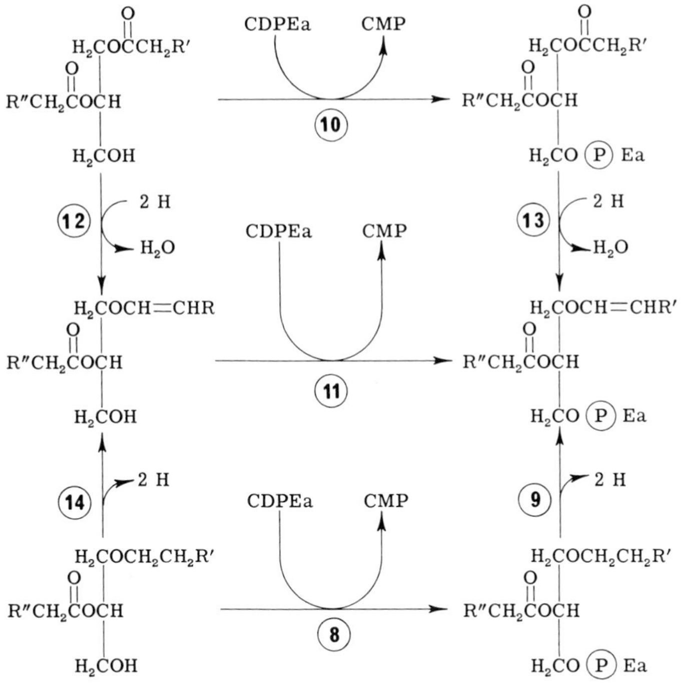

Fig. 3. The possible pathways for the synthesis of EPG from diradylglycerols.

The incorporation of CDPEa into alk-1-enylacyl-GPE was described by Kiyasu and Kennedy (1960), using a rat liver fraction that was primarily microsomal. A 10-fold stimulation of alk-1-enylacyl-GPE labeling was found with exogenous alk-1-enylacylglycerols, but along with a substantial stimulation of diacyl-GPE labeling.

Porcellati *et al.* (1970b) incubated chicken brain microsomes with labeled CDPEa and measured the radioactivity in the EPG. Without exogenous diradylglycerols, the proportions of alk-1-enylacyl-GPE and alkylacyl-GPE in the products were proportional to the content of these compounds in the microsomes. The rate of formation of the diacyl GPE was higher than that of the alk-1-enylacyl-GPE. The addition of alk-1-enylacylglycerols to the incubation mixture gave a sixfold stimulation of the formation of alk-1-enylacyl-GPE, but added 1,2-octadecylglycerols did not react with the CDPEa. Saturated dialkylglycerols are very insoluble and may not have been available to the enzyme, or it is possible that a 2-acyl group may be necessary for binding to the enzyme. The formation of alkylacyl-GPE from an endogenous substrate was depressed by the addition of the product.

The properties of the ethanolamine phosphotransferase from rat brain microsomes were described by Ansell and Metcalfe (1968, 1971). Very low rates of reaction were found without exogenous diradylglycerols. The stimulation was more than 70-fold with alk-1-enylacylglycerols and almost 300-fold with diacylglycerols. The K_m values were 1.7 mM for alk-1-enylacylglycerols and 1.5 mM for diacylglycerols. The highest specific activity of the enzyme was found in the microsomal fraction, but significant amounts of activity were also found in the myelin, nerve-ending, and cytosol fractions. The mitochondria contained very little enzyme activity. The enzyme activity of the brain microsomes was lower in 3-day-old and adult rats than in 16-day-old rats.

The SRA of the EPG after incubations of rat brain microsomes with labeled CDPEa were reported by Ansell (1968). The SRA of the alk-1-enylacyl-GPE was about one-half and the SRA of the alkylacyl-GPE was about one-tenth of the SRA of the diacyl-GPE when exogenous lipids were not present. Additional experiments of this type have been reported (Ansell *et al.*, 1965; Ansell and Metcalfe, 1966). The relative amounts of endogenous substrates (diacylglycerols, alk-1-enylacylglycerols, and alkylacylglcerols) in brain microsomes have not been reported.

Dispersions of brain tissue from young rats were incubated with labeled CDPEa by McMurray (1964). The diacyl-GPE had a higher SRA than the alk-1-enylacyl-GPE. The alkylacyl-GPE had a SRA more than 25 times that of the diacyl-GPE, but this result was due to a contaminant in the CDPEa preparation (McMurray, personal communication, cited by Ansell and Metcalfe, 1971). Only a small stimulation of alk-1-enylacyl-GPE labeling was found after additions of alk-1-enylacylglycerols. Saturated dialkylglycerols did not increase the labeling of the fraction that was resistant to hydrolysis. Reaction (8) (Figs. 2, 3) has recently been demonstrated *in vitro* with mouse and rat brain microsomes with a considerable stimulation of the ethanolaminephosphotransferase activity by alkylacylglycerols (Radominska-Pyrek and Horrocks, 1972).

Kiyasu and Kennedy (1960) also used labeled CDPCh in their experiments with rat liver; the results were similar to those obtained with CDPEa. With brain dispersions, McMurray (1964) found that the SRA of the alk-1-enylacyl-GPC was five times the SRA of the diacyl-GPC after an incubation with CDPCh. A small increase in the SRA was obtained with exogenous alk-1-enylacyl-GPC. The reaction catalyzed by phosphocholine transferase is reversible in liver but not in the brain (Ansell and Spanner, 1968b). Poulos *et al.* (1968) incubated the microsomal and mitochondrial fractions from bovine heart with CDPCh and alk-1-enylacylglycerols, and found that the content of alk-1-enylacyl-GPC in the mixture was almost doubled.

The formation of alkylacylglycerols *in vitro* is well established. The further conversion of these compounds to alkylacyl-GPE and alkylacyl-GPC by microsomes from preputial gland tumors required the addition of CDPEa and CDPCh (Snyder *et al.*, 1970e). Porcellati *et al.* (1970a) incubated chicken brain microsomes with labeled phosphorylethanolamine and found that the formation of CDPEa by the cytidyl transferase was the rate-limiting step for the synthesis of alk-1-enylacyl-GPE and diacyl-GPE in the presence of exogenous diacyl- and alk-1-enylacylglycerols. Under these conditions, the SRA of the alk-1-enylacyl-GPE was much lower than the SRA of the diacyl-GPE.

From experiments with the phosphotransferases, it is clear that they can form alk-1-enylacyl phosphoglycerides when alk-1-enylacylglycerols and the appropriate cytidine derivative are present. The physiological significance of these reactions for the *in vivo* biosynthesis of alk-1-enylacyl phosphoglycerides depends on the *in vivo* availability of alk-1-enylacylglycerols (Kiyasu and Kennedy, 1960). The only evidence supporting an endogenous supply of alk-1-enylacylglycerols is the limited *in vitro* incorporation of radioactivity from CDPEa into alk-1-enylacyl-GPE by brain microsomes, but an exchange of ethanolamine (Ansell and Spanner, 1966b) or the release of diradylglycerols by a phospholipase C reaction (Ansell and Spanner, 1966a) has not been ruled out in these investigations. No direct evidence for the presence of alk-1-enylacylglycerols has been found in lipid composition studies (Section II, B), but relatively small amounts of alkylacylglycerols may be present in brain tissue. Since cytidyl transferase activity is rate-limiting, *in vitro* comparisons of the preference for different types of diradylglycerols by the phosphotransferases may be of more physiological significance if phosphorylethanolamine and phosphorylcholine are used as the labeled precursors.

Studies on the incorporation of long-chain precursors into the alk-1-enyl groups of phosphoglycerides in cell-free systems from normal mammalian tissues have not been very successful. Wykle *et al.* (1970) and Snyder *et al.* (1971b) recently showed that dispersions of Ehrlich ascites cells can incorporate 1-^{14}C-hexadecanol into the alk-1-enyl groups of the alk-1-enylacyl-GPE in which the substituents on positions of the glycerol have been carefully characterized. The essential cofactors are $NADP^+$, CoA, ATP, and Mg^{2+}. With 20 nmoles of hexadecanol and 34 mg of protein, 2.4 to 5% of the hexadecanol was found in alk-1-enyl groups after 4 hours of incubation. In the EPG, 10% of the radioactivity was in alk-1-enyl groups. Most of the remaining radioactivity was found in alkyl groups from the EPG. CDPEa stimulates the incorporation of hexadecanol into alk-1-enylacyl-GPE (Wykle *et al.*, 1971). The SRA of the alk-1-enyl, alkyl, and acyl groups from this interesting system have not yet been reported.

In one experiment, Keenan (1960) incubated rat brain microsomes with ^{3}H-octadecanol, NAD^+, CoA, ATP, Mg^{2+}, Mn^{2+}, α-glycerol phosphate, CTP, and ethanolamine and then isolated radioactive 18:0 alk-1-enyl groups. The ratio of the 18:0 alk-1-enyl group SRA to the SRA of the total phospholipids was 0.10. The same ratio from a control incubation with ^{3}H-octadecanoic acid was 0.13, but a previous experiment with microsomes from younger rats had given a ratio of 0.35. These results are interesting, but no conclusions can be drawn since the EPG were not isolated and too few incubations were done. Gambal and Monty (1959) had earlier claimed that 5–15 nmoles of 1-^{14}C-hexadecanoic acid were incorporated into alk-1-enyl groups in a similar incubation mixture. Apparently, aldehyde impurities were present in the ^{14}C-hexadecanoic acid that was available at that time (Keenan, 1960). No incorporation of radioactivity into the alk-1-enyl groups was found after incubations of brain dispersions from young rats with ^{14}C-hexadecanal (Carr *et al.*, 1963).

If 1-alk-1′-enyl-2-acyl-GPC is hydrolyzed by phospholipase A, the product is 1-alk-1′-enyl-GPC. An acyl CoA: 1-alk-1′-enyl-GPC acyltransferase can then reacylate the 1-alk-1′-enyl-GPC (Waku and Lands, 1968). The activity of this enzyme in rabbits, especially for 18:2 and 20:4 acyl CoAs, is quite high in muscle sarcoplasmic reticulum, but lower in testis and human erythrocytes and very low in rabbit heart and brain. The activity was quite low with 1-alk-1′-enyl-GPE (Waku and Lands, 1968) and 1-alkyl-GPC (Waku and Nakazawa, 1970) as substrates.

2. *Catabolic Reactions*

The appreciable metabolic activity of alk-1-enyl and alkyl groups in the steady state (Section C, 3) requires that catabolic reactions must take place at a rate equal to the rate of the anabolic reactions. C. E. Anderson *et al.* (1960) incubated a rat liver dispersion and found that 70% of the endogenous plasmalogen was lost in 4 hours. Neither fluoride nor oxalate ions had any effect. Heating of the dispersion for 5 minutes at 100°C prevented the loss of plasmalogen.

Warner and Lands (1961) discovered a plasmalogenase enzyme, 1-alk-1′-enyl-GPC alk-1-enylhydrolase, in rat liver. Among the subcellular fractions, the light microsomes had the highest activity. No alk-1-enylhydrolase activity was found with alk-1-enylacyl-GPC or alk-1-enylacyl-GPE as the substrate. The specific activity of the 1-alk-1′-enyl-GPC alk-1-enylhydrolase in rat liver microsomes is between 0.4 and 2.7 mmoles/hr/g protein (Warner and Lands, 1963; Ellingson and Lands, 1968). Other substrates for this enzyme are 1-alk-1′-enyl-GPE, 1-alk-1′-enyl-GP, and 1-alk-1′-enylglycerol (Warner, 1962; Robertson and Lands, 1962). Disruption of the microsomal membrane structure reduces enzyme activity (Ellingson and Lands, 1968).

Ansell and Spanner (1965a) found a loss of 0.6 μmoles of endogenous alk-1-enyl groups per hour per gram of tissue during incubations of rat brain dispersions. With added alk-1-enylacyl-GPE, the value was 2.9. Alk-1-enyl-GPE was a poorer substrate. The plasmalogenase enzyme, alk-1-enylacyl-GPE alk-1-enylhydrolase, could be extracted from acetone powders. These investigators later measured the activity of this enzyme in tissues from the central nervous system of several species and discovered the enzyme activity in white matter to be more than tenfold that in gray matter (Ansell and Spanner, 1968a). In whole rat brain subcellular fractions, the highest activities are found in mitochondria and microsomes. Demyelinating conditions are associated with an increased activity of the alk-1-enylhydrolase in white matter (Ansell and Spanner, 1970). The alk-1-enylhydrolases from liver and brain appear to be different but in either tissue, however, alk-1-enyl groups are hydrolyzed to form free aldehydes.

An alk-1-enylhydrolase is probably present in ram spermatozoa because the plasmalogen content decreases during their incubation in sugarfree media (Hartree, 1964). The level of plasmalogens was not altered by prolonged incubations of bovine adrenal dispersions (Norton *et al.*, 1962).

An alkylhydrolase is present in rat liver microsomes (Tietz *et al.*, 1964). With alkylglycerols as the substrate, the reaction requires molecular oxygen and a tetrahydropteridine. Tetrahydrofolate has some activity. Reduced NADP is required for reduction of the pteridine. Since glycerol and aldehydes are released at the same rate, Tietz *et al.* (1964) proposed that the 1′-carbon atom was hydroxylated to form an unstable hemiacetal. The activity of the liver enzyme was higher in the rat than in six other species (Pfleger *et al.*, 1967). Lower activities of the enzyme were found in intestine, brain, and perirenal adipose tissue. The reported activities of the rat liver enzyme in micromoles per hour per gram of microsomal protein are 4.5–7.6 (Pfleger *et al.*, 1967), 22 (Soodsma *et al.*, 1970), and 360 (5 days of age, Snyder *et al.*, 1971a). Different times of incubation have been used in these studies. Soodsma (1970) reported a pH maximum of 8–9 for the alkylhydrolase and confirmed that aldehydes are the product. The information available on the cleavage of alkyl groups from other substrates is discussed in Chapter VII.

Alkylglycerols can also be acylated, particularly if a supply of acyl-CoA is available. Oswald *et al.* (1968) incubated 1-octadecylglycerol with a dispersion of intestinal tissue. A small amount of acylation but no significant alkylhydrolase activity was found. Gallo *et al.* (1968) included hexadecanoyl-CoA in incubations of 1-octadecylglycerol with microsomes from rat intestinal mucosa or dispersions of whole mucosa. Forty percent or more of the 1-octadecylglycerol was esterified, primarily at the 3-position. An active acylation of the 3-position and a much slower acylation of the 2-position of 1-alkyl-3-acylglycerols was also found by Snyder *et al.* (1970d) with 1-hexadecylglycerol incubated with rat liver dispersions or mouse preputial gland dispersions. The normal pathways for the catabolism of alkylacyl phosphoglycerides and alkyldiacylglycerols are unknown.

C. Intact Tissues

1. *Incorporation of Precursors into Water-Soluble Moieties*

The pathways shown in Fig. 2 can also be compared with the results from *in vivo* and tissue slice incubations. Generally, experiments with ^{32}P-phosphate have shown that the diacyl phosphoglycerides are formed at a faster rate than the alk-1-enylacyl phosphoglycerides. For example, 5 hr after the infusion of ^{32}P-phosphate into ram testes, the alk-1-enylacyl-GPC and the alk-1-enylacyl-GPE had SRA that were less than half of the SRA of the diacyl compounds (Scott and Setchell, 1968). Canine thyroid slices were incubated with ^{32}P-phosphate for 2 hr (Scott *et al.*, 1966) in media that contained glucose, and the SRA of the alk-1-enylacyl-GPC and the alk-1-enylacyl-GPE were 0.5 relative to diacyl-GPC and diacyl-GPE, respectively. After 3 hr of incubation, the relative SRA of the alk-1-enylacyl-GPE was only 0.4 but the relative SRA of the alk-1-enylacyl-GPC increased to 0.7 and 1.4 in two different experiments. Similar experi-

TABLE XXVII

The Uptake of ³²P-Phosphate into the 1-Alk-1′-enyl-2-acyl-GPE in Rat Brains after Systemic Injections[a]

Time after injection	SRA, 1-alk-1′-enyl-2-acyl-GPE / SRA, 1,2-diacyl-GPE
3 hr	0.3
12 hr	0.5
1 day	0.6
3 days	0.7
5 days	0.7

[a] The value of 12 hr is from Mandel *et al.* (1963). The remaining values were reported by Ivanova *et al.* (1967). Similar values were given by Freysz *et al.* (1965).

ments with 2 hour incubations of slices of bovine corpus luteum gave a relative SRA of 0.4 for alk-1-enylacyl-GPE and 1.2 for alk-1-enylacyl-GPC (Scott *et al.*, 1968). Broekhuyse and Veerkamp (1968) reported that lens slices incorporated phosphate into alk-1-enylacyl-GPE more slowly than into the diacyl-GPE. Rubel *et al.* (1967) compared the SRA of alk-1-enylacyl-GPE to the SRA of the diacyl-GPE in several organs of the rat after an administration of ^{32}P-phosphate. The lowest relative SRA was in liver and the highest relative SRA was for heart. Schmidt *et al.* (1961) reported that 3 hr after the injection of ^{32}P-phosphate, the relative SRA of the alk-1-enylacyl and diacyl phosphoglycerides from rabbits were 0.1 for heart and 0.2 for skeletal muscle.

A number of investigators have studied the uptake of ^{32}P-phosphate into brain phospholipids after systemic injections, which deliver only a small portion of the ^{32}P-phosphate to the brain and that over a long period of time (Ansell and Hawthorne, 1964). In the brain, as in other organs, the relative incorporation of ^{32}P-phosphate is slower in the alk-1-enylacyl-GPE than in the diacyl-GPE (Table XXVII). The incorporation is greater in younger rats (Rubel *et al.*, 1967). The SRA for alk-1-enylacyl-GPE relative to that for diacyl-GPE is greater in gray matter than in white matter (Ivanova *et al.*, 1967). The alk-1-enylacyl-GPC have a much higher SRA than do the alk-1-enylacyl-GPE according to Rubel and Ivanova (1965) and Freysz *et al.* (1969). In the latter report, the SRA for the alk-1-enyl-acyl-GPC was similar to the SRA for the phosphatidylinositol, which is one of the most metabolically active brain lipids.

Incorporation of ^{32}P-phosphate into the alkylacyl-GPE was first detected by Ansell and Spanner (1961). Mandel and Nussbaum (1966) reported the SRA of the alk-1-enylacyl-GPE and alkyl phosphoglycerides from rat brains after intraperitoneal injections of ^{32}P-phosphate. At 1 day after injection of the ^{32}P, these SRA were much higher for the microsomes than for the mitochondria or myelin. In similar experiments by Freysz *et al.* (1969), the diacyl-GPE had an SRA slightly higher than the SRA of the alkylacyl-GPE and the alk-1-enylacyl-GPE.

The relative metabolic activities of the diacyl, alk-1-enylacyl, and alkylacyl types of the EPG or CPG in the brain have been studied after intracerebral injections of ethanolamine or choline. Such studies partially avoid the problems of reutilization of catabolic products and lack of pulse labeling that are associated with intraperitoneal injections of ^{32}P-phosphate. The average ratio of the SRA for the alk-1-enylacyl-GPC and the diacyl-GPC from rat brain was 0.24 at 1, 2, 4, 6, and 8 hr after intracerebral injections of ^{14}C-choline (Ansell and Spanner, 1967, 1968b). There was a trend toward higher ratios at longer times. Ansell and Spanner (1967, 1968c) have also studied the EPG of rat brain after intracerebral injections of ^{14}C-ethanolamine. Here, the ratio of the SRA for the alk-1-enylacyl-GPE and the diacyl-GPE from whole brain increased from 0.14 at 1 hr to 0.47 at 18 hr after injection. The ratio of the SRA for the alk-1-enylacyl-GPE and the diacyl-GPE was higher in the microsomes and also increased from 0.23 at 1 hr to 0.54 at 1 day after injection. In mouse brain (Horrocks, 1969a), the ratio of SRA for these lipids was higher than in rat brain at 1 day after intracerebral injections of ^{14}C-ethanolamine. The ratios were lower for the myelin than for the microsomes in both studies. Comparisons of the SRA of the alk-1-enylacyl-GPE from the microsomal and myelin fractions show apparent precursor–product relationships. These data are consistent with a synthesis of the alk-1-enylacyl-GPE in the endoplasmic reticulum followed by an equilibration with the alk-1-enylacyl-GPE in the myelin.

Intracerebral injections of ^{14}C-ethanolamine have been used for studies of the metabolic relationships of the alkylacyl-GPE in rat brain (Horrocks and Ansell, 1967b) and in mouse brain (Horrocks, 1969b). In rat brain, the SRA of the alkylacyl-GPE was lower than the SRA of the diacyl-GPE, but the ratio was nearly the same from 30 min to 2 days after injection. At 30 min after injection, the ratio of the SRA of the alkylacyl-GPE to the SRA of the alk-1-enylacyl-GPE was greater than 5. This ratio declined with increasing time after injection, but was still greater than one at 2 days after injection. Subcellular fractions were separated from the brains of 5-month-old mice from 1 to 7 days after injection. The ratio of the SRA for the alkylacyl-GPE and the alk-1-enylacyl-GPE declined most rapidly in the microsomal fractions. At 3 days after injection, the ratio was less than 1 and at 7 days it had decreased to 0.3.

Very little is known about the *in vivo* incorporation of radioactivity into the glycerol moiety of lipids with alk-1-enyl or alkyl groups. Thompson and Hanahan (1963) incubated bovine bone marrow with glucose. They estimated that the alk-1-enyl groups from phosphoglycerides had twice the SRA of the alkyl groups because the alkylglycerols isolated after hydrogenation of the mixture had a higher SRA. These results led them to conclude that alk-1-enyl groups could not be produced from alkyl groups. In contrast, the alkylacyl-GPE from mouse brain are labeled much faster than the alk-1-enylacyl-GPE after intracerebral injections of ^{14}C-glycerol (Horrocks, 1972). At times from 2.5 to 60 min after injection, the highest SRA was found for the alkylacyl-GPE. At longer times, the SRA of the alkylacyl-GPE was less than the SRA of the diacyl-GPE. At all times up to 2 hr after injection, the SRA of the alkylacyl-GPE was greater than the SRA of the alk-1-enylacyl-GPE by a factor of at least 4. The ratios of these SRA decreased with time as they did after injections of ethanolamine. Waku and Nakazawa (1970) reported that the *in vivo* incorporation of ^{32}P-phosphate and ^{3}H-glycerol was more rapid into the alkylacyl-GPC than into the diacyl-GPC or alk-1-enylacyl-GPC in rabbit sarcoplasmic reticulum.

The sum of these results on the incorporation of radioactivity into the glycerol, phosphate, and nitrogen base moieties of diacyl-, alk-1-enylacyl-, and alkylacylglycerophosphoryl esters in intact mammalian cells is that the labeling of the alk-1-enylacyl type is almost always the slowest. More importantly, the SRA ratios involving the alk-1-enylacylglycerophosphoryl esters are not constant but are consistent with a precursor role for alkylacyl glycerophosphoryl esters. In one study (Horrocks, 1969b), a crossover of SRA values was found for the alkylacyl-GPE and the alk-1-enylacyl-GPE between 2 and 3 days after injections of ethanolamine into mouse brains.

2. *Incorporation of Precursors into Alkyl and Alk-1-enyl Groups*

Fatty acids, fatty aldehydes, and fatty alcohols have all been suggested as precursors of alkyl and alk-1-enyl groups. Studies of this question with intact cells are complicated by the presence of enzymes that oxidize or reduce these compounds and by the chemical instability of aldehydes. In brain tissue, the reduction of hexadecanoyl CoA to hexadecanol was described by Brady and Koval (1958) and Vignais and Zabin (1960). Tabakoff and Erwin (1970) reported that a soluble aldehyde reductase from brain reduces hexadecanal to hexadecanol with NADPH at a rate of 14 μmoles/min/g protein. Similar reactions with NADH were described by Day *et al.* (1970) in *Clostridium butyricum* and with NADH and NADPH by Kolattukudy (1970) in *Euglena gracilis*. In the reduction of hexadecanoyl-CoA to hexadecanol by microsomes from mouse preputial gland tumors, the cofactor is NADPH (Snyder and Malone, 1970). The reverse reactions also occur. The microsomes from preputial gland tumors contain an NAD^+-requiring enzyme that can oxidize hexadecanol to hexadecanoic acid (Snyder and Malone, 1970). A hexadecanol dehydrogenase was found in brain microsomes by Brady and Koval (1958) and Brady *et al.* (1958). Erwin and Dietrich (1966) partially purified an NAD^+-dependent aldehyde dehydrogenase from bovine brain mitochondria.

The evidence for a very low content or absence of free aldehydes and a low content of free alcohols is described in Section II, B. It appears that free aldehydes exist as transient intermediates but will be rapidly oxidized to acids or reduced to alcohols. Alcohols may be formed from injected acids and acids may be formed from injected alcohols as described above and as demonstrated *in vivo* by Schmid and Takahashi (1970).

The first evidence suggesting that alcohols might be the best precursors of alk-1-enyl groups was obtained with heart-lung preparations (Keenan, 1960; Keenan *et al.*, 1961). The precursors, 9,10-^{3}H-octadecanoate and 9,10-^{3}H-octadecanol, were suspended in the venous reservoir and allowed to recirculate for 2 hr. Alk-1-enyl and acyl groups were isolated from the cardiac phospholipids. The ratio of the alk-1-enyl group SRA to the acyl group SRA was 0.11 with octadecanoate and 0.35 with octadecanol. The ratio of the 18:0 alk-1-enyl group SRA to the 18:1 alk-1-enyl group SRA was also higher with octadecanol as the precursor. Both of these facts are consistent with the hypothesis that fatty alcohols are more direct precursors of the alk-1-enyl groups than the fatty acids. The results with 1-^{14}C-hexadecanoate were similar to the results with octadecanoate, except that no radioactivity was found in 18:0 alk-1-enyl groups after the administration of the hexadecanoate. Gilbertson *et al.* (1963) incubated rat epididymal fat pads with ^{14}C-acetate and ^{14}C-palmitate. The SRA of the alkyl groups was greater than the SRA of the alk-1-enyl groups that were isolated from the triradylglycerols.

Very recently, several papers have appeared which contain convincing evidence for the *in vivo* reaction sequence *alkanol*, *alkyl*acyl-GPE, *alk-1-enyl*acyl-GPE. The latter reaction has been studied by Paltauf (1971) in rat intestinal mucosa and by Stoffel and LeKim (1971) in rat brain. The incorporation of fatty alcohols into alkyl and alk-1-enyl groups of rat brain phosphoglycerides in the above sequence is supported by Bell *et al.* (1971) and Stoffel and LeKim (1971). Schmid and Takahashi (1970) measured the radioactivity in the alkyl, alk-1-enyl, and acyl groups from rat brain

EPG at 1, 2, 3, 6, and 22 hr after intracerebral injections of ^{14}C-hexadecanol. The highest radioactivity in alkyl groups was found at 6 hr and the highest radioactivity in alk-1-enyl and acyl groups was found at 22 hr after injection. Assuming that the relative contents of the alkylacyl-GPE and the alk-1-enylacyl-GPE did not change during the experiment, I have calculated the following ratios of the SRA for alkyl groups to the SRA for alk-1-enyl groups at the times given above: 8.1, 4.4, 3.2, 2.3, and 0.95. In my opinion, these data strongly suggest a precursor–product relationship for the alkyl and alk-1-enyl groups from the EPG, with a crossover point slightly before 22 hr after injection. All of the alkyl group radioactivity was found in 16:0 alkyl groups, but 5% of the alk-1-enyl group radioactivity was found in 18:0 and 18:1 alk-1-enyl groups at 22 hr after injection, perhaps because of problems of technique. If not, then about one-fifth of the radioactive alk-1-enyl groups did not arise directly from the alkyl groups. Acyl groups were excluded as a direct precursor of the alk-1-enyl groups because the alk-1-enyl groups had a much higher SRA than the 1-acyl groups at all times and much of the radioactivity in the acyl groups was found in 18:0 and 18:1 acyl groups. If an intermediate is formed in reaction (9) (Fig. 2), it is possible that the 1-radyl group of this intermediate could exchange with an unknown long-chain moiety. Such a possibility could explain the radioactivity in the 18:0 and 18:1 alk-1-enyl groups.

Further studies on the incorporation of ^{14}C-hexadecanol into alkyl and alk-1-enyl groups have been done by Horrocks (1971), using intracerebral injection of mice. The radioactivity in the acyl, alk-1-enyl, and alkyl groups from the EPG, CPG, and diradylglycerols was measured at 1, 5, and 24 hr and at 4, 8, and 16 days after injection. An apparent precursor–product relationship for the alkyl and alk-1-enyl groups from the EPG with a cross-over point near 24 hr was found, in confirmation of the study of Schmid and Takahashi (1970). The radioactivity in the alk-1-enyl groups reaches a peak in the CPG before the EPG. At 1 hr after injection, the amount of radioactivity in the alkylacylglycerols is about one-fourth of that in the alkyl groups from the EPG and CPG, but the radioactivity in the alk-1-enyl groups from the diradylglycerols is much less. In fact, there is no direct evidence for the existence of alk-1-enylacylglycerols in brain (see Section B, 2). Much of the radioactivity in the alkylacylglycerols was found in the alkylglycerols. These *in vivo* results are consistent with the pathway shown in Fig. 2.

Radioactivity in the alk-1-enyl groups has also been found after intracerebral injections of sphinganine into rats. Sphinganine is metabolized by formation of the phosphate derivative (Stoffel *et al.*, 1968) followed by a cleavage to give hexadecanal and ethanolamine phosphate (Stoffel *et al.*, 1968; Keenan and Maxam, 1969). Stoffel *et al.* (1970) later gave 3-^{3}H-sphinganine by intracerebral or intraperitoneal injection, and recovered a large portion of the radioactivity from brain and liver in alk-1-enyl groups. In a further experiment, 3-^{3}H-^{14}C-sphinganine was administered to young rats by intracerebral injections. Of the recovered ^{14}C, 8% was in the 16:0 alk-1-enyl groups. The ^{3}H/^{14}C ratios were 171 for sphinganine; 70 for 16:0 alk-1-enyl groups; and 4.5 for acyl groups. Since half of the ^{3}H was removed during the formation of the alk-1-enyl groups, the hexadecanal was probably reduced to hexadecanol before incorporation into phospholipids. In another experiment, 3-^{3}H-sphinganine and 1-^{14}C-hexadecanoate were administered together by intracerebral injection. After 70 hr, the alk-1-enyl and acyl groups were isolated from the CPG and EPG. The ^{3}H/^{14}C ratios were much higher in the alk-1-enyl groups than in the acyl groups. Thus, when available in excess, sphinganine is preferred as precursor of alk-1-enyl groups. Since the turnover of sphinganine is probably slower than the turnover of alk-1-enyl groups and alkyl groups, the alcohols used for alkyl and alk-1-enyl group biosynthesis are probably formed normally by reduction of endogenous fatty acids.

Another relevant study was reported by Bickerstaffe and Mead (1968), who gave an emulsion of U-^{3}H-hexadecylglycerol to rats by intracerebral injection. The hexadecylglycerol is apparently acylated to 1-hexadecyl-2-acylglycerol or it could be cleaved to yield hexadecanol (see Section B, 3). At 8 hr after injection, most of the radioactivity was in nonpolar lipids with more than 20% in diradyl- and triradylglycerols. Increasing proportions of the radioactivity were found in the EPG at 4, 8, and 16 hr after injection. At these times, the ratios of the 16:0 alkyl group SRA to the 16:0 alk-1-enyl group SRA were 11, 13, and 6. Further time periods were not examined. The large amount of radioactivity in the nonpolar lipids and the high SRA ratio at 16 hr after injection indicate that the injected material was metabolized rather slowly.

Bell and White (1968) gave 1-^{3}H-1-^{14}C-hexadecanal to young rats by intracerebral injection, removed the brains at 2 hr and isolated the EPG. The ^{3}H/^{14}C ratios of the alk-1-enyl and acyl groups were measured. In the first experiment, this ratio was 32 for the injected hexadecanal and 18 for the alk-1-enyl groups. In the second experiment, the ratios were 0.42 for the injected hexadecanal, 0.17 for the acyl groups, and 0.34 for the alk-1-enyl groups. If the aldehydes were incorporated directly into alk-1-enyl groups as postulated by Bell and White (1968), the expected ^{3}H/^{14}C ratios would be 32 and 0.42. If incorporation was by way of hexadecanol as outlined in Fig. 2, the ratios expected for the alk-1-enyl groups would be 16 and 0.295. In my opinion, the ratios that were found are more consistent with the latter possibility.

1-^{14}C-Hexadecanal was given to young rats by intracerebral injection by Bickerstaffe and Mead (1967). The SRA of 16:0 acyl, alkyl, and alk-1-enyl groups were measured at 2 hr, 6 hr, 2 days, and 3 days after injection. The SRA of the 16:0 alk-1-enyl groups was highest at 2 days after injection.At all times, the 16:0 acyl group SRA was higher than the 16:0 alk-1-enyl group SRA. The SRA of the 16:0 alk-1-enyl groups was less than half of the 16:0 alkyl group SRA at 2 hr after injection. At 3 days after injection, the 16:0 alkyl group SRA was lower than the SRA of the 16:0 acyl or alk-1-enyl groups. Bickerstaffe and Mead (1967) concluded that the hexadecanal was oxidized to hexadecanoic acid and that the 1-acyl groups of phosphoglycerides were the immediate precursors of the 1-alk-1′-enyl groups of phosphoglycerides. On the basis of present knowledge, the data are consistent with the proposed pathways (Fig. 2) except for the relatively low SRA of the 16:0 alkyl groups at 2 hr after injection.

The fate of palmitic acid after intracerebral injection has been studied quite thoroughly. Schmid and Takahashi (1970) found that the free hexadecanol had a very high SRA in rat brain at 3 hr after the injection of palmitic acid. Some radioactivity was also found in 18:0 and 18:1 alcohols. For the EPG, the ratio of the SRA of the alkyl groups to the SRA of the alk-1-enyl groups decreased with time from 12 at 1 hr to 1.9 at 22 hr after injection. The SRA of the alk-1-enyl groups was much less than the SRA of the 1-acyl groups from the EPG at all times up to 22 hr. This is also true in mouse brain at times from 12 hr to 8 days after injection (Sun and Horrocks 1969a). Sun and Horrocks (1972) have also measured the radioactivity in the alk-1-enylacyl-GPE from mouse brain subcellular fractions after intracerebral injections of ^{14}C-palmitic acid. The highest radioactivities in the 16:0 alk-1-enyl groups were found in the microsomes at about 3 days after injection, and in the myelin and mitochondria at 8–14 days after injection. The graphs for the SRA of the 16:0 alk-1-enyl groups as a function of time showed a good precursor–product relationship for the microsomes and myelin, with an intersection at 20–25 days after the injection. Similar curves for the microsomes and mitochondria did not intersect, indicating the possible presence of an unlabeled pool. The SRA of the 18:1 acyl groups from the alk-1-enylacyl-GPE were also plotted, showing that the microsomal curve intersected with both the myelin and mitochondrial curves near 8 days after injection. The highest radioactivity was found in the microsomal 18:1 acyl groups less than 1 day after injection. The delay in labeling of 16:0 alk-1-enyl groups from palmitic acid is longer than the delay in labeling 18:1 acyl groups. Since the latter transformation includes elongation and desaturation before esterification, the transformation of 16:0 acyl groups to 16:0 alk-1-enyl groups apparently requires at least the same number of steps.

Korey and Orchen (1959) studied the incorporation of ^{14}C-acetate and ^{14}C-palmitate into rat brain alk-1-enyl and acyl groups after intraperitoneal injections. At 16 hr after injection, they found that acetate but not palmitate was a better precursor of alk-1-enyl groups than of acyl groups. These results are difficult to interpret but they could indicate that a precursor of palmitic acid is the substrate for reduction to alcohols. The incorporation of acetate was not affected by the simultaneous administration of unlabeled fatty acids. Frosolono (1965) studied the uptake of ^{14}C-palmitic acid into the acyl and alk-1-enyl groups of liver lipids and found that the alk-1-enyl groups had very high SRA, but his results were not consistent.

Debuch (1964, 1966) and Etzrodt and Debuch (1970) have given intracerebral injections of ^{14}C-acetate to 14-day-old rats. One day after injection, the 16:0 and 18:0 acyl groups regularly have about twice the SRA of the corresponding alk-1-enyl groups. These results are consistent with a pathway of acetate to hexadecanoate (or hexadecanoyl CoA) to hexadecanol.

The relative SRA of 16:0 and 18:0 acyl, alk-1-enyl, and alkyl groups from rat brain EPG after intraperitoneal injections of ^{14}C-acetate were determined by Joffe (1969). The 17-day-old rats were killed at 1, 3, 6, 24, and 48 hr after injection. At all times, the 16:0 acyl groups had the highest SRA followed by 16:0 alkyl, 18:0 alkyl, 18:0 acyl, 16:0 alk-1-enyl, and 18:0 alk-1-enyl groups. The highest radioactivities were at 3 hr for the 16:0 acyl groups and at 6 hr for the alkyl groups. The 18:1 acyl groups from the alk-1-enylacyl-GPE were only slightly lower in SRA than the 18:1 acyl groups from the 2-position of the other EPG. These results are also consistent with the conclusion that alk-1-enyl groups are not formed directly from aldehydes but are most likely formed from alkyl groups. Proof of a precursor–product relationship between alkyl and alk-1-enyl groups was not demonstrated in this study since this requires pulse labeling. Peripheral injection and the high SRA of the acyl groups both contribute to a prolonged supply of alkyl group precursors.

The possible relationships between diradylglycerols and the three types of EPG are illustrated in Fig. 3. Several investigators have given intracerebral injections of intact EPG to rats in attempts to test for the presence of reactions 9 and 13 (Fig. 3). In each investigation, it was assumed that intact EPG were taken up by the cells. Bickerstaffe and Mead (1968) injected U-^{14}C-diacyl-GPE. The recovered radioactivity was found in all lipid classes with only 30% in the EPG at 4, 8, and 16 hr after injection. These results are similar to those found after intracerebral injections of fatty acids, so it is quite likely that the EPG were hydrolyzed before the acyl groups were metabolized by the cells. Similar experiments were reported by Stoffel *et al.* (1970), Debuch *et al.* (1970), and Segal and Wysocki (1970). Segal and Wysocki (1970) also injected EPG (from rat brain) that was labeled in the alk-1-enyl and acyl groups. Debuch *et al.* (1970) injected a hydrogenated alkylacyl-GPE preparation into rat brains and found a small amount of radioactivity in the alk-1-enyl groups. Most of the alkylacyl-GPE was recovered intact due to its insolubility.

Most of our knowledge of the *in vivo* biosynthesis of alkylacyl and alk-1-enylacyl phosphoglycerides in mammalian tissues is restricted to that concerning the EPG from rat and mouse brains, which now quite clearly indicates that fatty acids or their CoA derivatives are reduced to fatty alcohols. The latter are then incorporated into the alkyl groups of the alkylacylglycerols. These compounds appear to be the precursors of the alk-1-enylacyl-GPE, most likely by way of the alkylacyl-GPE. Reactions (14) and (11) (Fig. 3) for the biosynthesis of alk-1-enylacyl-GPE are unlikely because very little radioactivity can be found in the alk-1-enylacylglycerols when an active synthesis of radioactive alk-1-enylacyl-GPE is taking place. In addition, the lack of constancy with time in the SRA ratios with ^{14}C-ethanolamine as the precursor suggests that ethanolamine is not incorporated by reaction (11) (Fig. 3) but is incorporated in an earlier reaction such as reaction (8).

These reactions take place in the microsomes or cytosol and the products can then exchange with corresponding molecules in other membranes. The study of microsomal phosphoglycerides has been more fruitful than the study of whole brain phosphoglycerides because the latter include pools of alkylacyl phosphoglycerides that are not available to the desaturating enzyme [reaction (9)]. The rates of formation of the diacyl-GPE and the alkylacyl-GPE seem to be rather similar. The formation of the alk-1-enylacyl-GPE seems to be limited by the supply of alkylacyl-GPE. In contrast, in brain tissue the amounts of alk-1-enylacyl-GPC and alkylacyl-GPC are approximately the same. As a result, the SRA of the alk-1-enylacyl-GPC increases much faster than does the SRA of the alk-1-enylacyl-GPE after intracerebral injections of labeled precursors.

Important studies in other organisms have been reported by Thompson (1966, 1967) and Malins (1968). Excellent *in vivo* evidence for the incorporation of hexadecanol into alkyl and alk-1-enyl groups of EPG in Ehrlich ascites cells has been reported by Wood and Healy (1970a,b) and Wood *et al.* (1970).

Labeled alkylglycerols have been fed to animals in order to determine the normal pathways for intestinal digestion and absorption of triacylglycerols. An absorptive dose of 1′-^{14}C-1-octadecylglycerol in a test meal was fed to rats with a lymph-fistula by Swell *et al.* (1965). After 24 hr, 76% of the fed radioactivity had been recovered from the lymph, but only 55% of the recovered radioactivity was in alkyl groups, primarily in alkyldiacylglycerols. Very little of the radioactivity was found in the alkyl groups from the phospholipids. The remaining 45% of the recovered radioactivity was found in acyl groups. Comparable results were reported by Bergström and Blomstrand (1956) and Law *et al.* (1964). *In vitro* incubations have shown the presence of alkylhydrolase activity in rat intestinal contents (Oswald *et al.*, 1968) and hamster intestinal slices (Kern and Borgström, 1965). The SRA of tissue lipids after a meal containing 1′-^{14}C-1-hexadecylglycerol was reported by Snyder *et al.* (1969d). The highest SRA were in the spleen, liver, and plasma lipids at 27 hr after feeding. A small amount of radioactivity was found in the respired carbon dioxide. Studies with 2-alkylglycerols have been done by Sherr *et al.* (1963), Swell *et al.* (1965), Kern and Borgström (1965), Sherr and Treadwell (1965), Oswald *et al.* (1968), and Snyder *et al.* (1969d). Snyder *et al.* (1969d) found that *S*-alkylglycerols were metabolized differently than *O*-alkylglycerols.

Paltauf (1969) fed labeled dialkylglycerols and their phospholipid derivatives to rats. Most of the label was not absorbed and the absorbed label in the lymph was primarily in neutral lipids. At 24 hr after feeding, most of the absorbed label was found in the liver. Very little radioactivity from labeled trialkylglycerols was absorbed (Spener *et al.*, 1968; Carlson and Bayley, 1970).

Fat pads from rats, incubated in serum that contained 1-dodecylglycerol or 2-octadec-9′-enylglycerol (Belfrage, 1964), have shown a substantial uptake and acylation of the alkylglycerols. Rat liver slices in Tyrode's solution took up alkylglycerols rather slowly (Snyder and Pfleger, 1966). Most of the radioactivity in the phospholipids was in the acyl groups. Similar results were obtained with bone marrow cells and spleen slices.

Emulsions of alkylglycerols were given to lactating goats by intra-arterial injections (Bickerstaffe *et al.*, 1970). With the exception of a 2-alkylglycerol, very little uptake by mammary gland tissue was found. Hexadecylglycerols given to rats by intravenous injection (Snyder and Pfleger, 1966) were more active metabolically than were octadecylglycerols. At 6 hr after injection of the 1′-^{14}C-hexadecylglycerol, 30% of the radioactivity had been recovered in the respired CO_2, 6% in the urine, and 14% was found in the liver lipids. Only 1.6% of the radioactivity from the liver lipids could be recovered in alkylglycerols after acetolysis. Liver also rapidly removes plasmalogens from the blood after an intravenous injection of plasmalogen emulsion into rabbits (Robertson and Lands, 1962).

At 8 hr after intraperitoneal injections of 1-octadecylglycerol into rats (Oswald *et al.*, 1968), 25% of the radioactivity was in the liver and 8% was in the intestines. Results obtained with 1-hexadecylglycerol were similar. From 7 to 16% of the radioactivity in lipids at 4 and 8 hr after injection was found in the alkyl groups from phosphoglycerides, but significant amounts of radioactivity were not found in alk-1-enyl groups. In similar experiments, Blank *et al.* (1970) found that 20 to 32% of the total radioactivity in the liver lipids was in a previously unrecognized component of the EPG and CPG. After lithium aluminum hydride hydrogenolysis of the liver lipids, this component had TLC properties similar to a synthetic 2′-hydroxy-1-hexadecylglycerol, but GLC analyses of various derivatives proved that the two compounds were *not* the same. The presence of large amounts of radioactivity in this component suggests that it could be an intermediate in the biosynthesis of plasmalogens.

3. *Turnover*

Although turnover rates can be calculated from incorporation data, they are usually measured from graphs of the content of radioactivity against time and expressed in terms of half-life. In addition to problems with the reproducibility of injections and with the heterogeneity of pools, the interpretation of turnover studies is also dependent on the degree of reutilization of catabolic products. It is necessary to distinguish between the apparent turnover which includes reutilization and the true turnover. The likelihood of exchanges of intact lipids which are not catabolized should also be recognized as a form of turnover.

The values for apparent half-lives that are given in Table XXVIII are not comparable unless the same precursor was injected. In general, the degree of reutilization is lowest for ethanolamine, intermediate for phosphate, and highest for acetate and palmitate. For palmitate, the free fatty acids are rapidly esterified so that the precursors of the alk-1-enyl groups and the 18:1 acyl groups are primarily 16:0 acyl groups that had previously been a part of phosphoglyceride molecules (Sun and Horrocks, 1969a). Therefore, the assumption of pulse labeling is not valid for the experiments with acetate and palmitate precursors. The degree of recycling of hydrolyzed alk-1-enyl groups back into bound alk-1-enyl groups is not known. The ^{14}C-acetate was injected into rats that were 14 days of age and the half-lives were calculated from SRA values at 28 and 90 days of age. In this case, the concentration of alk-1-enyl groups increased considerably during the experiment. For these reasons, the apparent half-lives of the

TABLE XXVIII

Apparent Half-Lives of Ether-Linked Phosphoglycerides in Brains of Rats and Mice

Half-life (days)	Component measured[a]	Age of injection	Precursor injected[a]	Brain fraction	References
		1-Alk-1'-enyl-2-acyl-GPS			
13	Pi	Adult	Pi	Neurons	Freysz *et al.*, 1969
16	Pi	Adult	Pi	Glia	Freysz *et al.*, 1969
		1-Alk-1'-enyl-2-acyl-GPC			
6	Pi	Adult	Pi	Neurons	Freysz *et al.*, 1969
10	Pi	Adult	Pi	Glia	Freysz *et al.*, 1969
		1-Alk-1'-enyl-2-acyl-GPE			
11	Pi	Adult	Pi	Neurons	Freysz *et al.*, 1969
15	Pi	Adult	Pi	Glia	Freysz *et al.*, 1969
1.3[b]	E	40 days	E	Microsomes	Horrocks, 1969a
2.4	E	5 months	E	Microsomes	Horrocks, 1969b
10	E	2 years	E	Microsomes	Horrocks, 1969b
1.6[b]	E	40 days	E	Myelin	Horrocks, 1969a
3.2	E	5 months	E	Myelin	Horrocks, 1969b
5[b]	E	Adult	E	Myelin	Ansell and Spanner, 1968c
20	E	2 years	E	Myelin	Horrocks, 1969b
2.4	E	5 months	E	Mitochondria	Horrocks, 1969b
6	E	2 years	E	Mitochondria	Horrocks, 1969b

45[b]	18:0 Alk-1-enyl groups	14 days	A	Brain	Debuch, 1964, 1966
36[b]	16:0 Alk-1-enyl groups	14 days	A	Brain	Debuch, 1964, 1966
10	16:0 Alk-1-enyl groups	3 months	P	Microsomes	Sun, 1970
14	16:0 Alk-1-enyl groups	3 months	P	Myelin	Sun, 1970
8	16:0 Alk-1-enyl groups	3 months	P	Mitochondria	Sun, 1970
10	18:1 Acyl groups	3 months	P	Microsomes	Sun, 1970
16	18:1 Acyl groups	3 months	P	Myelin	Sun, 1970
12	18:1 Acyl groups	3 months	P	Mitochondria	Sun, 1970
		1-Alkyl-2-acyl-GPE			
11	Pi	Adult	Pi	Neurons	Freysz *et al.*, 1969
15	Pi	Adult	Pi	Glia	Freysz *et al.*, 1969
1.1[b]	E	Adult	E	Brain	Horrocks and Ansell, 1967b
1.3	E	5 months	E	Microsomes	Horrocks, 1969b
2.0	E	5 months	E	Myelin	Horrocks, 1969b
2.3	E	5 months	E	Mitochondria	Horrocks, 1969b

[a] A, ^{14}C-acetate; E, ^{14}C-ethanolamine; P, ^{14}C-palmitate; Pi, ^{32}P-phosphate. The latter was given by intraperitoneal injection, the remainder by intracerebral injection.

[b] Calculated from two SRA values.

alk-1-enyl and acyl groups of the alk-1-enylacyl-GPE may be much longer than the true half-lives. However, we can conclude that the turnovers of 1-alk-1′-enyl groups and 2-acyl groups are similar.

From the results with ^{32}P-phosphate as the precursor, several conclusions can be drawn although the assumption of pulse labeling is not valid. For each of the four lipids (Table XXVIII) the turnover was more rapid in neurons than in glia. The turnover of alk-1-enylacyl-GPC was more rapid than alk-1-enylacyl-GPE which was more rapid than alk-1-enylacyl-GPS.

With ^{14}C-ethanolamine as the precursor, the SRA of the alk-1-enyl acyl-GPE declines rapidly during the first week but then the decline of the SRA is much slower in both whole brain and myelin (Ansell and Spanner, 1968c). The apparent half-life for the alk-1-enylacyl-GPE from rat brain myelin was 30 days for the period from 2 to 9 weeks after injection, but only 5 days during the first week, which indicates that some recycling of ^{14}C-ethanolamine may have occurred. The turnover of alk-1-enylacyl-GPE slows down markedly with increasing age.

The most rapid turnover of the alkylacyl-GPE is in the microsomal fraction. Generally the turnover of alk-1-enylacyl-GPE and alkylacyl-GPE is somewhat slower in the myelin than in other subcellular fractions; this could be related to the glial origin of the myelin. The half-lives of alkylacyl-GPE and alk-1-enylacyl-GPE from rats (Horrocks and Ansell, 1967b; Ansell and Spanner, 1968c) are similar to those from mice (Horrocks, 1969a,b).

V. Summary

The limiting factor for research in this field has been the available methodology. Many improvements in the past decade have led to new discoveries, particularly in metabolism. A small but active pool of free long-chain alcohols is the most likely source of alkyl and alk-1-enyl groups as established by *in vivo* and *in vitro* studies. Aldehydes are also a good source if generated *in situ*. However, the aldehydes are transient because of the high activity of aldehyde reductase.

The most probable pathway for the biosynthesis of plasmalogens in the endoplasmic reticulum is shown in Fig. 2. The rate of formation of plasmalogens is slower than that of the corresponding diacyl or alkylacyl compounds with many different precursors. Studies *in vivo* with labeled hexadecanol, glycerol, and ethanolamine support the hypothesis that alkylacyl-GPE is the immediate precursor of alk-1-enylacyl-GPE. Among the EPG of brain, the alkylacyl-GPE has the highest turnover rate followed by diacyl-GPE and the alk-1-enylacyl-GPE. Alk-1-enylhydrolases (plasmalogenases) have been found in brain and liver. These organs also contain alkylglycerol alkylhydrolase activity.

Alk-1-enyl and alkyl groups have been found in the lipids from every mammalian tissue examined. In the phosphoglycerides, the concentration of alk-1-enyl groups is usually much greater than the concentration of alkyl groups but the two groups are always found in the same phosphoglyceride class, most often in the EPG. The CPG usually contain small amounts of alk-1-enyl and alkyl groups. In muscle tissues, the concentrations of both choline and ethanolamine plasmalogens are high. Small amounts of alk-1-enyl groups have been detected in IPG in several tissues. Numerous reports of serine plasmalogens have not yet firmly established their existence. Alk-1-enyl groups have also been reported in phosphatidic acids and cardiolipins.

Alkyldiacylglycerols and alk-1-enyldiacylglycerols are found in adipocytes and probably in several other cell types. The existence of alkylacylglycerols is quite probable but very little evidence for the existence of alk-1-enylacylglycerols has been reported.

The highest content of plasmalogens is found in the myelin sheath in which one-third of the phospholipids are alk-1-enylacyl-GPE. Human erythrocyte plasma membranes have a high concentration of alk-1-enylacyl-GPE whereas bovine erythrocytes have a high concentration of alkylacyl-GPE. A large amount of the latter is also found in red bone marrow and spleen. Liver and plasma phospholipids have the lowest concentrations of plasmalogens.

Alk-1-enyl and alkyl group compositions are quite simple with 16:0, 18:0, and 18:1 groups as the principal components. The 1-acyl groups of the corresponding diacyl phosphoglycerides have quantitatively different compositions. Branched-chain alk-1-enyl groups are of importance in ruminants only. The 2-acyl groups of phosphoglycerides with ether groups are highly unsaturated with compositions that are similar to the 2-acyl groups of the corresponding diacyl phosphoglycerides. Relatively large proportions of double bonds at the 11,12-position are found in monounsaturated alk-1-enyl groups from human tissues.

Acknowledgment

The preparation of this review was supported in part by PHS Research Grant NS-08291 from the National Institute of Neurological Diseases and Stroke.

REFERENCES

Adams, R. G. (1969). *J. Lipid Res.* **10,** 473.

Albro, P. W., and Dittmer, J. C. (1968). *J. Chromatogr.* **38,** 230.

Altrock, K., and Debuch, H. (1968). *J. Neurochem.* **15,** 1351.

Anderson, C. E., Williams, J. N., Jr., and Yarbro, C. L. (1960). *In* "Biochemistry of Lipids" (G. Popjak, ed.), pp. 105–113. Pergamon, Oxford.

Anderson, R. E. (1970). *Exp. Eye Res.* **10,** 339.

Anderson, R. E., Cumming, R. B., Walton, M., and Snyder, F. (1969a). *Biochim. Biophys. Acta* **176,** 491.

Anderson, R. E., Garrett, R. D., Blank, M. L., and Snyder, F. (1969b). *Lipids* **4,** 327.

Anderson, R. E., Maude, M. B., and Feldman, G. L. (1969c). *Biochim. Biophys. Acta* **187,** 345.

Anderson, R. E., Feldman, L. S., and Feldman, G. L. (1970a). *Biochim. Biophys. Acta* **202,** 367.

Anderson, R. E., Maude, M. B., and Feldman, G. L. (1970b). *Exp. Eye Res.* **9,** 281.

Ansell, G. B. (1968). *Sci. Basis Med.* p. 383.

Ansell, G. B., and Hawthorne, J. N. (1964). "Phospholipids—Chemistry, Metabolism, Function." Elsevier, Amsterdam.

Ansell, G. B., and Metcalfe, R. F. (1966). *Biochem. J.* **98,** 22P.

Ansell, G. B., and Metcalfe, R. F. (1968). *Biochem. J.* **109,** 29P.

Ansell, G. B., and Metcalfe, R. F. (1971). *J. Neurochem.* **18,** 647.

Ansell, G. B., and Spanner, S. (1961). *Biochem. J.* **81,** 36P.

Ansell, G. B., and Spanner, S. (1963a). *Biochem. J.* **88,** 56.

Ansell, G. B., and Spanner, S. (1963b). *J. Neurochem.* **10,** 941.

Ansell, G. B., and Spanner, S. (1965a). *Biochem. J.* **94,** 252.

Ansell, G. B., and Spanner, S. (1965b). *Biochem. J.* **97,** 375.

Ansell, G. B., and Spanner, S. (1966a). *Biochem. J.* **98,** 23P.

Ansell, G. B., and Spanner, S. (1966b). *Biochem. J.* **100,** 50P.

Ansell, G. B., and Spanner, S. (1967). *J. Neurochem.* **14,** 873.

Ansell, G. B., and Spanner, S. (1968a). *Biochem. J.* **108,** 207.

Ansell, G. B., and Spanner, S. (1968b). *Biochem. J.* **110,** 201.

Ansell, G. B., and Spanner, S. (1968c). *J. Neurochem.* **15,** 1371.

Ansell, G. B., and Spanner, S. (1970). *Biochem. J.* **117,** 11P.

Ansell, G. B., Chojnacki, T., and Metcalfe, R. F. (1965). *J. Neurochem.* **12,** 649.

Bandi, Z. L., Mangold, H. K., Hølmer, G., and Aaes-Jørgensen, E. (1971). *FEBS Lett.* **12,** 217.

Basinska, J., Sastry, P. S., and Stancer, H. C. (1969). *J. Neurochem.* **16,** 707.

Belfrage, P. (1964). *Biochem. J.* **92,** 41P.

Bell, O. E., Jr., and White, H. B., Jr. (1968). *Biochim. Biophys. Acta* **164,** 441.

Bell, O. E., Jr., Cain, C. E., Sulya, L. L., and White, H. B., Jr. (1967). *Biochim. Biophys. Acta* **144,** 481.

Bell, O. E., Jr., Blank, M. L., and Snyder, F. (1971). *Biochim. Biophys. Acta* **231,** 579.

Belsare, D. K., and Chowdhuri, D. R. (1968). *Lipids* **3,** 21.

Bergelson, L. D., Vaver, V. A., Prokazova, N. V., Ushakov, A. N., and Popkova, G. A. (1966). *Biochim. Biophys. Acta* **116,** 511.

Bergström, S., and Blomstrand, R. (1956). *Acta Physiol. Scand.* **38,** 166.

Berry, J. F., Cevallos, W. H., and Wade, R. R., Jr. (1965). *J. Amer. Oil Chem. Soc.* **42,** 492.

Bickerstaffe, R., and Mead, J. F. (1967). *Biochemistry* **6,** 655.

Bickerstaffe, R., and Mead, J. F. (1968). *Lipids* **3,** 317.
Bickerstaffe, R., Linzell, J. L., Morris L. J., and Annison, E. F. (1970). *Biochem. J.* **117,** 39P.
Bieth, R., Freysz, L., and Mandel, P. (1961). *Biochim. Biophys. Acta* **53,** 576.
Blank, M. L., and Snyder, F. (1970). *Lipids* **5,** 337.
Blank, M. L., Wykle, R. L., Piantadosi, C., and Snyder, F. (1970). *Biochim. Biophys. Acta* **210,** 442.
Blomstrand, R., Nakayama, F., and Nilsson, I. M. (1962). *J. Lab. Clin. Med.* **59,** 771.
Borggreven, J. M. P. M., Daemen, F. J. M., and Bonting, S. L. (1970). *Biochim. Biophys. Acta* **202,** 374.
Brady, R. O., and Koval, G. J. (1958). *J. Biol. Chem.* **233,** 26.
Brady, R. O., Formica, T. V., and Koval, G. J. (1958). *J. Biol. Chem.* **233,** 1072.
Brockerhoff, H., and Ackman, R. G. (1967). *J. Lipid Res.* **8,** 661.
Broderson, S. H. (1967). Ph.D. Dissertation, State University of New York at Buffalo (University Microfilms 67-13, 335).
Broekhuyse, R. M. (1968). *Biochim. Biophys. Acta* **152,** 307.
Broekhuyse, R. M. (1969). *Biochim. Biophys. Acta* **187,** 354.
Broekhuyse, R. M., and Veerkamp, J. H. (1968). *Biochim. Biophys. Acta* **152,** 316.
Brown, C. L., and Dittmer, J. C. (1968). *Biochem. Prep.* **12,** 73.
Buddecke, E., and Andresen, G. (1959). *Hoppe-Seyler's Z. Physiol. Chem.* **314,** 38.
Camejo, G., Rapport, M. M., and Morrill, G. A. (1964). *J. Lipid Res.* **5,** 75.
Carlson, W. E., and Bayley, H. S. (1970). *Fed. Proc., Fed. Amer. Soc. Exp. Biol.* **29,** 300.
Carr, H. G., Haerle, H., and Eiler, J. J. (1963). *Biochim. Biophys. Acta* **70,** 205.
Carter, H. E., Smith D. B., and Jones, D. N. (1958). *J. Biol. Chem.* **232,** 681.
Chang, T-C. Lo, and Sweeley, C. C. (1963). *Biochemistry* **2,** 592.
Cohen, P., and Derksen, A. (1969). *Brit. J. Haematol.* **17,** 359.
Colacicco, G., and Rapport, M. M. (1966). *J. Lipid Res.* **7,** 258.
Cornwell, D. G., and Horrocks, L. A. (1964). *In* "Symposium on Foods—Proteins and their Reactions" (H. W. Schultz and A. F. Anglemier, ed.), p. 117. Avi Publ. Co., Westport, Connecticut.
Cotman, C., Blank, M. L., Moehl, A., and Snyder, F. (1969). *Biochemistry* **8,** 4606.
Cumings, J. N., Grundt, I. K., and Yanagihara, T. (1968). *J. Neurol., Neurosurg. Psychiat.* **31,** 334.
Cuzner, M. L., and Davison, A. N. (1968). *Biochem. J.* **106,** 29.
Cuzner, M. L., Davison, A. N., and Gregson, N. A. (1965a). *Ann. N. Y. Acad. Sci.* **122,** 86.
Cuzner, M. L., Davison, A. N., and Gregson, N. A. (1965b). *J. Neurochem.* **12,** 469.
Das, M. L., and Rouser, G. (1967). *Lipids* **2,** 1.
Davenport, J. B. (1964). *Biochem. J.* **90,** 116.
Dawson, R. M. C. (1960). *Biochem. J.* **75,** 45.
Dawson, R. M. C. (1966). *Essays Biochem.* **2,** 69.
Dawson, R. M. C. (1967). *Lipid Chromatogr. Anal.* **1,** 163.
Dawson, R. M. C., and Scott, T. W. (1964). *Nature (London)* **202,** 292.
Dawson, R. M. C., Hemington, N., and Lindsay, D. B. (1960). *Biochem. J.* **77,** 226.
Dawson, R. M. C., Hemington, N., and Davenport, J. B. (1962). *Biochem. J.* **84,** 497.
Day, J. I. E., Goldfine, H., and Hagen, P. O. (1970). *Biochim. Biophys. Acta* **218,** 179.
Debuch, H. (1962). *Hoppe-Seyler's Z. Physiol. Chem.* **327,** 65.
Debuch, H. (1964). *Hoppe-Seyler's Z. Physiol. Chem.* **338,** 1.
Debuch, H. (1966). *Hoppe-Seyler's Z. Physiol. Chem.* **344,** 83.

Debuch, H., and Wendt, G. (1967). *Hoppe-Seyler's Z. Physiol. Chem.* **348,** 471.
Debuch, H., and Winterfeld, M. (1966). *Z. Klin. Chem.* **4,** 149.
Debuch, H., and Winterfeld, M. (1970). *Hoppe-Seyler's Z. Physiol. Chem.* **351**, 179.
Debuch, H., Friedemann, H., and Müller, J. (1970). *Hoppe-Seyler's Z. Physiol. Chem.* **351,** 613.
De Rooij, R. E., and Hooghwinkel, G. J. (1967). *Acta Physiol. Pharmacol. Neer.* **14,** 410.
Di Costanzo, G., and Clement, J. (1963). *Bull. Soc. Chim. Biol.* **45,** 137.
Di Costanzo, G., and Clement, J. (1965). *Bull. Soc. Chim. Biol.* **47,** 833.
Dodge, J. T., and Phillips, G. B. (1967). *J. Lipid Res.* **8,** 667.
Domonkos, J., and Heiner, L. (1968). *J. Neurochem.* **15,** 87.
Dvorkin, V. Y., and Gasteva, S. V. (1969). *Biokhimiya* **34,** 144.
Eichberg, J., and Hess, H. H. (1967). *Experientia* **23,** 993.
Eichberg, J. Whittaker, V. P., and Dawson, R. M. C. (1964). *Biochem. J.* **92,** 91.
Eichberg, J., Hauser, G., and Karnovsky, M. L. (1969). *In* "The Structure and Function of Nervous Tissue" (G. H. Bourne, ed.), Vol. 3, p. 185. Academic Press, New York.
Ellingson, J. S., and Lands, W. E. M. (1968). *Lipids* **3,** 111.
Eng, L. F., and Noble, E. P. (1968). *Lipids* **3,** 157.
Erickson, N. E., and Lands, W. E. M. (1959). *Proc. Soc. Exp. Biol. Med.* **102,** 512.
Erwin, V. G., and Dietrich, R. A. (1966). *J. Biol. Chem.* **241,** 3533.
Etzrodt, A., and Debuch, H. (1970). *Hoppe-Seyler's Z. Physiol. Chem.* **351,** 603.
Farquhar, J. W. (1962). *Biochim. Biophys. Acta* **60,** 80.
Farquhar, J. W., and Ahrens, E. H., Jr. (1963). *J. Clin. Invest.* **42,** 675.
Ferrans, V. J., Hack, M. H., and Borowitz, E. H. (1962). *J. Histochem. Cytochem.* **10,** 462.
Ferrell, W. J., and Radloff, D. M. (1970). *Physiol. Chem. Phys.* **2,** 551.
Ferrell, W. J., Radloff, J. F., and Jackiw, A. B. (1969). *Lipids* **4,** 278.
Ferrell, W. J., Radloff, D. M., and Radloff, J. F. (1970). *Anal. Biochem.* **37,** 227.
Fewster, M. E., and Mead, J. F. (1968a). *J. Neurochem.* **15,** 1041.
Fewster, M. E., and Mead, J. F. (1968b). *J. Neurochem.* **15,** 1303.
Fishman, M. A., Prensky, A. L., and Dodge, P. R. (1969). *Nature (London)* **221,** 552.
Fleischer, S., and Rouser, G. (1965). *J. Amer. Oil. Chem. Soc.* **42,** 588.
Fleischer, S., Klouwen, H., and Brierley, G. (1961). *J. Biol. Chem.* **236,** 2936.
Fleischer, S., Rouser, G., Fleischer, B., Casu, A., and Kritchevsky, G. (1967). *J. Lipid Res.* **8,** 170.
Flygare, R. E., Broderson, S. H., and Hayes, E. R. (1970). *Stain Technol.* **45,** 149.
Foltz, R. L., Ramachandran, S., and Cornwell, D. G. (1969). *Lipids* **4,** 225.
Forstner, G. G., Tanaka, K., and Isselbacher, K. J. (1968). *Biochem. J.* **109,** 51.
Freysz, L., Nussbaum, J. L., Bieth, R., and Mandel, P. (1963). *Bull. Soc. Chim. Biol.* **45,** 1019.
Freysz, L., Bieth, R., and Mandel, P. (1965). *Bull. Soc. Chim. Biol.* **47,** 1441.
Freysz, L., Bieth, R., Judes, C., Sensenbrenner, M., Jacob, M., and Mandel, P. (1968). *J. Neurochem.* **15,** 307.
Freysz, L., Bieth, R., and Mandel, P. (1969). *J. Neurochem.* **16,** 1417.
Frosolono, M. F. (1965). Ph.D. Dissertation, University of North Carolina (University Microfilms 67-985).
Funasaki, H., and Gilbertson, J. R. (1968). *J. Lipid Res.* **9,** 766.
Gallo, L., Vahouny, G. V., and Treadwell, C. R. (1968). *Proc. Soc. Exp. Biol. Med.* **127,** 156.
Gambal, D., and Monty, K. J. (1959). *Fed. Proc., Fed. Amer. Soc. Exp. Biol.* **18,** 232.

Gasteva, S. V., and Dvorkin, V. Y. (1967). *Dokl. Akad. Nauk SSSR* **176,** 1185.
Geison, R. L., and Waisman, H. A. (1970a). *J. Nutr.* **100,** 315.
Geison, R. L., and Waisman, H. A. (1970b). *J. Neurochem.* **17,** 469.
Gerstein, W. (1963). *J. Invest. Dermatol.* **40,** 105.
Gerstl, B., Tavaststjerna, M. G., Hayman, R. B., Smith, J. K., and Eng, L. F. (1963). *J. Neurochem.* **10,** 889.
Gerstl, B., Tavaststjerna, M. G., Hayman, R. B., Eng, L. F., and Smith, J. K. (1965). *Ann. N. Y. Acad. Sci.* **122,** 405.
Gerstl, B., Eng, L. F., Hayman, R. B., Tavaststjerna, M. G., and Bond, P. R. (1967). *J. Neurochem.* **14,** 661.
Gerstl, B., Eng, L. F., Hayman, R. B., and Bond, P. (1969). *Lipids* **4,** 428.
Getz, G. S., Bartley, W., Lurie, D., and Notton, G. M. (1968). *Biochim. Biophys. Acta* **152,** 325.
Gilbertson, J. R. (1969). *Metab., Clin. Exp.* **18,** 887.
Gilbertson, J. R., and Karnovsky, M. L. (1963). *J. Biol. Chem.* **238,** 893.
Gilbertson, J. R., Ellingboe, J., and Karnovsky, M. (1963). *Fed. Proc., Fed. Amer. Soc. Exp. Biol.* **22,** 304.
Gilbertson, J. R., Ferrell, W. J., and Gelman, R. A. (1967). *J. Lipid Res.* **8,** 38.
Gilbertson, J. R., Garlich, H. H., and Gelman, R. A. (1970). *J. Lipid Res.* **11,** 201.
Gluck, L., Kulovich, M. V., and Brody, S. J. (1966). *J. Lipid Res.* **7,** 570.
Goldfine, H. (1968). *Annu. Rev. Biochem.* **37,** 303.
Gottfried, E. L. (1967). *J. Lipid Res.* **8,** 321.
Gottfried, E. L., and Rapport, M. M. (1962). *J. Biol. Chem.* **237,** 329.
Gottfried, E. L., and Rapport, M. M. (1963). *Biochemistry* **2,** 646.
Gray, G. M. (1960a). *Biochem. J.* **74,** 1P.
Gray, G. M. (1960b). *Biochem. J.* **77,** 82.
Gray, G. M. (1967a). *Lipid Chromatogr. Anal.* **1,** 401.
Gray, G. M. (1967b). *Biochim. Biophys. Acta* **144,** 519.
Gray, G. M., and Macfarlane, M. G. (1958). *Biochem. J.* **70,** 409.
Gray, G. M., and Macfarlane, M. G. (1961). *Biochem. J.* **81,** 480.
Gray, G. M., and Macfarlane, M. G. (1964). *Biochem. J.* **91,** 16C.
Grogan, D. E., Mayer, D. T., and Sikes, J. D. (1966). *J. Reprod. Fert.* **12,** 431.
Gueuning, C., and Graff, G. L. A. (1967). *C. R. Soc. Biol.* **161,** 965.
Gurr, M. I., Finean, J. B., and Hawthorne, J. N. (1963). *Biochim. Biophys. Acta* **70,** 406.
Hack, M. H. (1961). *Fed. Proc., Fed. Amer. Soc. Exp. Biol.* **20,** 278.
Hack, M. H., and Ferrans, V. J. (1960). *Circ. Res.* **8,** 738.
Hack, M. H., and Helmy, F. M. (1965). *Comp. Biochem. Physiol.* **16,** 311.
Hack, M. H., and Helmy, F. M. (1967a). *Acta Histochem.* **27,** 291.
Hack, H. M., and Helmy, F. M. (1967b). *Comp. Biochem. Physiol.* **23,** 105.
Hajra, A. K. (1969). *Biochem. Biophys. Res. Commun.* **37,** 486.
Hajra, A. K., (1970a). *Fed. Proc., Fed. Amer. Soc. Exp. Biol.* **29,** 674.
Hajra, A. K. (1970b). *Biochem. Biophys. Res. Commun.* **39,** 1037.
Hajra, A. K. (1971). *Fed. Proc., Fed. Amer. Soc. Exp. Biol.* **30,** 1243Abs.
Hallgren, B., and Larsson, S. (1962). *J. Lipid Res.* **3,** 39.
Hanahan, D. J., and Thompson, G. A., Jr. (1963). *Annu. Rev. Biochem.* **32,** 215.
Hanahan, D. J., and Watts, R. (1961). *J. Biol. Chem.* **236,** PC59.
Hanahan, D. J., Ekholm J., and Jackson, C. M. (1963). *Biochemistry* **2,** 630.
Hartree, E. F. (1964). *In* "Metabolism and Physiological Significance of Lipids" (R. M. C. Dawson and D. N. Rhodes, eds.), pp. 207–217. Wiley, London.

Hartree, E. F., and Mann, T. (1959). *Biochem. J.* **71,** 423.
Hartree, E. F., and Mann, T. (1961). *Biochem. J.* **80,** 464.
Hayes, E. R. (1947). Ph.D. Dissertation, Ohio State University.
Helmy, F. M., and Hack, M. H. (1962). *Proc. Soc. Exp. Biol. Med.* **110,** 91.
Helmy, F. M., and Hack, M. H. (1963). *Proc. Soc. Exp. Biol. Med.* **114,** 361.
Helmy, F. M., and Hack, M. H. (1964). *Amer. J. Obstet. Gynecol.* **88,** 578.
Helmy, F. M., and Hack, M. H. (1966). *Lipids* **1,** 279.
Helmy, F. M., and Hack M. H. (1967). *Comp. Biochem. Physiol.* **23,** 329.
Helmy, F. M., and Hack, M. H. (1970). *Acta Histochem.* **35,** 253.
Helmy, F. M., and Longley, J. B. (1966). *Acta Histochem.* **25,** 300.
Hill, E. E., and Lands, W. E. M. (1970). *In* "Lipid Metabolism" (S. J. Wakil, ed.), Vol. 1, pp. 185–277. Academic Press, New York.
Hoevet, S. P., Viswanathan, C. V., and Lundberg, W. O. (1968). *J. Chromatogr.* **34,** 195.
Hogan, E. L., and Joseph, K. C. (1970). *J. Neurochem.* **17,** 1209.
Hogan, E. L., Joseph, K. C., and Schmidt, G. (1970). *J. Neurochem.* **17,** 75.
Hohenwald, H., and Braedel, C. (1967). *Acta Histochem.* **28,** 291.
Holloway, P. W., and Wakil, S. J. (1964). *J. Biol. Chem.* **239,** 2489.
Holub, B. J., and Kuksis, A. (1969). *Lipids* **4,** 466.
Horrocks, L. A. (1967a). *J. Lipid Res.* **8,** 569.
Horrocks, L. A. (1967b). *Fed. Eur. Biochem. Soc.* **4,** 45.
Horrocks, L. A. (1968a). *J. Neurochem.* **15,** 483.
Horrocks, L. A. (1968b). *J. Lipid Res.* **9,** 469.
Horrocks, L. A. (1968c). Unpublished data.
Horrocks, L. A. (1969a). *J. Neurochem.* **16,** 13.
Horrocks, L. A. (1969b). *In* "Second International Meeting of the International Society for Neurochemistry" (R. Paoletti, R. Fumagalli, and C. Galli, eds.), p. 221. Tamburini Editore, Milan.
Horrocks, L. A. (1970). Unpublished data.
Horrocks, L. A. (1971). *In* "Third International Meeting of the International Society for Neurochemistry" (J. Domonkos, A. Fonyó, I. Huszák, and J. Szentágothai, eds.), p. 312. Akadémiai Kiadó, Budapest.
Horrocks, L. A. (1972). Manuscript in preparation.
Horrocks, L. A., and Ansell, G. B. (1967a). *Biochim. Biophys. Acta* **137,** 90.
Horrocks, L. A., and Ansell, G. B. (1967b). *Lipids* **2,** 329.
Horrocks, L. A., and McConnell, D. (1969). Unpublished data.
Horrocks, L. A., and Sun, G. Y. (1972). *In* "Research Methods in Neurochemistry" (R. Rodnight and N. Marks, eds.) (in press). Plenum Press, New York.
Hübscher, G., and Clark, B. (1960). *Biochim. Biophys. Acta* **41,** 45.
Hughes, B. P., and Frais, F. F. (1967). *Nature (London)* **215,** 993.
Ivanova, T. N., Rubel, L. N., and Semenova, N. A. (1967). *J. Neurochem.* **14,** 653.
Joffe, S. (1969). *J. Neurochem.* **16,** 715.
Kates, M., and James, A. T. (1961). *Biochim. Biophys. Acta* **50,** 478.
Keenan, R. W. (1960). Ph. D. Dissertation, Ohio State University (University Microfilms 60-6381).
Keenan, R. W., and Maxam, A. (1969). *Biochim. Biophys. Acta* **176,** 348.
Keenan, R. W., Brown, J. B., and Marks, B. H. (1961). *Biochim. Biophys. Acta* **51,** 226.
Keenan, R. W., Schmidt, G., and Tanaka, T. (1968). *Anal. Biochem.* **23,** 555.
Kern, F., Jr., and Borgström, B. (1965). *Biochim. Biophys. Acta* **98,** 520.
Kishimoto, Y., and Radin, N. S. (1964). *J. Lipid Res.* **5,** 98.

Kishimoto, Y., Wajda, M., and Radin, N. S. (1968). *J. Lipid Res.* **9,** 27.
Kiyasu, J. Y., and Kennedy, E. P. (1960). *J. Biol. Chem.* **235,** 2590.
Klenk, E., and Debuch, H. (1963). *Progr. Chem. Fats Other Lipids* **6,** 3.
Klenk, E., and Doss, M. (1966). *Hoppe-Seyler's Z. Physiol. Chem.* **346,** 296.
Klenk, E., and Löhr, J. P. (1967). *Hoppe-Seyler's Z. Physiol. Chem.* **348,** 1712.
Klenk, E., Stoffel, W., and Eggers, H. J. (1952). *Hoppe-Seyler's Z. Physiol. Chem.* **290,** 246.
Kochetkov, N. K., Zhukova, I. G., and Glukhoded, I. S. (1963). *Biochim. Biophys. Acta* **70,** 716.
Kolattukudy, P. E. (1970). *Biochemistry* **9,** 1095.
Korey, S. R., and Orchen, M. (1959). *Arch. Biochem. Biophys.* **83,** 381.
Kuemmel, D. F., and Chapman, L. R. (1968). *Lipids* **3,** 313.
Kunze, D., and Olthoff, D. (1968). *Klin. Wochenschr.* **46,** 1308.
Lapetina, E. G., Soto, E. F., and DeRobertis, E. (1968). *J. Neurochem.* **15,** 437.
Laser, H. (1950), *J. Physiol. (London)* **110,** 338.
Law, M. D., Swell, L., and Treadwell, C. R. (1964). *Fed. Proc., Fed. Amer. Soc. Exp. Biol.* **23,** 340.
Leitch, G. J., Horrocks, L. A., and Samorajski, T. (1969). *J. Neurochem.* **16,** 1347.
Leupold, F. (1950). *Hoppe-Seyler's Z. Physiol. Chem.* **285,** 182.
MacBrinn, M. C., and O'Brien, J. S. (1969). *J. Neurochem.* **16,** 7.
McMartin, D. N., and Horrocks, L. A. (1969). Unpublished data.
McMurray, W. C. (1964). *J. Neurochem.* **11,** 315.
Mahadevan, V. (1970). *Progr. Chem. Fats Other Lipids* **10,** 81.
Mahadevan, V., Phillips, F., and Viswanathan, C. V. (1968). *Chem. Phys. Lipids* **2,** 183.
Malins, D. C. (1968). *J. Lipid Res.* **9,** 687.
Mandel, P., and Nussbaum, J. L. (1966). *J. Neurochem.* **13,** 629.
Mandel, P., Bieth, R., and Freysz, L. (1963). *J. Physiol. (Paris)* **55,** 293.
Marcus, A. J., Ullman, H. L., and Safier, L. B. (1969). *J. Lipid Res.* **10,** 108.
Marinetti, G. V., Temple, K., and Stotz, E. (1961). *J. Lipid Res.* **2,** 188.
Masoro, E. J. (1967). *J. Biol. Chem.* **242,** 1111.
Masoro, E. J., Rowell, L. B., and McDonald, R. M. (1964). *Biochim. Biophys. Acta* **84,** 493.
Masoro, E. J., Rowell, L. B., McDonald, R. M., and Steiert, B. (1966). *J. Biol. Chem.* **241,** 2626.
Matsumoto, M., Suzuki, Y., and Tamiya, K. (1967). *Jap. J. Exp. Med.* **37,** 355.
Miller, B., Anderson, C. E., and Piantadosi, C. (1964). *J. Gerontol.* **19,** 430.
Morgan, T. E., Tinker, D. O., and Hanahan, D. J. (1963). *Arch. Biochem. Biophys.* **103,** 54.
Morrill, G. A., and Rapport, M. M. (1964). *J. Biol. Chem.* **239,** 740.
Morton, I. D., and Todd, A. R. (1950). *Biochem. J.* **47,** 327.
Nakagawa, S., and McKibbin, J. M. (1962). *Proc. Soc. Exp. Biol. Med.* **111,** 634.
Nakayama, F., and Blomstrand, R. (1961). *Acta Chem. Scand.* **15,** 1595.
Neudoerffer, T. S. (1967). *Chem. Phys. Lipids,* **1,** 341.
Neudoerffer, T. S., and Lea, C. H. (1968). *Brit. J. Nutr.* **22,** 115.
Nordöy, A., and Lund, S. (1968). *Scand. J. Clin. Lab Invest.* **22,** 328.
Norton, W. T. (1960). *Biochim. Biophys. Acta* **38,** 340.
Norton, W. T. (1972). *In* "The Cellular and Molecular Basis of Neurologic Disease" (G. M. Shy, E. S. Goldensohn, and S. H. Appel, eds.) (in press). Lea & Febiger, Philadelphia, Pennsylvania.

Norton, W. T., and Autilio, L. A. (1965). *Ann. N. Y. Acad. Sci.* **122,** 77.
Norton, W. T., and Autilio, L. A. (1966). *J. Neurochem.* **13,** 213.
Norton, W. T., and Brotz, M. (1963). *Biochem. Biophys. Res. Commun.* **12,** 198.
Norton, W. T., and Brotz, M. (1967). *Fed. Proc., Fed. Amer. Soc. Exp. Biol.* **26,** 675.
Norton, W. T., and Poduslo, S. E. (1971). *J. Lipid Res.* **12,** 84.
Norton, W. T., Gelfand, M., and Brotz, M. (1962). *J. Histochem. Cytochem.* **10,** 375.
Norton, W. T., Poduslo, S. E., and Suzuki, K. (1966). *J. Neuropathol. Exp. Neurol.* **25,** 582.
O'Brien, J. S., and Sampson, E. L. (1965a). *J. Lipid Res.* **6,** 537.
O'Brien, J. S., and Sampson, E. L. (1965b). *J. Lipid Res.* **6,** 545.
O'Brien, J. S., Fillerup, D. L., and Mead, J. F. (1964). *J. Lipid Res.* **5,** 329.
O'Brien, J. S., Sampson, E. L., and Stern, M. B. (1967). *J. Neurochem.* **14,** 357.
Oshima, M., and Carpenter, M. P. (1968). *Biochim. Biophys. Acta* **152,** 479.
Oswald, E. O., Anderson, C. E., Piantadosi, C., and Lim, J. (1968). *Lipids* **3,** 51.
Owens, K. (1966). *Biochem. J.* **100,** 354.
Owens, K., and Hughes, B. P. (1970). *J. Lipid Res.* **11,** 486.
Paltauf, F. (1969). *Biochim. Biophys. Acta* **176,** 818.
Paltauf, F. (1971). *Biochim. Biophys. Acta* **239,** 38.
Paltauf, F., and Polheim, D. (1967). *Hoppe-Seyler's Z. Physiol. Chem.* **348,** 1551.
Paltauf, F., and Polheim, D. (1970). *Biochim. Biophys. Acta* **210,** 187.
Panganamala, R. V., Buntine, D. W., Geer, J. C., and Cornwell, D. G. (1969). *Chem. Phys. Lipids* **3,** 401.
Panganamala, R. V., Horrocks, L. A., Geer, J. C., and Cornwell, D. G. (1971). *Chem. Phys. Lipids* **6,** 97.
Parsons, P., and Basford, R. E. (1967). *J. Neurochem.* **14,** 823.
Patrikeeva, M. V. (1965). *Vopr. Med. Khim.* **11,** 99.
Pfleger, R. C., Piantadosi, C., and Snyder, F. (1967). *Biochim. Biophys. Acta* **144,** 633.
Pfleger, R. C., Anderson, N. G., and Snyder, F. (1968). *Biochemistry* **7,** 2826.
Piantadosi, C., and Snyder, F. (1970). *J. Pharm. Sci.* **59,** 283.
Pietruszko, R. (1962). *Biochim. Biophys. Acta* **64,** 562.
Pietruszko, R., and Gray, G. M. (1962). *Biochim. Biophys. Acta* **56,** 232.
Popović, M. (1965). *Hoppe-Seyler's Z. Physiol. Chem.* **340,** 18.
Porcellati, G., and Mastrantonio, M. A. (1964). *Ital. J. Biochem.* **13,** 332.
Porcellati, G., Biasion, M. G., and Arienti, G. (1970a). *Lipids* **5,** 725.
Porcellati, G., Biasion, M. G., and Pirotta, M. (1970b). *Lipids* **5,** 734.
Poulos, A. (1970). *J. Lipid Res.* **11,** 496.
Poulos, A., Hughes, B. P., and Cumings, J. N. (1968). *Biochim. Biophys. Acta* **152,** 629.
Pries, C., Aumont, A., and Böttcher, C. J. F. (1966). *Biochim. Biophys. Acta* **125,** 277.
Prostenik, M., and Popovic, M. (1963). *Nature (London)* **199,** 1285.
Prottey, C., and Hawthorne, J. N. (1966). *Biochem. J.* **101,** 191.
Pursel, V. G., and Graham, E. F. (1967). *J. Reprod. Fert.* **14,** 203.
Radloff, J. F., and Ferrell, W. J. (1970). *Physiol. Chem. Phys.* **2,** 105.
Radominska-Pyrek, A., and Horrocks, L. (1972). Submitted for publication.
Ramachandran, S., Sprecher, H. W., and Cornwell, D. G. (1968). *Lipids* **3,** 511.
Rao, P. V., Ramachandran, S., and Cornwell, D. G. (1967). *J. Lipid Res.* **8,** 380.
Rapport, M. M., and Lerner, B. (1959). *Biochim. Biophys. Acta* **33,** 319.
Rapport, M. M., and Norton, W. T. (1962). *Annu. Rev. Biochem.* **31,** 103.
Renkonen, O. (1962a). *Acta Chem. Scand.* **16,** 1288.
Renkonen, O. (1962b). *Biochim. Biophys. Acta* **59,** 497.

Renkonen, O. (1963a). *Acta Chem. Scand.* **17,** 634.
Renkonen, O. (1963b). *Acta Chem. Scand.* **17,** 1925.
Renkonen, O. (1965). *J. Amer. Oil Chem. Soc.* **42,** 298.
Renkonen, O. (1966). *Biochim. Biophys. Acta* **125,** 288.
Renkonen, O. (1967a). *Biochim. Biophys. Acta* **137,** 575.
Renkonen, O. (1967b). *Acta Chem. Scand.* **21,** 1108.
Renkonen, O. (1968). *J. Lipid Res.* **9,** 34.
Renkonen, O., and Varo, P. (1967). *Lipid Chromatogr. Anal.* **1,** 41.
Renkonen, O., Liusvaara, S., and Miettinen, A. (1965). *Ann. Med. Exp. Biol. Fenn.* **43,** 200.
Rhee, K. S., del Rosario, R. R., and Dugan, L. R., Jr. (1967). *Lipids* **2,** 334.
Robertson, A. F., and Lands, W. E. M. (1962). *J. Clin. Invest.* **41,** 2160.
Roots, B. I., and Johnston, P. V. (1965). *Biochem. J.* **94,** 61.
Rossiter, R. J. (1966). *In* "Nerve as a Tissue" (K. Rodahl and B. Issekutz, eds.), p. 175, Harper (Hoeber), New York.
Rossiter, R. J. (1967). *Lipids Lipidoses* p. 93.
Rouser, G. (1968). *Biochem. Prep.* **12,** 73.
Rouser, G., and Yamamoto, A. (1968). *Lipids* **3,** 284.
Rouser, G., Nelson, G. J., Fleischer, S., and Simon, G. (1968). *In* "Biological Membranes" (D. Chapman, ed.), pp. 5–69. Academic Press, New York.
Rubel, L. N., and Ivanova, T. N. (1965). *Dokl. Akad. Nauk SSSR* **165,** 943.
Rubel, L. N., Ivanova, T. N., and Semenova, N. A. (1967). *Zh. Evol. Biokhim. Fiziol.* **3,** 110.
Rumsby, M. G. (1967). *J. Neurochem.* **14,** 733.
Rumsby, M. G., and Rossiter, R. J. (1968). *J. Neurochem.* **15,** 1473.
Sanders, H. (1967). *Biochim. Biophys. Acta* **144,** 485.
Sansone, G., and Hamilton, J. G. (1969). *Lipids* **4,** 435.
Schmid, H. H. O., and Mangold, H. K. (1966a). *Biochem. Z.* **346,** 13.
Schmid, H. H. O., and Mangold, H. K. (1966b). *Biochim. Biophys. Acta* **125,** 182.
Schmid, H. H. O., and Takahashi, T. (1968). *Biochim. Biophys. Acta* **164,** 141.
Schmid, H. H. O., and Takahashi, T. (1970). *J. Lipid Res.* **11,** 412.
Schmid, H. H. O., Jones, L. L., and Mangold, H. K. (1967a). *J. Lipid Res.* **8,** 692.
Schmid, H. H. O., Tuna, N., and Mangold, H. K. (1967b). *Hoppe-Seyler's Z. Physiol. Chem.* **348,** 730.
Schmidt, G., Fingerman, L. H., Kreevoy, H. M., DeMarco, P., and Thannhauser, S. J. (1961). *Amer. J. Clin. Nutr.* **9,** 124.
Schogt, J. C. M., Begemann, P. H., and Koster, J. (1960). *J. Lipid Res.* **1,** 446.
Scott, T. W., and Setchell, B. P. (1968). *Biochem. J.* **107,** 273.
Scott, T. W., Dawson, R. M. C., and Rowlands, I. W. (1963). *Biochem. J.* **87,** 507.
Scott, T. W., Jay, S. M., and Freinkel, N. (1966). *Endocrinology* **79,** 591.
Scott, T. W., Voglmayr, J. K., and Setchell, B. P. (1967a). *Biochem. J.* **102,** 456.
Scott, T. W., Setchell, B. P., and Bassett, J. M. (1967b). *Biochem. J.* **104,** 1040.
Scott, T. W., Hansel, W., and Donaldson, L. E. (1968). *Biochem. J.* **108,** 317.
Segal, W., and Wysocki, S. J. (1970). *Biochem. J.* **119,** 43P.
Seminario, L. M., Hren, N., and Gómez, C. J. (1964). *J. Neurochem.* **11,** 197.
Shah, D. O., and Schulman, J. H. (1965). *J. Lipid Res.* **6,** 341.
Sheltawy, A., and Dawson, R. M. C. (1966). *Biochem. J.* **100,** 12.
Sheltawy, A., and Dawson, R. M. C. (1969). *Biochem. J.* **111,** 157.
Sherr, S. I., and Treadwell, C. R. (1965). *Biochim. Biophys. Acta* **98,** 539.

Sherr, S. I., Swell, L., and Treadwell, C. R. (1963). *Biochem. Biophys. Res. Commun.* **13**, 131.
Slagel, D. E., Dittmer, J. C., and Wilson, C. B. (1967). *J. Neurochem.* **14**, 789.
Snyder, F. (1969). *Prog. Chem. Fats Other Lipids* **10**, 287–337.
Snyder, F., and Blank, M. L. (1969). *Arch. Biochem. Biophys.* **130**, 101.
Snyder, F., and Malone, B. (1970). *Biochem. Biophys. Res. Commun.* **41**, 1382.
Snyder, F., and Pfleger, R. C. (1966). *Lipids* **1**, 328.
Snyder, F., Blank, M. L., and Morris, H. P. (1969a). *Biochim. Biophys. Acta* **176**, 502.
Snyder, F., Malone, B., and Wykle, R. L. (1969b). *Biochem. Biophys. Res. Commun.* **34**, 40.
Snyder, F., Malone, B., and Blank, M. L. (1969c). *Biochim. Biophys. Acta* **187**, 302.
Snyder, F., Piantadosi, C., and Wood, R. (1969d). *Proc. Soc. Exp. Biol. Med.* **130**, 1170.
Snyder, F., Wykle, R. L., and Malone, B. (1969e). *Biochem. Biophys. Res. Commun.* **34**, 315.
Snyder, F., Malone, B., and Cumming, R. B. (1970a). *Can. J. Biochem.* **48**, 212.
Snyder, F., Malone, B., and Blank, M. L. (1970b). *J. Biol. Chem.* **245**, 1790.
Snyder, F., Blank, M. L., Malone, B., and Wykle, R. L. (1970c). *J. Biol. Chem.* **245**, 1800.
Snyder, F., Piantadosi, C., and Malone, B. (1970d). *Biochim. Biophys. Acta* **202**, 244.
Snyder, F., Blank, M. L., and Malone, B. (1970e). *J. Biol. Chem.* **245**, 4016.
Snyder, F., Rainey, W. T., Jr., Blank, M. L., and Christie, W. H. (1970f). *J. Biol. Chem.* **245**, 5853.
Snyder, F., Hibbs, M., and Malone, B. (1971a). *Biochim. Biophys. Acta* **231**, 409.
Snyder, F., Blank, M. L., and Wykle, R. L. (1971b). *J. Biol. Chem.* **246**, 3639.
Song, M., Freysz, L., and Rebel, G. (1970). *Biochim. Biophys. Acta* **218**, 363.
Soodsma, J. F. (1970). *Fed. Proc., Fed. Amer. Soc. Exp. Biol.* **29**, 931.
Soodsma, J. F., Piantadosi, C., and Snyder, F. (1970). *Cancer Res.* **30**, 309.
Spanner, S. (1966). *Nature (London)* **210**, 637.
Spence, M. W. (1970). *Biochim. Biophys Acta* **218**, 347.
Spener, F., and Mangold, H. K. (1969). *J. Lipid Res.* **10**, 609.
Spener, F., Paltauf, F., and Holasek, A. (1968). *Biochim. Biophys. Acta* **152**, 368.
Spener, F., Mangold, H. K., Sansone, G., and Hamilton, J. G. (1969). *Biochim. Biophys. Acta* **192**, 516.
Stein, R. A., and Slawson, V. (1966). *J. Chromatogr.* **25**, 204.
Stoffel, W., and LeKim, D. (1971). *Hoppe-Seyler's Z. Physiol. Chem.* **352**, 501.
Stoffel, W., Sticht, G., and LeKim, D. (1968). *Hoppe-Seyler's Z. Physiol. Chem.* **349**, 1745.
Stoffel, W., LeKim, D., and Heyn, G. (1970). *Hoppe-Seyler's Z. Physiol. Chem.* **351**, 875.
Sun, G. Y. (1970). *Trans. Amer. Soc. Neurochem.* **1**, 70.
Sun, G. Y., and Horrocks, L. A. (1968). *Lipids* **3**, 79.
Sun, G. Y., and Horrocks, L. A. (1969a). *J. Neurochem.* **16**, 181.
Sun, G. Y., and Horrocks, L. A. (1969b). *J. Lipid Res.* **10**, 153.
Sun, G. Y., and Horrocks, L. A. (1970). *Lipids* **5**, 1006.
Sun, G. Y., and Horrocks, L. A. (1971). *Fed. Proc., Fed. Amer. Soc. Exp. Biol.* **30**, 1248 Abs.
Sun, G. Y., and Horrocks, L. A. (1972). Submitted for publication.
Svennerholm, L., and Thorin, H. (1960). *Biochim. Biophys. Acta* **41**, 371.
Swell, L., Law, M. D., and Treadwell, C. R. (1965). *Arch. Biochem. Biophys.* **110**, 231
Tabakoff, B., and Erwin, V. G. (1970). *J. Biol. Chem.* **245**, 3263.
Takahashi, T., and Schmid, H. H. O. (1968). *Chem. Phys. Lipids* **2**, 220.

Takahashi, T., and Schmid, H. H. O. (1970). *Chem. Phys. Lipids* **4**, 243.
Takeuchi, M., and Terayama, H. (1965). *Exp. Cell Res.* **40**, 32.
Tamai, Y. (1968). *Jap. J. Exp. Med.* **38**, 65.
Terner, J. Y., and Hayes, E. R. (1961). *Stain Technol.* **36**, 265.
Thiele, O. W. (1964). *Z. Klin. Chem.* **2**, 33.
Thiele, O. W., and Denden, A. (1967). *Hoppe-Seyler's Z. Physiol. Chem.* **348**, 1097.
Thompson, G. A., Jr. (1966). *Biochemistry* **5**, 1290.
Thompson, G. A., Jr. (1967). *Biochemistry* **6**, 2015.
Thompson, G. A., Jr., and Hanahan, D. J. (1963). *Biochemistry* **2**, 641.
Tietz, A., Lindberg, M., and Kennedy, E. P. (1964). *J. Biol. Chem.* **239**, 4081.
Todd, D., and Rizzi, G. P. (1964). *Proc. Soc. Exp. Biol. Med.* **115**, 218.
Vennart, G. P. (1963). *Lab Invest.* **12**, 327.
Vignais, P. V., and Zabin, I. (1960). *In* "Biochemistry of Lipids" (G. Popjak, ed.), pp. 78–84. Pergamon, Oxford.
Viswanathan, C. V., Basilio, M., Hoevet, S. P., and Lundberg, W. O. (1968a). *J. Chromatogr.* **34**, 241.
Viswanathan, C. V., Phillips, F., and Lundberg, W. O. (1968b). *J. Chromatogr.* **35**, 66.
Viswanathan, C. V., Phillips, F., and Lundberg, W. O. (1968c). *J. Chromatogr.* **38**, 267.
Viswanathan, C. V., Hoevet, S. P., Lundberg, W. O., White, J. M., and Muccini, G. A. (1969). *J. Chromatogr.* **40**, 225.
Waku, K., and Lands, W. E. M. (1968). *J. Biol. Chem.* **243**, 2654.
Waku, K., and Nakazawa, Y. (1970). *J. Biochem. (Tokyo)* **68**, 459.
Walker, B. L., and Kummerow, F. A. (1964). *J. Nutr.* **82**, 329.
Warner, H. R. (1962). Ph.D. Dissertation, University of Michigan (University Microfilms 63-470).
Warner, H. R., and Lands, W. E. M. (1961). *J. Biol. Chem.* **236**, 2404.
Warner, H. R., and Lands, W. E. M. (1963). *J. Lipid Res.* **4**, 216.
Ways, P., and Hanahan, D. J. (1964). *J. Lipid Res.* **5**, 318.
Webster, G. R. (1960). *Biochim. Biophys. Acta* **44**, 109.
Weinstein, D. B., Marsh, J. B., Glick, M. C., and Warren, L. (1969). *J. Biol. Chem.* **244**, 4103.
Wells, M. A., and Dittmer, J. C. (1966). *Biochemistry* **5**, 3405.
Wells, M. A., and Dittmer, J. C. (1967). *Biochemistry* **6**, 3169.
Wheeldon, L. W., Schumert, Z., and Turner, D. A. (1965). *J. Lipid Res.* **6**, **481**.
Wiley, R. A., Sumner, D. D., and Walaszek, E. J. (1970). *Lipids* **5**, 803.
Williams, J. H., Kuchmak, M., and Witter, R. F. (1966a). *Lipids* **1**, 89.
Williams, J. H., Kuchmak, M., and Witter, R. F. (1966b). *Lipids* **1**, 391.
Williams, J. N., Jr., Anderson, C. E., and Jasik, A. D. (1962). *J. Lipid Res.* **3**, 378.
Winterfeld, M., and Debuch, H. (1966). *Hoppe-Seyler's Z. Physiol. Chem.* **345**, 11.
Winterfeld, M., and Debuch, H. (1968). *Hoppe-Seyler's Z. Physiol. Chem.* **349**, 903.
Winterfeld, M., and Debuch, H. (1970). *Hoppe-Seyler's Z. Physiol. Chem.* **351**, 169.
Wittenberg, J. B., Korey, S. R., and Swenson, F. H. (1956). *J. Biol. Chem.* **219**, 39.
Wood, R., and Healy, K. (1970a). *Biochem. Biophys. Res. Commun.* **38**, 205.
Wood, R., and Healy, K. (1970b). *J. Biol. Chem.* **245**, 2640.
Wood, R., and Snyder, F. (1968). *Lipids* **3**, 129.
Wood, R., Harlow, R. D., and Snyder, F. (1969). *Biochim. Biophys. Acta* **176**, 641.
Wood, R., Walton, M., Healy, K., and Cumming, R. B. (1970). *J. Biol. Chem.* **245**, 4276.
Wykle, R. L., and Snyder, F. (1969). *Biochem. Biophys. Res. Commun.* **37**, 658.

Wykle, R. L., and Snyder, F. (1970). *J. Biol. Chem.* **245,** 3047.
Wykle, R. L., Blank, M. L., and Snyder, F. (1970). *FEBS Lett.* **12,** 57.
Wykle, R. L., Blank, M. L., and Snyder, F. (1971). *Fed. Proc., Fed. Amer. Soc. Exp. Biol.* **30,** 1243Abs.
Yabuuchi, H., and O'Brien, J. S. (1968). *J. Lipid Res.* **9,** 65.
Yanagihara, T., and Cumings, J. N. (1968). *Arch. Neurol.* **19,** 241.
Yanagihara, T., and Cumings, J. N. (1969). *Brain* **92,** 59.
Yarbro, C. L., and Anderson, C. E. (1956). *Proc. Soc. Exp. Biol. Med.* **91,** 408.
Yarbro, C. L., and Anderson, C. E. (1957). *Proc. Soc. Exp. Biol. Med.* **95,** 556.
Zilversmit, D. B., Entenman, C., and Fishler, M. C. (1943). *J. Gen. Physiol.* **26,** 325.
Zilversmit, R. D., Marcus, A. J., and Ullman, H. L. (1961). *J. Biol. Chem.* **236,** 47.

CHAPTER X

ETHER-LINKED LIPIDS AND FATTY ALCOHOL PRECURSORS IN NEOPLASMS

Fred Snyder

I. Introduction

A characteristic spot for alkyldiacylglycerols appears above triacylglycerols and below sterol esters when lipid extracts of tumors are chromatographed on silica gel layers in nonpolar solvent systems (Fig. 1). We first saw this spot in some experiments with tumor lipids in our labora-

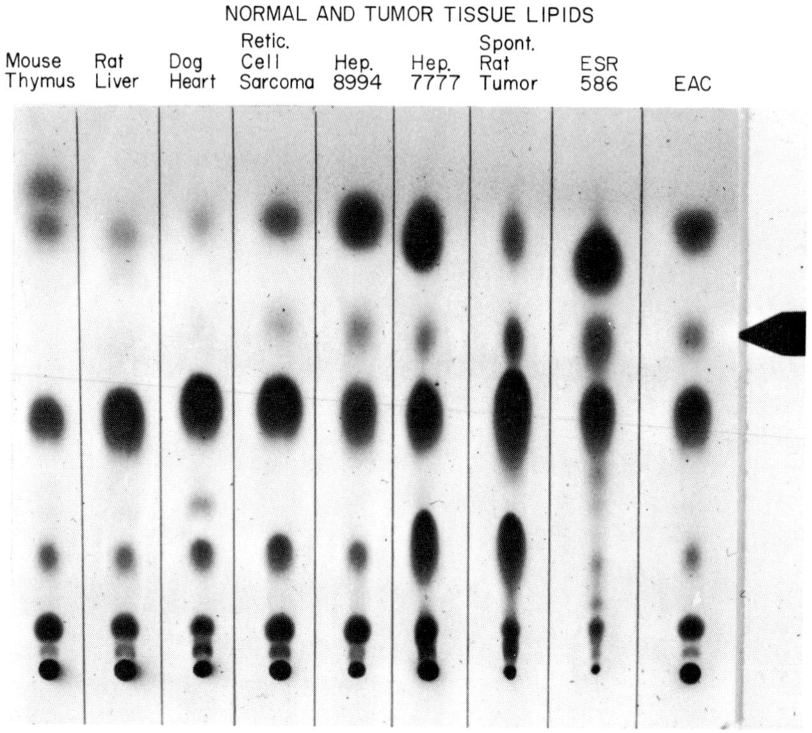

FIG. 1. Representative thin-layer chromatogram of total lipids from normal tissues and tumors. The abbreviated notations above each lane designate reticulum cell sarcoma (human), Morris hepatoma 8994 (rat), Morris hepatoma 7777 (rat), spontaneous rat tumor, preputial gland tumor ESR-586 (mouse), and Ehrlich ascites cells (mouse). The black arrow indicates the location of the alkyldiacylglycerols; the spots directly below the alkyldiacylglycerols are triacylglycerols. A solvent mixture of hexane:diethyl ether:acetic acid (90:10:1, v/v) was used for development, and Silica Gel G was the adsorbent.

tory about ten years ago, and did the preliminary chromatographic analysis in conjunction with a survey of lipid patterns from various tissues conducted to gain further experience with TLC.* At that time, Dr. Frank Comas, a radiotherapist on the Medical Division staff, was studying the effect of hypoxia in treating tumors with X-irradiation. The two tumors he was using (Walker-256 carcinosarcoma and a Fisher sarcoma R3259/96A) were made available to us for lipid studies. Malins and Mangold (1960) had just published illustrations of thin-layer chromatograms for

* Abbreviations used in this chapter: TLC, thin-layer chromatography; GLC, gas–liquid chromatography.

a variety of reference compounds, which revealed that alkyldiacylglycerols, methyl esters of fatty acids, and long-chain fatty aldehydes had similar chromatographic properties, each migrating directly above the triacylglycerols on Silica Gel G in hexane–diethyl ether–acetic acid (90:10:1, v/v). The unidentified spot above triacylglycerols that we found on thin-layer chromatograms was prominent for tumor lipids but not for healthy tissue lipids (Fig. 1). We first thought that it might be an artifact formed in lipid solvents, such as a methyl ester of a fatty acid. In fact, Lindlar and Wagener (1964) published a paper depicting a thin-layer chromatogram of "neutral" lipid patterns obtained for five different mouse tumors with a component above the triacylglycerols that the authors referred to as methyl esters of fatty acids, but which was probably alkyldiacylglycerols. Soon it became apparent that other transplantable and spontaneous tumors from animals and human tumors all showed the characteristic spot above triacylglycerols on thin-layer chromatograms, and we decided to identify its chemical structure (see Section II of this chapter).

In retrospect, these initial TLC experiments were important in changing the direction of our fledgling research program on glyceryl ethers in irradiated animals and their role in bone marrow (Snyder and Cress, 1963). The basis for our interest in alkyl glyceryl ethers was the radioprotective and hemopoietic-stimulating properties attributed to them (see Chapter VIII), but subsequent work in our laboratory with the ether-linked lipids from a variety of tumors and some normal cells eventually led to their complete characterization in neoplasms and to the elucidation of their enzymic pathways (see Chapter VII).

II. Nature of the Ether-Linked Aliphatic Moieties in Neoplastic Cells

A. Alkyl- and Alk-1-enyl-diacylglycerols

1. *General*

The alkyldiacylglycerols in tumors have been identified on the basis of the chemical, chromatographic, and physical behavior of the intact molecule and products (or derivatives) produced by $LiAlH_4$ reduction or saponification.

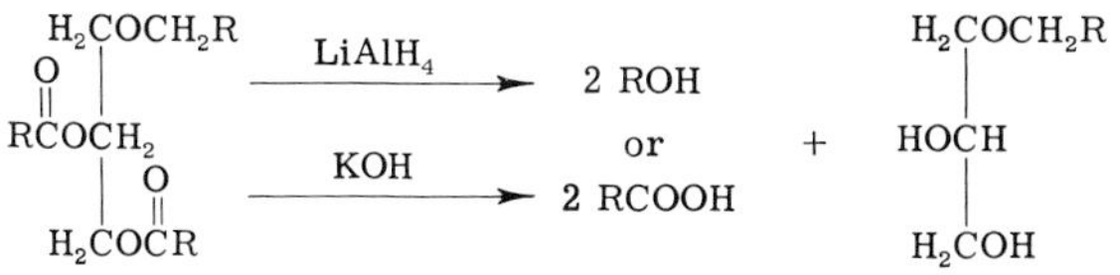

The first report (Snyder *et al.*, 1966) on alkyldiacylglycerols in neoplasms described their occurrence in two transplantable rat tumors (Walker-256 carcinosarcoma and R3259/96A giant cell sarcoma), a spontaneous rat tumor (fibroadenoma) that occurs in rats approximately 1 year after 800-R total-body irradiation, Ehrlich ascites cells grown in the peritoneal cavity of mice, and a lymphosarcoma from a 66-year-old man's auxillary lymph node. The alkyldiacylglycerols were not found in total lipid extracts from normal cells (liver, bone marrow, or plasma) of tumor-bearing animals, or from the peritoneal fluid containing Ehrlich ascites cells.

Chromatographic behavior of the "unidentified" tumor lipid class, later identified as alkyldiacylglycerols, was established for a wide range of solvent mixtures containing hexane:diethyl ether:acetic acid (Snyder *et al.*, 1966). The alkyldiacylglycerols migrated at a higher R_f than the methyl esters of fatty acids in solvent mixtures of high polarity, e.g., 70:30:1 v/v, but at a lower R_f in less polar systems, e.g., 95:5:1 v/v. We were able to show that saponification of the purified "unidentified" lipid from Ehrlich ascites cells produced alkylglycerols and fatty acids identifiable by TLC (Fig. 2); best results could be obtained by direct chromatography of the entire saponification mixture after acidification, since the usual washing and extraction of the nonsaponifiable fraction often resulted in a substantial loss of alkylglycerols into the water phase. We ruled out the presence of an alk-1-enyl moiety in the purified "unidentified" lipid fraction that corresponded to alkyldiacylglycerols, and argentation TLC was used to determine that the degree of unsaturation was less in the alkyldiacylglycerols than in the triacylglycerols.

The following year the alkyldiacylglycerols were unequivocally identified in Ehrlich ascites cells (Wood and Snyder, 1967) and in the Walker-256 carcinosarcoma and the human lymphosarcoma (Bollinger, 1967); however, Bollinger (1967) did not find alkyldiacylglycerols in a poorly differentiated sarcoma from the upper arm of another patient. These investigators used TLC, GLC, infrared spectroscopy, and nuclear magnetic resonance to prove that the "unidentified" tumor-lipid fraction consisted of 1-alkyl-2,3-diacylglycerols. Gas–liquid chromatography of the intact alkyldiacylglycerols from Ehrlich ascites cells indicated that their molecular weights ranged from 760 to 990 (Wood and Snyder, 1967).

2. *Alkyl Moieties*

Bollinger (1967) detected only the saturated alkyl moieties (16:0 and 18:0) in alkyldiacylglycerols from Walker-256 carcinosarcomas and a human lymphosarcoma; the 18:1 alkyl moiety was probably missed because of the poor GLC resolution obtained for the trimethylsilyl ether

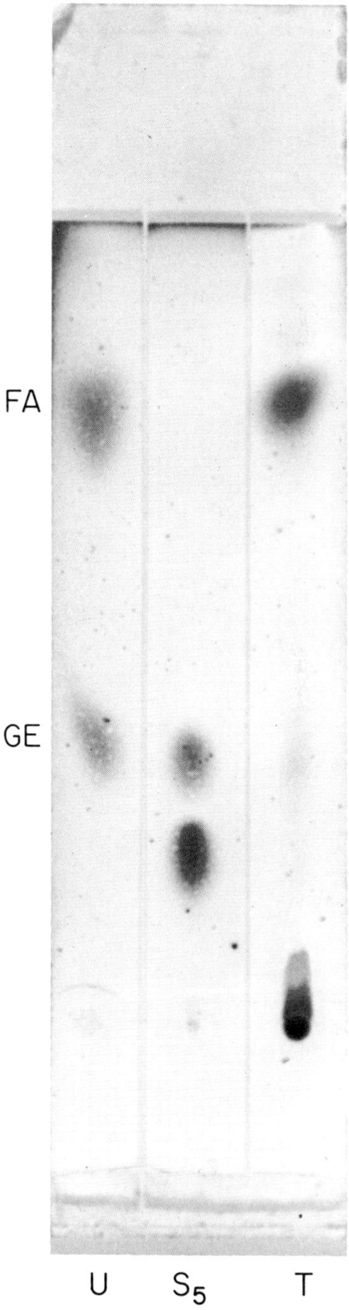

FIG. 2. Saponification products of purified unidentified lipid (U) from Ehrlich ascites carcinoma cells; T designates the lane for total lipid extract from the ascites cells. S_5 contained monopalmitin (lower R_f) and batyl alcohol (GE) (upper spot). FA refers to fatty acid position. Solvent development for TLC (Silica Gel G) was hexane:diethyl ether:methanol : acetic acid (80 : 20 : 10 : 1, v/v/v/v). Reproduced from F. Snyder, E. A. Cress, and N. Stephens, *Lipids* **1**, 381 (1966) by permission of the American Oil Chemists' Society, Chicago, Illinois.

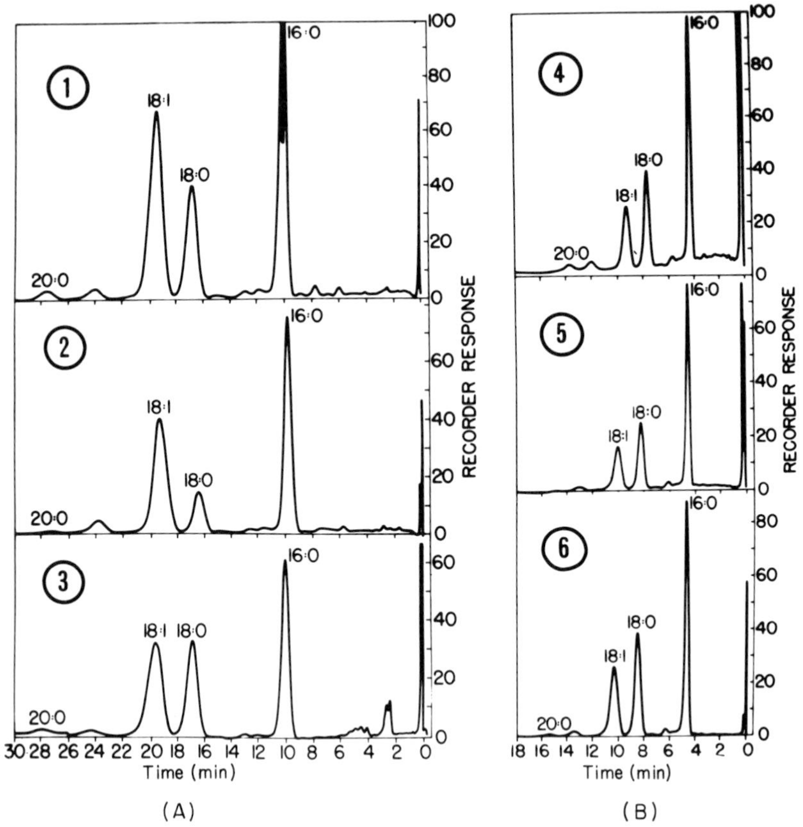

FIG. 3. Representative gas–liquid chromatograms showing the resolution of the (A) isopropylidene derivatives of alkylglycerols, and (B) fatty aldehydes derived from three different glycerolipid classes of Taper liver tumors from mice. The circled numbers designate tracings obtained for alkyl moieties from 1, alkyldiacylglycerols; 2, choline phospholipids; 3, ethanolamine phospholipids; and for alk-1-enyl moieties from 4, alk-1-enyldiacylglycerols; 5, choline phospholipids; and 6, ethanolamine phospholipids. The alkylglycerols were obtained from each lipid class after $LiAlH_4$ reduction, and the aldehydes were obtained by aqueous HCl treatment. Reproduced from Snyder (1969a) by permission of Plenum Publishing Corp., New York.

derivatives used in this study. In the investigation reported for Ehrlich ascites cells (Wood and Snyder, 1967), GLC of the trifluoroacetate derivatives of alkylglycerols derived from alkyldiacylglycerols demonstrated that the alkyl moieties were in the 1-position and that they were primarily hexadecyl (49%), octadecyl (21%), and octadecenyl (34%). In contrast, the acyl moieties contained 12 to 24 carbon atoms with varying degrees of unsaturation; dienes and polyenes were found in fatty acid chains $>C_{18}$ and only traces of odd chains were found.

Detailed studies of the alkyl moieties of the ether-linked glycerolipids have also been done on a number of other neoplasms: Walker-256 carcinosarcomas from rats, a human lymphosarcoma, and Taper liver tumors from mice (Snyder, 1969a); Morris hepatomas from rats (Snyder *et al.*, 1969a); Ehrlich ascites cells from mice (Wood and Snyder, 1969); preputial gland tumors from mice (Snyder *et al.*, 1970a); and in L-M cells (Anderson *et al.*, 1969), a subline of Earl's L cells that grows in chemically defined media and produces a high incidence of tumors when injected into mice. The data shown in Fig. 3 are typical and in general reflect the composite picture of analytical data that has emerged from our studies of the alkyldiacylglycerols and related lipid classes in tumors.

Itoh and Kasama (1970) found alkyldiacylglycerols in a transplantable lipid-producing tumor originating from harderian glands of mice. The harderian gland is located deep within the orbit of the eye and appears to function as a lacrimal gland in animals with a nictitating membrane. Itoh and Kasama (1970) identified the ether-linked lipid by infrared spectroscopy, TLC, and GLC before and after deacylation. Normal harderian glands of mice contain even higher quantities of alkyldiacylglycerols with 16:0, 18:0, 20:0, and 22:1 alkyl moieties (Kasama *et al.*, 1970). An unusual feature of the alkyl moieties in the harderian gland tumor was the high quantity of the 20:0 ($\simeq$22%), 20:1 ($\simeq$9%), and 22:1 ($\simeq$23%) moieties; 16:0 ($\simeq$13%) and 18:0 ($\simeq$22%) alkyl moieties accounted for the remainder. Paltauf and Polheim (1970) have reported an unusually high proportion of alkyl and alk-1-enyl moieties containing 20 carbon atoms in ether-linked glycerolipids isolated from rat intestinal mucosa. Recently, Cain and co-workers (1967) found long-chain *n*-alkanes in normal meninges and meningiomas in human brains. Although the qualitative patterns of the homologous series of alkanes were the same, those from the meningiomas contained a higher proportion of C_{24} and longer carbon chains than those from the normal meninges.

3. *Alk-1-enyl Moieties*

Alk-1-enyldiacylglycerols are not found in all tumor preparations (Wood and Snyder, 1967), but when they do occur the alk-1-enyl moieties (Fig. 3), like the alkyl moieties from the alkyldiacylglycerols from the same tumor preparations, consist mainly of 16:0, 18:0, and 18:1 hydrocarbon chains (Snyder, 1969a). The relative proportions of these moieties in the alk-1-enyldiacylglycerols were similar to those in the choline- and ethanolamine-containing phospholipids. In general, the alkyl moieties in

TABLE I

Nature of Acyl Chains (%) in Triacylglycerols (TA), Diacyl Glyceryl Ethers (DAGE), Choline Phospholipids (CP), and Ethanolamine Phospholipids (EP) of Neoplasms

	Walker-256 tumor				Taper liver tumor				Lymphosarcoma			
Carbon chain	TA	DAGE	CP	EP	TA	DAGE	CP	EP	TA	DAGE	CP	EP
14:0	1.8	1.5	2.2	0.2	1.7	2.0	0.9	t	3.7	4.5	1.1	t
14:1 (15:0)	t[a]	0.8	0.9	t	t	—	t	—	0.7	2.3	0.4	t
16:0	21.8	16.7	32.3	7.8	22.9	22.7	24.8	6.8	22.8	24.3	32.0	5.8
16:1 (17:0)	3.3	3.7	4.7	2.1	5.9	3.9	2.6	1.3	4.1	5.7	2.7	0.6
18:0	12.7	9.0	17.1	19.9	6.9	10.1	16.0	20.5	8.6	8.6	13.3	17.5
18:1	33.8	23.3	19.6	24.6	37.1	22.7	20.8	22.7	46.7	36.5	16.8	11.5
18:2 (19:0)	12.6	12.2	10.9	7.9	21.6	13.6	17.8	15.3	11.4	11.5	6.1	3.6
20:0	—	—	1.7	1.4	t	0.9	t	—	—	1.2	—	—
18:3 (20:1)	1.9	1.0	t	0.6	1.8	2.0	1.5	1.8	1.4	1.2	1.5	1.7
20:2	—	—	—	—	0.5	1.5	1.1	1.2	—	—	2.3	1.7
20:3	1.5	3.2	1.5	2.7	0.3	1.4	1.1	1.1	t	1.2	2.3	2.2
20:4	7.1	20.2	9.1	20.4	1.3	7.9	8.9	13.1	0.6	3.0	15.6	26.4
A[b]	t	0.1	t	1.4	t	—	1.2	4.0	t	t	2.6	9.2
B	1.4	3.4	t	4.1	t	3.7	t	1.1	t	t	t	3.7
C	0.7	1.9	t	2.6	t	1.6	t	1.7	t	t	1.2	3.6
22:6	1.4	3.0	t	4.3	t	6.0	3.3	9.4	t	—	2.1	12.5

[a]t indicates trace quantity.

[b]A, B, and C are unidentified peaks on a gas chromatogram having a retention time between 20:4 and 22:6. Reproduced from Snyder (1969a), by permission of Plenum Publishing Corp., New York.

glycerolipids contain larger quantities of 18:1 and lower quantities of 18:0 ether chains than the alk-1-enyl moieties of glycerolipids. References to the literature are the same as those listed for the alkyl studies (Section II, A, 2).

4. *Acyl Moieties*

Acyl moieties from the alkyldiacylglycerols of tumors are primarily palmitic, stearic, and oleic acids. However, linoleic, arachidonic, and other long-chain polyunsaturated fatty acids also occur, in marked contrast to the fact that fatty acids not normally synthesized *de novo* in mammalian cells are virtually absent in the *ether-linked* moieties of the neoplastic cells so far investigated. A complete analysis of the fatty acids present in alkyldiacylglycerols has been done on Ehrlich ascites cells (Wood and Snyder, 1967, 1969); Walker-256 carcinosarcoma, Taper liver tumor, and a human lymphosarcoma (Snyder, 1969a; see Table I); Morris hepatoma 7777 (Snyder *et al.*, 1969a); preputial gland tumors (Snyder *et al.*, 1970a); and L–M cells (Anderson *et al.*, 1969). In one series of experiments, the stereospecific analysis of triacylglycerols was compared to that of alkyldiacylglycerols from Ehrlich ascites cells (Wood and Snyder, 1969), demonstrating that the 2-position of alkyldiacylglycerols contained large quantities of 20:4 and 22:6 fatty acids, and that palmitic acid was essentially equally distributed between the 2- and 3-positions, whereas the 3-position contained most of the stearic acid present. Although the aliphatic composition of triacylglycerols differed from that of the alkyldiacylglycerols in Ehrlich ascites cells, they both had a 1-random-2-random-3-random distribution.

B. Alkyl and Alk-1-enyl Phospholipids

1. *Alkyl and Alk-1-enyl Moieties*

Gray (1963) used GLC to identify alk-1-enyl moieties in the choline and ethanolamine plasmalogens isolated in Landschutz ascites carcinoma and BP8/C3H ascites sarcoma. He found that the 16:0, 18:0, and 18:1 aldehydes, liberated from plasmalogens and measured as the dimethyl acetals, accounted for most of the alk-1-enyl moieties present. Odd-numbered and branched-chain alk-1-enyl moieties occurred only in trace amounts. These findings have now been substantiated for the alk-1-enyl moieties of plasmalogens from a wide variety of neoplastic tissues.

In 1968, the nature of both the alkyl and alk-1-enyl moieties of phospholipids containing choline and ethanolamine was reported at an international meeting (Drugs Affecting Lipid Metabolism) held in Milan,

Italy; the results of this study were published the following year in the proceedings of the meeting (Snyder, 1969a). These data (Fig. 3) demonstrated that the alkyl and alk-1-enyl moieties of both the choline and ethanolamine phospholipids from Walker-256 tumors, Taper liver tumors, and a human lymphosarcoma consisted primarily of 16:0, 18:0, and 18:1 moieties in varying proportions. Similar findings have since been reported for the ether-linked moieties of choline and ethanolamine phospholipids isolated from Ehrlich ascites cells (Wood and Snyder, 1969), Morris hepatomas 7794A and 7777 (Snyder *et al.*, 1969a), L–M cells (Anderson *et al.*, 1969), and the ethanolamine phospholipids from normal rat liver (Snyder *et al.*, 1969a). The alkyl and alk-1-enyl moieties of the ethanolamine phospholipids and the alkyl moieties of the choline phospholipids of preputial gland tumors are also mainly 16:0, 18:0, and 18:1 (Snyder *et al.*, 1970a). Essentially the same pattern has been found for the ether-linked moieties of the total phospholipids from muscle, bone marrow, spleen, heart, and brain of normal rats (Wood *et al.*, 1969). Bell and co-workers (1967) have reported a striking difference in the alk-1-enyl moieties from the choline plasmalogens of normal meninges and meningiomas of human brains, in that those from normal meninges consisted mainly of 16:0, 18:0, and 18:1 moieties and only traces ($<1\%$) of 18:2, 20:0, and 20:4 chain lengths, while those from meningiomas contained significant quantities of 18:2 (10%), 20:0 (2.2%), and 20:4 (2.9%) chains in addition to the predominating 16:0, 18:0, and 18:1 moieties.

2. *Acyl Moieties*

The acyl moieties of the choline and ethanolamine phospholipids were investigated in Morris hepatomas (Snyder *et al.*, 1969a); Walker-256 carcinosarcoma, Taper liver tumor, and a human lymphosarcoma (Snyder, 1969a; see Table I); Ehrlich ascites cells (Wood and Snyder, 1969); L–M cells (Anderson *et al.*, 1969); and preputial gland tumors (Snyder *et al.*, 1970a); they can be compared to those of normal rat livers (Snyder *et al.*, 1969a) and normal mouse preputial glands (Snyder and Blank, 1969). The ratio of 18:2 to 20:4 fatty acids in the choline and ethanolamine fractions was higher in hepatomas than in normal liver, indicating that essential fatty acids are poorly utilized by the hepatomas. In general, the acyl moieties of the choline phospholipids of preputial gland tumors are similar to those from the normal gland except for a large quantity of a fatty acid containing 22 carbon atoms and more than one double bond in the tumor. The acyl moieties of the ethanolamine phospholipids in the preputial gland tumors and normal glands were similar and consisted mainly of 16:0, 16:1, 18:0, 18:1, 18:2, and 20:4 or 20:1 fatty acids. The choline phospholipids contained more palmitic acid and less longer-chain polyunsaturated acids than the ethanolamine fractions in both normal and tumor-bearing preputial glands.

In Ehrlich ascites cells, the acyl moieties of 1-alkyl-2-acylglycerophosphorylcholine, 1-alkyl-2-acylglycerophosphorylethanolamine, and 1-alk-1-enyl-2-acylglycerophosphorylethanolamine consist mainly of 16:0, 18:1, 18:2, 20:4 + 22:1, and 22:6 fatty acids. The choline and ethanolamine phospholipid classes, except alk-1-enylacylglycerophosphorylethanolamine, contained a 1-random-2-random distribution (Wood and Snyder, 1969). In contrast to the ascites cells, Wood and Harlow (1969) found that normal rat liver had a nonrandom distribution of fatty acids in triglycerides, phosphatidylcholine, and phosphatidylethanolamine.

C. Substituted Groups on Alkyl Moieties

An unidentified glyceryl ether that appears to have a substituent on the alkyl moiety has been detected in metabolic experiments *in vitro* (Snyder *et al.*, 1970a) and *in vivo* (Blank *et al.*, 1970) with preparations from preputial gland tumors and Ehrlich ascites cells. K. Kasama, W. T. Rainey, Jr., and F. Snyder (unpublished data) recently discovered a hydroxy-substituted alkyl lipid derived from the triacylglycerol fraction of normal harderian glands (pink type) from rabbits. It is unknown whether the hydroxyl group is at the 11- or 12-position of the alkyl moiety and whether the hydroxyl group is esterified with acetate or a longer chain acyl moiety. β-Methoxy-substituted alkylglycerols occur in shark liver oil (Hallgren and Ställberg, 1967), but have not been reported in tumors.

III. Ether-Linked Lipid Content in Neoplastic Cells

A. General

Many investigations have dealt with the qualitative and quantitative differences in lipids that occur in normal and neoplastic cells (see Snyder, 1970a), and Snyder (1970b) has compiled an indexed listing of more than 800 references published on this subject between 1947 and 1970. A review of the methods used to quantitate ether-linked lipids is also available (Snyder, 1970a, 1971; see also Chapters II and IX). One of the characteristic features of most neoplastic cells is that they possess significantly higher levels of alkyldiacylglycerols than do healthy tissues (Fig. 1), and the level of alkyl and alk-1-enyl phospholipids is also higher in cancer cells than in most normal cells.

In comparing quantitative studies of alkyl and alk-1-enyl glycerolipids, it is important to consider how the values are expressed. Are calculations based on only a portion of the native form, e.g., the alkylglycerols or alk-1-enylglycerols liberated by $LiAlH_4$, or are they based on the native lipid molecules containing substituents at the other two positions of glycerol? Such values can differ by a factor of 2 to 3. The method of analysis can also have an important bearing on the results. For example, $LiAlH_4$ reduction is quantitative for nonpolar alkyl glycerolipids but not for polar alk-1-enyl glycerolipids (Wykle *et al.*, 1970; Snyder *et al.*, 1971b; see Chapter IX). Better procedures for quantitation of plasmalogens include the preparation of dimethylacetals of the fatty aldehydes (Farquhar, 1962; Morrison and Smith, 1964), reduction with $NaAlH_2(OCH_2CH_2\text{-}OCH_3)_2$ (Snyder *et al.*, 1971a), and the preparation of dioxolane derivatives (Rao *et al.*, 1967).

B. Concentration in Whole Cells

Rapport and Lerner (1959) conducted the first systematic survey of the quantity of alk-1-enyl glycerolipids in normal and neoplastic tissues from rabbits, rats, and humans. The plasmalogen content was measured by methods based on the formation of the aldehyde *p*-nitrophenylhydrazone and on the addition of iodine to the α,β-unsaturated bond in the alk-1-enyl moiety. This investigation revealed that neoplastic tissues had plasmalogen concentrations similar to those in many normal tissues, with the exception of liver, which contains only negligible quantities.

Ohnishi (1960) found substantial quantities of choline plasmalogen in rat primary hepatomas (induced by *p*-dimethylaminoazobenzene), rat ascites hepatomas (AH7974), and rat fibrosarcomas (Umeda rhodamine sarcoma). He reported that the choline plasmalogens from the tumors were profoundly less unsaturated than those in normal tissues (liver, lung, heart). Subsequent investigations by Veerkamp *et al.* (1961), Gray (1963), and Gerstl *et al.* (1965) also demonstrated that neoplastic tissues contained significant quantities of plasmalogens. Veerkamp *et al.* (1961) were the first to point out that primary hepatomas (in rats fed *p*-dimethylaminoazobenzene) and Ehrlich ascites cells (Veerkamp, 1960) contained a relatively high level of plasmalogens in comparison to many normal cells. However, Gerstl and co-workers (1965) found no difference in the plasmalogen content of five pulmonary carcinomas and three normal specimens of lungs.

Snyder *et al.* (1966) reported in their initial paper that the alkyldiacylglycerols in Ehrlich ascites cells represented 1–2% of their total lipids. Bollinger (1967) found that the level of alkyldiacylglycerols ranged be-

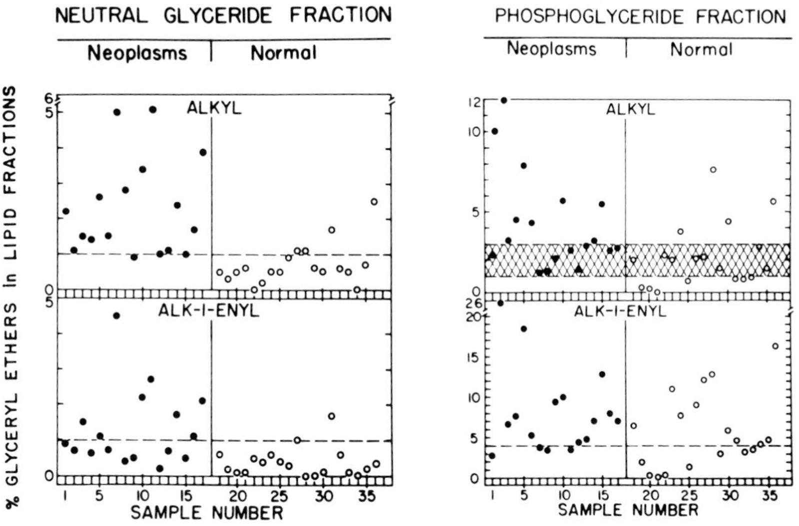

FIG. 4. Percentage distribution of alkyl and alk-1-enyl glycerolipids in the neutral lipid fraction and in the phospholipid fraction of normal (○) and neoplastic (●) human cells. Sample numbers along the abscissa refer to 1, metastatic adenocarcinoma, liver; 2, neurofibrosarcoma, retroperitoneal tissue; 3, squamous cell carcinoma, larynx; 4, adenocarcinoma, sigmoid colon; 5, adenoma, rectum; 6, metastatic undifferentiated carcinoma, liver; 7, metastatic embryonal carcinoma, omentum; 8, metastatic liposarcoma, peritoneal; 9, metastatic undifferentiated carcinoma, omentum; 10, metastatic undifferentiated adenocarcinoma, ovary; 11, metastatic poorly differentiated transitional cell carcinoma, abdomen; 12, chronic myelocytic leukemia, lymph; 13, adenocarcinoma, breast; 14, astrocytoma, brain; 15, metastatic squamous cell carcinoma, lymph node; 16, squamous cell carcinoma, lung; 17, undifferentiated carcinoma, prostate; 18, lung; 19, adipose tissue (peritoneal); 20, adipose tissue (omental); 21, adipose tissue (subcutaneous); 22, liver; 23, brain; 24, spleen; 25, lymph; 26, larynx; 27, heart; 28, colon; 29, rectum; 30, testes; 31, kidney; 32, prostate; 33, pancreas; 34, testes; 35, spleen; 36, heart. The horizontal lines designate an arbitrary zone of demarcation between the values for neoplasms and normal samples. All values are expressed as esterified glycerolipids. Reproduced in part from Snyder and Wood (1969) by permission of Cancer Research, Inc.

tween 1% and 3% of the total lipids in five Walker-256 tumors, and approximately 3.5% of the total lipids from the human lymphosarcoma. More extensive investigations were finally done with a variety of transplantable animal tumors (Snyder and Wood, 1968) and human tumors (Snyder and Wood, 1969), and also with a number of normal tissues (e.g., Gilbertson and Karnovsky, 1963; Schmid and Mangold, 1966; Wood and Snyder, 1968; Snyder and Wood, 1969). Figure 4 illustrates typical results for a variety of tumors and normal tissues from humans and is also in-

TABLE II

Choline and Ethanolamine Phospholipids of Normal Liver and Two Morris Hepatomas from Rats[a]

Lipid class	Normal liver	Hepatoma 7794A	Hepatoma 7777
	(% of total lipid class)		
Choline phospholipids			
Diacylglycerophosphorylcholine	99	99	97
Alkylacylglycerophosphorylcholine	0.83	0.70	2.8
Alk-1-enylacylglycerophosphorylcholine	0.14	0.14	0.46
Ethanolamine phospholipids			
Diacylglycerophosphorylethanolamine	97	93	92
Alkylacylglycerophosphorylethanolamine	0.50	0.56	1.2
Alk-1-enylacylglycerophosphoryl-ethanolamine	2.2	6.3	7.1

[a] Based on data published by Snyder *et al.* (1969a). Alkyldiacylglycerols represented 2.2% of the total lipid weight.

dicative of the results obtained for a group of 13 transplantable animal tumors (Snyder and Wood, 1968), and for a variety of other neoplasms (F. Snyder, unpublished data).

However, comparisons of ether-linked lipids in normal and cancer cells are not too meaningful unless the results obtained with tumor tissue are compared to normal cells of the same origin and functional activity. This type of comparison has been made between two Morris hepatomas (7777 and 7794A) of varying growth rate, and between livers from hepatoma-bearing and normal rats of the same strain (Snyder *et al.*, 1969a). Table II summarizes the data obtained for the phospholipids. The increase in alkyldiacylglycerols ($\simeq 2\%$ of the total lipids) in the 7777 hepatoma over that in the 7794A hepatoma and normal liver indicates that the levels of the alkyldiacylglycerols might be related to tumor growth rate. Both the 7777 and 7794A hepatomas also contained relatively high quantities of choline and ethanolamine plasmalogens, whereas only the fast-growing tumor contained higher quantities of the alkyl phospholipids. The hepatoma data thus reflect the same general trend for the concentration of ether-linked phospholipids as that observed in the earlier survey investigations of different types of tumors.

In general, the alkyl glycerolipids predominate in the neutral lipid fraction and the alk-1-enyl glycerolipids in the phospholipids of both normal and neoplastic cells. Usually, the alkyl type is highest in the choline phospholipids and the alk-1-enyl type highest in the ethanolamine fraction. This conclusion is based on quantitative measurements of alkyldiacylglycerols, alkylacylglycerophosphorylcholine, alkylacylglycerophosphorylethanolamine, alk-1-enylacylglycerophosphorylcholine, and alk-1-enylacylglycerophosphorylethanolamine that have been published for lipid extracts from Morris hepatomas (Snyder *et al.*, 1969a), Walker-256 carcinosarcomas, Taper liver tumors, and a human lymphosarcoma (Snyder, 1969a), Ehrlich ascites cells (Wood and Snyder, 1969), L–M cells (Anderson *et al.*, 1969), preputial gland tumors (Snyder *et al.*, 1970a), and normal preputial glands from mice (Snyder and Blank, 1969).

Fibroblasts (L–M cells) grown in suspension cultures in synthetic media accumulate large quantities of alkyldiacylglycerols ($\simeq$20% of the neutral lipids), whereas fibroblasts (L-strain) grown as monolayers in media containing horse serum (Cheng *et al.*, 1967) are devoid of them. These findings could be related to the influence of oxygen tensions, pH, media, or the surface properties of growing cells. Whether the difference between the cells grown as monolayers and suspension cultures is related to cell transformation is unknown; however, it is interesting that the L–M cells grown on the synthetic media have a much higher incidence of tumor development than normal L cells when they are reinjected into mice (Hellman *et al.*, 1968).

C. Subcellular Distribution

Little is known about the subcellular distribution of ether-linked lipids in neoplastic cells, but Snyder and Wood (1968) found that the alkyldiacylglycerols that accumulate in Ehrlich ascites cells were highest in the mitochondrial fraction. Later, Wood *et al.* (1970b) reported that alkyl and alk-1-enyl glycerolipids are virtually absent from a complex membranous mass (rough-surfaced microsomes, lysosomal-like structures, virus-like particles, few intact mitochondria, and unidentified amorphous material), but they did not examine purified membranes of known organelles. The only facts obtained so far on ether-linked lipids from other neoplastic tissues are that microsomes from hepatomas (Theise and Bielka, 1968) and preputial gland tumors (T-C. Lee and F. Snyder, unpublished data) contain higher quantities of plasmalogens than other portions of the cell.

IV. Fatty Alcohols in Neoplasms

Long-chain fatty alcohols are most prominent in plants and marine animals (Hilditch and Williams, 1964). However, fatty alcohols do exist in both free and esterified forms in mammalian cells (Blank and Snyder, 1970; Takahashi and Schmid, 1970). Previous work on the occurrence of fatty alcohols in mammalian tissues was primarily limited to wax esters in pig liver (Gershbein and Singh, 1969), certain glands (Sansone and Hamilton, 1969; Snyder and Blank, 1969), and skin lipids (see Nicolaides, 1965; Nikkari, 1965).

In Gershbein and Singh's investigation (1969) on the occurrence of small quantities of wax esters in pig liver, the alcohols liberated varied in chain length between 12:0 and 30 carbon atoms with a considerable degree of unsaturation. In wax esters of skin lipids, the fatty alcohols appear to be similar to the fatty acids (Nicolaides, 1965; Nikkari, 1965) but this is not true in preputial glands. The principle chain lengths of the fatty alcohols (14:1, 16:0, 16:1) that form the wax esters in normal preputial glands of mice differ significantly from the acyl moieties (14:1, 16:0, 16:1, 17:0, 18:0, 18:1, 18:2, 20:0, and 20:3 or 20:4) to which they are esterified (Sansone and Hamilton, 1969; Snyder and Blank, 1969). A detailed structural investigation of these fatty alcohols showed that the 14:1 monoene contained 74% $\Delta5$ and 26% $\Delta6$ isomers, and that the 16:1 contained 4% $\Delta5$, 67% $\Delta6$, 17% $\Delta7$, and 12% $\Delta9$ isomers (Snyder and Blank, 1969). The major isomeric forms of the fatty acids in the wax esters were 14:1 (4% $\Delta4$, 29% $\Delta5$, 41% $\Delta6$, 2% $\Delta7$, 4% $\Delta8$, 20% $\Delta9$), 16:1 (2% $\Delta5$, 42% $\Delta6$, 5% $\Delta7$, 49% $\Delta9$), and 18:1 (92% $\Delta9$, 8% $\Delta11$).

In metabolic studies, Friedberg and Greene (1967) found that rat liver enzymes could catalyze the formation of significant quantities of waxes from 1-^{14}C-hexadecanol and fatty acids. Such reactions also appear to occur in other tissues that are essentially devoid of wax esters (Snyder and Malone, 1970). Friedberg and Greene (1967) suggested that under their experimental conditions the two aliphatic moieties and enzymes reacted in a miceller state without any need for activation. However, in *Euglena* (Kolattukudy, 1970) and preputial gland tumors (Snyder and Malone, 1970), ATP, CoA, and Mg^{2+} are required to activate the fatty acids before they can react with the fatty alcohols to form esters.

The levels of fatty alcohols in neoplastic cells (Walker-256 carcinosarcoma, Ehrlich ascites cells, and Morris hepatoma 7777) are considerably elevated ($\simeq$0.3% of the neutral lipid fraction) when compared to levels in normal tissues ($\simeq$0.01% of the neutral lipid fraction in beef brain, beef heart, rat liver, and mouse preputial glands) (Blank and Snyder, 1970). Takahashi and Schmid (1970) reported that fatty alcohols represented about 0.002% (w/w) of the total lipids isolated from brain and heart of pigs and beef. We (Blank and Snyder, 1970) found hexadecanol, octadecanol, and octadecenol (free and esterified wax esters) were

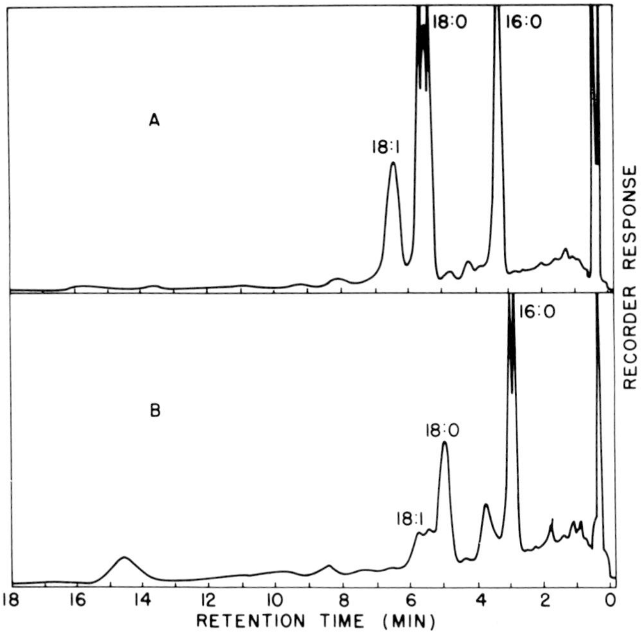

FIG. 5. Gas–liquid chromatograms of the acetate derivatives of fatty alcohols isolated from the total lipids of Walker-256 carcinosarcoma (A) and normal beef heart (B). Reproduced from M. L. Blank and F. Snyder, *Lipids* **5**, 337 (1970) by permission of the American Oil Chemists' Society, Chicago, Illinois.

the major fatty alcohols (>90%) in lipid extracts of normal tissues (beef brain, beef heart, rat liver, and mouse preputial glands) and three transplantable tumors (Walker-256 carcinosarcoma, Ehrlich ascites cells, and Morris hepatoma 7777). Figure 5 illustrates typical GLC data obtained for fatty alcohols in lipid extracts from Walker-256 carcinosarcoma and normal beef heart. Takahashi and Schmid (1970) found the same major alcohols in heart and brain from pigs and beef, but the beef also contained a significant amount of docosanol (12–15%).

V. Metabolism of Ether-Linked Glycerolipids in Neoplasms

A. Enzymic Studies

1. *Biosynthesis*

Enzymic systems for the biosynthesis of alkyl and alk-1-enyl glycerolipids were first detected in preputial gland tumors (Snyder *et al.*, 1969b,c,

1970a,b,c,e,f; Wykle and Snyder, 1969) and Ehrlich ascites cells (Wykle and Snyder, 1970; Wykle *et al.*, 1970). The initial products formed in the biosynthesis of ether-linked glycerolipids, alkyldihydroxyacetone phosphate and alkyldihydroxyacetone, have also been synthesized chemically (Piantadosi *et al.*, 1970, 1971). Similar systems for the biosynthesis of alkyl lipids have been found in normal cells from mammals and lower forms, but activities of the alkyl-synthesizing enzymes appear to be much higher in the tumor preparations than in most normal cells (see Chapter VII), although the enzymic activities vary according to the age of the animal (Snyder *et al.*, 1971b) and the age and type of tumor used (Snyder *et al.*, 1970a). Even the maximum activities reported for microsomal enzymes from normal brain and liver are from 6 to 20 times less than similarly prepared preparations from tumors (Snyder *et al.*, 1971b). However, microsomal preparations from fibroblasts (L–M cells) grown in suspension cultures (Snyder *et al.*, 1970d) and starfish (Snyder *et al.*, 1969d) contain alkyl-synthesizing activities in the same range as the tumor preparations.

The tumor preparations also contain enzymes that can interconvert fatty acids and fatty alcohols (Snyder and Malone, 1970). Our experiments with unwashed microsomes from preputial gland tumors have demonstrated that a microsomal reductase requiring NADPH (and NADH to some extent) reduces fatty acids to the corresponding fatty alcohols and that the reverse reaction is mediated by an NAD^+-linked enzyme. These enzymes appear to be of special significance in the cancer cell, since the reduction of fatty acids to fatty alcohols provides one of the precursors for the abnormal production of ether-linked lipids in neoplasms. Details of the biosynthetic and degradative enzymic systems for alkyl and alk-1-enyl glycerolipids and long-chain fatty alcohols isolated from neoplastic cells and normal cells, and a complete list of references citing the original papers can be found in Chapter VII of this book and in earlier reviews (Snyder, 1969b; Piantadosi and Snyder, 1970).

2. *Biocleavage*

A hydroxylase that requires tetrahydropteridine cleaves alkylglycerols (Tietz *et al.*, 1964), but it is essentially absent in fast-growing tumors. Soodsma and co-workers (1970) demonstrated that the ether cleavage system was most active in normal rat liver and a slow-growing Morris hepatoma (7794A), much less active in a fast-growing Morris hepatoma (7777), and barely detectable or absent in 12 other mouse and rat tumors tested. Neoplasms that contained high levels of alkyldiacylglycerols did not have an active alkyl ether cleavage system. Since the 7777 Morris hepatoma was originally derived from a tissue that contains an active alkyl ether cleavage enzyme, it is possible that the hepatoma has lost a functional component of this enzymic system through genetic aberration (Soodsma *et al.*, 1970). At the present time, it is not known whether an alteration in a specific enzyme protein or some other factor is responsible for the nonexpression of enzymic activity in the fast-growing tumors.

The suggestion by Pfleger and co-workers (1967) that the "highest activities of glyceryl ether-cleaving enzymes occur in cells that contain the lowest concentrations of ether-containing lipids" appears to be a general rule of thumb. Similar conclusions have been reached in studies of enzymes involved in the biosynthesis and biocleavage of alkyl lipids in brain and liver during fetal and postnatal development (Snyder *et al.*, 1971b). Microsomes from rat brain, a tissue rich in ether lipids, had an activity of 0.25 nmoles per 10 min per mg protein, whereas microsomes from rat liver, a tissue that normally contains little ether-linked lipids, had an activity of 58 nmoles per 10 min per mg protein.

B. *In Vivo* Studies

The first metabolic experiments with the ether-linked lipids in tumors involved the administration of 1-^{14}C-palmitic acid and 1-^{14}C-sodium acetate to irradiated rats bearing fibroadenomas (Snyder *et al.*, 1966). Six hours after intravenous injections of these labeled compounds, essentially no radioactivity had been incorporated into the alkyldiacylglycerols, although ^{14}C was incorporated into phospholipids. In these initial experiments, we did not measure the radioactivity in ether-linked aliphatic moieties.

Shortly after, Cheng and co-workers (1967), using Ehrlich ascites cells grown as monolayers in tissue cultures, were able to demonstrate that these cells were capable of synthesizing the alkyl ether bond in the absence of other tissues and in the absence of alkylglycerols in the media used for growth of the cells. Their data ruled out the possibility that the tumor cells merely incorporated and concentrated the alkyldiacylglycerols from other tissues or from dietary intake. Essentially the same level of alkyldiacylglycerols was found in the Ehrlich ascites cells even after four to six transfers in tissue culture. The data also showed that the tumor cells did not release any of the alkyl lipids since they were never observed in lipid extracts of the media either before or after cell growth; similar results were reported for media used to grow L–M cells that accumulate ether-linked glycerolipids (Anderson *et al.*, 1969).

Snyder and Wood (1968) measured the incorporation of ^{14}C into alkyldiacylglycerols of Ehrlich ascites cells from a wide variety of potential lipid and carbohydrate-derived precursors. The labeled tracers were given as single or eight daily intravenous injections; analysis was done 3 hr after the single injection or 6 hr after the eighth injection. More radioactivity was incorporated into the alkyldiacylglycerols from the long-chain 1-^{14}C-fatty alcohols than from the 1-^{14}C-labeled fatty acids, but the experiments were of a survey nature and lacked sufficient evidence to take into account interconversions of the labeled lipids tested.

After long-chain fatty alcohols had been established as the precursors of alkyl lipids in enzymic systems (see Chapter VII), *in vivo* experiments were conducted to determine the incorporation of fatty alcohols into the alkyl and alk-1-enyl lipids in Ehrlich ascites cells grown in the peritoneal cavities of mice and in tissue culture. The long-chain fatty alcohols were labeled on the 1-position with ^{14}C and ^{3}H (Wood and Healy, 1970a,b; Wood *et al.*, 1970a) or on the oxygen with ^{18}O (Snyder *et al.*, 1970f). The data obtained support the results established in the enzymic systems for the biosynthesis of alkyl and alk-1-enyl glycerolipids.

Wood and Healy (1970a,b) demonstrated that 1-^{14}C,1-^{3}H-hexadecanol was incorporated better than the corresponding acid or aldehyde into alkyl glycerolipids of Ehrlich ascites cells grown in mice. Furthermore, their data revealed that the radioactivity from hexadecanol was also incorporated into alk-1-enyl glycerolipids and that half of the tritium was lost. The authors' data led them to conclude that plasmalogens were derived from intact alkylacyl phospholipids. These experiments also clearly documented that fatty acids could be reduced to fatty alcohols in the intact tumor cells and that fatty aldehydes were primarily oxidized to fatty acids, although some were also reduced to fatty alcohols.

Similar results and conclusions were reached by Wood and co-workers (1970a) when they repeated this type of experiment with Ehrlich ascites cells grown in tissue culture. In the tissue culture experiments, three or more unidentified polar lipids containing radioactivity were isolated from the media used to grow the cells. Since the ^{3}H to ^{14}C ratios of these unidentified compounds were the same as in the alkylglycerols derived from the various lipid classes, the authors felt that the unidentified lipids might be intermediates formed during the biosynthesis of ether-linked glycerolipids. The unidentified components were stable to alkali treatment but not to acid. Since the acetates could be formed, they apparently contained a hydroxy group(s) but only a small portion reacted with acetone to form the isopropylidene derivative.

Subsequent *in vivo* experiments with ^{18}O-labeled hexadecanol (Snyder *et al.*, 1970f) demonstrated that the oxygen of the alcohol is retained in both the alkyl and alk-1-enyl lipids of Ehrlich ascites cells grown in the peritoneal cavity of mice. The ^{18}O content of the alkylglycerols and alk-1-enylglycerols isolated 48 hr after an intraperitoneal injection of the ^{18}O-labeled hexadecanol to mice bearing Ehrlich ascites cells was 34% and 16%, respectively.

The direct incorporation of alkylglycerols into ethanolamine phospholipids containing alkyl and alk-1-enyl moieties has also been shown in tumors. *rac*-1-^{14}C-Hexadecyl-2-^{3}H-glycerol was incorporated into the ether-containing phospholipids of Ehrlich ascites cells, including the plasmalogens, without any change in the $^{3}H/^{14}C$ ratio, demonstrating that phosphorylation of alkylglycerols must occur (Blank *et al.*, 1970). The alkylglycerol phosphate formed can then be acylated and subsequently incorporated into the more complex lipids; enzymes that catalyze these steps are located in the microsomal fraction of neoplastic cells (Wykle and Snyder, 1970; Snyder *et al.*, 1970c). Blank *et al.* (1970) found a labeled unidentified lipid in the tracer experiments with Ehrlich ascites cells (after $LiAlH_4$ reduction of the choline or ethanolamine lipid fraction) that had the same $^{3}H/^{14}C$ ratio as the alkylglycerol precursor and the alkyl and alk-1-enyl glycerolipid products. The chromatographic behavior of the unidentified lipid in the *in vivo* experiments was identical to that of an unidentified lipid isolated from the enzymic system that forms alkyl glycerolipids (Snyder *et al.*, 1970a). It had a number of chromatographic properties comparable to those of 1-hexadecyl(β-hydroxy)glycerol, but GLC of the acetate derivatives demonstrated that they were not identical. These data suggested that substitution of a functional group occurs on the alkyl moiety of a glycerolipid and may be involved in the formation of the alk-1-enyl moiety in plasmalogens of tumor cells (Snyder *et al.*, 1970a).

REFERENCES

Anderson, R. E., Cumming, R. B., Walton, M., and Snyder, F. (1969). *Biochim. Biophys. Acta* **176,** 491.

Bell, O. E., Jr., Cain, C. E., Sulya, L. L., and White, H. B., Jr. (1967). *Biochim. Biophys. Acta* **144,** 481.

Blank, M. L., and Snyder, F. (1970). *Lipids* **5,** 337.

Blank, M. L., Wykle, R. L., Piantadosi, C., and Snyder, F. (1970). *Biochim. Biophys. Acta* **210,** 442.

Bollinger, J. N. (1967). *Lipids* **2,** 143.

Cain, C. E., Bell, O. E., Jr., White, H. B., Jr., Sulya, L. L., and Smith, R. R. (1967). *Biochim. Biophys. Acta* **144,** 493.

Cheng, S., Piantadosi, C., and Snyder, F. (1967). *Lipids* **2,** 193.

Farquhar, J. W. (1962). *J. Lipid Res.* **3,** 21.

Friedberg, S. J., and Greene, R. C. (1967). *J. Biol. Chem.* **242,** 234.

Gershbein, L. L., and Singh, E. J. (1969). *J. Amer. Oil Chem. Soc.* **46,** 34.
Gerstl, B., Hayman, R. B., Ramorino, P., Tavaststjerna, M. G., and Smith, J. K. (1965). *Amer. J. Clin. Pathol.* **43,** 314.
Gilbertson, J. R., and Karnovsky, M. L. (1963). *J. Biol. Chem.* **238,** 893.
Gray, G. M. (1963). *Biochem. J.* **86,** 350.
Hallgren, B., and Ställberg, G. (1967). *Acta Chem. Scand.* **21,** 1519.
Hellman, K. B., Hellman, A., Cumming, R. B., and Novelli, G. D. (1968). *J. Nat. Cancer Inst.* **41,** 653.
Hilditch, T. P., and Williams, P. N. (1964). "The Chemical Constitution of Natural Fats," 4th ed., pp. 666–670. Wiley, New York.
Itoh, K., and Kasama, K. (1970). *Gann.* **61,** 271.
Kasama, K., Uezumi, N., and Itoh, K. (1970). *Biochim. Biophys. Acta* **202,** 56.
Kolattukudy, P. E. (1970). *Biochemistry* **9,** 1095.
Lindlar, F., and Wagener, H. (1964). *Schweiz. Med. Wochenschr.* **94,** 243.
Malins, D. C., and Mangold, H. K. (1960). *J. Amer. Oil Chem. Soc.* **37,** 576.
Morrison, W. R., and Smith, L. M. (1964). *J. Lipid Res.* **5,** 600.
Nicolaides, N. (1965). *J. Amer. Oil Chem. Soc.* **42,** 708.
Nikkari, T. (1965). *Scand. J. Clin. Lab. Invest.* **17,** Suppl. 85.
Ohnishi, T. (1960). *Gann* **51,** 11.
Paltauf, F., and Polheim, D. (1970). *Biochim. Biophys. Acta* **210,** 187.
Pfleger, R. C., Piantadosi, C., and Snyder, F. (1967). *Biochim. Biophys. Acta* **144,** 633.
Piantadosi, C., and Snyder, F. (1970). *J. Pharm. Sci.* **59,** 283.
Piantadosi, C., Ishaq, K. S., and Snyder, F. (1970). *J. Pharm. Sci.* **59,** 1201.
Piantadosi, C., Ishaq, K. S., Wykle, R. L., and Snyder, F. (1971). *Biochemistry* **10,** 1417.
Rao, P. V., Ramachandran, S., and Cornwell, D. G. (1967). *J. Lipid Res.* **8,** 380.
Rapport, M. M., and Lerner, B. (1959). *Biochim. Biophys. Acta* **33,** 319.
Sansone, G., and Hamilton, J. G. (1969). *Lipids* **4,** 435.
Schmid, H. H. O., and Mangold, H. K. (1966). *Biochem. Z.* **346,** 13.
Snyder, F. (1969a). *Advan. Exp. Med. Biol.* **4,** 609–621.
Snyder, F. (1969b). *Progr. Chem. Fats Other Lipids* **10,** 287–335.
Snyder, F. (1970a). *Methods Cancer Res.* **6,** 399–436.
Snyder, F. (1970b). *U.S. At. Energy Comm., Rep.* **ORAU 111.**
Snyder, F. (1971). *In* "Progress in Thin-Layer Chromatography and Related Methods" (A. Niederwieser and G. Pataki, eds.), Vol. II, pp. 105–141. Ann Arbor Sci. Publ., Ann Arbor, Michigan.
Snyder, F., and Blank, M. L. (1969). *Arch. Biochem. Biophys.* **130,** 101.
Snyder, F., and Cress, E. A. (1963). *Radiat. Res.* **19,** 129.
Snyder, F., and Malone, B. (1970). *Biochem. Biophys. Res. Commun.* **41,** 1382.
Snyder, F., and Wood, R. (1968). *Cancer Res.* **28,** 972.
Snyder, F., and Wood, R. (1969). *Cancer Res.* **29,** 251.
Snyder, F., Cress, E. A., and Stephens, N. (1966). *Lipids* **1,** 381.
Snyder, F., Blank, M. L., and Morris, H. P. (1969a). *Biochim. Biophys. Acta* **176,** 502.
Snyder, F., Malone, B., and Wykle, R. L. (1969b). *Biochem. Biophys. Res. Commun.* **34,** 40.
Snyder, F., Wykle, R. L., and Malone, B. (1969c). *Biochem. Biophys. Res. Commun.* **34,** 315.
Snyder, F., Malone, B., and Blank, M. L. (1969d). *Biochim. Biophys. Acta* **187,** 302.
Snyder, F., Malone, B., and Blank, M. L. (1970a). *J. Biol. Chem.* **245,** 1790.

Snyder, F., Blank, M. L., Malone, B., and Wykle, R. L. (1970b). *J. Biol. Chem.* **245,** 1800.
Snyder, F., Blank, M. L., and Malone, B. (1970c). *J. Biol. Chem.* **245,** 4016.
Snyder, F., Malone, B., and Cumming, R. B. (1970d). *Can. J. Biochem.* **48,** 212.
Snyder, F., Piantadosi, C., and Malone, B. (1970e). *Biochim. Biophys. Acta* **202,** 244.
Snyder, F., Rainey, W. T., Jr., Blank, M. L., and Christie, W. H. (1970f). *J. Biol. Chem.* **245,** 5853.
Snyder, F., Blank, M. L., and Wykle, R. L. (1971a). *J. Biol. Chem.* **246,** 3639.
Snyder, F., Hibbs, M., and Malone, B. (1971b). *Biochim. Biophys. Acta* **231,** 409.
Soodsma, J. F., Piantadosi, C., and Snyder, F. (1970). *Cancer Res.* **30,** 309.
Takahashi, T., and Schmid, H. H. O. (1970). *Chem. Phys. Lipids* **4,** 243.
Theise, H., and Bielka, H. (1968). *Arch. Geschwulstforsch.* **32,** 11.
Tietz, A., Lindberg, M., and Kennedy, E. P., (1964). *J. Biol. Chem.* **239,** 4081.
Veerkamp, J. H. (1960). Ph.D Thesis, University of Utrecht, Netherlands.
Veerkamp, J. H., Mulder, I., and van Deenen, L. L. M. (1961). *Z. Krebsforsch.* **64,** 137.
Wood, R., and Harlow, R. D. (1969). *Arch. Biochem. Biophys.* **131,** 495.
Wood, R., and Healy, K. (1970a). *Biochem. Biophys. Res. Commun.* **38,** 205.
Wood, R., and Healy, K. (1970b). *J. Biol. Chem.* **245,** 2640.
Wood, R., and Snyder, F. (1967). *J. Lipid Res.* **8,** 494.
Wood, R., and Snyder, F. (1968). *Lipids* **3,** 129.
Wood, R., and Snyder, F. (1969). *Arch. Biochem. Biophys.* **131,** 478.
Wood, R., Harlow, R. D., and Snyder, F. (1969). *Biochim. Biophys. Acta* **176,** 641.
Wood, R., Walton, M., Healy, K., and Cumming, R. B. (1970a). *J. Biol. Chem.* **245,** 4276.
Wood, R., Anderson, N. G., and Swartzendruber, D. C. (1970b). *Arch. Biochem. Biophys.* **141,** 190.
Wykle, R. L., and Snyder, F. (1969). *Biochem. Biophys. Res. Commun.* **37,** 658.
Wykle, R. L., and Snyder, F. (1970). *J. Biol. Chem.* **245,** 3047.
Wykle, R. L., Blank, M. L., and Snyder, F. (1970). *FEBS Lett.* **12,** 57.

CHAPTER XI

*THE ETHER BOND IN MARINE LIPIDS**

Donald C. Malins and Usha Varanasi

I. Introduction

In 1922 Tsujimoto and Toyama reported the isolation of two dihydric alcohols from sharks and other elasmobranch fish. The discovery and further classification of these compounds (Toyama, 1924a,b,c) focused attention on the glycerol ethers, a new class of lipids. Although interest in the ether-linked lipids evolved from these early studies with marine

* This work was conducted under Office of Naval Research contract N00014-69-C0404, in cooperation with the Oceanic Institute, Hawaii.

animals, investigations developed that reflected the traditional concern with terrestrial animals, such as the rat, and select microorganisms. However, recent emphasis on the viability of our imperiled marine resources will undoubtedly stimulate a renewed interest in the ubiquitous ether-linked lipids of aquatic species. Reviews on the structure, composition, and biochemistry of marine lipids have been published (Lovern, 1964; Malins, 1967; Malins and Wekell, 1969), and a complete review on the ether-linked lipids in general (Snyder, 1969) is available. Accordingly, this chapter will focus attention on recent developments.

II. Alkyl Glycerolipids

A. Occurrence

The alkyl glycerolipids are found in the organ and body lipids of a wide variety of marine organisms. This extensive distribution has been discussed (Lovern, 1964; Malins, 1967; Malins and Wekell, 1969; Snyder, 1969) and pinpointed in marine fish (Malins *et al.*, 1965; Schmid and Mangold, 1966; Lewis, 1969a; Kayama *et al.*, 1971), mammals (Varanasi and Malins, 1970b), birds (Lewis, 1966, 1969b), and worms (Pocock *et al.*, 1969).

The alkyldiacylglycerols have received the most attention in studies on the ether-linked lipids of marine organisms. Their importance was recognized after chromatography of marine lipids on thin layers of silicic acid revealed spots just above triacylglycerols, particularly when weakly polar developing solvents were used (Mangold and Malins, 1960). The availability of this simple chromatographic technique, thin-layer chromatography (TLC), led to a keen interest in the alkyldiacylglycerols of both marine and terrestrial species (Malins and Wekell, 1969; Snyder, 1969).

The ether linkage of alkyldiacylglycerols from marine species is located on position 1 of glycerol (Malins and Wekell, 1969). We have not succeeded in finding alkyldiacylglycerols containing the 2-*O*-alkyl moiety in marine fish (Malins and Barone, 1969), but the 2-*O*-alkyl structure does exist in the marine biota, as demonstrated by the discovery of a diether analog of phosphatidylglycerophosphate in halophilic bacteria (Kates *et al.*, 1963, 1965) (see Chapter XV). Studies on the physical properties of the alkyldiacylglycerols and analogous alk-1-enyldiacylglycerols of ratfish (*Hydrolagus colliei*) clearly show that these ether-linked lipids occur in the D-configuration (Baumann *et al.*, 1966).

Recently, a few studies have been carried out on the composition and structure of the alkyldiacylglycerols (Wood and Snyder, 1969; Lewis,

1969a; Baumann *et al.*, 1970; Kasama *et al.*, 1970; Malins and Robisch, 1971), so that it is now possible to compare these compounds with other lipid classes, such as the triacylglycerols, and speculate upon their biosynthesis.

Hanahan *et al.* (1963) determined the position of the double bond in selachyl alcohol (18:1) from dogfish (*Squalus acanthias*) liver. The principal isomer was a Δ9 structure. Spener and Mangold (1971) showed that *O*-alkyl chains in alkyldiacylglycerols of *S. acanthias* and soupfin shark (*Galeorhinus galeus*) also contain about 60% octadecenyl moieties, and that the main isomer in the hepatic lipids of *S. acanthias* was indeed the Δ9 octadecenyl structure. However, they also detected substantial amounts of Δ7 and lesser amounts of Δ6, Δ8, Δ10, and Δ11 isomers. In *G. galeus*, Δ5, Δ6, Δ7, Δ8, Δ11, and Δ13 octadecenyl moieties were present in addition to the predominant Δ9 isomer. The isomeric *O*-octadecenyl chains in the alkyldiacylglycerols from the silky shark (*Carcharhinus falciformis*) were essentially the same as those in the alkyldiacylglycerols from *G. galeus* (Spener and Mangold, 1971). Schmid *et al.* (1969) studied the composition of the monounsaturated *O*-alkyl chains from the alkyldiacylglycerols of *H. colliei* liver. The Δ9 isomer predominated in C_{18} chains, whereas the Δ7 isomer was the major constituent of the C_{16} and C_{20} chains.

The Δ9 isomer predominates in the isomeric monoenoic fatty acids of alkyldiacylglycerols from *H. colliei* when the chains are 18 to 24 carbons long. However, the major component of the C_{16} acids is the Δ7 isomer (Schmid *et al.*, 1969). In addition, compositional studies on isomeric structures (Malins and Houle, 1961; Hanahan *et al.*, 1963; Spener and Mangold, 1971) show that certain positional isomers are preferred in the biosynthesis of *O*-alkyl chains of alkyldiacylglycerols. Preferential synthesis probably takes place via enzymes controlling the conversions, fatty acid → fatty alcohol → *O*-alkylglycerols (Malins and Sargent, 1971). The occurrence of substantial amounts of Δ7 octadecenyl chain in *O*-alkyl moieties of *S. acanthias* alkyldiacylglycerols may reflect the selective incorporation of the Δ7 isomer through such interconversions.

Such selective incorporation of hydrocarbon chains in the synthesis of alkylglycerolipids is indicated by a recent study on porpoises (Varanasi and Malins, 1970b). Particularly noteworthy was the apparent absence of the isopentyl chain in the neutral alkylglycerolipids, despite the occurrence of high percentages of isovaleric acid in the acyl groups (Table I).

A comparative discussion of the molecular structures of alkyldiacylglycerols from two marine species, *H. colliei* (Baumann *et al.*, 1970) and *S. acanthias* (Malins and Robisch, 1971), and from Ehrlich ascites carcinoma cells (Wood and Snyder, 1969) will be given special attention

TABLE I

A Comparison of Hydrocarbon Chains of Triacylglycerols and Alkylglycerols from the Mandible Lipids of the Porpoise, Phocoena phocoena[a]

Chain	Triacylglycerols (mole %)	Alkylglycerols (mole %)
5:0 br	40.5[b]	—
8:0	1.2[b]	—
9:0	—	0.4
10:0 br	0.3[b]	0.7[c]
10:0	0.5	0.4
11:C br	0.7[b]	—
11:0	0.3	—
12:0 br	1.6[b]	4.3[c]
12:0	—	0.1
12:1	—	0.5
Unknown	1.7	—
13:0 br	1.3[b]	—
13:0	—	3.3
14:0 br	1.5[b]	—
14:0	8.0	2.4
14:1	1.0	—
15:0 br	4.8[b]	—
15:0	2.2	1.7
16:0 br	2.8[b]	4.3[c]
16:0	8.4	43.0
16:1	13.1	—
17:0	—	2.8
17:?	1.7	—
18:0	1.0	7.5
18:1	7.4	27.0

[a] From the work of Varanasi and Malins (1970b); br designates branched chains. Copyright (1970) by the American Chemical Society. Reprinted by permission of the copyright owner.

[b] Iso structure.

[c] Specific branched structure not determined.

because the necessary data have only recently become available. Alkyldiacylglycerols derived from the livers of the two marine species are very similar to each other with respect to *O*-alkyl chains of position 1 in that this position attracts 18:1 and lesser amounts of 16:0 and 16:1 chains. However, the Ehrlich ascites carcinoma cells contain mainly 16:0

and substantial amounts of 18:0 and 18:1 chains but no detectable 16:1 *O*-alkyl moieties.

Significant differences also exist between the acyl moieties on position 2 of the alkyldiacylglycerols of *S. acanthias* and *H. colliei* liver and those of the Ehrlich ascites carcinoma cells, in that the marine ether-linked lipids do not contain the 18:2 and 20:4 acids common to terrestrial food chains. In *S. acanthias* liver, the acyl moieties on position 2 are characterized by high proportions of 20:1 and 22:1 acids, and also by the presence of some 22:5 and 22:6 acids in the triacylglycerols and alkyldiacylglycerols (Malins and Robisch, 1971). Baumann *et al.* (1970) did not report the presence of C_{20} and C_{22} polyenoic acids in the alkyldiacylglycerols of *H. colliei*. Polyenoic C_{20} and C_{22} acids are present in appreciable amounts in the acyl moieties of position 2 of Ehrlich ascites carcinoma cells. However, it should be remembered that the polyenoic acids of marine fish are primarily members of the linolenic family of acids (Gruger, 1967), whereas the polyenoic acids of carcinoma cells (Wood and Snyder, 1969) belong mainly to the linoleic family.

Large amounts of 16:0 acid are present on both positions 2 and 3 of alkyldiacylglycerols derived from *S. acanthias*, *H. colliei*, and Ehrlich ascites carcinoma cells, with no distinct preference being shown for this acid by either acyl position. High proportions of 20:1 and 22:1 acids also accumulate on position 3 as well as on position 2 of the alkyldiacylglycerols derived from *S. acanthias*. In contrast, *H. colliei* accumulates almost all 20:1 and 22:1 acyl moieties on position 3. The small percentages of these structures in Ehrlich ascites carcinoma cells show no distinct tendency for either acyl position of glycerol.

Polyenoic acids of the C_{20} and C_{22} series were not found in more than trace amounts on position-3 of the ether-linked lipids from the marine fish, but they were present on position 3 of the Ehrlich ascites carcinoma cells, which also accumulated significant amounts of 22:5 and 22:6 acids on position 2 and small amounts of 22:4 acids on both acyl positions. The limited studies undertaken so far indicate that future investigations will reveal great diversity in the molecular structures of alkyldiacylglycerols from natural sources.

Brockerhoff and co-workers (1964a) postulated that marine fish may synthesize triacylglycerols from a 2-acylglycerol structure containing high proportions of C_{20} and C_{22} polyenoic acids that is carried through the food chain essentially unchanged. This structure may have its origin in the phytoplankton (Brockerhoff *et al.*, 1964b), where synthesis of the polyenoic chains takes place. The alkyldiacylglycerols in *S. acanthias* liver exhibit a 2-acylglycerol backbone similar to that found in triacylglycerols from marine fish. Baumann *et al.* (1970) have postulated that the 2-acylglycerol moieties in alkyldiacylglycerols of *H. colliei* liver are derived from dietary lipids, which may also form the backbone of the analogous alk-1-enyldiacylglycerols, whose 2-position acyl moieties are similar to those of the alkyl compounds. However, conclusions drawn from molecular structure about the *in vivo* formation of alkyldiacylglycerols are best evaluated in the light of complementary metabolic studies.

B. Cellular Distribution

Unfortunately, little attention has been given to the cellular distribution of lipids in the fatty tissues of marine fish. In the livers of sharks, such as *S. acanthias* (Malins and Wekell, 1969) and *Scyliorhinus caniculus* (Wardle, 1967), lipids are stored in specialized fat cells that are presumably unique in structure and function. In the livers of *S. acanthias*, for example, the intracellular fat contains more than 99 mole % of the total ether-linked lipids. The intracellular fat is 37 mole % alkyldiacylglycerols and 57 mole % triacylglycerols (Malins and Wekell, 1969). Analogous alk-1-enyldiacylglycerols also tend to be in the intracellular fat in livers of *S. acanthias*. Karnovsky *et al.* (1955) have shown that the intracellular lipid of starfish (*Asterias forbesi*) contains most of the neutral ether-linked lipids. Metabolic interrelations between biomembranes and intracellular lipid of specialized fat cells should be a stimulating and fruitful area for research. The presence of a high percentage of alkyldiacylglycerols in the intracellular fat strongly suggests that they play a vital role in the metabolism of specialized fat cells such as that suggested by Prop (1965) for the pineal metabolic processes in the rat. For example, a high lipase and non-specific esterase activity was found in the pineal cells; the centers of esterase activity were on the surface of fat droplets. The findings of Prop (1965) and of Friedberg and Greene (1967b), who suggested that wax ester synthesis in *S. acanthias* takes place under hydrophobic conditions, should rightfully stimulate an interest in metabolism at hydrophilic–hydrophobic interfaces of specialized fat cells, with particular attention to the metabolism of the alkyldiacylglycerols.

C. Metabolism

Recently, a keen interest has developed in the mechanism of biosynthesis of the *O*-alkyl bond in lipids. Notable contributions have been made by Snyder *et al.* (1969a,b, 1970a,b) and Hajra (1969) in studies on cell-free systems. Dihydroxyacetone phosphate and a fatty alcohol were found to be obligatory for the enzymic formation of the ether bonds

(Snyder *et al.*, 1969a,b, 1970a,b). Hajra (1969) showed that the ether bonds are biosynthesized by direct reaction of a fatty alcohol and acyldihydroxyacetone phosphate. Friedberg and Greene (1967a) found that isolated stomachs of dogfish (*S. acanthias*) and the shark (*Raja erinacea*) incorporated long-chain alcohols, but not long-chain fatty acids, into alkyl glycerolipids. In livers of *S. acanthias* 1-^{14}C-palmitic acid is reductively incorporated *in vivo* into alkyl chains of alkyl glycerolipids, although the incorporation is not extensive (Malins, 1968). In digestive glands of the starfish (*A. forbesi*), fatty alcohols are more efficiently incorporated into alkyl glycerolipids than are fatty acids of the same chain length (Ellingboe and Karnovsky, 1967). Recent studies with a cell-free preparation from *S. acanthias* liver also revealed that in the biosynthesis of *O*-alkyl chains of alkyldiacylglycerols, fatty alcohol was favored over fatty acid (Malins and Sargent, 1971). An obvious possible precursor of the *O*-alkyl bond, fatty aldehyde, had lower rates of incorporation than fatty alcohols in studies with *A. forbesi* (Ellingboe and Karnovsky, 1967).

These studies from independent laboratories clearly indicate that fatty alcohols are important precursors in the biosynthesis of *O*-alkyl bonds in marine life. Other questions now arise: What, for example, are the relative rates of biosynthesis of acyl and alkyl moieties in alkyldiacylglycerols? *S. acanthias* liver incorporated 1-^{14}C-palmitic acid, *in vivo*, into alkyldiacylglycerols (Malins and Robisch, 1971). In these investigations, a distinct preference was shown for incorporation of radioactivity into acyl moieties. In studies with 1-^{14}C-oleic acid and 9,10-^{3}H-oleyl alcohol in a cell-free preparation from *S. acanthias* liver radioactivity from both precursors was also decidedly higher in acyl rather than *O*-alkyl chains (Malins and Sargent, 1971). The preferential incorporation of both alcohol and acid into acyl moieties of alkyldiacylglycerols led to a study of the relative rates of incorporation into each acyl position of glycerol. Experiments on *S. acanthias* liver with labeled fatty acid in intact cells (Malins and Robisch, 1971) and in a cell-free system (Malins and Sargent, 1971), revealed that the radioactivity ratio of position 2 to that of position 3 of alkyldiacylglycerols increased substantially with time, thereby indicating that the acyl groups on position 2 are metabolized at the expense of the acyl groups on position 3. These results are not compatible with a significant role for either acyl migration or transacylation reactions in the biosynthesis of alkyldiacylglycerols from an exogenous 2-acylglycerol structure (Baumann *et al.*, 1970), but are consistent with the well-known pathway for alkyldiacylglycerol biosynthesis proposed by Snyder *et al.* (1969a,b, 1970a,b). It is interesting from a comparative point of view that in trout, cod, and lobsters a high proportion of triacylglycerols containing 9,10-

^3H-oleic acid on position 2 and 1-^{14}C-oleic acid on positions 1 and 3 is completely metabolized when included in the diet (Brockerhoff and Hoyle, 1967). These data, together with findings of Malins and Robisch (1971) showing extensive metabolism of acyl groups in position 2 in triacylglycerols of *S. acanthias* liver, do not support the previous hypothesis (Brockerhoff *et al.*, 1964a,b) that biosynthesis takes place via a 2-acylglycerol structure.

D. Biological Role

Marine organisms are continually exposed to fluctuating environmental conditions requiring unique adaptions. For example, in the aquatic environment, the attainment of neutral or near neutral buoyancy is of utmost importance in survival. Species that make vertical migrations may require specialized organs for the delicate attenuation of buoyancy. The attainment of near weightlessness at various depths can be accomplished by such devices as gas-filled swim bladders (Jones and Marshall, 1953), coeloms filled with a solution of ammonium chloride (Denton *et al.*, 1969), or livers containing high proportions of fat (Marshall, 1960; Corner *et al.*, 1969; Malins and Barone, 1970).

Corner *et al.* (1969) showed that lipids of relatively low specific gravity, such as squalene (0.858), are important in the maintenance of close-to-neutral buoyancy, and Lewis (1970) observed the significant differences in the specific gravities of wax esters (0.858), alkyldiacylglycerols (0.891), and triacylglycerols (0.915), hypothesizing that these compounds, when present in appreciable amounts, contribute to the maintenance of neutral or near neutral buoyancy in certain fish. It must be remembered that an aquatic animal such as *S. acanthias*, weighing 3 kg in air, weighs only 80 g in seawater (Bone and Roberts, 1969). Thus, small differences in specific gravities of lipids in fatty fish can produce significant alterations in buoyancy. For example, 1 g of alkyldiacylglycerols (0.908) gives 14% more lift in seawater than 1 g of triacylglycerols (0.922) (Malins and Barone, 1970). Squalene in the liver of the small shark (*Scyliorhinus caniculus*) does not appear to turn over very rapidly (Sargent *et al.*, 1970), suggesting that squalene, a common major constituent of the lipids of deep sea shark livers, may be progressively deposited as the animal grows. A very slow turnover may imply that squalene serves to provide a constant lift. In contrast, the rapid turnover of alkyldiacylglycerols (0.908) and triacylglycerols (0.922) in the liver of *S. acanthias*, an organ known to be hydrostatic, suggests that their metabolism is intimately related to the attainment of near neutral buoyancy at various depths (Malins and Barone, 1970). When the body weight of *S. acanthias* was artificially increased, the hepatic ratio

TABLE II

Alterations in Levels of Alkyldiacylglycerols and Triacylglycerols in Dogfish (S. acanthias) Liver Resulting from an Artificial Increase in Body Weight[a]

Fish	Weight		Lipid in	Ratio
	Fish (kg)	Liver (g)	liver (%)	(ADG/TG)[b]
		Experimenal group		
A	2.7	189	57.2	1.25
B	3.4	337	66.8	1.37
C	3.4	266	64.7	1.58
D	2.7	177	59.3	1.10
E	2.7	196	62.2	1.59
F	2.7	246	62.6	1.03
G	3.0	460	72.2	1.08
		Control group		
H	2.7	282	75.8	0.90
I	3.0	299	64.5	0.72
J	2.7	295	62.4	0.90
K	2.7	229	62.4	0.49
L	2.3	193	62.5	0.49
M	3.2	318	68.6	0.88

[a] From the work of Malins and Barone (1970). Copyright (1970) by the American Association for the Advancement of Science.

[b] The ratios of alkyldiacylglycerols (ADG) to triacylglycerols (TG) are less than 1 in the control group (unweighted fish) and greater than 1 in the experimental group in which 4 oz (114 g) weights were added to each fish.

of alkyldiacylglycerols to triacylglycerols changed (Table II). Because the change in the ratio strongly favored the less dense alkyldiacylglycerols, Malins and Barone (1970) postulated that these ether-linked lipids were deposited at the expense of triacylglycerols to offset the increase in body weight. Thus, a regulatory mechanism involving the selective metabolism of alkyldiacylglycerols and triacylglycerols may be used by *S. acanthias* for the delicate attenuation of near neutral buoyancy. Further investigation of this hypothesis may reveal much about a biological role for alkyldiacylglycerols in fatty livers. The presence of large amounts of alkyldiacylglycerols in marine birds (Lewis, 1966, 1969b) still remains perplexing, but it is possible that these compounds, present in the stomach, may be ejected as a defense mechanism (Kritzler, 1948). Even more obscure is the function of unusually high contents of alkyldiacylglycerols in coelomic fluids and body wall, but not the gut, of the marine worm (*N. virens*) (Pocock *et al.*, 1969).

III. Alk-1-enyl Glycerolipids

A. Occurrence

Data on the alk-1-enyl glycerolipids of marine origin are not as extensive as those on the analogous alkyl derivatives. Nevertheless, the alk-1-enyl glycerolipids are found in a variety of aquatic organisms (Malins and Wekell, 1969; Snyder, 1969). The wide distribution of these ether-linked lipids is exemplified by their occurrence in *S. acanthias* (Malins *et al.*, 1965; Malins, 1968), *H. colliei* (Schmid and Mangold, 1966; Schmid *et al.*, 1967), *A. forbesi* (Ellingboe and Karnovsky, 1967; Snyder *et al.*, 1969c), petrel stomachs (Lewis, 1966), and marine worms (Pocock *et al.*, 1969).

The ether-linked lipids occur as alk-1-enyldiacylglycerols in starfish (*A. forbesi*) diverticulum (0.3% of total lipid) (Ellingboe and Karnovsky, 1967; Snyder *et al.*, 1969c), the liver of *S. acanthias* (0.2% of the total lipid) (Malins *et al.*, 1965), and the liver of *H. colliei* (Schmid *et al.*, 1967). Alk-1-enyl glycerolipids were not detected in the spermatazoa of the alewife (*Alosa pseudoharengus*) or in the rainbow trout (*Salmo gairdnerii*) (Minassian and Terner, 1966).

Schmid *et al.* (1969) investigated the composition of the monounsaturated *O*-alk-1-enyl chains of alk-1-enyldiacylglycerols from *H. celliei* liver. The major C_{18} isomer was $\Delta 9$, whereas the $\Delta 7$ isomer predominated in the C_{20} and C_{22} chains. The monoenoic acids of these compounds were mainly the $\Delta 9$ isomer in the C_{18} through C_{24} acids, but the $\Delta 7$ isomer predominated in the C_{16} acids. Baumann *et al.* (1970) have shown that the *O*-alk-1-enyl chains in alk-1-enyldiacylglycerols of *H. colliei* are mostly saturated and monounsaturated, with 16.4% branched structures. The acyl moieties on position 2 contain largely 18:1 chains, whereas those on position 3 contain a wide range of saturated and monounsaturated chains, including 2.6% branched structures (Baumann *et al.*, 1970).

B. Metabolism

Recently, precursor roles for fatty aldehydes (Bickerstaffe and Mead, 1967; Ellingboe and Karnovsky, 1967; Hagen and Goldfine, 1967; Bell and White, 1968) and alkylglycerols (Thompson, 1968; Malins, 1968; Friedberg and Greene, 1967a; Wood and Healy, 1970; Blank *et al.*, 1970) have been considered in the biosynthesis of alk-1-enyl glycerolipids. The fatty aldehydes have not been extensively studied in the biosynthesis of alk-1-enyl glycerolipids of marine origin, but Ellingboe and Karnovsky (1967) have shown that aldehyde was a better precursor than fatty alcohol and fatty acid in the formation of alk-1-enyl glycerolipids of the starfish (*A. forbesi*). This study failed to find a direct precursor–product relation between alkyl and alk-1-enyl glycerolipids. Studies on the metabolism of *S. acanthias* liver (Malins, 1968), however, indicated a probable conversion of alkyl glycerolipids to alk-1-enyl glycerolipids in the live animal, while interconversion of *O*-alk-1-enyl and *O*-alkyl moieties in ratfish (*H. colliei*) liver was not indicated in Baumann and co-workers' studies (1970) on molecular structure. Recent evidence strongly favoring the conversion of *O*-alkyl to *O*-alk-1-enyl moieties in various tissues is discussed (see Chapters VII and XII).

C. Environmental Effects

There is an increasing body of evidence implicating alk-1-enyl glycerolipids in regulating the degree of plasticity and permeability of the biological membrane: Levels of alk-1-enyl glycerolipids undergo significant alterations during myelination in the rat (Korey and Orchen, 1959; Debuch, 1966), a 150% increase in alk-1-enyl glycerolipids takes place in rat spermatazoa as they pass through the epididymis (Scott *et al.*, 1963), and the alk-1-enyl glycerolipid content of rat adipose tissue decreases to a very low value in the 1- to 21-day postnatal period (Yarbro and Anderson, 1956). It is possible that in the marine environment, the alk-1-enyl glycerolipids are metabolized in response to environmental stresses so as to maintain the integrity of cellular architecture. Roots and Johnston (1968)

TABLE III

Molar Proportions of Octadecanal and Octadecenal in Ethanolamine Plasmalogens Extracted from the Brains of Goldfish Acclimated to Different Temperatures[a]

Acclimation temperature (°C)			
5°	15°	25°	30°
0.68 ± 0.11	0.77 1.01	0.81 ± 0.3	1.69 ± 0.53

[a] From the work of Roots and Johnston (1968).

recently showed that environmental temperature has a striking effect on levels of alk-1-enyl glycerolipids in goldfish (*Carassius auratus* L.) brains. Plasmalogen values (moles aldehyde/moles phosphorus $\times$ 100) for fish acclimated at 5°C were 35.9 $\pm$ 1.37, but rose to 45.73 $\pm$ 3.51 at 30°C. Alteration in temperature also produced significant changes in fatty aldehydes of phosphatide plasmalogens, e.g., more unsaturated aldehydes were synthesized at lower temperatures (Table III).

IV. Ether-Linked Phosphatides and Miscellaneous Structures

Alkyl glycerophosphatides, such as the analogs of phosphatidylethanolamine and phosphatidylcholine, occur in several marine organisms (Thompson and Lee, 1965; Lewis, 1966; Ellingboe and Karnovsky, 1967; Snyder *et al.*, 1969c; Pocock *et al.*, 1969). Lewis' observation (1966) that alkyl ether analogs of phosphatidic acid occur in significant amounts in squid (*Loligo* species), shrimp (*Peneus* species), and anchovies (*Engraulis mordax*) is particularly interesting (Table IV), because he did not find alkyl glycerophosphatides containing an amino base. Although the liver of *S. acanthias* contains large amounts of alkyldiacylglycerols, Malins (1968) found no alkyl moieties in the phospholipid fraction. Nevertheless, when he administered 1-^{14}C-chimyl alcohol to live animals, there was considerable radioactivity in the alkylglycerols derived from the phospholipids, demonstrating the capability of this species to phosphorylate alkylglycerols.

Phosphorylated alk-1-enyl glycerolipids were detected in the hepatic lipids of *S. acanthias* (Malins, 1968) and in the diverticulum and gonads of *A. forbesi* (Ellingboe and Karnovsky, 1967; Snyder *et al.*, 1969c). The marine worm (*N. virens*) contains alk-1-enyl glycerophosphatides complexed with ethanolamine and choline (Pocock *et al.*, 1969), and the phospholipids of goldfish brains contain alk-1-enyl glycerolipids with ethanolamine, choline, and serine residues (Roots and Johnston, 1968). The occurrence of alk-1-enyl glycerolipids in mollusks is discussed in Chapter XII.

Recently, alkyl ethers of cholesterol were established as naturally occurring lipids in bovine cardiac muscle (Funasaki and Gilbertson, 1968). Gilbertson *et al.* (1970) then found that alk-1-enyl ethers of cholesterol are present in bovine and porcine cardiac muscle. Sterol ethers may occur in marine species, as is suggested by the work of Ellingboe and Karnovsky (1967) with the starfish (*A. forbesi*), which also contains small amounts of cyclic acetals, compounds previously reported to be present in high percentages in the sea anemone (*Anthropleura elegantissima*) (Bergmann

TABLE IV

Alkylglycerols Derived from Total Phospholipids of Three Marine Organisms[a]

Chain[b]	Anchovy (%)	Shrimp (%)	Squid (%)
10	—	1.6	—
11	0.9	—	—
12	2.3	4.3	0.4
12:1	—	—	0.5
13	1.4	2.0	—
13:1	—	—	0.7
14:br[c]	—	0.9	0.5
14	5.7	6.4	7.2
14:1	3.9	5.4	1.3
15:br	1.1	4.4	1.0
15	2.6	7.3	2.6
16:br	0.7	0.9	0.5
16	30.5	36.2	59.0
16:1	4.1	11.0	4.8
17:br	2.5	3.6	4.1
17	3.1	2.0	3.1
17:1	—	—	0.3
18:br	—	—	1.0
18	15.7	9.7	10.6
18:1	15.7	3.4	4.6
19	7.6	Trace	Trace
19:1	0.7	—	—
20	1.4	Trace	—

[a] From the work of Lewis (1969b).
[b] Analyzed as trimethyl silyl ether derivatives.
[c] br designates branched chains.

and Landowne, 1958). However, the occurrence of these compounds in the sea anemone has been questioned (Frosolono *et al.*, 1967).

V. Diol Lipids

Bergelson and co-workers (1966) inferred the presence of alk-1-enyl ether and acyl derivatives of ethane, propane, and butane diols in various animal and vegetable products. Evidence now suggests that the diol lipids are widely distributed in nature (Bergelson, 1969). So far the *O*-alk-1-enyl moiety is the only ether-linked structure that has been found in the diol lipids of terrestrial species. Recently, however, the mandibular canal lipids of the porpoise (*Phocoena phocoena*) yielded a new class of alkyl ethers, the di-*O*-alkyl ethers of pentane diol (Varanasi and Malins, 1969). Further investigations of the dialkoxypentane fraction showed that the principal structure was the 1,5-isomer (Varanasi and Malins, 1970b). The *O*-alkyl chains were 86.4% C_{18}, 7.2% C_{16}, and 2.7% C_{14}. Despite the fact that the isovaleroyl moiety is characteristic of the main lipid classes no isopentyloxy structure was detected in dialkoxypentanes. These findings are consistent with results obtained from the analysis of wax esters (Varanasi and Malins, 1970a) and neutral glycerolipids (Varanasi and Malins, 1970b), which indicate that the isopentyl chain is primarily associated with acyl moieties. Isovaleric acid, presumably derived from L-leucine metabolism (Christophe, 1963), is therefore not reduced and incorporated into alkyl moieties to an appreciable extent. The selective incorporation of the isopentyl moiety into glycerolipids and wax esters indicated by these data suggests a unique relation between leucine metabolism and lipid metabolism in marine porpoises.

REFERENCES

Baumann, W. J., Mahadevan, V., and Mangold, H. K. (1966). *Hoppe-Seyler's Z. Physiol. Chem.* **347**, 52.

Baumann, W. J., Takahashi, T., Mangold, H. K., and Schmid, H. H. O. (1970). *Biochim. Biophys. Acta* **202**, 468.

Bell, O. E., Jr., and White, H. B., Jr. (1968). *Biochim. Biophys. Acta* **164**, 441.

Bergelson, L. D. (1969). *Progr. Chem. Fats Other Lipids* **10**, 241.

Bergelson, L. D., Vaver, V. A., Prokazova, N. V., Ushakov, A. N., and Popkova, G. A. (1966). *Biochim. Biophys. Acta* **116**, 511.

Bergmann, W., and Landowne, R. H. (1958). *J. Org. Chem.* **23**, 1241.

Bickerstaffe, R., and Mead, J. F. (1967). *Biochemistry* **6**, 655.

Blank, M. L., Wykle, R. L., Piantadosi, C., and Snyder, F. (1970). *Biochim. Biophys. Acta* **210**, 442.

Bone, Q., and Roberts, B. L. (1969). *J. Mar. Biol. Ass. U.K.* **49**, 913.

Brockerhoff, H., and Hoyle, R. J. (1967). *Can. J. Biochem.* **45**, 1365.

Brockerhoff, H., Hoyle, R. J., and Ronald, K. (1964a). *J. Biol. Chem.* **239**, 735.

Brockerhoff, H., Yurkowski, H., Hoyle, R. J., and Ackman, R. G. (1964b). *J. Fish. Res. Bd. Can.* **21**, 1379.

Christophe, J. (1963). *In* "Biochemical Problems of Lipids" (A. C. Frazer, ed.), p. 373. American Elsevier, New York.

Corner, E. D. S., Denton, E. J., and Forster, G. R. (1969). *Proc. Roy. Soc., Ser. B* **171**, 415.

Debuch, H. (1966). *Hoppe-Seyler's Z. Physiol. Chem.* **344**, 83.

Denton, E. J., Gilpin-Brown, J. B., and Shaw, T. I. (1969). *Proc. Roy. Soc., Ser. B* **174**, 271.

Ellingboe, J., and Karnovsky, M. L. (1967). *J. Biol. Chem.* **242**, 5693.

Friedberg, S. J., and Greene, R. C. (1967a). *J. Biol. Chem.* **242**, 5709.

Friedberg, S. J., and Greene, R. C. (1967b). *J. Biol. Chem.* **242,** 234.
Frosolono. M. F., Kisic, A., and Rapport, M. M. (1967). *J. Org. Chem.* **32,** 3998.
Funasaki, H., and Gilbertson, J. R. (1968). *J. Lipid Res* **9,** 766.
Gilbertson, J. R., Garlich, H. H., and Gelman, R. A. (1970). *J. Lipid Res.* **11,** 201.
Gruger, E. H., Jr. (1967). *In* "Fish Oils" (M. E. Stansby, ed.), p. 3. Avi Publ. Co., Westport, Connecticut.
Hagen, P. O., and Goldfine, H. (1967). *J. Biol. Chem.* **242,** 5700.
Hajra, A. K. (1969). *Biochem. Biophys. Res. Commun.* **37,** 486.
Hanahan, D. J., Ekholm, J., and Jackson, C. M. (1963). *Biochemistry* **2,** 630.
Jones, F. R. H., and Marshall, N. B. (1953). *Biol. Rev.* **28,** 16.
Karnovsky, M. L., Jeffery, S. S., Thompson, M. S., and Deanae, H. W. (1955). *J. Biophys. Biochem. Cytol.* **1,** 173.
Kasama, K., Nayao, U., and Katsuya, I. (1970). *Biochim. Biophys. Acta* **202,** 56.
Kates, M., Sastry, P. S., and Yengoyan, L. S. (1963). *Biochim. Biophys. Acta* **70,** 705.
Kates, M., Yengoyan, L. S., and Sastry, P. S. (1965). *Biochim. Biophys. Acta* **98,** 252.
Kayama, M., Tsuchiya, Y., and Nevenzel, J. C. (1971). *Bull. Jap. Soc. Scient. Fisheries* **37,** 111.
Korey, S. R., and Orchen, M. (1959). *Arch. Biochem. Biophys.* **83,** 381.
Kritzler, H. (1948). *Condor* **50,** 5.
Lewis, R. W. (1966). *Comp. Biochem. Physiol.* **19,** 363.
Lewis, R. W. (1969a). *Comp. Biochem. Physiol.* **31,** 715.
Lewis, R. W. (1969b). *Comp. Biochem. Physiol.* **31,** 725.
Lewis, R. W. (1970). *Lipids* **5,** 151.
Lovern, J. A. (1964). *Oceangr. Mar. Biol. Annu. Rev.* **2,** 169.
Malins, D. C. (1967). *In* "Fish Oils" (M. E. Stansby, ed), p. 31. Avi Publ. Co., Westport, Connecticut.
Malins, D. C. (1968). *J. Lipid Res.* **9,** 687.
Malins, D. C., and Barone, A. (1969). Unpublished results.
Malins, D. C., and Barone, A. (1970). *Science* **167,** 79.
Malins, D. C., and Houle, C. R. (1961). *Proc. Soc. Exp. Biol. Med.* **108,** 126.
Malins, D. C., and Robisch, P. A. (1971). *Biochim. Biophys. Acta* **248,** 430.
Malins, D. C., and Sargent, J. R. (1971). *Biochemistry* **10,** 1107.
Malins, D. C., and Wekell, J. C. (1969). *Progr. Chem. Fats Other Lipids* **10,** 339.
Malins, D. C., Wekell, J. C., and Houle, C. R. (1965). *J. Lipid Res.* **6,** 100.
Mangold, H. K., and Malins, D. C. (1960). *J. Amer. Oil Chem. Soc.* **37,** 383.
Marshall, N. B. (1960). *'Discovery' Rep.* **31,** 1.
Minassian, E. S., and Terner, C. (1966). *Amer. J. Physiol.* **210,** 615.
Pocock, D. M. E., Marsden, J. R., and Hamilton, J. G. (1969). *Comp. Biochem. Physiol.* **30,** 133.
Prop, N. (1965). *Progr. Brain Res.* **10,** 454.
Roots, B. I., and Johnston, P. V. (1968). *Comp. Biochem. Physiol.* **26,** 553.
Sargent, J. R., Williamson, I. P., and Towse, J. B. (1970). *Biochem. J.* **117,** 26P.
Schmid, H. H. O., and Mangold, H. K. (1966). *Biochim. Biophys. Acta* **125,** 182.
Schmid, H. H. O., Baumann, W. J., and Mangold, H. K. (1967). *J. Amer. Chem. Soc.* **89,** 4797.
Schmid, H. H. O., Bandi, P. C., Mangold, H. K., and Baumann, W. J. (1969). *Biochim. Biophys. Acta* **187,** 208.
Scott, T. W., Dawson, R. M. C., and Rowlands, I. W. (1963). *Biochem. J.* **87,** 507.
Snyder, F. (1969). *Progr. Chem. Fats Other Lipids* **10,** 287.

Snyder, F., Malone, B., and Wykle, R. L. (1969a). *Biochem. Biophys. Res. Commun.* **34,** 40.
Snyder, F., Malone, B., and Wykle, R. L. (1969b). *Biochem. Biophys. Res. Commun.* **34,** 315.
Snyder, F., Malone, B., and Blank, M. L. (1969c). *Biochim. Biophys. Acta* **187,** 302.
Snyder, F., Blank, M. L., Malone, B., and Wykle, R. L. (1970a). *J. Biol. Chem.* **245,** 1800.
Snyder, F., Malone, B., and Blank, M. L. (1970b). *J. Biol. Chem.* **245,** 1790.
Spener, F., and Mangold, H. K. (1971). *J. Lipid Res.* **12,** 12.
Thompson, G. A., Jr. (1968). *Biochim. Biophys. Acta* **152,** 409.
Thompson, G. A., Jr., and Lee, P. (1965). *Biochim. Biophys. Acta* **98,** 151.
Toyama, Y. (1924a). *Chem. Umsch. Geb. Fette, Oele, Wachse Harze* **31,** 13.
Toyama, Y. (1924b). *Chem. Umsch. Geb. Fette, Oele, Wachse Harze* **31,** 61.
Toyama, Y. (1924c). *Chem. Umsch. Geb. Fette, Oele, Wachse Harze* **31,** 153.
Tsujimoto, M., and Toyama, Y. (1922). *Chem. Umsch. Geb. Fette, Oele, Wachse Harze* **29,** 27.
Varanasi, U., and Malins, D. C. (1969). *Science* **166,** 1158.
Varanasi, U., and Malins, D. C. (1970a). *Biochemistry* **9,** 3629.
Varanasi, U., and Malins, D. C. (1970b). *Biochemistry* **9,** 4576.
Wardle, C. R. (1967). Unpublished results.
Wood, R., and Healy, K. (1970). *Biochem. Biophys. Res. Commun.* **38,** 205.
Wood, R., and Snyder, F. (1969). *Arch. Biochem. Biophys.* **131,** 378.
Yarbro, C. L., and Anderson, C. E. (1956). *Proc. Soc. Exp. Biol. Med.* **91,** 408.

CHAPTER XII

ETHER-LINKED LIPIDS IN MOLLUSCS

*Guy A. Thompson, Jr.**

I. Introduction

We have known for many years that molluscs are rich in ether-containing lipids. Reports of glycerolipids high in alkyl and alk-1-enyl ether content in molluscan tissues were among the first to appear after the development of reliable analytical methods for these compounds, and members of the phylum Mollusca have since been the subject of numerous studies of ether lipid metabolism.

* Research cited in this review which originated in the author's laboratory was supported in part by the United States Public Health Service and the National Science Foundation.

II. Distribution

A. Alkyl Ethers

The first broad survey for alkyl glyceryl ethers in organisms was conducted by Karnovsky *et al.* (1946). They estimated the alkyl ether content of tissue glycerolipids by oxidizing the nonsaponifiable fraction with periodate and quantifying any formaldehyde produced. A number of molluscs representing Gastropoda, Lamellibranchia, and Cephalopoda, three of the four major classes of the phylum, were found to contain alkyl glyceryl ethers. One species, the rock octopus (*Octopus rugosus*), contained as much as 30% of its nonsaponifiable lipid as alkyl glyceryl ethers.

Because of the saponification technique employed by Karnovsky *et al.* (1946), it was impossible to determine the nature of the parent lipid giving rise to alkyl glyceryl ethers. Much later, Thompson and Hanahan (1963b) examined the lipids of *Ariolimax columbianus* and *Arion ater*, two species of terrestrial slug, in more detail. Alkyl glyceryl ethers were found in the neutral lipids as the diacyl derivatives and also in the phospholipid fraction. Analysis of the individual phospholipids by acetolysis and saponification revealed that the alkyl ethers are localized mainly in the two major fractions of glycerolipids, those containing ethanolamine (13 mole % alkyl ethers) and those containing choline (49 mole % alkyl ethers). Plasmalogens are also present, but mainly in the ethanolamine phospholipids, where they account for 20 mole % of the total fraction. Alkyl ethers occur in similar concentrations in the ethanolamine and choline phosphatides of the water snail *Lymnaea stagnalis* (Liang and Strickland, 1969); plasmalogens account for less than 4% of either fraction.

Analysis of the chain length distribution of alkyl glyceryl ethers by gas–liquid chromatography demonstrated a predominance of chimyl alcohol in *Ariolimax* (Thompson and Hanahan, 1963b). This homolog accounts for 94 mole % of the total alkyl ethers in the choline phosphatides while the other phosphatides contain considerable amounts of batyl alcohol and some branched-chain alkyl ethers. No unsaturated alkyl ethers were detected.

The alkyl glyceryl ether analysis was later extended to other representative molluscs (Thompson and Lee, 1965), including marine species of the four principal classes: Amphineura (*Katherina tunicata*), Gastropoda (*Thais lamellosa*), Lamellibranchia (*Protothaca staminea*), and Cephalopoda (*Octopus dofleini*). All species examined are rich in alkyl glyceryl ethers (Table I), both in the phospholipid and the neutral lipid fractions. In fact, when the sizable amounts of plasmalogens present in these organisms are included, the figures for total ether-containing phospholipids would in some

TABLE I

The Content of Various Lipid Constituents in Molluscan Tissues

Species	Fresh wt. of tissue (g)	Total lipid (mg)	Phospholipid (mg)	Alkyl glyceryl ethers in neutral lipids (mg)	Alkyl glyceryl ethers in phospholipids (mg)	Alkyl glyceryl ethers in phospholipids (average mole %)[a]
K. tunicata	205	2125	829	52.0	60.1	24.5
				54.2	70.4	
T. lamellosa	117	5280	1990	63.6[b]	140.2	20.0
				47.8	—	
P. staminea	226	2340	1540	15.1	56.0[c]	9.4
				14.7	44.6	
O. dofleini (hepatopancreas)	87	3380	1580	34.2	104.0	17.8
				43.0	102.8	
O. dofleini (tentacle)	480	2250	1650	Trace	159	24.8
				Trace	138	

[a] Based on total lipid phosphorus.
[b] Small amount of nonglyceryl ether contaminant.
[c] Small amount lost.

cases reach 35 mole % of the lipid phosphorus, certainly among the highest levels ever reported in biological samples.

Although there is some variation from one species to another, the four molluscs described above generally have as their principal alkyl ether components 16:0, 18:0, and (in *K. tunicata*) 18:1 side chains. A similar pattern was reported by Lewis (1966), who found mainly alkyl ethers with chain lengths of 16:0 (59%) and 18:0 (11%), in the squid *Loligo*.

On the basis of limited evidence, it would appear that there is no marked localization of alkyl ether glycerolipids in any particular part on the molluscan body. Somewhat more is present in the viscera than in the mantle of the slug *Arion ater* (Thompson and Hanahan, 1963b), and among the internal organs, alkyl ethers varied in content from 10 mole % of the phospholipids in the crop and intestine to 40% in the sexual organs (Thompson, 1966). A smaller difference was noted between hepatopancreas and tentacle of the octopus (Thompson and Lee, 1965).

Despite the fact that no exhaustive survey has been made of phylum Mollusca for the presence of alkyl ethers in glycerolipids, existing evidence leads one to assume that these lipids are of widespread, if not universal, occurrence, often in high concentration. They are found in marine and freshwater species as well as in terrestrial species. They occur as components of phospholipids in all major portions of the body and in storage fats. Further studies on additional species will be necessary to decide whether the variations noted within the phylum indicate significant trends in alkyl ether levels.

B. Alk-1-enyl Ethers

The first comparative analysis of molluscan plasmalogens made with modern analytical techniques was by Rapport and Alonzo (1960). The whelk (*Busycon canaliculatum*), squid (*Loligo pealeii*), mussel (*Mytilus edulis*), softshell clam (*Mya arenaria*), scallop (*Pecten irradians*), and quahog (*Venus mercenaria*) were all found to contain large amounts of plasmalogens. Many of the species examined have plasmalogen levels higher than that found in mammalian brain, which is usually regarded as being extremely rich in the alk-1-enyl ether glycerolipids. Representatives from other phyla, namely sea anemone and sea cucumber, also were shown to abound in plasmalogens.

The alk-1-enyl ethers are found in both neutral lipids and phospholipids, as are the alkyl ethers. Rapport and Alonzo detected a general pattern in which the phosphorus-bound alk-1-enyl ethers of glycerol occur mainly in the ethanolamine-containing lipids. Very little exists as serine phosphatides and even less as choline glycerolipids. The high plasmalogen levels in marine molluscs were confirmed by Thompson and Lee (1965), who reported that more than 10% of the phospholipids of *Protothaca staminea* and tentacles of *Octopus dofleini* have the alk-1-enyl group.

Thiele (1959) carried out plasmalogen analyses on the terrestrial snail *Helix pomatia* throughout the year, finding the level fairly constant at approximately 7–9% of the lipid phosphorus except for a rise to 12.8% in March and a drop to 4.4% in June (breeding season). The terrestrial slug *Arion ater* also contains plasmalogens in the range of 9–13% of total lipid phosphorus (Thompson and Hanahan, 1963b). Just as in marine species (Rapport and Alonzo, 1960), most of the plasmalogens in slugs are ethanolamine phospholipids. The chain-length distribution of aldehydes from the ethanolamine plasmalogens is remarkably similar to that of the alkyl ethers isolated from the same ethanolamine lipid fraction; both are mainly the 16:0 and 18:0 homologs (Thompson, 1965).

Like the alkyl ether glycerolipids, plasmalogens are distributed throughout all tissues of the species that have been analyzed. Rapport (1961) discovered that in a squid and six types of clam, gill tissue is highest in plasmalogen content, having up to 400 μmoles/g lipid (as compared to

140 μmoles/g lipid for rat brain using a similar method of analysis). Thompson and Hanahan (1963b) noted little difference in the plasmalogen content of the mantle and the combined viscera of the slug *Arion ater*.

We may conclude from the data available that molluscs contain ether glycerolipids of both the alkyl and alk-1-enyl types, together accounting for approximately 15–35 mole % of the phospholipids. This is far more than in most other organisms that have been examined. The fairly uniform distribution of the ether-containing phospholipids suggests they they play some general structural or metabolic function in cells of the organism.

III. Metabolism

A. Biosynthesis

Since they contain such high levels of ether-linked lipids, it is not surprising that molluscs have served a useful role in the elucidation of ether lipid biochemistry. When the high content of plasmalogens and alkyl ether phospholipids was found in slugs (Thompson and Hanahan, 1963b) these organisms became likely condidates for studies of the metabolic relationship, if any, existing between the two structurally similar molecules. Synthesis of alkyl glyceryl ethers from glucose-6-^{14}C and palmitic acid-1-^{14}C was shown to be rapid, with the alkyl ether phosphatides having approximately the same specific radioactivity as the diacyl phosphatides (Thompson, 1965). After the administration of 1-^{14}C-palmitate, alk-1-enylglycerols derived from intact plasmalogens were surprisingly low in specific radioactivity compared to either alkylglycerols or fatty acids recovered from the phospholipids. The differences were most striking in feeding experiments extending over a 1 hr or a 4 hr time span, but plasmalogens were low in radioactivity even 16 hr after feeding ^{14}C-palmitate. Similarly, alk-1-enylglycerols isolated from slugs 18 hr after feeding ^{3}H-chimyl alcohol contained only traces of tritium, although the alkyl ether phospholipids were very radioactive.

While those labeling studies confirmed the observations made earlier with other tissues, i.e., that alkylglyceryl ethers can be synthesized from glucose (Thompson and Hanahan, 1962) and palmitic acid (Gilbertson *et al.*, 1963; Thompson and Hanahan, 1963a), they also suggested that the metabolic pathways of alkyl ethers and alk-1-enyl ethers are not closely related. However, subsequent investigations proved that this latter conclusion is false.

Hoping to find some condition under which appreciable *in vivo* biosynthesis of plasmalogens could be detected, a number of experimental variables, including incubation time, temperature, and oxygen tension, were tested. The incorporation of radioactivity into alk-1-enyl glycerolipids was shown to depend upon only one of these variables—the passage of time (Thompson, 1966). If slugs were allowed to metabolize fed ^{14}C-palmitate or ^{3}H-chimyl alcohol for 48 hr, the alk-1-enylglycerols isolated from intact plasmalogens had nearly one-half the specific radioactivity of the alkylglycerols. After 72 hr the alk-1-enyl ether derivatives were even nearer to the alkyl ethers in specific radioactivity. It was clear from the ^{3}H-labeled chimyl alcohol experiments, where low radioactivity in fatty acids indicated little metabolic cleavage of the ether moiety, that the alk-1-enyl side chain was not being derived from an acyl group.

Evidence of a different sort from the same study further implicated alkyl ethers as precursors of alk-1-enyl ethers in glycerolipids. Slugs fed selachyl alcohol (1-*O*-*cis*-9′-octadecenylglycerol) for 10 days were found to have incorporated this uncharacteristic alkyl ether for slugs into their phospholipids. The octadecenyl side chain was a prominent component, not only of the isolated alkyl ether but also of the alk-1-enyl glycerolipids. However, no increase in the corresponding acyl group, oleic acid, was detected.

The experiments described above provided support for the hypothesis that the alk-1-enyl side chain arises through enzymatic desaturation of the alkyl ether moiety, probably in phospholipid molecules. It seemed important to learn whether the ether bond remained intact during this transformation. In order to accomplish this, the long-term chimyl alcohol feeding experiments were repeated, this time using a doubly labeled substrate (Thompson, 1968). The $^{14}C/^{3}H$ ratios obtained after feeding 1-*O*-(1-^{14}C)-hexadecyl-2(^{3}H)-glycerol (Table II) strongly support the concept that the glyceryl

TABLE II

Incorporation of Radioactivity into Components of the Phospholipid Fraction of Arion ater after Feeding [^{14}C]-, [^{3}H]-Chimyl Alcohol[a]

	Specific radioactivity of ^{14}C (counts/min/mmole, $\times 10^5$)		
Days fed	Alcohols[b]	Alk-1-enyl glyceryl ethers	Alkyl glyceryl ethers
3	0.48 [310]	8.45 [5.2]	25.8 [4.2]
3	0.43 [554]	8.78 [4.8]	26.4 [4.0]
7	0.40 [19]	7.10 [4.6]	20.2 [4.0]
7	0.43 [45]	6.80 [4.8]	20.0 [3.6]

[a] Approx. 2×10^6 counts/min ^{14}C and 5×10^5 counts/min ^{3}H administered. $^{14}C/^{3}H$ ratios are given in brackets.

[b] Derived from fatty acids by $LiAlH_4$ reduction.

moiety as well as the alkyl side chain of alkyl glyceryl ethers is converted to the analogous alk-1-enyl glyceryl ether without appreciable scission of the ether bond.

In summary, the picture emerging from the *in vivo* slug experiments is depicted in reaction (1). While the nature

(1)

of R′ and R″ could not be determined with certainty, all available evidence points to a fatty acyl group for R′ and phosphorylethanolamine for R″. It would seem reasonable to postulate that the reaction is a general one in which any alkyl ether-containing phospholipid can be dehydrogenated to yield the corresponding plasmalogen. Additional evidence is needed to support this point; it is also possible that the transformation takes place only at the level of an alkyl ether analog of phosphatidic acid. At any rate the role of alkyl ethers as precursors of alk-1-enyl ethers seems well established in molluscs. Other evidence supporting this pathway has come from recent studies of such widely different organisms as ascites tumor cells (Wood and Healy, 1970; Blank *et al.*, 1970) and the Myxomycete, *Physarum polycephalum* (Poulos and Thompson, 1971). Similar studies done in mammalian tissues are discussed in Chapters VII, IX, and X.

The metabolic origin of "neutral plasmalogens" (alk-1-enyldiacylglycerols) has received less intensive study. In the slug experiments, these compounds were always very low in radioactivity, and it is thought likely that they arise by esterification of alk-1-enyl ether compounds liberated from phospholipids.

Little attention has been directed to the metabolic control of ether lipid biosynthesis. Some finely controlled mechanism must act to maintain the proper balance of diacyl and alkylacyl phospholipids in tissues. An interesting indication that synthesis of alkyl ether glycerolipids is subject to feedback inhibition was noted in studies on *Arion ater* (Thompson, 1965). The prefeeding of large amounts of alkylglyceryl ethers depressed alkyl ether biosynthesis in glycerolipids from ^{14}C-palmitic to 10% of the normal rate. Much more experimentation is called for in this important area.

B. Catabolism

No systematic studies have been reported on the degradation of ether-linked lipids in molluscs. Radioactive chimyl alcohol ingested by the slug *Arion ater* is partly degraded, as evidenced by the recovery of variable percentages of the label in the form of fatty acids (Thompson, 1965, 1966). The limited data available would point to some kind of direct relationship between the amount of alkyl glyceryl ether fed and the amount degraded although the stimulation of ether bond cleavage is not particularly striking.

Pfleger *et al.* (1967) compared the ether-cleaving ability of homogenates of the slug *Lima maximus* with that of liver homogenates from various mammals. Cleavage by the slug enzyme was markedly less than observed in liver. The authors speculated that high levels of tissue ether lipids may go hand in hand with low levels of cleavage activity (see Chapter VII).

REFERENCES

Blank, M. L., Wykle, R. L., Piantadosi, C., and Snyder, F. (1970). *Biochim. Biophys. Acta* **210**, 442.

Gilbertson, J. R., Ellingboe, J., and Karnovsky, M. L. (1963). *Fed. Proc., Fed. Amer. Soc. Exp. Biol.* **22**, 304.

Karnovsky, M. L., Rapson, W. S., and Black, M. (1946). *J. Soc. Chem. Ind., London* **65**, 425.

Lewis, R. W. (1966). *Comp. Biochem. Physiol.* **19**, 363.

Liang, C.-R., and Strickland, K. P. (1969). *Can. J. Biochem.* **47**, 85.

Pfleger, R. C., Piantadosi, C., and Snyder, F. (1967). *Biochim. Biophys. Acta* **144**, 633.

Poulos, A., and Thompson, G. A., Jr. (1971). *Lipids* **6**, 470.

Rapport, M. M. (1961). *Biol. Bull.* **121**, 376.

Rapport, M. M., and Alonzo, N. F. (1960). *J. Biol. Chem.* **235**, 1953.

Thiele, O. W. (1959). *Z. Vergl. Physiol.* **42**, 484.

Thompson, G. A., Jr. (1965). *J. Biol. Chem.* **240**, 1912.

Thompson, G. A., Jr. (1966). *Biochemistry* **5**, 1290.

Thompson, G. A., Jr. (1968). *Biochim. Biophys. Acta* **152**, 409.

Thompson, G. A., Jr., and Hanahan, D. J. (1962). *Arch. Biochem. Biophys.* **96**, 671.

Thompson, G. A., Jr., and Hanahan, D. J. (1963a). *Biochemistry* **2**, 641.

Thompson, G. A., Jr., and Hanahan, D. J. (1963b). *J. Biol. Chem.* **238**, 2628.

Thompson, G. A., Jr., and Lee, P. (1965). *Biochim. Biophys. Acta* **98**, 151.

Wood, R., and Healy, K. (1970). *J. Biol. Chem.* **245**, 2640.

CHAPTER XIII

ETHER-LINKED LIPIDS IN PROTOZOA

*Guy A. Thompson, Jr.**

I. Introduction

The amount of information pertaining to ether lipids in members of the phylum Protozoa is not great, but justification for writing a chapter on the subject arises from these considerations. First, one of the few protozoa examined in detail, *Tetrahymena pyriformis*, has an unusually high concentration of alkyl glyceryl ethers distributed among its phospholipids in a novel way. Second, some recent investigations suggest that other protozoa are equally rich in ether-containing lipids. The future value of this chapter depends upon the accuracy of my prediction that protozoa will be uniquely useful tools in studying ether lipid metabolism and function.

* Research cited in this review which originated in the author's laboratory was supported in part by the United States Public Health Service, the National Science Foundation, and the Robert A. Welch Foundation.

II. Distribution

Discouragingly little attention has been devoted to the lipid composition of the protozoa. This apparent neglect stems mainly from the fact that very few species have been grown successfully in axenic culture. Homogeneous material for study has, until recently, simply not been available for analysis. In view of this state of affairs, it is ironic that a protozoan was one of the first cells shown to contain plasmalogens. Feulgen and Voit (1924) observed that the entire protoplasm of *Nyctotherus* was stained when subjected to the reaction that now bears Feulgen's name. After alcohol extraction, only the nucleus stained, indicating that lipid material was responsible for the cytoplasmic coloration (see Chapter I).

More refined techniques have been employed to detect protozoan and other plasmalogens in recent years. Hack *et al.* (1962) analyzed the lipids extracted from several algae and protozoa by carrying out the periodic acid-Schiff reaction on developed chromatograms. While none of the algal lipids gave a positive test for plasmalogen, lipids from all the animal flagellates, including *Crithidia fasciculata*, *C. luciliae*, *Herpetomonas culicis*, *Trypanosoma cruzi*, and *Leishmania donovani*, had a strong, positive reaction for ethanolamine plasmalogen and also contained traces of choline plasmalogen. The presence of plasmalogens was confirmed by $HgCl_2$ hydrolysis of the lipids on chromatograms, followed by chromatography of the resulting fatty aldehydes in a second dimension. Two other protozoa, *Acanthamoeba* sp. and *Tetrahymena pyriformis*, had lipids giving only a faint positive test for plasmalogen. *Tetrahymena* did have a lipid with chromatographic properties suggestive of an alkyl glyceryl ether. This agreed with the conclusions of Taketomi (1961) that although plasmalogens were but minor components of *Tetrahymena*, it was likely that the organism contained a sizable amount of alkyl ether glycerol lipids. Somewhat later, Helmy *et al.* (1967), using the periodic acid-Schiff color reaction, identified ethanolamine plasmalogen as the principal lipid of *Blastocrithidia culicis*.

Data of a semiquantitative nature thus began to accumulate in the literature linking high levels of ether-containing lipids with protozoa. A few analyses contradicted these indications. For example, Meyer and Holz (1966) reported very little plasmalogen in *Crithidia fasiculata*. This apparent contradiction could have resulted from a degradative loss of plasmalogen during silicic acid column chromatography at room temperature.

The first detailed analysis of protozoan lipids for ether derivatives involved the ciliate *Tetrahymena pyriformis* (Thompson, 1967). Alkyl glyceryl ethers, almost exclusively chimyl alcohol, were found in both choline- and ethanolamine-containing phospholipids. Alkyl ether derivatives account for 60 mole % of the choline lipid fraction and 22 mole % of the ethanolamine fraction. The latter fraction includes 2-aminoethyl phosphonate-containing analogs of ethanolamine phosphatides. Further analysis of this mixed fraction revealed an unexpected preference of association between alkyl ether and 2-aminoethyl phosphonate (AEP) derivatives of glycerol. Whereas the diacyl phospholipids contain only 12–20 mole % of AEP, the alkylacyl phospholipids contain 70–80 mole % AEP. Thompson (1967) found no trace of plasmalogens, although Taketomi (1961) reported their existence in low concentrations in *Tetrahymena*.

Plasmalogens are more prevalent than their alkyl glyceryl ether analogs in rumen protozoa, making up approximately 12% of their phospholipids, while alkyl ether phospholipids account for only a third as much (Dawson and Kemp, 1967). An aminoethyl phosphonate-containing plasmalogen was detected in these preparations. Although *Entodinium* was the major protozoan present, pure cultures of *E. caudatum* contained considerably less plasmalogen than the mixed cultures.

Ether lipids have recently been reported in high concentrations in plasmodia of the true slime mold, *Physarum polycephalum* (Poulos *et al.*, 1971). This organism and other members of phylum Myxophyta are considered by some taxonomists to belong with the protozoa, although others classify them among the fungi (Alexopoulos, 1962). *P. polycephalum* contains as much as one-quarter of its phospholipids as plasmalogens, while another 12 mole % are alkyl ether phospholipids (Poulos *et al.*, 1971). As in most protozoa, the plasmalogens in the slime mold are principally in the ethanolamine lipid fraction, whereas the alkyl ethers are distributed more widely among all the various phosphatides. Both the alkyl and the alk-1-enyl ethers of glycerol usually have a saturated ether-linked side chain of 16 carbon atoms. This close biochemical resemblance to protozoa may prove useful in a taxonomic reevaluation of the slime molds, especially since ether lipids have never been reported in fungi.

III. Metabolism

Tetrahymena is capable of synthesizing the alkyl ether side chain of glycerolipids from 1-^{14}C-palmitic acid (Thompson, 1967). Radioactive chimyl alcohol administered in tracer amounts can also be taken up by the cells and incorporated intact into the ether-linked phospholipids. Only a small amount of the chimyl alcohol was degraded to fatty acids during the period of incorporation.

Later experiments tracing the incorporation of ^{3}H-chimyl alcohol uncovered a peculiar pattern (Thompson, 1969). The choline phospholipids became labeled very rapidly and then lost a significant fraction of their radioactivity, this loss in radioactivity of the choline lipid fraction being offset by a corresponding rise in labeling of the AEP-containing lipids. Although the details have not been clarified, some curious metabolic interrelationship seems to exist between these two lipid classes, perhaps involving the alkylglycerol backbone as a participant in the synthesis of the carbon–phosphorus bond of the phosphonoglycerolipid.

Tetrahymena has been used to investigate the mechanism of ether bond formation. Friedberg and Greene (1968) measured the $^{3}H/^{14}C$ ratios of alkyl glyceryl ethers synthesized by cells grown with uniformly ^{14}C-labeled glycerol and [1,3-^{3}H]-glycerol. The $^{3}H/^{14}C$ ratio of the alkyl glyceryl ethers isolated was 6:1 as compared to a ratio of 20:1 in the glycerol substrate incubated and in the glycerol recovered after hydrolysis of esterified glycerolipids. The results suggested to the authors that the glycerol backbone of alkylglyceryl ethers is derived not from free glycerol or glycerophosphate, but from phosphoglyceraldehyde or dihydroxyacetone phosphate, since these intermediates might be expected to lose some tritium from the 1-position. Further work with other experimental systems subsequently proved this indication to be correct. (See Chapter VII.)

Cell-free preparations of *Tetrahymena* are also capable of synthesizing alkyl glyceryl ethers (Kapoulas and Thompson, 1969). The properties of the system are generally similar to those of the preputial gland tumor enzymes described by Snyder *et al.* (1969) Friedberg and co-workers have used cell-free preparations to study details of the ether bond formation. When *Tetrahymena* microsomes were incubated with hexadecanol and [1,3-^{3}H]-dihydroxyacetone phosphate (DHAP) plus [1,3-^{14}C]-DHAP, alkyl glyceryl ether synthesis was shown to involve the loss of one proton from carbon 3 of DHAP (Friedberg *et al.*, 1971). Later experiments demonstrated that the proton is extracted after a stereochemical labilization, and that the specific proton lost is the same one labilized by an unrelated enzyme, triosephosphate isomerase (Friedberg *et al.*, 1972).

As one would expect for an organism rich in alkyl ether glycerolipids, *Tetrahymena* has the capacity to enzymically cleave the ether bond. However, the degradative system is activated only under certain conditions. When traces of radioactive chimyl alcohol are fed to growing *Tetrahymena* cultures, the molecules are promptly incorporated into phospholipids without appreciable ether bond cleavage (Thompson, 1967). On the other hand, when chimyl alcohol is administered in an amount equalling or exceeding that naturally present in lipids of the cells, a large proportion of the absorbed substrate is degraded to fatty acid, and presumably, glycerol (Kapoulas *et al.*, 1969a). At the same time that the cells are engaged in degrading one portion of the chimyl alcohol, they are actively utilizing another for the formation of phospholipids. Indeed, essentially all phospholipid molecules synthesized by cells growing in the presence of excess chimyl alcohol are of the alkyl ether type.

The mechanism of ether bond cleavage is probably quite similar to that described for rat liver. (See Chapter VII.) Kapoulas *et al.* (1969b) studied the degradative enzymes of cell-free preparations of *Tetrahymena*. As in *in vivo* experiments, fatty acids are the principal cleavage products, but smaller amounts of fatty aldehydes and fatty alcohols were tentatively identified as intermediates in the breakdown process. The enzyme is principally in the postmicrosomal supernatant, from which it may be sedimented by centrifugation at 250,000*g* for 150 min. NAD^+, an NADPH-generating system, and molecular O_2 are needed for maximal activity. Contrary to findings with the rat liver enzyme system (Tietz *et al.*, 1964), no requirement for a pteridine cofactor could be demonstrated, and unlike the situation in the living cells, even trace amounts of chimyl alcohol incubated *in vitro* with the enzyme were rapidly cleaved.

Little experimentation has been conducted on the intermediary metabolism of plasmalogens in protozoa, but Poulos and Thompson (1971) measured the incorporation of radioactive precursors into plasmalogens of the slime mold *Physarum polycephalum*. The authors were especially interested in investigating any possible metabolic interrelationship between the alkyl and alk-1-enyl ether-linked glycerolipids. A comparison of the specific radioactivities of both ether types isolated from plasmodia grown in the presence of ^{14}C-palmitate or ^{3}H-chimyl alcohol strongly suggested that plasmalogens arise from the enzymic desaturation of alkyl ether phospholipids. This pathway has also been proposed for other cells. (See Chapters VII, IX, X, XII.) There were some indications that the desaturation in *Physarum* may take place at the level of an alkyl ether analog of phosphatidic acid.

IV. Biological Function

There is no reason to doubt that each specific class of lipids has some special functions for which its structure makes it uniquely suited. Yet there is not enough experimental evidence available to construct a sound hypothesis relating detailed structure and biological function. Among the suggestive correlations that have been recognized is one involving ether lipids of the protozoan, *Tetrahymena pyriformis*. As described earlier, the total lipids of this ciliate are rich in alkyl ether phospholipids, which amount to approximately 25 mole % of the total phospholipids. Preliminary evidence has been reported to the effect that the ether lipids are localized to some extent in the surface membranes providing the interface between the cell contents and the environment (Nozawa and Thompson, 1971). This evidence has been confirmed by quantitative analysis demonstrating that as much as 50 mole % of the isolated ciliary membrane phospholipids contain alkyl ethers (Thompson *et al.*, 1971). It so happens that the ether derivatives in this organism show a preferential association with molecules containing 2-aminoethyl phosphonate, as mentioned earlier in the chapter, and more than 60 mole % of the phosphorus-containing lipids of ciliary membranes are phosphonolipids (Kennedy and Thompson, 1970). It may be more than a coincidence that lipids containing the ether linkage (Thompson, 1967) and/or the carbon–phosphorus bond (Thompson, 1969) are markedly resistant to attack by lipolytic enzymes, including those endogenous to *Tetrahymena*. An exciting hypothesis, which remains to be tested, is that the stability conferred upon the *Tetrahymena* surface membrane by the presence of the ether bond and the direct carbon–phosphorus bond in glycerolipids serves as a specialized protective device for a cell that, in its natural habitat, is frequently exposed to attack by chemical and enzymic agents.

V. Conclusions

The data available pertaining to lipids of protozoa indicate that members of this phylum generally contain high concentrations of ether-linked lipids. There is considerable diversity within the phylum, with some species possessing both alkyl and alk-1-enyl derivatives and others containing only alkyl ethers. With the advent of more refined culture techniques, an increasing variety of organisms is becoming available in pure culture for analysis. Because of their rapid growth rate and ease of experimental manipulation, it is likely that protozoa will serve as useful subjects for further investigation of the metabolism and biological function of ether lipids.

REFERENCES

Alexopoulos, C. J. (1962). "Introductory Mycology," 2nd ed., p. 67. Wiley, New York.
Dawson, R. M. C., and Kemp, P. (1967). *Biochem. J.* **105**, 837.
Feulgen, R., and Voit, K. (1924). *Pfluegers Arch. Gesamte Physiol. Menschen Tiere* **206**, 389.
Friedberg, S. J., and Greene, R. C. (1968). *Biochim. Biophys. Acta* **164**, 602.
Friedberg, S. J., Heifetz, A., and Greene, R. C. (1971). *J. Biol. Chem.* **246**, 5822.

Friedberg, S. J., Heifetz, A., and Greene, R. C. (1972). *Biochemistry* **11,** 297.
Hack, M. H., Yaeger, R. G., and McCaffery, T. D. (1962). *Comp. Biochem. Physiol.* **6,** 247.
Helmy, F. M., Hack, M. H., and Yaeger, R. G. (1967). *Comp. Biochem. Physiol.* **23,** 565.
Kapoulas, V. M., and Thompson, G. A., Jr. (1969). *Biochim. Biophys. Acta* **187,** 594.
Kapoulas, V. M., Thompson, G. A., Jr., and Hanahan, D. J. (1969a). *Biochim. Biophys. Acta* **176,** 237.
Kapoulas, V. M., Thompson, G. A., Jr., and Hanahan, D. J. (1969b). *Biochim. Biophys. Acta* **176,** 250.
Kennedy, K. E., and Thompson, G. A., Jr. (1970). *Science* **168,** 989.
Meyer, H., and Holz, G. G., Jr. (1966). *J. Biol. Chem.* **241,** 5000.
Nozawa, Y., and Thompson, G. A., Jr. (1971). *J. Cell Biol.* **49,** 712.
Poulos, A., and Thompson, G. A., Jr. (1971). *Lipids* **6,** 470.
Poulos, A., LeStourgeon, W. M., and Thompson, G. A., Jr. (1971). *Lipids* **6,** 466.
Snyder, F., Wykle, R. L., and Malone, B. (1969). *Biochem. Biophys. Res. Commun.* **34,** 315.
Taketomi, T. (1961). *Z. Allg. Mikrobiol.* **1,** 331.
Thompson, G. A., Jr. (1967). *Biochemistry* **6,** 2015.
Thompson, G. A., Jr. (1969). *Biochim. Biophys. Acta* **176,** 330.
Thompson, G. A., Jr., Bambery, R. J., and Nozawa, Y. (1971). *Biochemistry* **10,** 4441.
Tietz, A., Lindberg, M., and Kennedy, E. P. (1964). *J. Biol. Chem.* **239,** 4081.

CHAPTER XIV

BACTERIAL PLASMALOGENS

Howard Goldfine and Per-Otto Hagen

Before 1962 reports on the presence of plasmalogens in bacteria were scattered and generally unconfirmed (Rapport and Norton, 1962), but during the past decade there has been increasing interest in microbial lipids. With the development of modern tools such as gas–liquid chromatography (GLC) and thin-layer chromatography (TLC), especially when used in conjunction with radioisotopes, the study of microbial lipids received tremendous impetus, and reliable evidence for the presence of plasmalogens in bacteria was quickly obtained.

I. Distribution of Plasmalogens among Bacteria

The work of Allison *et al.* (1962), Wegner and Foster (1963), and Katz and Keeney (1964) firmly established the presence of plasmalogens in several species of anaerobic rumen bacteria of cows and sheep. Allison *et al.* (1962) investigated two species of rumen organisms that require volatile, branched-chain fatty acids for growth. By use of ^{14}C-isovalerate, they demonstrated the incorporation of this precursor into C_{15} branched-chain fatty acids and fatty aldehydes by growing cells of *Ruminococcus flavefaciens.* They also demonstrated that the labeled fatty aldehydes were components of the cellular phospholipids and thus presumably present in plasmalogens. About 7.5% of the lipid ^{14}C was recovered in the branched-chain C_{15} aldehyde. Experiments on *R. albus* grown with 1-^{14}C-isobutyrate were also reported. Approximately 11% of the lipid ^{14}C was recovered as the 2,4-dinitrophenylhydrazone derivatives of 14- and 16-carbon aldehydes which were presumably branched, but not further identified.

According to Allison *et al.* (1962), work along similar lines was simultaneously being carried out by Wegner and Foster at Wisconsin but their results were published in the following year (1963). Working with a different species of rumen bacteria, *Bacteroides succinogenes*, they showed that the isobutyrate and isovalerate required for the growth of this species were incorporated almost exclusively into the cellular lipids, 98.7% and 94%, respectively, and that most of the label was in the phospholipid fraction of the cells grown with 1-^{14}C-isobutyrate. The major, if not sole, plasmalogen was an alk-1-enylacylglycerophosphorylethanolamine (ethanolamine plasmalogen), which accounted for at least 80% of the total phospholipids. The fatty acids derived from labeled isobutyrate were mainly iso-C_{14} and iso-C_{16}, but the labeled aldehydes so derived were not identified. ^{14}C-valerate was incorporated into 13:0 and 15:0 fatty acids and unidentified fatty aldehydes.

These reports were followed by another paper by Katz and Keeney (1964), which showed that mixed rumen bacteria from cows were a rich source of aldehydogenic lipids. The ratio of aldehyde to phosphorus was 0.19 in the polar lipid fraction, implying about 20% plasmalogen content. The nonpolar lipids also contained aldehydes, but approximately one-eighth as much per gram of lipid as the polar lipid fraction. The aldehydes derived from both fractions were analyzed and were found to contain large amounts of 15:0 branched, 14:0 branched, 16:0, 14:0, and 18:1 chains.

All of these authors suggested that the branched-chain aldehydes of ruminant milk and tissues might be derived from the plasmalogens of anaerobic rumen bacteria. The recent work on the positional isomers of bacterial aldehydes that will be discussed below has led to similar conclusions on the origin of animal aldehydes.

The presence of plasmalogens in the lipids of a saprophytic soil anaerobe was first revealed in studies on the complex lipids of *Clostridium butyricum* (Goldfine, 1963). A complete analysis of the polar lipids of this organism showed that the glycerophosphatides represented about 70% of the total extractable lipid by weight and that three of the major components of the phospholipid fraction of log phase cells were *N*-methylethanolamine phosphatides (38%), ethanolamine phosphatides (14%), and phosphatidylglycerol (26%). When each of these components was assayed for plasmalogens, it was found that 78% of the *N*-methylethanolamine phosphatides, 55% of the ethanolamine phosphatides, and 9% of the phosphatidylglycerol were of the alk-1-enylacylglycerophosphate type (Goldfine, 1962, 1963; Baumann *et al.*, 1965). A fraction representing about 15% of the total phospholipid had *N*-methylethanolamine as the predominant "base" and a ratio of 1.0 ester/P, but no alk-1-enyl ether. The structure of this component(s) remains unresolved, but it is clearly neither a saturated ether lipid nor a lysophosphatide (Baumann *et al.*, 1965).

Until recently the lipids of only one other *Clostridium*, *C. welchii* (*perfringens*), had been analyzed. Macfarlane (1962) had found polyglycerol phosphatides and *O*-amino acid esters of phosphatidylglycerol, but had not reported the presence of plasmalogens in the lipids of this organism.

The lipids of one other rumen bacterial species have also been investigated in depth (Kanegasaki and Takahashi, 1968). *Selenomonas ruminantium*, a strict anaerobe of sheep rumen, when grown on glucose medium, requires a normal volatile fatty acid of chain length from C_3 to C_{10} for growth. 1-^{14}C-Valerate or 1-^{14}C-caproate was incorporated into the extractable lipids of these cells and the incorporated isotope migrated essentially as one spot on silicic acid-impregnated paper chromatography. The radioactivity was in a phosphatide with a high amino nitrogen and aldehyde content. When the total ethanol-ether extracts from glucose-valerate-grown cells were analyzed, ratios of 0.96 amino nitrogen/P and 0.25 aldehyde/P were found. When cells were grown on lactate, which abolishes the voltatile fatty acid requirement seen with glucose-grown cells, the amino nitrogen/P ratio was 0.90 and aldehyde/P was 0.47.

The wider distribution of plasmalogens in anaerobic bacteria was recently studied by Kamio *et al.* (1969). They examined the mixed bacteria of sheep rumen and found, as did Katz and Keeney (1964) in cow's rumen, that bacterial fractions obtained by differential centrifugation had phospholipids with aldehyde/P ratios ranging from 0.16 to 0.59. The most slowly sedimenting bacterial fraction, which consisted mainly of small cocci, had lipids with an aldehyde/P ratio of 0.16. Some of the heavier fractions, which were rich in plasmalogens, also contained protozoa. Kamio *et al.* (1969) also grew enrichment cultures of anaerobic rumen bacteria on various energy sources and obtained morphologically different cell types on different carbon sources. The molar ratios of aldehyde to phosphorus in the lipids obtained from these enrichment cultures ranged from 0.19 for a culture rich in diplococci to 0.89 for a culture containing a mixture of rods and diplococci. Enrichment cultures of anaerobic soil bacteria were also grown on various energy sources and the ratios of aldehyde/P in the lipids of these organisms, mostly rods, ranged from 0.25 to 1.38.

In the most definitive part of their study, Kamio *et al.* (1969) examined the lipids of a variety of pure cultures of aerobic, anaerobic, and facultative bacteria for the presence of plasmalogens. None of the aerobic or facultative species, whether grown aerobically or anaerobically, contained detectable amounts of plasmalogens. The strictly anaerobic bacteria, from a variety of families and genera, contained plasmalogens in varying amounts. The results obtained by this group, along with those obtained by other workers, are summarized in Table I.

In agreement with the results of Macfarlane (1962), the amount of plasmalogen in *C. welchii (perfringens)* is very low. Similarly, the amounts in *Desulfovibro* sp. and *Bacteroides ruminicola* are of doubtful significance. With these exceptions, all of the strictly anaerobic bacteria examined thus far contain substantial amounts of plasmalogens. In addition to the strict anaerobes surveyed by Kamio *et al.* (1969) and the organisms discussed above, two other anaerobic bacteria have recently been found to contain plasmalogens: *Sphaerophorus ridiculosus* (Hagen, 1970), and *Treponema pallidum* (Reiter) (Meyer and Meyer, 1971) (Table I).

As mentioned above, none of the microaerophilic, aerobic, and facultative organisms examined by Kamio *et al.* (1969) contained plasmalogens. These included *Escherichia coli* (grown aerobically or anaerobically); *Rhodospirillum rubrum, Streptococcus faecalis, Leuconostoc mesenteroides, Lactobacillus casei*, and *L. delbrueckii* (all grown anaerobically); and *Bacillus subtilis, Corynebacterium sepedonicum, Pseudomonas fluorescens*, and *Streptomyces aureofaciens* (all grown aerobically). These results confirm the work done in many laboratories on the lipids of aerobic, micro-

TABLE I

Plasmalogens and Alkyl Glyceryl Ethers in Bacteria

Bacterial strains	Molar ratios	
	Aldehyde/P[a]	Alkyl glyceryl ether/P[e]
Desulfovibrio sp.	0.09	—
Selenomonas ruminantium		
Lactate-grown	0.50	0.034
Glucose-grown	0.25	0.036
Bacteroides ruminicola	0.004	—
Bacteroides succinogenes	0.71[b]	—
Ruminococcus flavefaciens	0.56–0.80[c]	—
Veillonella gazogenes	0.69	0.002
Peptostreptococcus elsdenii	1.04	—
Propionibacterium freudenreichii		
Anaerobic culture	0.68	0.014
Standing culture	0.60	0.002
Prop. shermanii		
Anaerobic culture	0.37	0.015
Standing culture	0.45	0.003
Clostridium butyricum	0.4[d]	0.002[f]
Cl. saccharoperbutylacetonicum	0.80	0.007
Cl. acetobutylicum 179-121	0.72	0.002
Cl. acetobutylicum 314-48	0.92	—
Cl. perfringens	0.04	0.007
Cl. kaneboi	0.82	0.009
Cl. kainantoi	0.86	N.D.[g]
Sphaerophorus ridiculosus	0.25[f]	—
Treponema pallidum (Reiter)	0.14[h]	—

[a] All data from Kamio *et al.* (1969) except where otherwise noted.
[b] Wegner and Foster (1963).
[c] Allison *et al.* (1962).
[d] Baumann *et al.* (1965).
[e] All data from Kim *et al.* (1970) except where otherwise noted.
[f] Hagen (1970).
[g] Not detectable.
[h] Meyer and Meyer (1971).

aerophilic, and facultative bacteria. To site a few specific examples: Gray and Wilkinson (1965) found no plasmalogens in *Pseudomonas aeruginosa, Proteus mirabilis, Alcaligenes faecalis,* and *E. coli.* Randle *et al.* (1969) examined the lipids of eight species of gram-negative aerobic and facultative bacteria and found no plasmalogens and White (1968) reported no plasmalogens in *Haemophilus parainfluenzae.*

The structures of the aldehydogenic phospholipids of the anaerobic bacteria studied by Kamio *et al.* (1969) were not determined. These investigators did, however, subject the phospholipids from enrichment cultures and from five pure cultures to thin-layer chromatography on silica gel H in chloroform-methanol-water (65:25:4 v/v). In each case a major spot, R_f 0.45–0.50, was found that gave positive reactions for fatty aldehyde, phosphate, and amino nitrogen. These data are consistent with the presence of an ethanolamine plasmalogen but are only indicative, since *N*-methylethanolamine plasmalogen shows somewhat similar behavior on thin-layer chromatography (Baumann *et al.*, 1965).

II. The Aldehyde Chains of Bacterial Plasmalogens

A. Position of the Alk-1-enyl Chain

The 1-alk-1-enyl-2-acyl-*sn*-glycerol-3-phosphate structure of the plasmalogens derived from animal sources has been known for some time (Chapter I). An attempt to confirm this structure for bacterial plasmalogens was carried out with the phospholipids from *Clostridium butyricum* (Hagen and Goldfine, 1967). The mixed ethanolamine and *N*-methylethanolamine phosphatides were catalytically hydrogenated in order to convert the alk-1-enyl ethers to alkyl ethers. After saponification to remove the ester residues from the hydrogenated plasmalogens as well as from the diacylphosphatides, the lipids were subjected to acetolysis. The alkyl ether diacetates were examined by nuclear magnetic resonance spectroscopy and gave the spectrum characteristic of the diacetates of 1(3)-alkylglycerol. The acetate moieties were then removed and the yield of formaldehyde obtained on periodate oxidation confirmed the 1(3)-linkage.

B. Composition of the Aldehydes Derived from Bacterial Plasmalogens

Even with the scant information presently available, it is clear that there is wide species variation in the types of fatty aldehyde chains found in bacteria. Indeed, this variation appears to mirror that seen in examination of the fatty acids of bacteria, which have definite familial relationships (Kates, 1964; O'Leary, 1967).

1. *Ruminococcus flavefaciens*

Allison and co-workers (1962) examined the aldehydes from *Ruminococcus flavefaciens* by isolating them as the 2,4-dinitrophenylhydrazones, regenerating them, and then analyzing them by GLC. Most of the radioactivity from isovalerate-1-^{14}C-grown cells was coincident with a branched-chain C_{15} peak. Branched-chain iso-C_{15} fatty acid was a major component. A complete analysis of the aldehydes was not reported.

2. *Ruminococcus albus*

Again, Allison *et al.* (1962) give data only for the aldehydes derived from a labeled precursor, in this case 1-^{14}C-isobutyrate, in their study of *R. albus*. The labeled aldehydes were obtained as the 2,4-dinitrophenylhydrazones and subjected to reverse-phase paper chromatography. They migrated with known 14- and 16-carbon aldehyde derivatives and were presumed to be iso-C_{14} and iso-C_{16}, but no further work was reported. The major fatty acids of *R. albus* grown on isobutyrate appeared to be 16:0, iso-14:0, and iso-16:0.

3. *Selenomonas ruminantium*

The influence of volatile fatty acid nutrients on the fatty aldehyde and fatty acid composition of *S. ruminantium* has been studied by Kanegasaki and Takahashi (1968) and Kamio *et al.* (1970a). When grown on glucose, this organism requires a volatile, saturated fatty acid, 3:0 to 10:0, for growth. When grown on lactate, no volatile fatty acid is required. When it was grown on lactate or glucose plus valerate, the predominant fatty aldehydes in the phospholipids were 17:1 > 15:0 > 15:1 > 12:0 > 17:0. Similar results were obtained with cells grown on glucose plus heptanoate, with the exception that there was less 12:0 than 17:0 fatty aldehydes. The fatty acids from these cells were also examined and were found to be somewhat different in composition from the aldehydes. There was more 13:0, 15:0, and 15:1 acids than aldehydes, and less 17:1 and 17:0. There was very little 12:0 in the fatty acids of lactate-, valerate-, and heptanoate-grown cells (Kamio *et al.*, 1970a).

When *S. ruminantium* was grown on glucose plus even-numbered volatile fatty acids (4:0 to 8:0), the fatty aldehydes and acids isolated from the phospholipids reflected this by having a greater proportion of even-numbered chains. The predominant aldehydes were 16:1 (53–72%), 16:0 (7–19%), 18:1 (3.5–8.1%), 14:0 (1–4%), and an unidentified, pre-presumably branched, 12:0 in the caproate-grown cells. Again, the fatty acids were qualitatively similar to the aldehydes but quantitatively dissimilar. Hexadecenoate was less predominant (18–43%) and there was somewhat less 16:0 (3.6–8.3%); 18:1 was lower (trace–5.0%) and 14:0 much higher (25–55%) (Kamio *et al.*, 1970a).

4. *Mixed Rumen Bacteria*

Katz and Keeney (1964) determined the fatty aldehyde composition of the plasmalogens from the bacteria of cow-rumen digesta. The major components were 16:0 (31.5%), 15:0 br (28%), 14:0 br (8.4%), 18:1 (6.7%), 14:0 (6.3%), 15:0 (5.5%), and 16:0 br (4.7%). These were strikingly similar to those of the rumen-bacterial phospholipid fatty acids investigated by Keeney *et al.* (1962).

5. *Clostridium butyricum*

Aldehydes obtained from the total lipids of *C. butyricum* were similar in composition to the total fatty acids. The major aldehydes were 16:0 (44%), 17:cyc (25.5%), 19:cyc (11%), 16:1(9.1%), 18:0 (3.2%), and 18.1 (2.1%). The unusual C_{17} and C_{19} cyclopropane aldehydes were identified by NMR and infrared spectroscopy and by oxidation to the corresponding fatty acids for which standards were available for comparison on GLC (Goldfine, 1964). When the aldehydes derived from the phospholipids alone were examined, somewhat more 16:1 plus 17:cyc (45%) and 18:1 plus 19:cyc (18%) and less 16:0(35%) were seen (Hagen and Goldfine, 1967). Comparison to the reported composition of the 1-linked fatty acids of the diacyl phosphatides of this organism (Hildebrand and Law, 1964) revealed a remarkable similarity. The sums of the monounsaturated precursors and the cyclopropane products were given because the proportion of cyclopropane to unsaturated fatty acids increases as cells go from log-phase to stationary phase, but the sum tends to remain relatively constant (Chung and Goldfine, 1965; Hagen and Goldfine, 1967).

Recently the positional isomers of the monoenoic and cyclopropane fatty acids and aldehydes derived from the phospholipids of *C. butyricum* were analyzed by open capillary column GLC (Goldfine and Panos, 1971). The analyses of these components obtained from cells harvested near the end of the log-phase are given in Table II. The proportions of positional isomers of the monoenoic fatty acids and aldehydes are quite similar, but there are considerably higher proportions of 17:cyc 7,8 and 19:cyc 9,10 aldehydes than the corresponding acids. Apparently there is not quite as strong a specificity in the formation of the cyclopropane fatty aldehydes from the $(n - 7)$ monoenoic precursors as there is in the formation of the cyclopropane fatty acids from their monoenoic precursors. This is especially striking in the case of the 17:cyc, which is almost entirely (>99%) the 9,10-isomer in the fatty acids but only 76% 9,10-isomer in the fatty aldehydes; the remaining 24% is the 17:cyc 7,8 aldehyde isomer.

TABLE II

Positional Isomers of Apolar Chains from Clostridium butyricum Phospholipids

	Fatty acids (wt. %)	Fatty aldehydes (wt. %)
16:1 Δ^7	16	15
16:1 Δ^9	7.0	7.7
17:cyc 7,8	0.07	4.4
17:cyc 9,10	12	14
18:1 Δ^9	1.3	1.2
18:1 Δ^{11}	1.1	0.7
19:cyc 9,10	0.90	2.6
19:cyc 11,12	3.4	3.6

There have been two reports on the presence of the $(n - 7)$ monoenoic series in the alk-1-enyl ethers of higher organisms. Schmid *et al.* (1969) reported large amounts of 18:1 Δ^{11}, 20:1 Δ^{13}, and 22:1 Δ^{15} in the diacylalk-1-enyl glycerols of the ratfish (*Hydrolagus colliei*). In the fatty acids, however, the usual $(n - 9)$ isomers were the major constituents. Debuch and Winterfeld (1970) have found almost equal mixtures of 18:1 Δ^9 and 18:1 Δ^{11} in the alk-1-enylglycerols from human placental plasmalogens. Small amounts of 20:1 Δ^{13} were also found. The finding of the $(n - 7)$ series of aldehydes in *Clostridium butyricum* led to the suggestion that some animal alk-1-enyl glycerolipids may arise from bacterial sources (Goldfine and Panos, 1971). The monoenoic aldehydes of other bacteria will have to be analyzed to confirm the generality of the $(n - 7)$ series, already established in bacterial fatty acids by many workers (Kates, 1964).

III. Alkyl Glycerolipids in Bacteria

The diether phospholipids containing the polyisoprenoid side chains found in the extremely halophilic bacteria are discussed in Chapter XV. Alkyl glycerolipids have recently been reported in the same species of bacteria that contain the plasmalogen type. Hagen and Blank (1970) discovered the presence of small amounts of alkylglycerols derived from lipids of *C. butyricum*. An extensive survey was recently reported by Kim *et al.* (1970), who found small amounts of alkyl glycerolipids in a variety of strictly anaerobic and moderately aerotolerant species (Table I). The ratios of glyceryl ethers to phosphorus are low, ranging from 0.002 to 0.034, in contrast to ratios of aldehydes (derived from alk-1-enyl glycerolipids) to phosphorus, ranging up to 1.04 (Table I). A detailed account of the isolation and characterization of the alkylglycerols derived from lipids of *Selenomonas ruminantium* has also appeared (Kamio *et al.*, 1970b). In both this organism and in *C. butyricum* the alcohol chains have been analyzed and their lack of correspondence to the fatty acids and aldehyde chains of these organisms has been noted. This will be discussed in relation to plasmalogen biosynthesis (Section IV, D).

IV. Bacterial Plasmalogen Biosynthesis

A number of pathways for plasmalogen biosynthesis have been proposed. Among them are: (a) the direct conversion of an ester linkage to an alk-1-enyl glyceryl ether through dehydration of a hemiacetal, (b) the reaction of a free hydroxyl group in the 1-position of a glyceride or phosphoglyceride with a long-chain fatty aldehyde to form a hemiacetal which is then dehydrated, (c) the conversion of alkyl to alk-1-enyl glyceryl ethers. (See Chapters VII and IX.) Much of the experimental evidence for and against these pathways has come from studies with mammalian cells and tissues. Until very recently, *Clostridium butyricum* was the only bacterial species in which plasmalogen biosynthesis had been studied in depth.

A. Fatty Acids as the Precursors of the Aldehyde Chains of Bacterial Plasmalogens

Long-chain fatty acids appear to be precursors for the vinyl ether-linked aldehydic chains in the plasmalogens of growing cultures of *Clostridium butyricum*. Exogenous long-chain fatty acids were readily incorporated into cellular phosphatides both as esterified fatty acids and as the alk-1-enyl ether-linked, aldehydogenic chains of plasmalogens (Baumann *et al.*, 1965). Labeled octanoic and decanoic acids were incorporated into saturated, unsaturated, and cyclopropane aldehydes and fatty acids. On the other hand, longer chain length saturated precursors were incorporated primarily into saturated fatty aldehydes and acids, either unchanged or after chain elongation. The relative specific activities of the long-chain aldehydes separated as the dimethylacetals by GLC were, with few exceptions, the same as those of the corresponding acids.

The monounsaturated fatty acid precursors of *C. butyricum*, *cis*-Δ^3-decenoic acid and *cis*-Δ^3-dodecenoic acid are synthesized by dehydration of the corresponding D(—)-β-hydroxy fatty acids. Chain elongation of these monounsaturated fatty acids leads to the mixture of long-chain monounsaturated fatty acids, 16:1 Δ^7, 16:1 Δ^9, 18:1 Δ^9, and 18:1 Δ^{11} (Goldfine and Bloch, 1961; Scheuerbrandt *et al.*, 1961). It has recently been shown that the monounsaturated aldehyde chains isolated from *C. butyricum* plasmalogens have the same mixtures of positional isomers as the corresponding chain length fatty acids, in approximately the same ratios (Table II) (Goldfine and Panos, 1971). This finding, along with the incorporation data cited above, can reasonably be interpreted to argue for the origin of the long-chain aldehydes by way of the same pathway, and using the same enzymes, as those involved in the *de novo* synthesis of long-chain fatty acids.

In these experiments, the precursors were added when the cultures were inoculated, and the products were isolated after the cells ceased to grow logarithmically. In order to study the kinetics of formation of the aldehyde chains relative to the corresponding fatty acids, Hagen and Goldfine (1967) examined the time course of incorporation of ^{14}C-acetate into the fatty acid and aldehyde chains in both neutral lipids and phospholipids of *C. butyricum.* The fatty acids derived from the phospholipids were found to have specific activities six times those of the corresponding phospholipid aldehydes after a 5 or 10 min exposure to the labeled acetate. Even 90 min after addition of the labeled acetate, the specific activities of the phospholipid fatty acids were two to three times higher than those of the corresponding aldehydes. The esterified fatty acids and the aldehydes derived from the neutral lipid fraction had consistently lower specific activities than the corresponding phosphatide fatty acids. The pattern of specific activities indicated that the long-chain fatty aldehydes of the phosphatides were derived from fatty acids. The specific activities of the fatty aldehydes derived from the neutral lipids and those derived from the phospholipids were essentially the same during the first 45 min after the addition of ^{14}C-acetate. Since 2.5 to 5 times as many total counts were found in the phosphatide-bound aldehydes as in the neutral lipid fraction during the entire incubation period, it seems unlikely that the alk-1-enyl glyceryl ethers of the phosphatides were derived from those of the neutral lipid fraction. It appears that either exogenous fatty acids or those synthesized *de novo* can serve as precursors of the aldehydogenic chains of complex lipids in *C. butyricum.* Whether reduction occurs before or after incorporation of fatty acids into complex lipids is, however, not clear.

B. Conversion of Ester to Alk-1-enyl Glyceryl Linkage

There are several lines of evidence from experiments with mammalian tissues suggesting that diacyl phosphatides undergo transformation to plasmalogens; perhaps by reduction of the ester bond (Debuch, 1966; Bickerstaffe and Mead, 1967). However, later work (Bickerstaffe and Mead, 1968; Debuch *et al.*, 1970) on the incorporation of labeled phosphatidylethanolamine after intracerebral injection in rat brain failed to demonstrate the transformation of exogenous phosphatidylethanolamine to plasmalogen. Studies on the kinetics of incorporation of labeled inorganic phosphate into the phospholipids of logarithmically growing cultures of *C. butyricum* were reported by Baumann *et al.* in 1965. The phosphatides were separated by TLC and subjected to mild alkaline hydrolysis, and their specific activities were determined. Incorporation of $^{32}P_i$ into diacylphosphatidylethanolamine occurred without a detectable lag, whereas incorporation into ethanolamine plasmalogen was barely detectable in the first sample and did not become linear for at least 10 min. Incorporation into diacylphosphatidyl-*N*-methylethanolamine also preceded incorporation into the corresponding plasmalogen.

Hagen (1970) obtained supporting evidence for the conversion of a diacyl phosphatide to a plasmalogen by growing cultures of *C. butyricum* in the presence of $^{32}P_i$ until early logarithmic phase. The $^{32}P_i$ was removed and the turnover of the phosphatides measured at time intervals during logarithmic growth (Fig. 1). There was a reciprocal redistribution of radioactivity from the diacylphosphatidylethanolamine into the 1-alk-1-enyl-2-acylglycerylphosphoryl-*N*-methylethanolamine. Although specific activity data are lacking in these experiments, it seems reasonable to conclude, considering the known composition of the phosphatides of this organism (Baumann *et al.*, 1965), that diacyl phosphatides undergo transformation to plasmalogens. If such a conversion does take place, phosphatidic acid, the proposed intermediate for polar lipid synthesis in *C. butyricum* (Goldfine, 1966), should be devoid of alk-1-enyl ether-linked moieties. After sequential alkaline and acid hydrolysis of lipids from logarithmically growing cultures of *C. butyricum*, about 2% of the lipid phosphorus was associated with the glycerophosphate released by alkali, and no glycerophosphate was detected in the water-soluble products of acid hydrolysis. Thus, no plasmalogen, or at most only trace amounts, could be present in the phosphatidic acid fraction (Hagen, 1970).

The results so far described suggest a conversion of the diacylphosphatides to the plasmalogens. Supporting evidence for this contention is that there is a remarkable similarity between the fatty acid composition of the fatty acids esterified in the 1-position of the diacylphosphatides and those of the aldehydic chains of the total phospholipid plasmalogens of *C. butyricum* (Hagen and Goldfine, 1967). These results cannot, however, be unequivocally interpreted to mean that plasmalogens are syn-

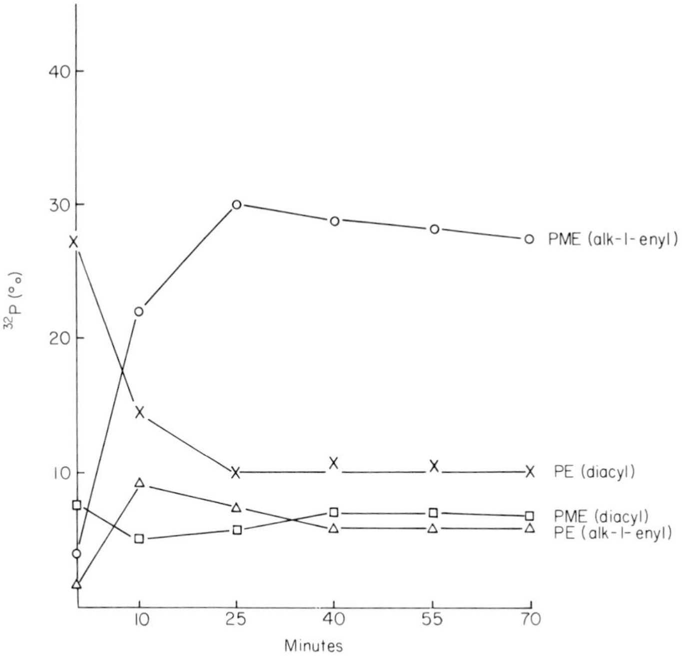

FIG. 1. Distribution of ^{32}P after dilution of isotope. *C. butyricum* was grown in the presence of $^{32}P_i$ until early logarithmic phase, then rapidly transferred to a nonradioactive medium. Aliquots were harvested at time of transfer and at intervals thereafter. The lipids were isolated and separated into phosphatidyl-*N*-methylethanolamine (PME) and phosphatidylethanolamine (PE) by TLC. The radioactivity in diacyl phosphatides and plasmalogens (alk-1-enyl) was measured after mild alkaline hydrolyses of the individual phosphatides. The percent values on the ordinate refer to the distribution of ^{32}P in the lipids after removal of $^{32}P_i$ from the medium.

thesized in *C. butyricum* by reduction of the ester linkage of the diacyl phosphatides. Attempts to show such a transformation in cell-free preparations from *C. butyricum* have so far failed (Hagen and Goldfine, 1966). Other evidence against direct reduction is presented below.

C. Aldehydes as Direct Precursors of Alk-1-enyl Ethers

Evidence for the incorporation of aldehydes into the plasmalogens of various tissues has also accumulated (Baumann *et al.*, 1965; Ellingboe and Karnovsky, 1967; Hagen and Goldfine, 1967; Bell and White, 1968; Stoffel *et al.*, 1970). The incorporation of fatty acids and their precursors into both acid and aldehyde chains of the lipids of growing cells of *C.*

butyricum led to experiments to determine whether exogenous long-chain aldehydes could also be utilized by growing cells, and, if they were, whether they would be preferentially incorporated into the alk-1-enyl chains of glycerolipids (Baumann *et al.*, 1965). Cells were grown in the presence of palmitaldehyde labeled in the carboxyl carbon, and the specific activities of the aldehydic and fatty acid moieties of the lipids were determined. Long-chain aldehydes were readily incorporated into the lipids of *C. butyricum*, and in several experiments more than 20% of the radioactive material found in the washed cells was associated with compounds other than free fatty aldehydes. These compounds included phospholipids, free fatty acids, and esterified fatty acids. The data indicated little, if any, preference for the incorporation of labeled aldehydes into the alk-1-enyl chain of glyceryl ethers as opposed to the fatty acid moieties. When unlabeled palmitaldehyde was added to cultures growing in the presence of trace amounts of 1-^{14}C-palmitate, there was a marked dilution of radioactivity in both the acyl and alk-1-enyl chains of phospholipids. Apparently, relatively large amounts of palmitaldehyde in the medium were utilized efficiently in competition with palmitate. When unlabeled palmitic acid was added to the medium of cells growing in the presence of 1-^{14}C-palmitaldehyde, the dilution of radioactivity in the alk-1-enyl and acyl chains of phospholipids was also approximately equal, but not as marked.

Equilibration of exogenous fatty acids and aldehydes was also indicated in experiments in which labeled free fatty acids and labeled non-phospholipid-bound aldehydes were detected in cells grown in the presence of labeled palmitaldehyde and palmitate, respectively. Although no evidence for a special role for free long-chain aldehydes as precursors of the alk-1-enyl ether-linked chain in plasmalogens was obtained, a rapid equilibration of free fatty acids and aldehydes along with incorporation of the aldehyde chain into alk-1-enyl glyceryl ethers could still account for these results. If the latter were true, one might see some incorporation of tritium into the phosphatide-bound aldehyde chains when cells are grown in the presence of 1-^{3}H-1-^{14}C-palmitaldehyde. On the other hand, if oxidation of the aldehyde is required, all of the tritium should be lost in the phosphatide-bound aldehyde chains.

C. butyricum was grown in the presence of 1-^{3}H-1-^{14}C-palmitaldehyde, and the isolated phospholipids, after initial purification by acetone precipitation, were analyzed by two methods (Hagen and Goldfine, 1967). In the first experiments, the phospholipids were purified by repeated TLC. The zone corresponding to the mixed ethanolamine and *N*-methylethanolamine phosphatides was eluted, and the ^{3}H/^{14}C ratio in the alk-1-enyl ethers was determined. In the second experiment, the phospholipids were separated by repeated column chromatography on silicic acid using, gradients of chloroform and methanol. A fraction that contained 0.96 mole of alk-1-enyl ether equivalents per mole of phosphorus was obtained and subjected to mild alkaline hydrolysis designed to remove the remaining diacyl phosphatides and the fatty acids esterified in the 2-position of the plasmalogens. Approximately 15% of the tritium in the starting material was retained in the alk-1-enyl ether-linked chains; therefore, added long-chain aldehydes can be incorporated into the alk-1-enyl glyceryl ethers of *C. butyricum* without necessarily undergoing oxidation to fatty acids.

Ellingboe and Karnovsky (1967) obtained better evidence by showing that 70–90% of the tritium from 1-^{3}H-1-^{14}C-steraldehyde was retained in the aldehydes of plasmalogens after incubation of labeled aldehyde with starfish digestive gland. Contrary to the report of Bickerstaffe and Mead (1967), Bell and White (1968) have found that long-chain aldehydes were precursors of alk-1-enyl glyceryl ethers in developing rat brain without prior oxidation. Their conclusion was also based on experiments with double-labeled aldehydes. In none of these experiments with double-labeled aldehydes was further reduction, for example to alcohols, prior to incorporation excluded.

D. Conversion of Alkyl Glyceryl Ethers to Alk-1-enyl Glyceryl Ethers

In vivo studies by Thompson (1966, 1968) in slugs and Blank *et al.* (1970) in mammals on the synthesis of alk-1-enyl ethers of glycerol from ^{14}C/^{3}H-labeled alkylglycerols, suggested that alkyl glycerolipids are the precursors of plasmalogens. Recently, Hajra (1969), Wykle and Snyder (1969, 1970), and Snyder *et al.* (1970b,c,d) have shown that dihydroxyacetone and long-chain fatty alcohols are the precursors of alkyl glycerolipids in several animal tissues. (See Chapters VII and IX.) In agreement with Thompson, Wood and Healy (1970) postulate that alkyl glycerolipids are the immediate precursors of alk-1-enyl glycerolipids in Ehrlich ascites cells, whereas the work of Ellingboe and Karnovsky (1967) and Snyder *et al.* (1969c) with starfish tissue suggested that fatty alcohols were the precursor of alkyl glycerolipids but probably not of alk-1-enyl glyceryl ethers. Recently, Snyder *et al.* (1970a) have shown that the oxygen of ^{18}O-labeled hexadecanol is incorporated into *O*-alkyl lipids by microsomal enzymes of preputial gland tumors, and the ^{18}O from hexadecanol is recovered in the plasmalogens in experiments *in vivo* with Ehrlich ascites cells. Similar findings have been reported for rat brain (Bell *et al.*, 1971). The results from a cell-free system isolated from neoplastic cells that synthesizes ethanolamine plasmalogens is consistent with the *in vivo* results (Wykle *et al.*, 1970; Snyder *et al.*, 1971).

Hill and Lands (1970) have tested the hypothesis that dihydroxyacetone phosphate may serve as a precursor for plasmalogens in *C. butyricum* by growing the cells in the presence of both ^{14}C-glycerol and 2-^{3}H-glycerol. If alkyldihydroxyacetone-P or a subsequent intermediate (Snyder *et al.*, 1969a,b, 1970c; Hajra, 1970) is a precursor of plasmalogens, the $^{3}H/^{14}C$ ratio in the plasmalogens would be essentially zero, since the ^{3}H at position 2 of glycerol would be labilized during formation of the keto derivative. Hill and Lands (1970) found that when *C. butyricum* was grown to mid or late logarithmic phase in the presence of ^{14}C-glycerol plus 2-^{3}H-glycerol, both the diacyl and alk-1-enylacyl phosphatides were *enriched* in tritium compared to the tritium-to-carbon ratio of the glycerol added. The ratio of ^{3}H to ^{14}C of diacyl phosphatides and plasmalogens was found to be similar to that of the total lipids. They therefore concluded that the hydrogen at the 2-position of glycerol was not labilized during the incorporation of glycerol into plasmalogen, which precludes a 2-keto intermediate in plasmalogen synthesis in *C. butyricum*. Attempts to achieve synthesis of plasmalogens in cell-free preparations of *C. butyricum* from dihydroxyacetone phosphate or glyceraldehyde 3-phosphate have also been unsuccessful (Hagen, 1970).

The results obtained from the bacterial system discussed so far cannot distinguish between a pathway that involves the alkyl ethers as intermediates in the formation of alk-1-enyl glyceryl ethers and a pathway in which a fatty aldehyde becomes linked to a 2-acyl lysophosphatide to form an alk-1-enyl ether bond directly.

When *C. butyricum* was grown in the presence of ^{14}C-labeled batyl or chimyl alcohol, most of the radioactivity was recovered as unchanged glyceryl ethers (Hagen and Goldfine, 1967). With ^{14}C-chimyl alcohol, some of the radioactivity was recovered in the phospholipid fraction, and it migrated on thin-layer chromatograms with the ethanolamine and *N*-methylethanolamine phosphatides. No radioactive dimethyl acetals were recovered, however, when this material was subjected to methanolysis and GLC. Similarly, no formation of plasmalogens could be detected when a crude cell-free preparation of *C. butyricum* obtained by sonication and low-speed centrifugation of logarithmically growing cells was incubated in the presence of preputial tumor microsomes, glyceraldehyde 3-phosphate, 1-^{14}C-cetyl alcohol, ATP, CoA, and Mg^{2+} in a reconstruction of the system described by Snyder *et al.* (1969a). Under these conditions alkyl glycerolipids are formed, but the *C. butyricum* extract lacked the ability to convert them to plasmalogens (Hagen, 1970).

When *C. butyricum* was grown in the presence of 1-^{3}H-1-^{14}C-cetyl alcohol, slight incorporation into plasmalogens was seen in some experiments (Hagen, 1970), but not in others (Goldfine, 1970). In one such experiment, 2% of the added radioactivity was recovered from the cells when they were allowed to grow to the late logarithmic phase and about 98% of this radioactivity was in the neutral lipid fraction. The added cetyl alcohol had a ^{3}H/^{14}C ratio of 4.4; the ratios in the neutral lipids and phospholipids were found to be 4.3 and 2.3, respectively. When the polar lipids were reduced with $LiAlH_4$ and the reduction products separated by TLC, 42% of the ^{14}C chromatographed with alcohols (derived from esterified fatty acids), 20% with alkylglycerols, and 7% with alk-1-enylglycerols. These bands had ^{3}H/^{14}C ratios of 1.2, 4.4, and 3.0, respectively. It therefore appears that although the incorporation into growing cells of *C. butyricum* of exogenous long-chain fatty alcohols is small compared to the incorporation of long-chain fatty acids and aldehydes (Baumann *et al.*, 1965), fatty alcohols can be utilized for complex lipid synthesis. A portion of the alcohols is oxidized to fatty acids. Some radioactivity is found with a ^{3}H/^{14}C ratio identical to that of the starting material in compounds that chromatograph with alkylglycerols. The fraction of the polar lipid radioactivity that chromatographed with alk-1-enylglycerols had a ^{3}H/^{14}C ratio approximately two-thirds of that of the starting material. From these results, it seems possible that plasmalogens are synthesized in *C. butyricum* from alkyl glycerolipids but incorporation of an aldehyde into plasmalogens cannot be ruled out.

In another experiment (Hagen, 1970) in which *C. butyricum* was grown in the presence of 1-^{14}C-cetyl alcohol, the alk-1-enylglycerol and alkylglycerol fractions were isolated by TLC after $LiAlH_4$ reduction of the polar lipids. About 95% of the radioactivity in the alk-1-enylglycerol fraction after acid hydrolysis chromatographed with palmitaldehyde on GLC. However, only 26% of the radioactivity in the alkylglycerol fraction chromatographed with the 2,3-isopropylidene derivative of chimyl alcohol. TLC, GLC, and periodate oxidation showed that the remaining radioactivity was 51% alkane-1,2-diol and 23% alkane-1,3-diol. Neither a 16-carbon saturated alkane-1,2-diol nor a 16-carbon saturated 2-hydroxy acid served as precursors for plasmalogen biosynthesis in *C. butyricum* (Hagen, 1970). Both precursors were incorporated into polar lipids, but no radioactivity could be detected in the alk-1-enyl ether-linked chains.

About 0.14% of the total polar lipids of *C. butyricum* exist as alkyl glycerolipids (Hagen and Blank, 1970). The compositions of the aldehydes, alkyl ethers, and fatty acids in the polar lipids of *C. butyricum* are given in Table III. The aldehydic chains of the polar lipids were iso-

TABLE III

Comparison of the Chain Composition of Phospholipid-Bound Fatty Aldehydes, Fatty Acids, and Alkylglycerols of C. butyricum[a]

Chain length, no. of double bonds	Wt. %		
	Aldehydes	Alkylglycerols	Fatty acids
14:0	1	7.0	5.0
16:0	37	50	54
18:0	2.7	7.4	2.4
16:1 + 17:cyc	44	19	31
18:1 + 19:cyc	16	17	8.2
Total saturated	39	64	61
Total unsaturated	60	36	39

[a] Polar lipids were treated with hydrochloric acid and the aldehydes liberated were isolated by TLC. The phosphatides were reduced with $LiAlH_4$ and the resulting alkylglycerols and alcohols were isolated by TLC. The isopropylidene derivatives of the alkylglycerols and the acetate derivatives of the alcohols were further purified by TLC. The aldehydes, alcohol acetates, and isopropylidenes of the alkylglycerols were analyzed by GLC on a 6 ft × ⅛ in. column packed with 10% EGSS-X on 100–120 mesh Gas-Chrom P at column oven temperatures of 175°, 190°, and 200°C, respectively.

lated after hydrochloric acid hydrolysis (Anderson *et al.*, 1969) and the remaining lipids were reduced by $LiAlH_4$. The chain compositions of the alkylglycerols and the alcohols derived from the esterified fatty acids were determined by GLC after separation of these compounds by TLC and preparation of their 2,3-isopropylidene and acetate derivatives, respectively. The difference in composition between the aldehydogenic chains and the alcohol chains of alkyl glycerolipids argues against transformation of the alkyl moiety to the alk-1-enyl moiety in plasmalogens by dehydrogenation in this organism, unless a selected pool of these precursors is utilized.

Recently, Kim *et al.* (1970) and Kamio *et al.* (1970b) have reported the presence of alkyl moieties in the polar glycerolipid fraction of a number of anaerobic bacteria. The percentage alkyl phosphatides varied between 0.2 and 3%. Alkyl glycerolipids were detected in all the organisms examined that contain plasmalogens with the exception of *Clostridium kainantoi* 182−72 (Kim *et al.*, 1970). *Selenomonas ruminantium* was shown by Kamio *et al.* (1970b) to incorporate radioactivity mainly into the $C_{11:1}$, $C_{9:0}$, and C_{13} fatty alcohol side chains of the alkyl glycerolipids when grown in the presence of 1-^{14}C-sodium valerate. When this organism is grown in the presence of 1-^{14}C-sodium caproate, the major labeled fatty alcohol components were $C_{12:0}$, $C_{12:1}$, and $C_{10:0}$. Previously, Kamio *et al.* (1970a) had shown that the major fatty acid and aldehyde components of plasmalogens from *S. ruminantium* grown under similar conditions in the presence of valerate were $C_{15:0}$, $C_{15:1}$, and $C_{17:1}$. In caproate-grown cells, the major fatty acid components in the total phosphatides were $C_{14:0}$ and $C_{15:1}$, while the aldehydic side chain was mainly $C_{16:1}$. In *S. ruminantium*, as in *C. butyricum*, there appears to be no precursor–product relationship between the alcohol side chains of the alkyl glycerolipids and the fatty aldehyde side chains of plasmalogens, but the existence of a small pool of alkyl glycerolipids that serve as precursors of plasmalogens cannot be ruled out.

E. Interconversion of Long-Chain Fatty Acids, Fatty Aldehydes, and Fatty Alcohols

The experiments with long-chain fatty acids, fatty aldehydes, and fatty alcohols described above suggested an interconversion of these compounds in growing cultures of *C. butyricum*. Recently, Day *et al.* (1970) reported that a 100,000*g* supernatant fraction of extracts from *C. butyricum* can convert palmityl CoA to palmitaldehyde and cetyl alcohol. NADH appears to be required for both reduction steps. Partial reversibility of this reaction was shown by incubation of a crude extract of *C. butyricum* with 1-^{3}H-1-^{14}C-palmitaldehyde. In one experiment, 15.2% of the added aldehyde was oxidized to fatty acid and 11.8% reduced to alcohol. Oxidation of cetyl alcohol has so far not been demonstrated in cell-free preparations of this organism. Free alcohols were present in the neutral lipid fraction of cells harvested in the late logarithmic phase (Day *et al.*, 1970).

V. Summary

It is becoming apparent that plasmalogens and alkyl glycerolipids are not confined to animal tissues and a few bacteria, but these lipids are present in varying quantities in most anaerobic bacteria. Ether lipids are presumably located in the membranes of anaerobic bacteria, but their specific functions are not clear.

With the animal cells and tissues studied, evidence has accumulated suggesting that dihydroxyacetone phosphate and long-chain fatty alcohols are the precursors for alkyl glycerolipids. The mechanism of synthesis is not fully understood, but the postulated reaction sequence involves the formation of a 1-acyl dihydroxyacetone phosphate, which is replaced by a long-chain fatty alcohol with the formation of a 1-alkyl dihydroxyacetone phosphate. The keto group is reduced with NADPH or NADH followed by acylation. The resulting 1-alkyl-2-acylglycerol 3-phosphate, or phospholipids derived from it, may then be dehydrogenated, with the formation of the corresponding plasmalogens.

In bacterial systems, however, evidence against such a pathway is accumulating. The work of Hill and Lands (1970) showed that a keto intermediate in the formation of plasmalogens from glycerol is highly unlikely in *C. butyricum.* This points to a noninvolvement of dihydroxyacetone phosphate in the synthesis of these lipids in this organism. The observation that phosphatidic acid is virtually plasmalogen-free in *C. butyricum* makes it further unlikely that 1-alk-1-enyl-2-acylglycerol 3-phosphate is a direct precursor of its plasmalogens.

Direct dehydrogenation of alkyl glycerolipids to the corresponding plasmalogens also appears unlikely in the bacterial systems so far studied. No incorporation of label from either chimyl alcohol or batyl alcohol was observed when *C. butyricum* was grown in the presence of these compounds, and the marked differences in the chain compositions of alkyl and alk-1-enyl moieties in glycerolipids of both *C. butyricum* and *S. ruminantium* indicate that a precursor–product relationship between these lipids is unlikely.

The evidence obtained from the studies on bacterial plasmalogen biosynthesis points, however, to a direct involvement of preformed diacyl phosphatides. The precursor–product relationship between diacyl phosphatides and plasmalogens seen in incorporation studies with $^{32}P_i$ and ^{14}C-acetate, and turnover experiments with $^{32}P_i$ support this contention. The mechanism of synthesis is, however, not understood, but it seems unlikely that the conversion of the diacyl phosphatide to the plasmalogens occurs by direct conversion of an acyl to an alk-1-enyl linkage. Existing data suggest that the synthesis of bacterial plasmalogens occurs by a replacement reaction between the acyl chain in the 1-position of the phosphatides and a long-chain fatty aldehyde.

In view of the large evolutionary gap between the anaerobic bacteria, the only primitive organisms in which plasmalogens are found, and animals, it is likely that different mechanisms for plasmalogen biosynthesis have evolved. With few exceptions, plasmalogens have not been found in fungi, algae, or plants. It seems probable that the biosynthetic pathway to the plasmalogens in the primitive anaerobic bacteria was lost during the stages of evolution leading up to the animals. In these organisms, which could have obtained long-chain alcohols and aldehydes from the anaerobic bacteria by ingestion or later via the rumen bacteria, new pathways may have developed to utilize these alcohols and aldehydes for the formation of alkyl and alk-1-enyl glycerolipids. *De novo* synthesis may also have arisen as the utility of these lipid types in the animal cell membrane became established. These speculations, while somewhat premature at this stage of our ignorance of the detailed mechanisms of alk-1-enyl ether lipid synthesis in anaerobic bacteria and in animals, are not totally unfounded, since similar alterations in biosynthetic pathways during evolution have already been proven for unsaturated fatty acids, tyrosine, and nicotinic acid (Goldfine and Bloch, 1963), and lysine (Vogel, 1965).

REFERENCES

Allison, M. J., Bryant, M. P., Katz, I., and Keeney, M. (1962). *J. Bacteriol.* **83,** 1084.

Anderson, R. E., Garrett, R. D., Blank, M. L., and Snyder, F. (1969). *Lipids* **4,** 327.

Baumann, N. A., Hagen, P-O., and Goldfine, H. (1965). *J. Biol. Chem.* **240,** 1559.

Bell, O. E., Jr., and White, H. B., Jr. (1968). *Biochim. Biophys. Acta* **164,** 441.

Bell, O. E., Jr., Blank, M. L., and Snyder, F. (1971). *Biochim. Biophys. Acta.* **231,** 579.

Bickerstaffe, R., and Mead, J. F. (1967). *Biochemistry* **6,** 655.

Bickerstaffe, R., and Mead, J. F. (1968). *Lipids* **3,** 317.

Blank, M. L., Wykle, R. L., Piantadosi, C., and Snyder, F. (1970). *Biochim. Biophys. Acta* **210,** 442.

Chung, A. E., and Goldfine, H. (1965). *Nature (London)* **206,** 1253.

Day, J. I. E., Goldfine, H., and Hagen, P-O. (1970). *Biochim. Biophys. Acta* **218,** 179.

Debuch, H. (1966). *Hoppe-Seyler's Z. Physiol. Chem.* **344,** 83.

Debuch, H., and Winterfeld, M. (1970). *Hoppe-Seyler's Z. Physiol. Chem.* **351,** 179.

Debuch, H., Friedemann, H., and Müller, J. (1970). *Hoppe-Seyler's Z. Physiol. Chem.* **351,** 613.

Ellingboe, J., and Karnovsky, M. L. (1967). *J. Biol. Chem.* **242,** 5693.

Goldfine, H. (1962). *Biochim. Biophys. Acta* **59,** 504.

Goldfine, H. (1963). *Fed. Proc., Fed. Amer. Soc. Exp. Biol.* **22,** 415.

Goldfine, H. (1964). *J. Biol. Chem.* **239,** 2130.

Goldfine, H. (1966). *J. Biol. Chem.* **241,** 3864.

Goldfine, H. (1970). Unpublished results.

Goldfine, H., and Bloch, K. (1961). *J. Biol. Chem.* **236,** 2596.

Goldfine, H., and Bloch, K. (1963). *In* "Control Mechanisms in Respiration and Fermentation" (B. Wright, ed.), pp. 81–103. Ronald Press, New York.

Goldfine, H., and Panos, C. (1971). *J. Lipid Res.* **12,** 214.

Gray, G. W., and Wilkinson, S. G. (1965). *J. Gen. Microbiol.* **39,** 385.

Hagen, P-O. (1970). Unpublished results.

Hagen, P-O., and Blank, M. L. (1970). *J. Amer. Oil Chem. Soc.* **47,** 83A.

Hagen, P-O., and Goldfine, H. (1967). *J. Biol. Chem.* **242,** 5700.

Hagen, P-O., and Goldfine, H. (1966). Unpublished experiments.

Hajra, A. K. (1969). *Biochem. Biophys. Res. Commun.* **37,** 486.

Hajra, A. K. (1970). *Biochem. Biophys. Res. Commun.* **39,** 1037.

Hildebrand, J. G., and Law, J. H. (1964). *Biochemistry* **3,** 1304.

Hill, E. E., and Lands, W. E. M. (1970). *Biochim. Biophys. Acta* **202,** 209.

Kamio, Y., Kanegasaki, S., and Takahashi, H. (1969). *J. Gen. Appl. Microbiol.* **15,** 439.
Kamio, Y., Kanegasaki, S., and Takahashi, H. (1970a). *J. Gen. Appl. Microbiol.* **16,** 29.
Kamio, Y., Kim, K. C., and Takahashi, H. (1970b). *J. Gen. Appl. Microbiol.* **16,** 291.
Kanegasaki, S., and Takahashi, H. (1968). *Biochim. Biophys. Acta* **152,** 40.
Kates, M. (1964). *Advan. Lipid Res.* **2,** 17.
Katz, I., and Keeney, M. (1964). *Biochim. Biophys. Acta* **84,** 128.
Keeney, M., Katz, I., and Allison, M. J. (1962). *J. Amer. Oil Chem. Soc.* **39,** 198.
Kim, K. C., Kamio, Y., and Takahashi, H. (1970). *J. Gen. Appl. Microbiol.* **16,** 321.
Macfarlane, M. G. (1962). *Nature (London)* **196,** 136.
Meyer, H., and Meyer, F. (1971). *Biochim. Biophys. Acta* **231,** 93.
O'Leary, W. M. (1967). "The Chemistry and Metabolism of Microbial Lipids." World Publ. Co., Cleveland, Ohio.
Randle, C. L., Albro, P. W., and Dittmer, J. C. (1969). *Biochim. Biophys. Acta* **187,** 214.
Rapport, M. M., and Norton, W. T. (1962). *Annu. Rev. Biochem.* **31,** 103.
Scheuerbrandt, G., Goldfine, H., Baronowsky, P. E., and Bloch, K. (1961). *J. Biol. Chem.* **236,** PC70.
Schmid, H. H. O., Bandi, P. C., Mangold, H. K., and Baumann, W. J. (1969). *Biochim. Biophys. Acta* **187,** 208.
Snyder, F., Malone, B., and Wykle, R. L. (1969a). *Biochem. Biophys. Res. Commun.* **34,** 40.
Snyder, F., Wykle, R. L., and Malone, B. (1969b). *Biochem. Biophys. Res. Commun.* **34,** 315.
Snyder, F., Malone, B., and Blank, M. L. (1969c). *Biochim. Biophys. Acta* **187,** 302.
Snyder, F., Rainey, W. T., Jr., Blank, M. L., and Christie, W. H. (1970a). *J. Biol. Chem.* **245,** 5853.
Snyder, F., Blank, M. L., and Malone, B. (1970b). *J. Biol. Chem.* **245,** 4016.
Snyder, F., Malone, B., and Blank, M. L. (1970c). *J. Biol. Chem.* **245,** 1790.
Snyder, F., Malone, B., and Cumming, R. B. (1970d). *Can. J. Biochem.* **48,** 212.
Snyder, F., Blank, M. L., and Wykle, R. L. (1971). *J. Biol. Chem.* **246,** 3639.
Stoffel, W., Le Kim, D., and Heyn, G. (1970). *Hoppe-Seyler's Z. Physiol. Chem.* **351,** 875.
Thompson, G. A., Jr. (1966). *Biochemistry* **5,** 1290.
Thompson, G. A., Jr. (1968). *Biochim. Biophys. Acta* **152,** 409.
Vogel, H. J. (1965). *In* "Evolving Genes and Proteins" (V. Bryson and H. J. Vogel, eds.), p. 25. Academic Press, New York.
Wegner, G. H., and Foster, E. M. (1963). *J. Bacteriol.* **85,** 53.
White, D. C. (1968). *J. Bacteriol.* **96,** 1159.
Wood, R., and Healy, K. (1970). *J. Biol. Chem.* **245,** 2640.
Wykle, R. L., and Snyder, F. (1969). *Biochem. Biophys. Res. Commun.* **37,** 658.
Wykle, R. L., and Snyder, F. (1970). *J. Biol. Chem.* **245,** 3047.
Wykle, R. L., Blank, M. L., and Snyder, F. (1970). *FEBS Let.* **12,** 57.

CHAPTER XV

ETHER-LINKED LIPIDS IN EXTREMELY HALOPHILIC BACTERIA

Morris Kates

I. Introduction

Investigation of the lipids of halophilic bacteria was initiated in 1958 at the National Research Council, Ottawa, on the suggestion of Dr. N. E.

Gibbons, then head of the Microbiology Section of the Division of Biosciences. Dr. Gibbons and his group had been engaged for many years in physiological and biochemical studies of halophilic bacteria, particularly their salt requirements, and suspected that these organisms might possess an unusual membrane structure to help them cope with high salt concentrations in their environment.

An early study by his group (Smithies *et al.*, 1955) had dealt only with the total lipid content of the cell envelopes of several species of halophiles. A detailed analysis of the lipids of a moderate halophile, *Micrococcus halodenitrificans,* grown in salt concentrations ranging from 0.55 to 1.0 *M* was subsequently undertaken by the author and a post-doctoral fellow, Dr. S. N. Sehgal, from Dr. Gibbons' group. The results of this study (Kates *et al.*, 1961) were rather disappointing since the lipids turned out to be not very different from those of nonhalophiles [e.g., *Bacillus cereus* (see Kates, 1964)] and little or no influence of salt concentration on lipid composition, and hence on membrane composition, was detected.

It became apparent that we would have to examine the lipids of a halophile growing in much higher concentrations of salt if we were to find any unusual membrane lipids. Such organisms, the extremely halophilic bacteria that grow optimally in 25–30% salt, were no strangers to Dr. Gibbons' laboratory, and indeed had been studied there for many years; furthermore, a number of these extreme halophiles were readily available in the type-culture collection of the National Research Council, Ottawa.

Dr. Sehgal and the present author then proceeded, early in 1959, to extract and analyze the lipids of a typical extremely halophilic bacterium, *Halobacterium cutirubrum*; we soon realized that we were dealing with strange and unusual lipids (Sehgal *et al.*, 1962). First, the total lipids extracted had a deep red color due to the presence of large amounts of red carotenoid pigments; the polar lipids could, however, be easily separated from these pigments by precipitation with acetone. Secondly, chromatography revealed several phosphatide components, none of which corresponded to any normal bacterial phosphatides. Finally, after methanolysis and saponification of the lipids, large amounts of unsaponifiable material, but only traces of fatty acids were obtained. The absence of fatty acid esters was confirmed by the negative hydroxamate—$FeCl_3$ test for ester groups and the lack of an ester C=O band at 1730 cm^{-1} in the infrared spectrum of the total lipids (Fig. 1). The first positive clue to the identity of these lipids came from the infrared spectrum of the unsaponifiable material, which showed a strong C—O—C ether band at 1110 cm^{-1} (Fig. 4).

These observations led to the hypothesis, fully validated by subsequent investigations, that *H. cutirubrum* cells contain a new class of lipids having ether-linked alkyl groups instead of ester-bound fatty acids. This chapter will deal with the elucidation of the complete structures of the main alkyl ether-derived lipids of extremely halophilic bacteria, and will also describe studies bearing on the metabolism of these lipids. However, before proceeding, a brief description of the characteristics of extremely halophilic bacteria will be given.

II. The Extremely Halophilic Bacteria

A. General Characteristics and Classification

The extremely halophilic bacteria are a group of unusual organisms that *require* saturated or nearly saturated salt solutions for their growth and survival (Larsen, 1962, 1967; Kushner, 1964, 1968). These organisms are commonly found in salted or preserved foods, in salted hides, in salt lakes such as the Great Salt Lake, and in the Dead Sea. They also occur in concentrated brines, in salt flats formed by evaporation of seawater, and in the common salt obtained from these salt flats, to all of which these organisms impart a characteristic red or pink color. The coloration is attributed to the presence in most extreme halophiles of red, orange, or pink carotenoid pigments, one of which, bacterioruberin, has recently been identified as a C_{50}-tetrahydroxy carotenoid (Kelly and Liaaen Jensen, 1967; Kelly *et al.*, 1970).

Two distinct groups of extremely halophilic bacteria are known: the rod-shaped "*Halobacterium* group" assigned to the family Pseudomonadaceae, and the coccoid "Sarcina-Micrococcus group" assigned to the family Micrococcaceae [see Larsen (1962) and Kushner (1968) for a discussion of the taxonomy of halophilic bacteria]. Bacteria in the first group are gram-negative, obligate aerobes, requiring at least 12% salt for growth, the optimal salt concentration being 25–30%. They are non-spore-formers and, if motile, possess polar flagella. The halophilic cocci are gram-negative or gram-variable spheres, 1–1.5 μ in diameter, occurring singly, in pairs, or in packets (sarcinae). They are all obligate aerobes, non-spore-forming, and nonmotile, and require at least 5–10% NaCl for growth, the optimal concentration being 20–25%, with growth occurring in up to 30% salt. It has been suggested that the extremely halophilic cocci be accorded generic rank with the name *Halococcus* (see Larsen, 1962, 1967).

B. Salt Requirements

The extremely halophilic bacteria require high concentrations of salt for growth and preservation of their structure. In cultures of halobacteria, growth of the cells ceases when the salt concentration is decreased below 3 *M*, and the rods gradually lose their regular shape and become spherical; below 1 *M* NaCl the cell envelopes begin to disintegrate and the cells lyse rapidly (Larsen, 1962, 1967). The concentration of NaCl required for growth and maintenance of shape of the extreme halophiles ranges between 3 and 5 *M*. Concentrations of 0.1–0.5 *M* Mg^{2+} and 1.3–2.5 $\times 10^{-3}$ *M* K^+ are also required for optimal growth. No salt has been found that replaces NaCl to any significant extent for optimal growth (Kushner, 1968).

The loss of structural rigidity and lysis of intact cells and cell envelopes upon salt depletion is not completely understood at present, but it is known that it does not depend on enzymic or osmotic processes. It is generally assumed that a high charge density exists on the cell envelope and that removal or reduction of the shielding counterions leads to disruption of the surface structure (Brown, 1963; Kushner *et al.*, 1964; Kushner, 1968). However, this mechanisms of lysis does not explain the specific NaCl requirement for growth and maintenance of the rod shape of the cells. Stoeckenius and Rowen (1967) have shown by electron microscopy that the cell envelope consists of an outer cell wall layer as well as a cytoplasmic membrane, and that stepwise reduction in salt concentration caused lysis of the cell wall before, and independent of, membrane lysis.

It is interesting that the extreme halophiles possess very high internal salt concentrations, as high as that of their external environment. However, these cells are not freely permeable to all ions but are able to actively concentrate K^+ and to exclude Na^+. For example, the internal concentrations of Na^+ and K^+ in *H. salinarium* are 1.37 *M* and 4.57 *M*, respectively, whereas the external concentrations are 4.0 *M* Na^+ and 0.03 *M* K^+ (Christian and Waltho, 1962). The high internal concentration of K^+ may be related to the fact that this ion (as well as Mg^{2+} is required for maintaining the ribosomes in their physiologically active 70 S form (Bayley and Kushner, 1964); also, most, if not all, of the known enzymes in extreme halophiles require high concentrations of salts for optimum activity, and K^+ is usually more effective than Na^+ (Larsen, 1962; Kushner, 1968).

C. Cellular Localization of Lipids

A detailed discussion of cellular structure of extreme halophiles and localization of the lipids is beyond the scope of this article; the reader may, however, consult the reviews by Larsen (1962, 1967) and Kushner (1964, 1968) for further details. One of the difficulties in discussing this subject is that the structure of the cell envelope of extreme halophiles has not yet been established unambiguously. It would appear reasonable to assume that the external layer of the envelope, which in the electron microscope shows a regular hexagonal arrangement of spherical particles, is the cell wall and that beneath this is the cytoplasmic membrane (see Stoeckenius and Rowen, 1967; Stoeckenius and Kunau, 1968; Kushner, 1968). The cell wall, however, is unusual in that it contains no muramic acid or diaminopimelic acid, but only glucosamine and hexoses (Kushner, 1968). The cell membrane is undoubtedly lipoprotein in nature (Kushner *et al.*, 1964), but it is not known whether the lipids in the cell envelope, which in *H. cutirubrum* account for more than 93% of the total cellular lipids (Kushner *et al.*, 1964), are associated entirely with the cytoplasmic membrane or also partly with the cell wall. The studies of Stoeckenius and Kunau (1968) on *H. halobium* cell fractions show, however, that the cell wall in this organism is lipid-free, as are the intracytoplasmic gas vacuole membranes, and that the cellular lipids are associated largely, if not entirely, with the cytoplasmic membrane.

III. Chemistry of the Ether-Linked Lipids of Extremely Halophilic Bacteria

A. Characterization of Total Cellular Lipids

1. *Overall Composition*

The total lipid material extracted from *Halobacterium cutirubrum*, the first extremely halophilic bacterium examined, was deep red, due to the presence of carotenoid pigments (chiefly bacterioruberin), and accounted for about 2% of the cell's dry weight (Sehgal *et al.*, 1962); in subsequent studies total lipid contents were found to be in the range 2.5–4% (Kates *et al.*, 1965a; Joo *et al.*, 1968). The total lipid-P amounted to 3.9–4.6% (Sehgal *et al.*, 1962; Kates *et al.*, 1965a; Joo *et al.*, 1968), indicating a high phosphatide content. Precipitation of the lipids with cold acetone yielded an acetone-insoluble tan precipitate of polar lipids accounting for about 90% by weight of the total lipids, and an acetone-soluble, "neutral lipid" fraction amounting to about 10% of the total lipids (Sehgal *et al.*, 1962; Joo *et al.*, 1968; Tornabene *et al.*, 1969).

The total lipids and the polar lipid fraction had low nitrogen contents (%N, 0.24 and 0.18, respectively; N/P atomic ratio, 0.13 and 0.09, respectively; Sehgal *et al.*, 1962) indicating the presence of only traces of nitrogenous lipids. A lipopeptide has, however, been detected in small amounts in *Halobacterium halobium*, along with traces of two unidentified nitro-

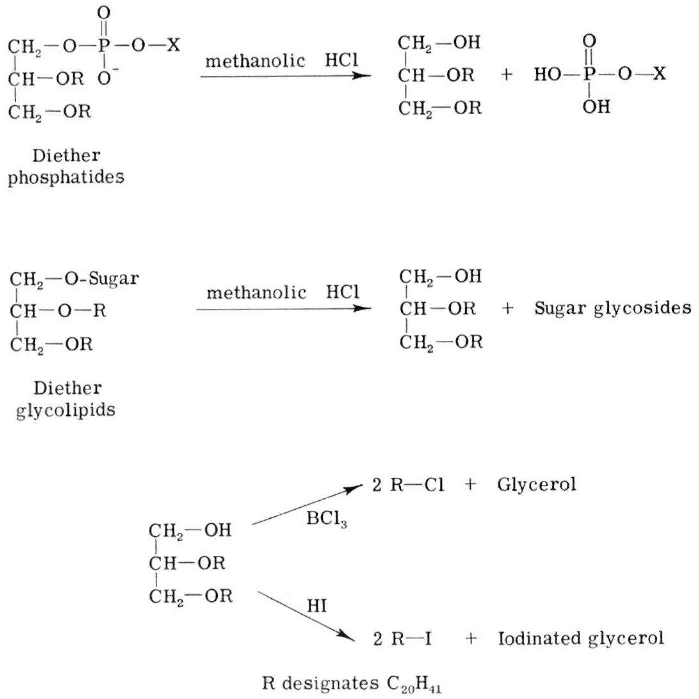

SCHEME I. Chemical degradation of diether phosphatides and glycolipids and of the dialkyl glycerol ether moiety.

genous phosphatides (Marshall and Brown, 1968). Nevertheless, the low content or virtual absence of nitrogenous lipids in extreme halophiles is unusual, since all nonhalophilic Eubacteria examined contain major amounts of nitrogenous phosphatides, usually phosphatidylethanolamine or lipoamino acids (Kates, 1964).

Another unusual feature of the lipids of *H. cutirubrum* became apparent when it was found (Sehgal *et al.*, 1962) that no fatty acids or water-soluble phosphate esters were released after mild alkaline hydrolysis or "deacylation" by Dawson's (1954) procedure. Even more drastic hydrolysis procedures, such as heating in boiling 0.7 *N* methanolic HCl for 2–4 hr and then in boiling 0.1 *N* NaOH for 1 hr, released only traces of fatty acids and long-chain aldehydes; surprisingly, about 80% of the lipids was recovered as a phosphorus-free unsaponifiable material in the ether extract of this alkaline hydrolysate, and 96% of the lipid-P was recovered in the methanol-water phase as organically bound phosphorus (Sehgal *et al.*, 1962; Kates *et al.*, 1965a). These findings indicated that the lipids of this extreme halophile were not derivatives of diacylglycerols, as in normal bacteria, but were probably derived from the long-chain "unsaponifiable" compound (see Scheme I).

2. *Infrared Spectrum*

The absence of fatty acid ester groups was completely confirmed by the infrared absorption spectrum of the total lipids (Fig. 1, Spectrum 1), which showed no ester absorption in the expected range (1730–1750 cm^{-1}). However, the spectrum did show strong absorption bands indicative of long-chain groups (2920, 2850, and 1460 cm^{-1}), OH groups (3300 cm^{-1}, broad), and phosphate ester groups (P=O, 1260–1230 cm^{-1}; P—O—C, 1060–1090; P—O^-, 1100 cm^{-1}). The band at 1100–1110 cm^{-1} might have been interpreted as indicating an ether C—O—C group, as it indeed later turned out to be, but overlapping of the P—O^- band at 1100 cm^{-1} precluded the identification of an ether group at this stage of investigation. Another band that proved to be important, but the signi-

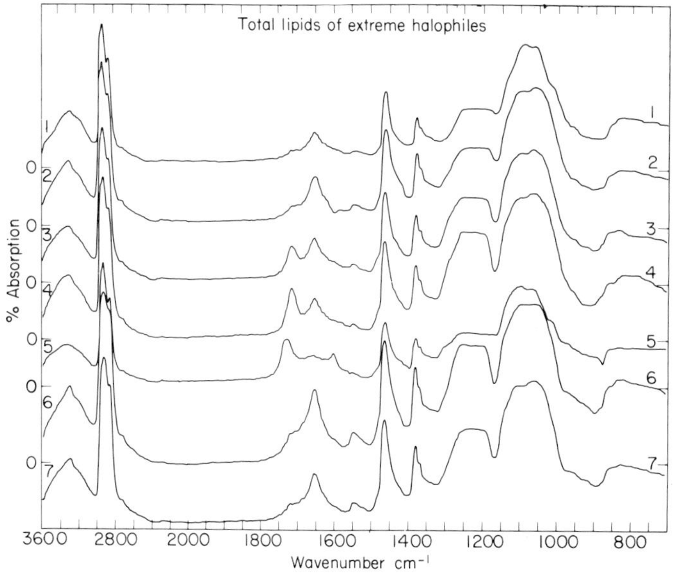

FIG. 1. Infrared spectra, in chloroform solution, of total lipids of extremely halophilic bacteria: (1) *H. cutirubrum*, (2) *H. halobium* M, (3) *H. halobium* P, (4) *H. salinarium*, (5) *S. litoralis*, (6) halophilic *Sarcina* sp., and (7) halophile A-2C. [From Kates *et al.* (1966b). Copyright (1966) by the American Chemical Society. Reprinted by permission of the copyright owner.]

TABLE I

Lipid Analysis of Halophilic and Nonhalophilic Bacteria[a]

Sample No.	Organism	Color of cells	NaCl conc. (%)	Lipid content (% of cell dry wt.)	Lipid analyses (% of total lipids): P	Unsaponifiable material	Fatty acids
	Extreme Halophiles						
1	*H. cutirubrum*	Red	25	3.5	4.3	70	0.6
2	*H. halobium* M	Pink	25	4.1	4.3	74	0.4
3	*H. halobium* P	Red	25	2.6	3.8	69	0.3
4	*H. salinarium*	Red	25	3.6	3.4	67	0.7
5	*S. literalis*	Yellow	25	—	3.7	65	2.3
6	*Sarcina* sp.	Colorless	25	3.5	4.2	68	0.3
7	Halophile A-2C	Colorless	25	3.5	4.4	71	0.3
	Moderate Halophiles						
8	Halophile A-31C	Colorless	5.85	2.3	3.8	14	60
9	*M. halodenitrificans* (48 hr culture)	Colorless	5.85	—	3.3	0.5	63
10	*M. halodenitrificans* (20 hr culture)	Colorless	5.85	6.0	3.7	9	68
11	*V. costicolus*	Colorless	5.85	9.4	3.8	5	61
	Nonhalophiles						
12	*S. lutea*	Yellow	0	7.2	2.9	14	68
13	*S. flava*	Yellow	0	5.9	3.3	14	60

[a] Data taken from Kates *et al.* (1966b). All cells were grown in the medium for extreme halophiles (Sehgal *et al.*, 1962), except for the 48 hr culture of *M. halodenitrificans*, which was grown in a proteose peptone-tryptone medium (Kates *et al.*, 1961).

ficance of which was not at first recognized, was the doublet at 1365–1375 cm^{-1} indicative of an isopropyl group.

3. *Comparison of Lipids of Extreme and Moderate Halophiles and Nonhalophiles*

The question whether *H. cutirubrum* was unique in having lipids deficient in fatty acid ester groups, or whether other extreme halophiles, moderate halophiles, or even nonhalophiles also contained such lipids arose at this point and was investigated in detail (Kates *et al.*, 1966b). Surveying the lipid composition of several extreme halophiles (halobacteria and halococci) and a few moderately halophilic bacteria, we found that all of the extreme halophiles had rather lower total lipid contents (2–4%) and higher lipid-P contents than the moderate or nonhalophiles (Table I). More striking, however, was the finding that the lipids of all of the extreme halophiles yielded only traces of fatty acids (0.3–2%) and very high amounts of unsaponifiable material (65–75%) after hydrolysis; in contrast, the moderate and nonhalophiles released large amounts of fatty acids and only small amounts of unsaponifiable material (Table I). The trace amount of fatty acids in *H. halobium* was also reported at about the same time by Cho and Salton (1966).

The low fatty acid content of the extreme halophiles was again apparent from the infrared spectra of their total lipids (Fig. 1), which showed little or no ester carbonyl absorption at 1730 cm^{-1}; the latter absorption band was, however, very prominent in the spectra of the ester-containing lipids of moderate and nonhalophiles (Fig. 2). Furthermore, the spectra

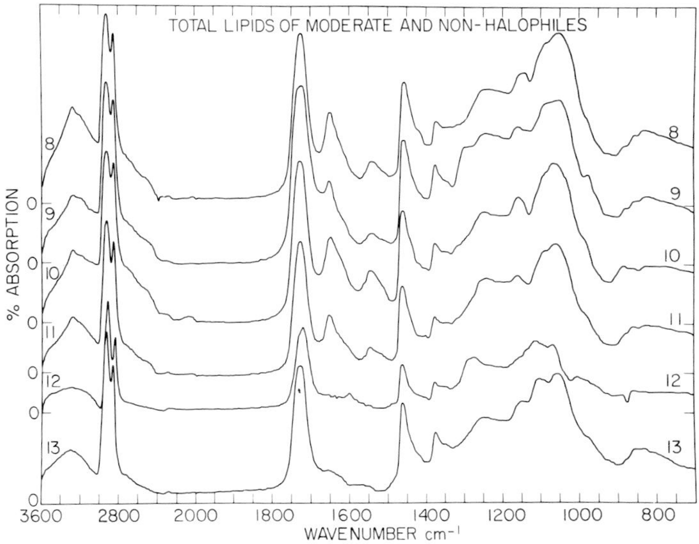

FIG. 2. Infrared spectra, in chloroform solution, of total lipids of moderately and nonhalophilic bacteria: (8) halophile A-31C, (9) *M. halodenitrificans*, 48 hr culture, (10) *M. halodenitrificans*, 20 hr culture, (11) *V. costicolus*, (12) *S. lutea*, and (13) *S. flava*. [From Kates *et al.* (1966b). Copyright (1966) by the American Chemical Society. Reprinted by permission of the copyright owner.]

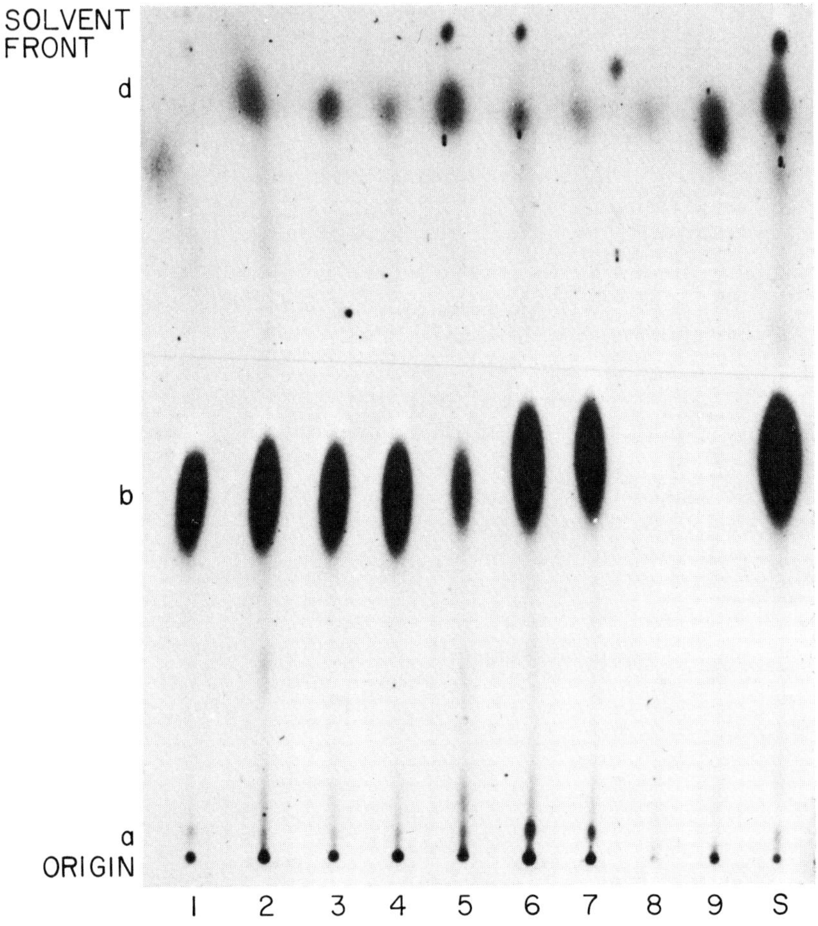

FIG. 3. Thin-layer chromatogram of unsaponifiable material from extremely halophilic (1–7), and moderately halophilic (8–9), bacteria; identity of bacteria 1–9 is given in Figs. 1 and 2. S, synthetic 2,3-di-*O*-phytanylglycerol; a, glycerol monoalkyl ethers; b, di-*O*-phytanylglycerol; d, nonpolar components (pigments, squalenes). Solvent: chloroform–ethyl ether (9:1).

of the lipids of the extreme halophiles were virtually identical to that of the *H. cutirubrum* lipids described above (Fig. 1, Spectrum 1). These results strongly suggested that the lipids of extremely halophilic bacteria, as a group, have unique and novel structural features that are not present in those of any other bacteria examined. It may be noted that the colorless rod, designated "halophile A-31C", which had been maintained in a 25% NaCl medium (Table I), was first taken to be an extreme halophile; however, after it was shown to contain high amounts of fatty acids and little unsaponifiable material (Table I), a reexamination of its salt requirements showed that it grew best in 7.5–15% salt and hence was a moderate halophile. The other colorless rod, designated "halophile A-2C," was found to grow best in 25% NaCl and not at all below 15% NaCl, and hence was an extreme halophile, consistent with the absence of fatty acid esters in its lipids (Kates *et al.*, 1966b).

B. Identification of the Ether-Containing Lipid Moiety as 2,3-Di-*O*-phytanyl-*sn*-glycerol

1. *Isolation and Characterization of the Unsaponifiable Material*

The ether-soluble unsaponifiable material obtained after methanolic HCl hydrolysis of the total lipids of the extreme halophiles listed in Table

TABLE II

Analytical and Physical Data for the Glyceryl Diether Moiety of Lipids of Extreme Halophiles

Data	Found				Calc. for $C_{43}H_{88}O_3$
	Sample 1[a]	Sample 2[b]	Sample 3[c]	Sample 4[d]	
% C	78.67	79.60	79.43	—	79.07
% H	12.98	13.44	13.87	—	13.58
Molecular weight	—	660	—	—	653.1
Iodine value	—	1.5	—	8	0
Alkyl group: Glycerol, mole ratio[e]	—	2.0:0.9	—	—	2.0:1.0
Alkyl group: free OH, mole ratio[f]	—	—	—	2.0:1.15	2.0:1.0
$[\alpha]_D^{20}$ (in $CHCl_3$)	—	+7.8°	+8.4°	+7.5°, +8°	—

[a] Data from Sehgal *et al.* (1962).
[b] Data from Kates *et al.* (1963b, 1965a)
[c] Data from Joo *et al.* (1968).
[d] Data of Faure *et al.* (1963, 1964).
[e] Estimated after cleavage with BCl_3 (see Scheme I).
[f] Estimated by quantitative acetylation of the intact unsaponifiable compound followed by colorimetric ester analysis.

I was a viscous colorless oil which revealed one major component* on TLC in chloroform-ether (9:1) with an R_f corresponding to that of a long-chain monohydroxy compound (Fig. 3; Table II); this compound was present *only* in the unsaponifiable material from the extreme halophiles. After elution from the TLC plate, or purification by chromatography on a column of silicic acid (Kates *et al.*, 1965a; Joo *et al.*, 1968) the unsaponifiable compound had an infrared spectrum (Fig. 4) showing the presence of OH group(s) (3580 cm^{-1}), long-chain groups with terminal isopropyl groups (2920, 2850, 1465, 1385–1375 cm^{-1}), ether C—O—C groups (1110 cm^{-1}), and primary alcoholic C—O groups (1040 cm^{-1});

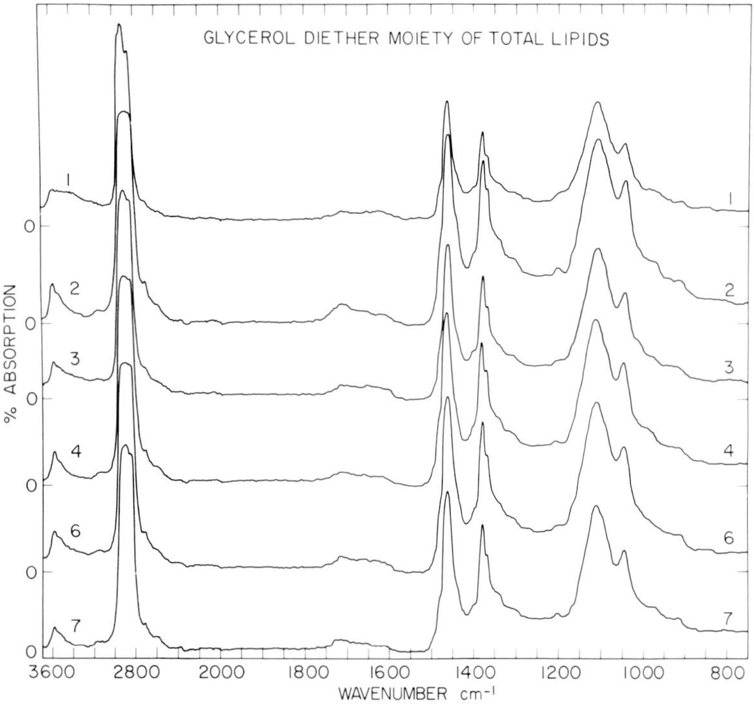

FIG. 4. Infrared spectra, in carbon tetrachloride solution, of purified glycerol diether moiety of total lipids from (1) *H. cutirubrum*, (2) *H. halobium* M, (3) *H. halobium* P, (4) *H. salinarium*, (6) *Sarcina* sp., and (7) halophile A-2C. Insufficient material from *S. litoralis* was available for spectral determination.

* A minor slow-moving component was also observed in the unsaponifiable material (Fig. 3), and was later identified as a mixture of 2-*O*- and *sn*-3-*O*-phytanylglycerol (Joo *et al.*, 1968). These monoethers are now known to be formed during methanolic-HCl hydrolysis of the lipids by cleavage of one of the ether linkages (Kates *et al.*, 1971).

no bands indicative of C=C double bonds were present. This spectrum was very similar to that of synthetic α,β-dioctadecyl glycerol ether (Kates *et al.*, 1963a) except for the presence of the isopropyl band at 1385–1375 cm^{-1}. Analytical data showed that the unsaponifiable compound had a low iodine value and possessed a free, acetylatable OH group (Faure *et al.*, 1963, 1964; Table II). Furthermore, treatment of the compound with BCl_3 to cleave the ether groups (Scheme I), yielded an alkyl chloride and free glycerol in the mole ratio of 2:1 (Table II). These

TABLE III

Gas–Liquid Chromatographic Data for Derivatives of Alkyl Group in Glyceryl Diether of H. cutirubrum[a]

Compound	Apiezon L at 197° C		Butanediol succinate polyester at 197° C	
	Relative retention	Carbon number	Relative retention	Carbon number
1-Octadecyl chloride[b]	1.00	18.00	—	—
Bacterial alkyl chloride	0.852	17.62	—	—
1-Octadecyl iodide[b]	—	—	1.00	18.00
3,7,11,15-Tetramethyl-1-hexadecyl iodide	—	—	0.818	17.33
Bacterial alkyl iodide	—	—	0.818	17.33
Octadecan-1-ol[b]	1.00	18.00	1.00	18.00
3,7,11,15-Tetramethyl-hexadecan-1-ol	0.878	17.69	0.843	17.44
Bacterial alcohol	0.875	17.69	0.842	17.44
Octadecyl acetate[b]	1.00	18.00	1.00	18.00
3,7,11,15-Tetramethyl-hexadecyl 1-acetate	0.816	17.52	0.791	17.24
Bacterial acetate	0.818	17.52	0.791	17.24
Methyl octadecanoate[b]	1.00	18.00	1.00	18.00
Methyl 3,7,11,15-tetramethyl-hexadecanoate	0.785	17.43	0.739	17.01
Bacterial methyl ester	0.785	17.43	0.740	17.01
Octadecane[b]	1.00	18.00	1.00	18.00
3,7,11,15-Tetramethylhexadecane	0.965	17.92	0.960	17.86
Bacterial hydrocarbon	0.964	17.92	0.960	17.86

[a] Data from Kates *et al.* (1965a).

[b] Retention of reference compounds relative to methyl octadecanoate, on Apiezon L and butanediol succinate polyester, respectively: 1-octadecyl chloride, 1.28, –; 1-octadecyl iodide, –, 1.80; octadecan-1-ol, 1.12, 1.78; 1-octadecyl acetate, 1.42, 1.36; octadecane, 0.37, 0.14.

findings suggested that the unsaponifiable compound was a saturated long-chain dialkyl ether of glycerol, and the analytical data (Table II) revealed a molecular formula of $C_{43}H_{88}O_3$, corresponding to a di-C_{20}-alkyl ether of glycerol.

The presence of an absorption band in the infrared spectrum at 1040 cm^{-1} was attributed to a primary C—O– bond and the high optical rotation (Table II) further suggested that the glycerol was asymmetrically substituted and that the unsaponifiable compound was an α,β-di-*O*-alkylglycerol. The dextrorotatory rotation of the diether, as was pointed out by Faure *et al.* (1963), suggested that it had the L-α,β- or *sn*-2,3-configuration since synthetic D-α,β-glycerol diethers, as well as D-α,β-diglycerides are levorotatory (Kates *et al.*, 1963a). This was an unusual situation since the diglyceride moieties of naturally occurring glycerophosphatides were known to have the D-α,β- or *sn*-1,2-configuration.

2. *Identification of Alkyl Groups as Phytanyl Groups*

At this stage the identity of the ether-linked alkyl groups was not yet established: the C and H analyses of partially purified samples suggested C_{17} or C_{18} chains (Sehgal *et al.*, 1962; Table II), while later, more highly

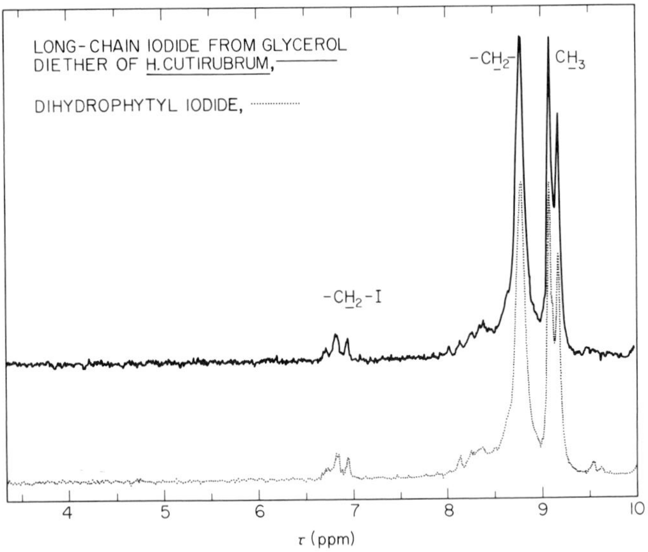

FIG. 5. Nuclear magnetic resonance spectrum of alkyl iodide derived from the glycerol diether of *H. cutirubrum*, compared with that of phytanyl iodide (in CCl_4). [From Kates *et al.* (1965a). Copyright (1965) by *Biochimica Biophysica Acta*. Reprinted by permission of the copyright owner.]

purified samples gave us data for C_{20} chains (Kates *et al.*, 1963b, 1965a); GLC relative retention data of the diether on SE-52 at 220°C suggested an average carbon number of 17.2 for the chains (Kates *et al.*, 1963a,b). The diether was then subjected to cleavage with HI (Scheme I), which yields the alkyl groups as alkyl iodides. The alkyl iodide fraction gave only one peak on GLC with a carbon number of 17.3 (Table III), as did the alkyl iodides obtained from the glycerol diethers of each of the extremely halophilic bacteria examined (Kates *et al.*, 1966b).

The nonintegral carbon number obtained suggested that the alkyl group was probably branched, and this was confirmed by the NMR spectrum of the alkyl iodide (Fig. 5) which showed a doublet at 9.1–9.2 ppm (τ) indicative of C—CH_3 groups. A signal at 8.8 ppm for aliphatic CH_2 groups, and a triplet at 6.8 ppm indicative of a primary CH_2—I group were also prominent but no signal at 4.9 ppm for –C═C—H was detected. Integration of the areas under these peaks showed that the ratio of CH_3 to CH_2 to CH_2I groups was 5:10:1, suggesting that the alkyl iodide was a saturated C_{20} primary iodide with four C—CH_3 branches. The infrared spectrum of the iodide (Fig. 6A) was consistent with this structure, and also showed the characteristic bands for a *gem*-dimethyl group: a doublet at 1385–1375 cm^{-1} and a band at 1170 cm^{-1}.

A saturated isoprenoid arrangement of the methyl branches was therefore considered to be most probable, and a comparison was undertaken between the bacterial alkyl iodide and dihydrophytyl iodide. The latter was prepared by hydrogenation of the plant C_{20}-isoprenoid alcohol, phytol, and conversion of the resulting dihydrophytol to the iodide by treatment with HI. The two were completely identical with respect to TLC mobility, GLC relative retention (Table III), NMR spectrum (Fig. 5), and infrared spectrum (Fig. 6A). Chain lengths of C_{25} (Faure *et al.*, 1963) and C_{24} (Faure *et al.*, 1964) have been reported for the alkyl iodide obtained from the lipids of *H. cutirubrum*, on the basis of analysis for iodine only. It is doubtful whether such analyses are sufficiently sensitive to establish the chain length of the alkyl group within a few carbon atoms. Furthermore, we have never observed peaks corresponding to C_{24}, C_{25}, or longer chains on GLC chromatograms of the alkyl iodides from any of the extreme halophiles studied. The influence of culture conditions can probably be eliminated, since the strain of organism (obtained from the NRC collection) and the culture medium and growth conditions used by Faure *et al.* (1963, 1964) were the same as those in our studies.

To confirm the identification of the alkyl group as a dihydrophytyl (or phytanyl) group the alkyl iodide was converted to the corresponding acetate, free alcohol, carboxylic acid and methyl ester, and finally hydrocarbon, as shown in Scheme II, and these were compared to the corresponding phytanyl derivatives prepared from dihydrophytol (Kates *et al.*, 1965a). All derivatives of the bacterial alkyl group proved to be identical to the corresponding dihydrophytyl derivatives with respect to infrared spectra (Fig. 6) and GLC retentions on two liquid phases (Table III).

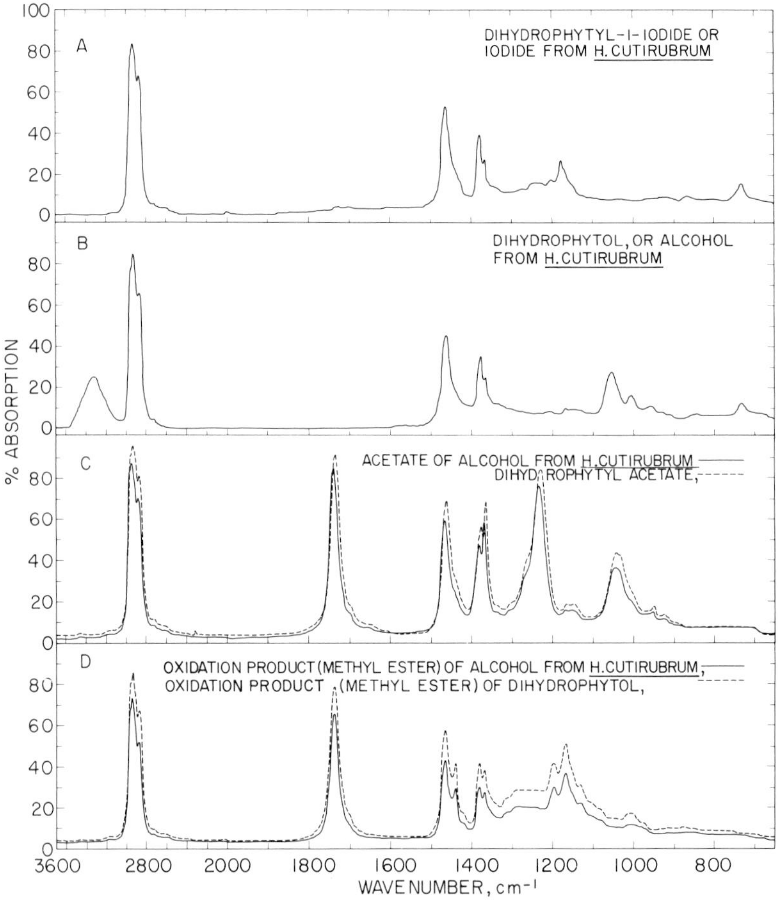

FIG. 6. Infrared spectra of derivatives of the alkyl group in the glycerol diether of *H. cutirubrum* lipids compared to the corresponding phytanyl derivatives. A, the bacterial alkyl iodide, or phytanyl iodide; B, the bacterial alcohol or phytanol; C, the bacterial alcohol acetate or phytanyl acetate; and D, the bacterial methyl ester or methyl phytanate. A and B as liquid films, C and D in CCl_4. [From Kates *et al.* (1965a). Copyright (1965) by *Biochimica Biophysica Acta*. Reprinted by permission of the copyright owner.]

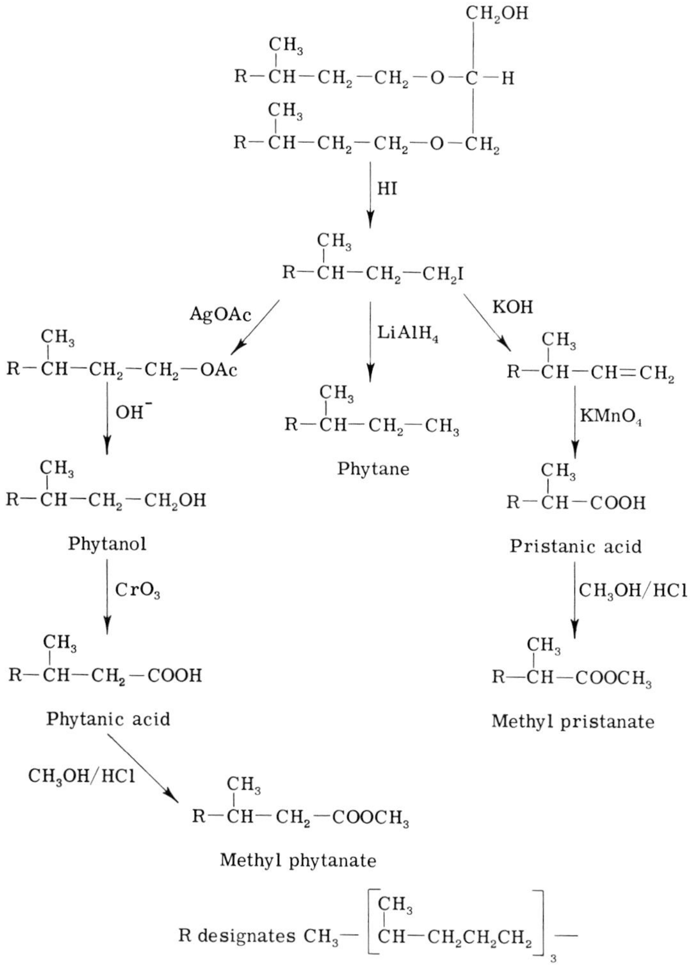

SCHEME II. Conversion of ether-linked phytanyl groups to various phytanyl derivatives.

Finally, the mass spectrum of the free alcohol was shown to be identical to that of dihydrophytol (Fig. 7). These results, therefore, establish the structure of both alkyl groups in the bacterial dialkyl glycerol ether as 3,7,11,15-tetramethylhexadecyl (phytanyl) groups.

3. *Absolute Stereochemical Configuration of Phytanyl Groups*

Since carbon atoms 3, 7, and 11 in the phytanyl group are asymmetrically substituted, the question now arose as to their configuration. The

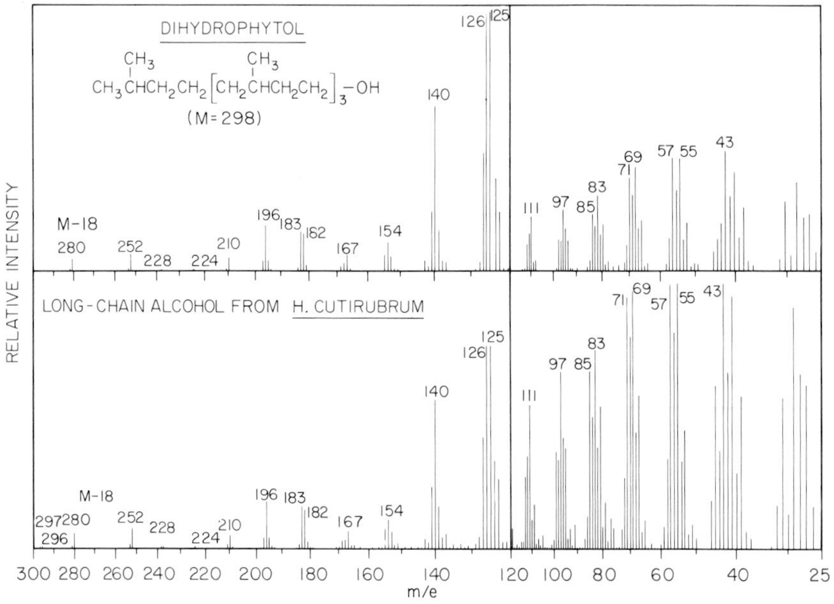

FIG. 7. Mass spectrum of the long-chain alcohol derived from the glycerol diether of *H. cutirubrum* lipids compared with that of authentic phytanol. [From Kates *et al.* (1965a). Copyright (1965) by *Biochimica Biophysica Acta*. Reprinted by permission of the copyright owner.]

approach used to determine the configuration at C-3 was based on the fact that in monomethyl-branched carboxylic acids of the same absolute configuration the molecular rotations alternate in sign and decrease in magnitude as the methyl branch is moved from the 2-position down the chain. Thus, in the D-series, a 2-methyl alkanoic acid has $M_D - 28°$, a 3-methyl acid has $+13°$, a 4-methyl acid has $-2°$, and the 5-, 6-, and 7-methyl acids have rotations close to zero; these correlations are essentially independent of chain length (Abrahamson *et al.*, 1963). Thus, the rotation of a multibranched acid will be determined by the configuration of the asymmetric center closest to the carboxyl group, the contributions from the asymmetric centers further down the chain being negligible.

With this in mind, the bacterial phytanyl group was converted (Kates *et al.*, 1967b) to the corresponding phytanic (3,7,11,15-tetramethylhexadecanoic) and pristanic (2,6,10,14-tetramethylpentadecanoic) acids and their respective methyl esters by the reactions shown in Scheme II. Each methyl ester was shown to consist of a single stereoisomer by GLC on an open tubular column with polyester coating (Ackman and Hansen,

TABLE IV

Molecular Rotations and Configurations of Bacterial Phytanyl Group Derivatives[a]

Compound	M_D	Configuration assigned
Phytanic acid (bacterial)	+11°	3-D (3*R*)
D-3-Methylnonadecanoic acid (synthetic)[b]	+13°	3-D (3*R*)
Methyl pristanate (bacterial)	−37°	2-D (2*R*)
Methyl D-2-methyloctadecanoate (synthetic)[b]	−38°	2-D (2*R*)
C_{18}-Ketone (bacterial)	+1.6°	6-D, 10-D (6*R*, 10*R*)
6-D, 10-D, 14-Trimethylpentadecan-2-one (from phytol)	+1.7°	6-D, 10-D (6*R*, 10*S*)
Pristane (bacterial)	0.0°	6-D, 10-D (6*R*, 10*S*)

[a] Data from Kates *et al.* (1967b).
[b] Data from Abrahamsson *et al.* (1963).

1967). The phytanic acid had $M_D + 11°$, a value close to that of synthetic D-3-methylnonadecanoic acid, and the methyl pristanate had $M_D - 37°$, almost identical to that of synthetic methyl D-2-methyloctadecanoate (Table IV). These results show that the C-3 and C-2 positions in phytanic and pristanic acids, respectively, have the D-configuration, and hence that the C-3 in the bacterial phytanyl group has the D or in absolute terms, the *R* configuration.

The approach used to establish the configurations at C-7 and C-11 was to convert the bacterial phytanyl group to the C_{18}-ketone, 6,10,14-trimethylpentadecanone-2. The two diastereoisomers of this ketone (the 6*R*,10*R*- and 6*S*,10*R*-isomers) had previously been synthesized and their optical rotations were known; furthermore it was known that the C_{18}-ketone obtained by oxidation of phytol was the 6*R*,10*R*-isomer (Burrell *et al.*, 1959). Conversion of the bacterial phytanyl group to the C_{18}-ketone was achieved by Barbier-Wieland degradation of the bacterial methyl pristanate as shown in Scheme III. The ketone was isolated as the bisulfite addition compound and proved to have a specific rotation identical to that of the 6*R*,10*R*-isomer derived from phytol and isolated in the same way (Table IV). These results show that the C-7 and C-11 centers in the bacterial phytanyl group have the *R* (or D) configuration. In confirmation of this, the pristane derived from the methyl pristanate as shown in Scheme III had no measurable rotation and was therefore the *meso*, 6*R*,10*S*-isomer. These results thus establish the bacterial phy-

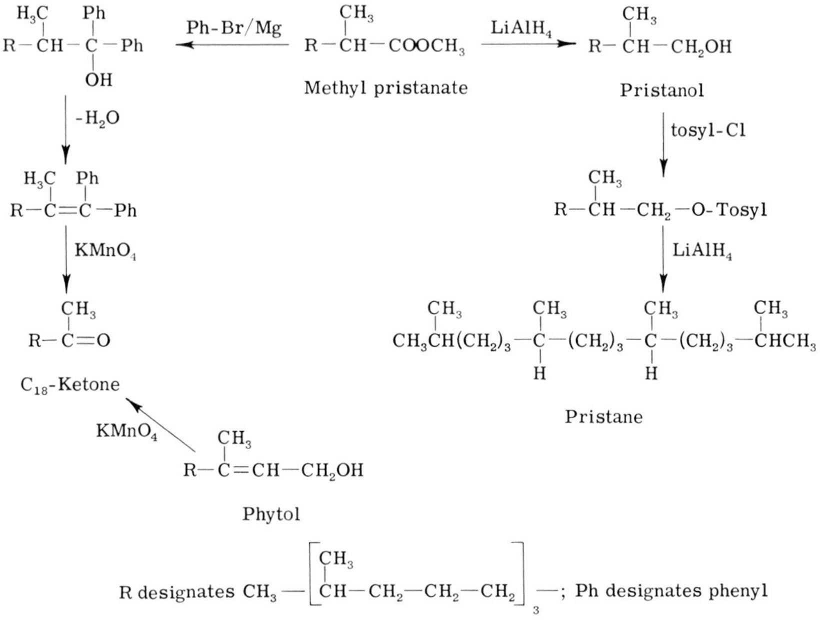

SCHEME III. Conversion of methyl pristanate to pristane and the C_{18}-ketone.

tanyl group as being 3*R*,7*R*,11*R*,15 (or 3-D,7-D,11-D,15)-tetramethylhexadecyl (**1**):

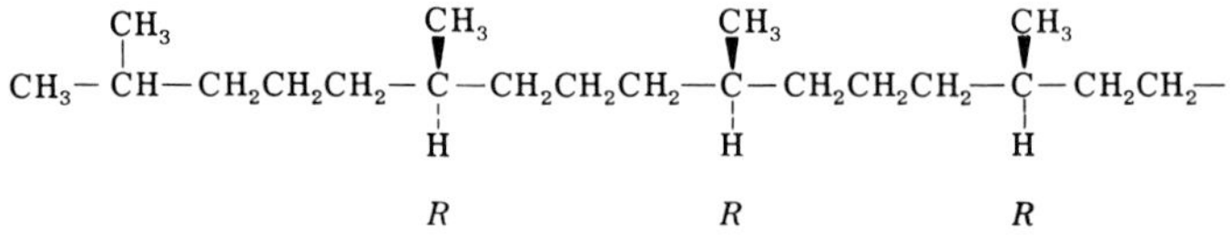

(**1**)

4. *Structure and Configuration of the Diphytanyl Glycerol Ether*

The evidence so far has established the structure of the ether-containing unsaponifiable moiety of the lipids of extreme halophiles as a di-*O*-phytanylglycerol. By analogy with the diglyceride moiety of natural phosphatides and glycosyl diglycerides it was assumed that the diether would have the D-α,β(*sn*-1,2-)-structure and configuration. However, the bacterial diether was found to be dextrorotatory (Table II), whereas synthetic D-α,β-(*sn*-1,2-)-dialkyl glycerol ethers are levorotatory (Kates *et al.*, 1963a). The natural diether was therefore considered to have the L-α,β- or *sn*-2,3-structure and configuration.

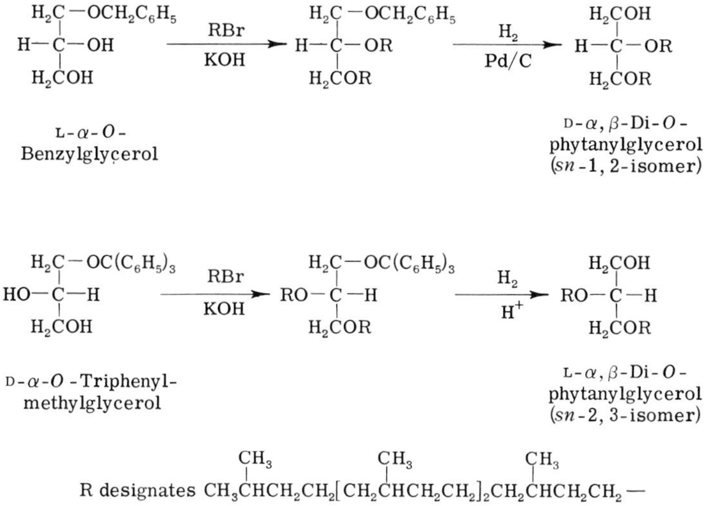

SCHEME IV. Synthesis of *sn*-1,2- and *sn*-2,3-di-*O*-phytanylglycerol.

To confirm this assignment, the synthesis of both the D- and L-stereoisomers of α,β-di-*O*-phytanylglycerol was carried out by the unambiguous procedures described for the synthesis of straight-chain dialkyl glycerol ethers (Kates *et al.*, 1963a), as shown in Scheme IV. It is important to note that in our first synthesis (Kates *et al.*, 1965b) the phy-

TABLE V

Properties of Synthetic and Natural α,β-Di-O-phytanylglycerols

Isomer	$[\alpha]_D$[a]	TLC, R_f[b] 1	2	3	GLC, relative retention[c]
Natural	+8.4°	0.24	0.40	0.55	2.11
Synthetic					
L-α,β (*sn*-2,3)[d]	+8.5°	0.24	0.40	0.55	2.10
L-α,β (*sn*-2,3)[e]	+7.6°	0.24	0.40	0.55	2.10
D-α,β (*sn*-1,2)[e]	−7.0°	0.24	0.40	0.55	2.15

[a] In chloroform. Data from Kates *et al.* (1965b) and Joo *et al.* (1968).

[b] In solvents: 1, chloroform; 2, chloroform-ethyl ether (20:1); 3, chloroform-ethyl ether (9:1); data from Kates *et al.* (1965b, 1966b) and Joo *et al.* (1968).

[c] Relative to 1,2-di-*O*-hexadecylglycerol on SE-52 at 220° C. Data from Kates *et al.* (1965b).

[d] Contains bacterial 3-D, 7-D, 11-D-phytanyl group (Joo *et al.*, 1968).

[e] Contains 3-DL, 7-D, 11-D-phytanyl group derived from phytol (Kates *et al.*, 1965b).

tanyl group used was the mixture of 3-D,7-D,11-D and 3-L,7-D,11-D isomers derived from phytol, while in a later study (Joo *et al.*, 1968) the bacterial phytanyl group (consisting only of the 3-D,7-D,11-D-isomer) was used. Both the D- and the L-stereoisomers of the synthetic α,β-diphytanyl glycerol ether had the same mobilities on TLC in several solvents, the same relative retention on GLC (Table V), and the same infrared spectrum (Fig. 6) as the bacterial diether. However, only the L- or *sn*-2,3-isomer containing the D,D,D-phytanyl groups had an optical rotation identical in magnitude and sign to that of the natural diether (Table V). Therefore the bacterial diether has the structure and configuration, 2,3-di-*O*-(3′*R*,7′*R*,11′*R*,15-tetramethylhexadecyl)-*sn*-glycerol (**2**):

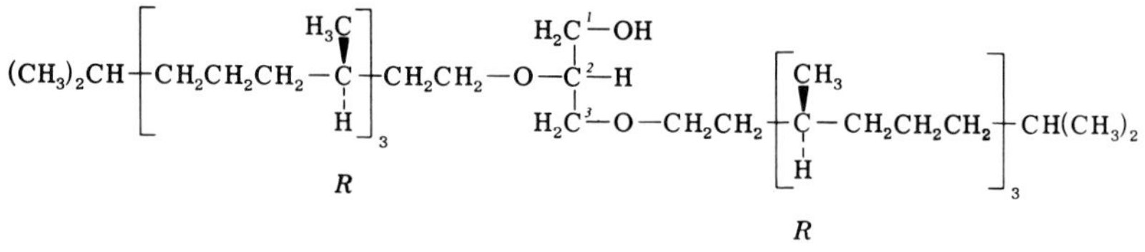

(2)

C. Identification of Polar Lipid Components

1. *Composition and Characterization of the Polar Lipids*

Chromatography of the lipids of *H. cutirubrum* on silica-gel-loaded paper in the solvent system diisobutyl ketone–acetic acid–water (40:25:5) revealed the presence of two major components (spots 2 and 6, Fig. 8) and four or five minor components (spots 1, 3, 4, 5 and 8, Fig. 8) (Sehgal *et al.*, 1962; Kates *et al.*, 1966b). The two major components and one or more of the minor components were later found to be present in all of the extremely halophilic bacteria examined (Fig. 8), but not in the moderately halophilic or nonhalophilic bacteria (Kates *et al.*, 1966b). The same major components were found by Marshall and Brown (1968) in *H. halobium*, but several additional minor components, including two unidentified nitrogenous lipids, were also detected.

The components were characterized by their R_f values on silica-loaded paper and on TLC in several solvents, by their staining behavior towards specific reagents, and by their lipid-P content (Table VI). In all of the extreme halophiles, the major component (spot 6, Fig. 8) was a non-nitrogenous, acidic phosphatide that accounted for more than 85% of the lipid-P and 60% of the weight of polar lipids; as described below, it proved to be the diphytanyl ether analog of phosphatidylglycerophosphate. The second most abundant component (spot 2, Fig. 8) was a non-

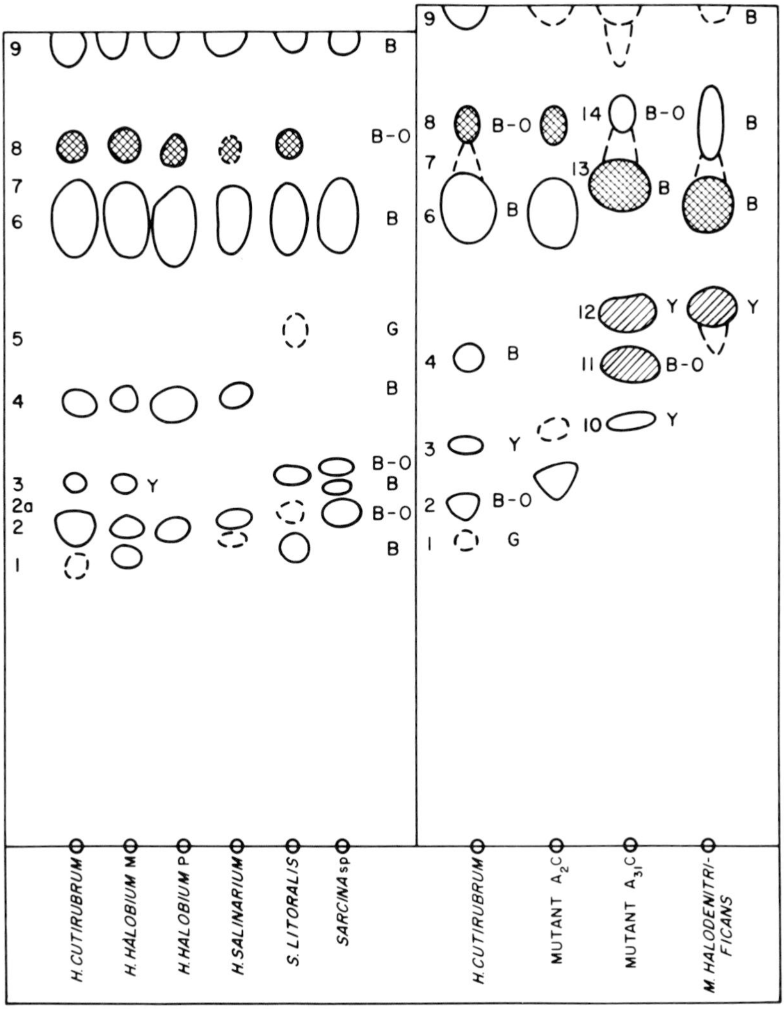

FIG. 8. Chromatogram of total lipids of extremely halophilic and moderately halophilic bacteria on silicic acid-impregnated paper; solvent, diisobutyl ketone–acetic acid–water (40:25:5; v/v). Identity of components (see Table VI): (1, 2a, 4, 5, 7, 10) unidentified; (2) glycolipid sulfate; (3) glycolipid; (6) phosphatidylglycerophosphate (diether form); (8) phosphatidylglycerol (diether form); (9) pigments and neutral lipids; (11) phosphatidylaminoacylglycerol (diester form); (12) phosphatidylethanolamine (diester form); (13) phosphatidylglycerol (diester form); and (14) diphosphatidylglycerol (diester form). Chromatograms were stained with Rhodamine 6G (Y, yellow; B, blue; O, orange, G, gray), ninhydrin (singly hatched spots), and periodate-Schiff reagent (doubly hatched spots). [From Kates *et al.* (1966b). Copyright (1966) by the American Chemical Society. Reprinted by permission of the copyright owner.]

TABLE VI

Composition and Characterization of Polar Lipids

Spot No. (Fig. 8)	R_f[a] 1	R_f[a] 2	R_f[a] 3	R_f[a] 4	Staining behavior[b] Rhodamine 6G	Staining behavior[b] HIO_4-Schiff	Staining behavior[b] Sugar stain	Lipid-P (% of total)[c]	Wt. % of total polar lipid[c]	Identity of components
1	0.25	0.03	0.08	0.12	Blue	−	+	—	Trace	Unidentified sulfolipid
2	0.35	0.10	0.16	0.18	Blue-orange	−	+	—	25	Glycolipid sulfate
3	0.45	0.27	—	0.23	Yellow	+	+	—	Trace	Glycolipid
4	0.55	0.30	0.36	0.30	Blue	−	−	6	4	Phosphatidylglycerol sulfate (diether analog[d])
5	0.65	—	—	—	Blue	−	−	Trace	Trace	Unidentified phosphatide
6	0.75	0.46	0.25	0.44	Blue	−	−	86	64	Phosphatidylglycerophosphate (diether analog)
8	0.85	0.60	0.70	0.34	Blue	+	−	5	4	Phosphatidylglycerol (diether analog)
9	0.95	0.95	0.95	0.95	Orange	−	−	Trace	—	Pigments + neutral lipids

[a] R_f values are for : 1, chromatography on silicic acid impregnated Whatman 3MM paper in diisobutyl ketone–acetic acid–water (40:25:5) (Kates *et al.*, 1966b); 2, 3, and 4, TLC on silica gel H in chloroform–methanol–90% acetic acid (30:4:20), chloroform–methanol–NH_4OH (65:35:5), and chloroform–methanol–water (65:35:5), respectively (Kates and Hancock, 1972).

[b] No ninhydrin-positive spot was detected; sugars were detected with the α-naphthol H_2SO_4 stain of Siakotos and Rouser (1965) on TLC plates.

[c] Analysis of components separated by TLC in solvent 3 (Kates and Hancock, 1972).

[d] Hancock and Kates (1972).

phosphatide which gave a positive sugar test and could be strongly labeled with ^{35}S (see Table XIII); this sulfur-containing glycolipid accounted for 35% by weight of the polar lipids, and proved to be a sulfate ester of a triglycosyl diphytanyl glycerol ether (see Section III, C,4).

Of the remaining components, spot 8 (Fig. 8) was a minor, acidic, periodate-Schiff-positive phosphatide that proved to be the diphytanyl ether analog of phosphatidylglycerol (Faure *et al.*, 1964; Kates *et al.*, 1966b). It is interesting that it had a higher mobility on silicic acid-impregnated paper (and also on TLC) than the corresponding straight-chain diacyl form, as is evident by comparing the R_f of the diether PG (spot 8) with that of the diester PG (spot 13) in Fig. 8. In general, all of the diphytanyl ether analogs appear to have higher mobilities than their ester counterparts, most likely because of the presence of the highly branched phytanyl groups.

Spot 4 is another minor acidic phosphatide which appeared to be associated with traces of ^{35}S (see Table XIII). Recent studies have, however, shown that this compound is actually a sulfate ester of the diether PG (Hancock and Kates, 1972). Spots 1, 3, and 5 are trace components, spot 1 being another as yet unidentified sulfolipid; spot 3 a glycolipid probably identical to the triglycosyl diether formed by desulfation of the glycolipid sulfate (spot 2); and spot 5 an unidentified acidic phosphatide (Table VI).

In our early studies (Sehgal *et al.*, 1962), some of the minor components were tentatively identified as phosphatidyl inositol, lysolecithin, and lecithin, on the basis of their R_f values and staining behavior only. None of these phosphatides has subsequently been identified unambiguously in the lipids of extreme halophiles, and it is highly unlikely that they actually exist in these bacteria.

2. *Phosphatidylglycerophosphate (Diphytanyl Ether Analog)*

Early studies of the structure of the major phosphatide (Sehgal *et al.*, 1962) of *H. cutirubrum* were carried out on material corresponding to spot 6 (Fig. 8) eluted from several chromatograms. Because of the small quantities of material obtained only qualitative analyses and tests could be performed, and these suggested that the major phosphatide was a polyglycerol phosphatide, probably a diether analog of cardiolipin. This conclusion was based largely on the fact that the major phosphatide was acidic, had a high R_f value similar to that of cardiolipin, and yielded the dialkyl glycerol ether and glycerol diphosphate after acid hydrolysis in a manner analogous to the hydrolysis of cardiolipin.

The major bacterial phosphatide component was then isolated (Kates

TABLE VII

Analytical Data for the Diphytanyl Glycerol Ether Analog of Phosphatidylglycerophosphate and Its Salts

Data	Free acid		Sodium salt		Potassium salt		Barium salt	
	Found[a]	Calc. for $C_{46}H_{96}O_{11}P_2$	Found[a]	Calc. for $C_{46}H_{94}O_{11}P_2Na_2$	Found[b]	Calc. for $C_{46}H_{94}O_{11}P_2K_2$	Found[a,b]	Calc. for $C_{46}H_{93}O_{11}P_2Ba_{3/2}$
Mol. wt.	—	887.2	—	930.7	—	962.8	—	1089
C (%)	—	62.27	59.50	59.35	57.11	57.40	—	50.6
H (%)	—	10.91	10.24	10.18	10.02	9.85	—	8.54
P (%)	6.80	6.98	6.66	6.66	6.43	6.45	5.61	5.69
M (%)	—	—	4.87	4.95	8.07	8.13	18.4	18.9
M/P (atomic ratio)	—	—	0.98	1.00	0.995	1.00	0.74	0.75
Diphytanyl glycerol ether (%)	—	73.5	67.7	69.8	—	—	—	—
Diether/P (mole ratio)	—	0.50	0.49	0.50	—	—	—	—
H^+ strong/P (atomic ratio)	0.96[a,c]	1.00	—	—	—	—	—	—
H^+ weak/P (atomic ratio)	0.42[a] 0.0[c]	0.50	—	—	—	—	—	—
OH/P (mole ratio)	—	0.50	—	0.50	0.51[c]	0.50	—	0.50

[a] Data from Kates *et al.* (1963b, 1965a).
[b] Data from Joo and Kates (1969).
[c] Data from Faure *et al.* (1963).

et al., 1963b, 1965a) by a procedure similar to that of Faure and Morelec-Coulon (1958) for preparation of cardiolipin. This procedure involved its precipitation as the barium salt by addition of a $BaCl_2$ solution to a methanol solution of the total lipids, followed by reprecipitation from chloroform solution by addition of acetone. The barium salt obtained was chromatographically homogeneous and was converted to the free acid and then to the sodium salt. Essentially the same procedure was used by Faure *et al.* (1963) to prepare the pure major bacterial phosphatide in the form of its potassium salt. In a later study (Joo and Kates, 1968, 1969), the major phosphatide was isolated by preparative TLC in chloroform–methanol–conc. ammonia (65:35:5) followed by chloroform–methanol–water (80:20:2), and converted to the sodium, potassium, and barium salts.

Analytical data for these salts (Table VII) were in excellent agreement with those expected for a phosphatidylglycerophosphate-type structure (**3**), $C_{46}H_{96}O_{11}P_2$, in which 1 mole of 2,3-di-*O*-phytanylglycerol is esterified with one of the phosphate groups of glycerol 1,3-diphosphate. Results of degradative studies (Scheme I) were consistent with this structure: hydrolysis with methanolic HCl (or 90% acetic acid) yielded 1 mole each of 2,3-di-*O*-phytanyl-*sn*-glycerol and a water-soluble phosphate ester that was isolated as the barium salt and identified as 1,3-glycerol diphosphate (Table VIII).

So far, however, the data did not establish unambiguously the phos-

TABLE VIII

Analytical and Physical Data for Water-Soluble Phosphate Ester Degradation Product of Phosphatidylglycerophosphate (Diphytanyl Ether Analog)[a]

Data	Barium salt of phosphate ester degradation product	Authentic 1,3-glycerol diphosphate ($C_3H_6O_9P_2Ba_2 \cdot 2H_2O$)[b]	1,2-Glycerol diphosphate[c]
P (%)	10.95	11.09	—
P/glycerol (mole ratio)	2.06	2.00	—
R_f in phenol-water (satd.)	0.17	0.17	0.15
R_f in butanol-acetic acid-water (5:3:1)	0.42	0.43	0.33

[a] Data from Kates *et al.* (1963b, 1965a).
[b] Synthesized according to Kates *et al.* (1966a).
[c] Synthesized according to Michelson (1959).

phatidylglycerophosphate structure (**3**) for the major phosphatide. Unexpectedly, the sodium and the potassium salts isolated had a cation:P atomic ratio of 1:1 (Table VII). Furthermore, only one ionizable acid group per atom of P (i.e., two acid groups per molecule of PGP) could be detected by titration with indicator dyes such as phenolphthalein (Faure *et al.*, 1963; Kates *et al.*, 1963b), although three acid groups per phosphatidylglycerophosphate molecule (**3**) were required. It was for this reason that Faure *et al.* (1963), who had also arrived essentially at a phosphatidylglycerophosphate structure, concluded that the bacterial phosphatide consisted of two molecules of phosphatidylglycerophosphate joined by a pyrophosphate bond between the terminal phosphate groups (**4**).

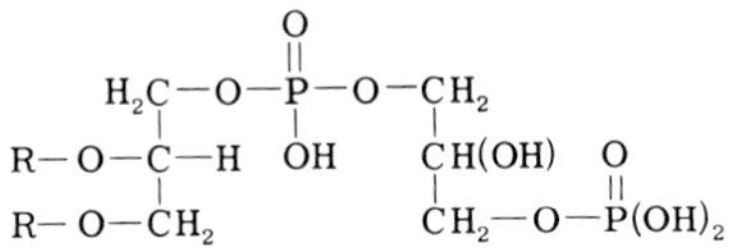

R = $C_{20}H_{41}$ (phytanyl)

(**3**)

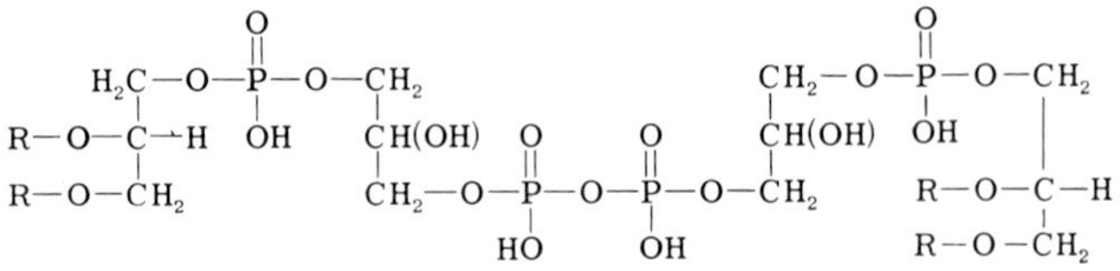

R = $C_{20}H_{41}$ (phytanyl)

(**4**)

However, potentiometric titration did reveal the presence of two strongly ionized acid groups with p*K* 3.25 and one weakly ionized group with p*K* 7.5–8 (Kates *et al.*, 1965a); furthermore, analysis of the barium salt showed it to contain three equivalents of Ba per two atoms of P (Table VII). The presence of three ionizable acid groups per two atoms of P has now been verified with the finding that, after treatment of the free acid form of the bacterial phosphatide with diazomethane, the methylated product contained three methyl ester groups per two atoms of P, as shown by its NMR spectrum, elementary analysis and analysis for $-OCH_3$ (Kates and Hancock, 1971). These data favor the phosphatidylglycerophosphate structure (**3**) rather than the pyrophosphate structure (**4**). The latter is further considered unlikely because of the absence of a P—O—P absorption band at 900–980 cm^{-1} in the infrared spectrum of the bacterial phosphatide (Fig. 9). Formation of only the disodium or

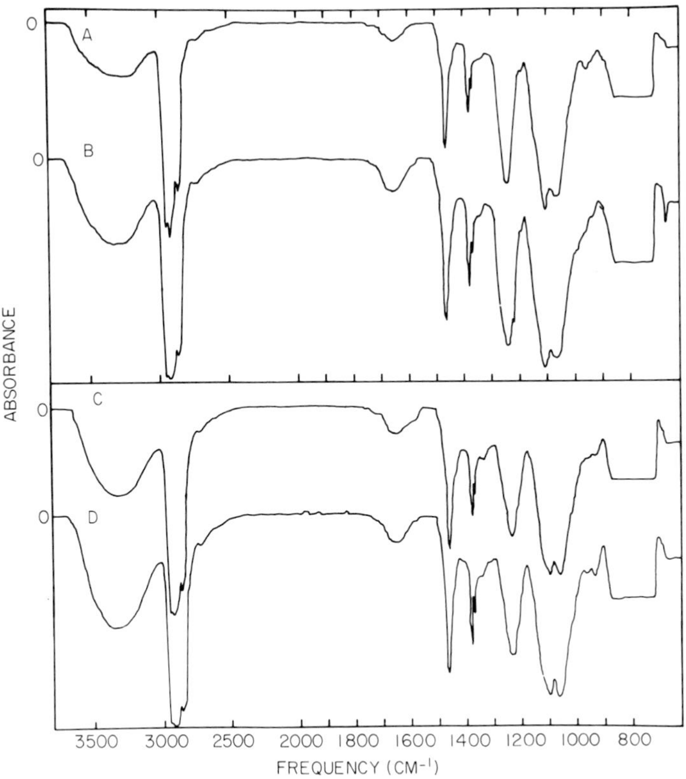

FIG. 9. Infrared spectra (in CCl_4) of A, natural and B, synthetic stereoisomers of the diphytanyl ether analog of phosphatidylglycerol (disodium salt); C, natural and D, synthetic stereoisomers of the diphytanyl ether analog of phosphatidylglycerol (monosodium salt). [From Joo and Kates (1969). Copyright (1969) by *Biochimica Biophysica Acta*. Reprinted by permission of the copyright owner.]

dipotassium salts by titration to the phenolphthalein endpoint can now be explained by the fact that the endpoint for the third acid group is above the pH range of this indicator. The trisodium or tripotassium salts can, however, be obtained after addition of three equivalents of the corresponding base per mole of free acid PGP (M. Kates and A. J. Hancock, unpublished results).

Although the configuration of the glycerol diether moiety of (**3**) was known, the configuration of the glycerophosphate moiety still remained to be determined. For this reason and also to establish the phosphatidylglycerophosphate structure (**3**) unambiguously, the synthesis of the two diastereomeric forms of (**3**) was undertaken (Joo and Kates, 1968, 1969)

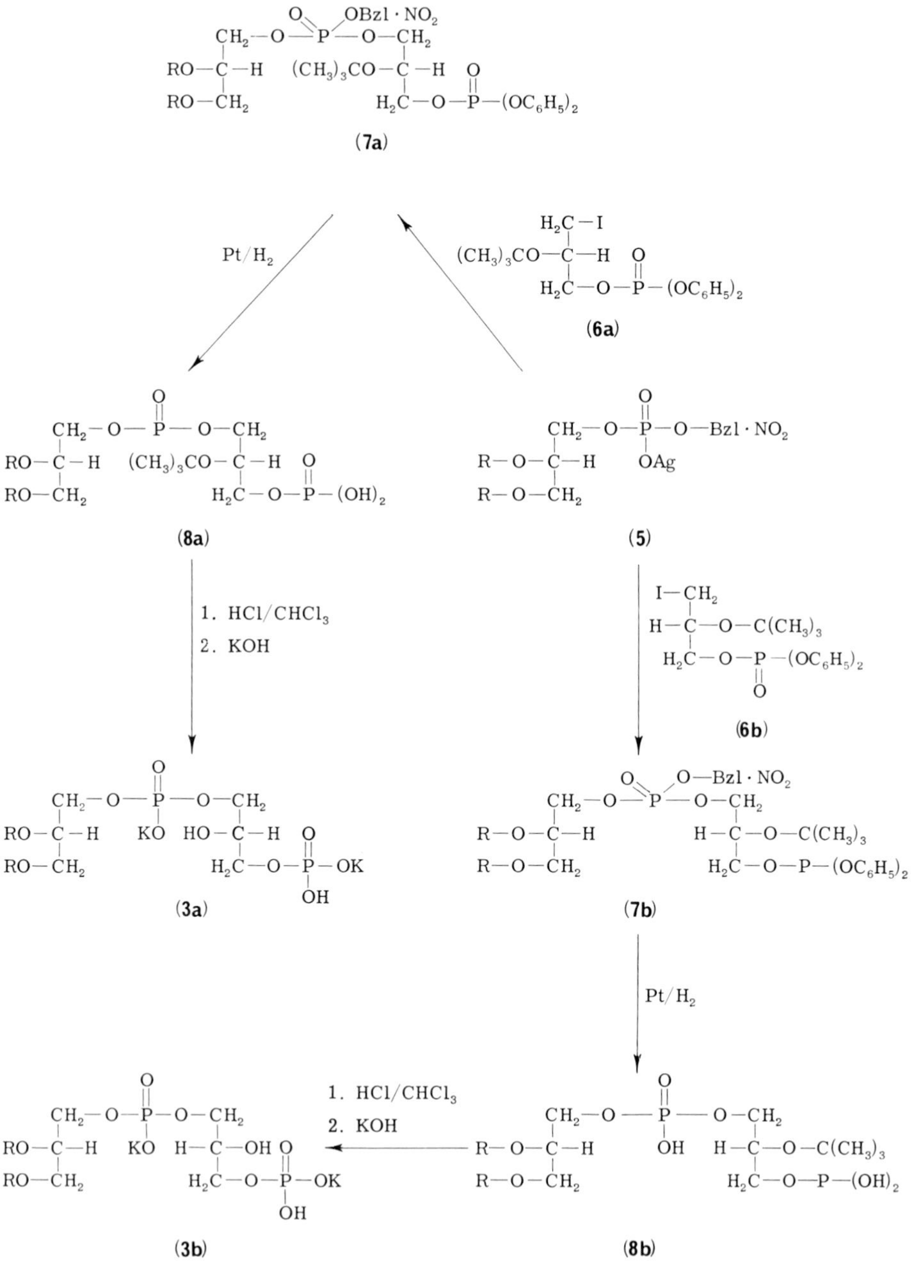

R designates 3*R*, 7*R*, 11*R*, 15-tetramethylhexadecyl;
Bzl designates benzyl

SCHEME V. Synthesis of diastereomeric phosphatidylglycerophosphates (diphytanyl ether analogs).

and the products obtained were compared with the natural isomer isolated from *H. cutirubrum*. The synthesis was carried out by the sequence of reactions shown in Scheme V. Briefly, the monosilver salt mono-*p*-nitrobenzyl ester of 2,3-di-*O*-phytanyl-*sn*-glycerol-1-phosphate [compound (**5**) prepared from the bacterial diphytanylglycerol] was condensed with either of the two enantiomers of a suitably blocked iodoglycerol derivative, namely, 1-iodo-2-*tert*-butyl-3-diphenylphosphorylglycerol [(**6a**) and (**6b**)] to give the fully blocked diastereomeric phosphotriesters [(**7a**) and (**7b**)]. Removal of nitrobenzyl and phenyl groups was effected by catalytic hydrogenolysis, and the *tert*-butyl group was removed from the resulting compounds (**8a**) and (**8b**) by treatment with anhydrous HCl in chloroform at 0°C. The resulting diastereomeric phosphatidylglycerophosphoric acids [(**3a**) and (**3b**)] were converted to the sodium or potassium salts and purified by preparative TLC as described for the natural isomer. The salts of the two synthetic diastereoisomers had the same elementary analyses (Table VII), the same staining behavior and chromatographic mobilities (Table VI), and the same infrared spectra (Fig. 9) as the natural isomer. However, only the stereoisomer synthesized from 1-diphenylphosphoryl-2-*tert*-butyl-3-iodo-*sn*-glycerol (**6b**) and having the

TABLE IX

Specific Rotations of Diastereomeric Phosphatidylglycerophosphates and Phosphatidylglycerols (Phytanyl Diether Analogs)[a]

Compound	$[\alpha]_D$ in chloroform		
	Potassium salt	Sodium salt	Barium salt
Phosphatidylglycerophosphates			
Natural isomer	+1.89°, + 2.5°[b]	0.0°	−1.94°
1-*sn*-Phosphatidyl-3′-*sn*-glycero-1′-phosphate (**3b**)	+1.93°	0.0°	—
1-*sn*-Phosphatidyl-1′-*sn*-glycero-3′-phosphate (**3a**)	−2.24°	−1.8°	—
Phosphatidylglycerols			
Natural isomer	+2.6°[b]	+3.46°	—
1-*sn*-Phosphatidyl-3′-*sn*-glycerol (**9b**)	—	+3.43°	—
1-*sn*-Phosphatidyl-1′-*sn*-glycerol (**9a**)	—	−1.13°	—

[a] Data from Joo and Kates (1969).
[b] Faure *et al.* (1964).

TABLE X

Analytical Data for the Diphytanyl Glycerol Ether Analog of Phosphatidylglycerol

Data	Potassium salt		Sodium salt	
	Found[a]	Calc. for $C_{46}H_{94}O_8PK$	Found[b]	Calc. for $C_{46}H_{94}O_8PNa$
Mol. wt.	—	845.7	—	829.6
C (%)	—	65.32	66.89	66.70
H (%)	—	11.18	11.00	11.41
P (%)	3.30	3.66	3.67	3.74
M (%)	—	4.63	2.53	2.77
M/P (atomic ratio)	—	1.00	0.93	1.00
H^+ strong/P (atomic ratio)	1.0[c]	1.00	—	1.00
H^+ weak/P (atomic ratio)	0.0[c]	0.00	—	0.00
Vicinal OH/P (mole ratio)	0.98	1.00	—	1.00

[a] Data of Faure *et al.* (1964).
[b] Data of Joo and Kates (1969).
[c] Analyses for the free acid (Faure *et al.*, 1964).

configuration 1-*sn*-phosphatidyl-3′-*sn*-glycerol-1′-phosphate (**3b**) had a specific rotation identical to that of the natural isomer (Table IX).

These results therefore establish unambiguously the structure and configuration of the phosphatidylglycerophosphate in extreme halophiles as 2,3-di-*O*-phytanyl-*sn*-glycerol-1-phosphoryl-3′-*sn*-glycerol-1′-phosphate [structure (**3b**), Scheme V]. This structure is unusual stereochemically since both glycerol moieties have the opposite configurations to those in the corresponding diester form found in rat liver and in *E. coli*.

3. *Phosphatidylglycerol (Diphytanyl Ether Analog)*

Isolation of the phosphatidylglycerol component of the lipids of *H. cutirubrum* was first achieved by Faure *et al.* (1964) by chromatography on a column of silicic acid eluted with ethyl ether containing increasing amounts of methanol. The phosphatidylglycerol appeared in the 30% methanol fraction and was obtained as the pure potassium salt. Later, this phosphatide was isolated (Joo and Kates, 1969) by preparative TLC in the same solvent systems as described for phosphatidylglycerophosphate, and was converted to the sodium salt.

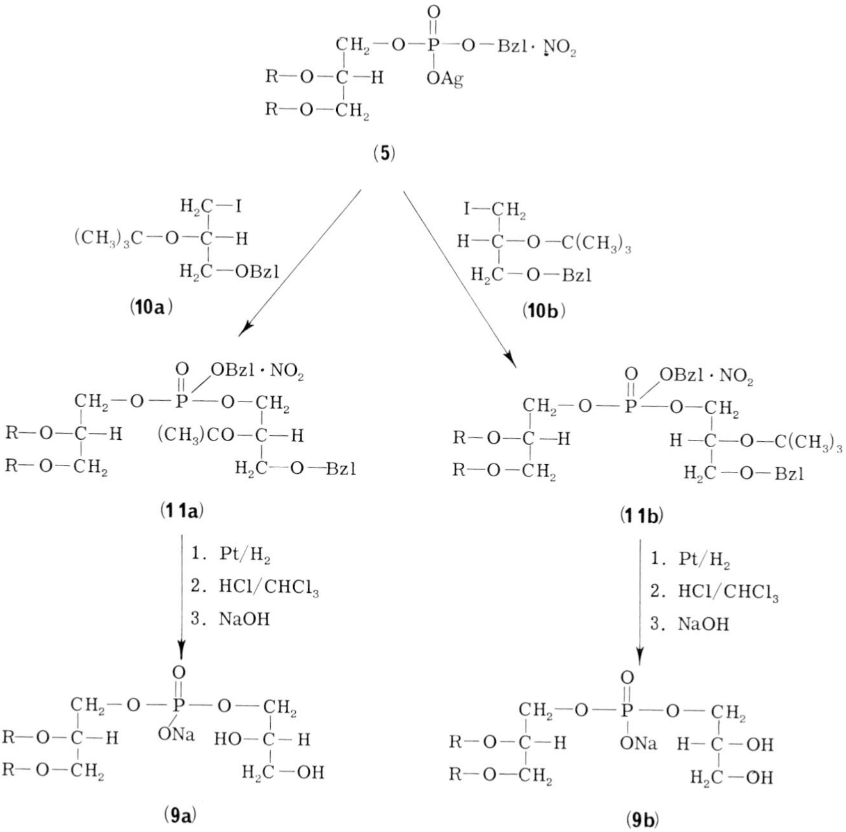

R designates 3*R*, 7*R*, 11*R*, 15-tetramethylhexadecyl;
Bzl designates benzyl

Scheme VI. Synthesis of diastereomeric phosphatidylglycerols (diphytanyl ether analogs).

Analytical data for the sodium and potassium salts (Table X) were in excellent agreement for the α-isomeric form of a phosphatidylglycerol in which 2,3-di-*O*-phytanylglycerol is esterified with α-glycerophosphoric acid (**9**).

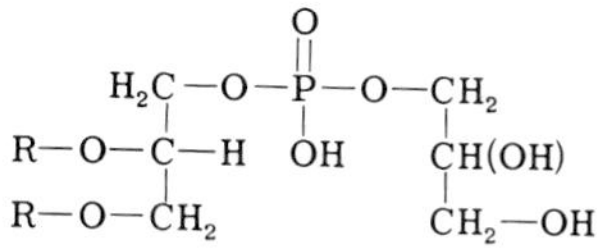

(**9**)

Consistent with this structure, hydrolytic degradation of the phosphatide with hot 90% acetic acid or with methanolic HCl yielded the 2,3-di-*O*-phytanyl-*sn*-glycerol and α-glycerophosphate (together with some cyclic glycerophosphate) (Faure *et al.*, 1964; Kates and Hancock, 1972). However, there still remained the question of the configuration of the glycerophosphate moiety of (**9**); for this reason, and also to establish structure (**9**) unambiguously, the synthesis of the two diastereoisomeric forms of (**9**) was undertaken, using the sequence of reactions shown in Scheme VI (Joo and Kates, 1969). As in the synthesis of phosphatidylglycerophosphate (Scheme V), the starting material was the monosilver salt mono-*p*-nitrobenzyl phosphate ester of the bacterial 2,3-di-*O*-phytanyl-*sn*-glycerol (**5**), which was condensed with either of the two enantiomers of 1-iodo-2-*tert*-butyl-3-*O*-benzylglycerol [(**10a**) and (**10b**)] to give the fully blocked diastereoisomeric triesters [(**11a**) and (**11b**)]. Removal of the nitrobenzyl and benzyl groups by catalytic hydrogenolysis, followed by removal of the *tert*-butyl group with anhydrous HCl in chloroform yielded the diastereoisomeric phosphatidylglycerols [(**9a**) and (**9b**)] which were isolated as the sodium salts. Both diastereoisomers had the same elementary analyses, staining behavior, chromatographic mobilities, and infrared spectra as the natural isomer (Tables X and VI, Fig. 9), but only the isomer synthesized from 1-*O*-benzyl-2-*tert*-butyl-3-iodo-*sn*-glycerol (**10b**), with the configuration 1-*sn*-phosphatidyl-3′-*sn*-glycerol (**9b**), had the same specific rotation as the natural isomer (Table IX).

These results thus establish the structure and configuration of the phosphatidylglycerol in extreme halophiles as 2,3-di-*O*-phytanyl-*sn*-glycerophosphoryl-3′-*sn*-glycerol (**9b**). It is noteworthy that the configuration of the two glycerol moieties in (**9b**) is the same as in the phosphatidylglycerophosphate (**3b**), and that it is stereochemically the mirror image of the diester phosphatidylglycerol in all other organisms.

4. *Glycolipid Sulfate*

Isolation of the second major component, the glycolipid sulfate (spot 2, Fig. 8), was achieved by taking advantage of its sparing solubility in methanol (Kates *et al.*, 1967a). The total acetone-insoluble polar lipids of *H. cutirubrum* were triturated with several portions of methanol, and the insoluble residue was dissolved in a minimum of chloroform and precipitated with acetone. The precipitate was greatly enriched in the sulfolipid, which was further purified by chromatography on a silicic acid column eluted with gradually increasing concentrations of methanol in chloroform. The glycolipid sulfate appeared in the 20–40% methanol–chloroform eluates and was finally purified by reprecipitation from chloroform solu-

TABLE XI

Analytical Data for Glycolipid Sulfate[a]

Data	Found	Calc. for $C_{61}H_{117}O_{21}SNa$
Mol. wt.	—	1241.6
C (%)	58.2	58.95
H (%)	10.1	9.50
S (%)	2.20	2.58
Na (%)	1.09	1.85
K (%)	0.31	—
Mg (%)	0.37	—
Total sugar (%)	41.7	43.6
Di-*O*-phytanylglycerol (%)	52.5	52.6
Na/S, atomic ratio	0.7	1.0
Sugar/S, mole ratio	3.3	3.0
Di-*O*-phytanylglycerol/S (mole ratio)	1.1	1.0

[a] Data of Kates *et al.* (1967a).

tion by addition of methanol. The product obtained was not converted to any one salt form but was kept in the form (largely the sodium salt) representative of its natural ionic state. The glycolipid sulfate was a colorless, hygroscopic powder and was chromatographically pure (R_f values given in Table VI); it was strongly dextrorotatory ($[\alpha]_D$ +36.8°, in $CHCl_3$).

The analytical data (Table XI) indicated that the glycolipid sulfate was a sulfate ester of a glycolipid containing 3 moles of sugar and 1 mole of diphytanylglycerol ether per sulfate group, and that it was largely in the form of a monosodium salt (small amounts of potassium and magnesium salts were also present). The infrared spectrum (Fig. 10) showed a strong OH absorption band at 3400 cm^{-1}; strong alcoholic C—O– and glycosidic C—O– bands at 1010 and 1100 cm^{-1}, indicative of sugar groups; strong CH_2 and CH_3 absorption bands at 2930, 2880, and 1465 cm^{-1}, and an isopropyl doublet at 1380–1365 cm^{-1}, indicative of phytanyl groups; and ester sulfate absorption bands at 1235 and 815 cm^{-1}. The spectrum was thus consistent with that of a glycolipid sulfate ester structure.

Hydrolysis of the sulfolipid with methanolic HCl followed by aqueous 1 *N* HCl yielded 1 mole of 2,3-di-*O*-phytanyl-*sn*-glycerol, identical in structure and configuration to the diether obtained from the total lipids,

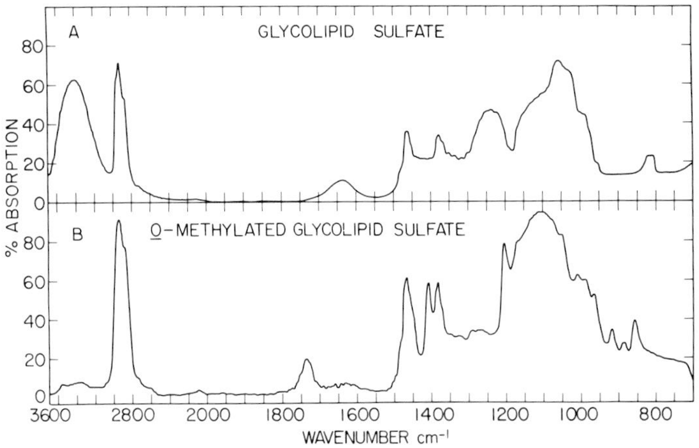

FIG. 10. Infrared spectra of A, the glycolipid sulfate (in KBr) and B, the fully methylated glycolipid sulfate (in CCl_4).

and 3 moles of sugar, identified chromatographically as glucose, galactose, and mannose, in equimolar proportions. These findings showed that the sulfolipid probably consisted of a triglycosyl sulfate linked glycosidically to the diphytanyl glycerol ether.

To determine the position of attachment of the sulfate group and the linkages between the sugars, the sulfolipid was converted to the free acid and methylated with methyl iodide and silver oxide under reflux. The methylated product had no OH absorption in the infrared (Fig. 10), but had a strong C—O—C ether band at 1105 cm^{-1} and sulfate bands at 1200 and 855 cm^{-1}, as expected for a methyl sulfate ester. After methanolysis of the fully methylated product in methanolic-HCl, the methyl glycosides of the methylated sugars were analyzed by GLC on Carbowax 6000 at 175°C. The identified products (Table XII) indicated that (1) galactose was most likely the terminal sugar (the small amount of tetramethyl galactose probably arose from partial hydrolysis of the sulfate group during the methylation procedure) and was esterified with sulfate at the 3-position; (2) mannose was linked at position 6; and (3) glucose was linked at position 2.

To confirm the identity of the terminal sugar and the position of attachment of the sulfate group, the glycolipid sulfate was subjected to mild acid hydrolysis (0.05 *N* HCl in methanol-chloroform, at 25°C for 20 hr) to hydrolyze the sulfate group. The desulfated glycolipid was then methylated and methanolyzed as described above, and the methylated sugars

TABLE XII

Products of Methylation of Glycolipid Sulfate and Free Glycolipid[a]

Substance methylated	Methylated sugars found[b]
Free glycolipid sulfate	2,4,6-Trimethyl galactose + 2,3,4,6-tetramethyl galactose (traces)
	2,3,4-Trimethyl mannose
	3,4,6-Trimethyl glucose
Desulfated glycolipid	2,3,4,6-Tetramethyl galactose
	2,3,4-Trimethyl mannose
	3,4,6-Trimethyl glucose

[a] Data of Kates *et al.* (1967a).
[b] Analyzed by GLC on Carbowax 6000 at 175° C.

again identified by GLC (Table XII). The only tetramethyl sugar found was 2,3,4,6-tetramethyl galactose, and no trimethyl galactose was detected, thus confirming that galactose was indeed the terminal sugar and that it was esterified with sulfate at the 3-position.

At this stage, two equally probable structures, (**12**) and (**13**), could be postulated for the glycolipid sulfate ester of *H. cutirubrum* (Kates *et al.*, 1967a).

[SO_3^--*O*-3′-Gal-(1′→6′)-Man-(1′→2′)-Glu(1′→1)] -
2,3-di-*O*-phytanyl-*sn*-glycerol

(**12**)

[SO_3^--*O*-3′-Gal-(1′→2′)-Glu-(1′→6′)-Man-(1′→1)]-
2,3-di-*O*-phytanyl-*sn*-glycerol

(**13**)

To determine which of these structures was the correct one, the glycolipid sulfate (as ammonium salt) was subjected to prolonged mild acid hydrolysis in 0.1 *N* HCl in methanol-chloroform (3:4) at 25°C for 72 hr (Kates and Deroo, 1972). The products formed consisted of almost equal amounts of the triglycosyl, diglycosyl, and monoglycosyl diphytanyl glycerol ethers, which were separated by preparative TLC in chloroform–methanol–90% acetic acid (30:4:20). Each glycolipid was hydrolyzed with methanolic HCl followed by aqueous 1 *N* HCl, and the free sugars identified by paper chromatography in pyridine–ethyl acetate–water (2:5:5, upper phase). The monoglycosyl diether was found to contain only glucose, the diglycosyl diether only glucose and mannose, and the triglycosyl diether glucose, mannose, and galactose. These results clearly establish the order of the sugars as that present in structure (**12**). Assuming the D-configuration for the sugars, the mono- and diglycosyl diethers had molecular rotations close to those calculated for α-glycosidically linked sugars; the rotation of the triglycosyl diether, however, was much lower than that expected for an α-linked galactose, but agreed well with that calculated for a β-linked galactose (Kates and Deroo, 1972). The glycolipid sulfate in *H. cutirubrum* may thus be assigned the complete structure (**12a**): 2,3-di-*O*-phytanyl-1-*O*-[β-D-galactopyranosyl-3′-sulfate-(1′→6′)-*O*-α-D-mannopyranosyl-(1′→2′)-*O*-α-D-glucopyranosyl]-*sn*-glycerol.

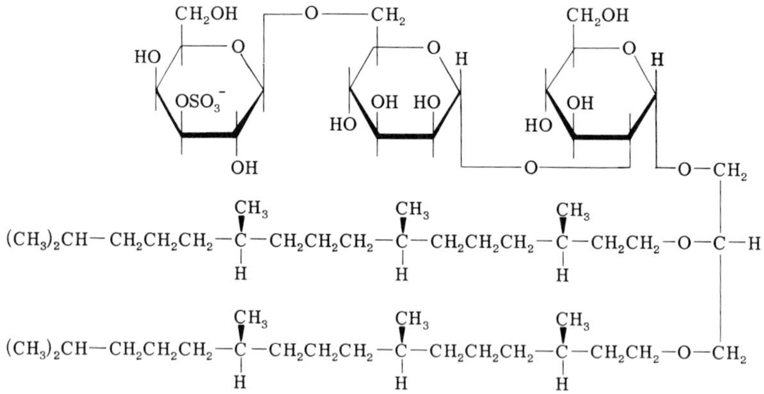

(**12a**)

Final proof of this structure by chemical synthesis has not yet been achieved.

IV. Metabolic Studies

A. Incorporation of Labeled Precursors into Ether-Linked Lipids by Whole Cells

As a first approach to the study of ether-linked lipid metabolism in *H. cutirubrum* (Kates *et al.*, 1968a,b), cells were grown to the stationary phase in the presence of the simple labeled phosphate and sulfate precursors, ^{32}P-orthophosphate and ^{35}S-sulfate, and of precursors of long-chain hydrocarbon groups, 1-^{14}C-acetate, 2-^{14}C-mevalonate, and 1,3-^{14}C-glycerol (2-^{14}C-malonate was also tried but only traces were incorporated into the

TABLE XIII

Distribution of Radioisotopes among Lipid Components[a]

Spot No. (see Fig. 8)	Component	Percentage of total radioisotope in total lipids				
		^{14}C-Acetate	^{14}C-Mevalonate	^{14}C-Glycerol	^{32}P-Phosphate	^{35}S-Sulfate
Origin	—	3	2	2	0.3	0.2
1	Unidentified sulfolipid	3	1	4	0.4	13
2	Glycolipid sulfate	37	13	44	0.5	80
3	Glycolipid	1	1	2	0.4	0
4	Phosphatidylglycerol sulfate	15	7	13	11	6
5	Unidentified phosphatide	2	3	1	2	0
6	Phosphatidylglycerophosphate	27	44	24	77	0
8	Phosphatidylglycerol	7	13	6	8	0
9	Pigments + neutral lipids[b]	6	16	4	0.5	0

[a] Data of Kates *et al.* (1968b).

[b] Includes squalene, dihydro- and tetrahydrosqualenes, vitamin MK_8 and carotenoid pigments (Tornabene *et al.*, 1969).

lipids). Relatively low incorporations of ^{32}P-phosphate and ^{35}S-sulfate into the lipids were obtained (1.5 and 0.1%, respectively), but this was attributed to the presence of inorganic phosphate and sulfate in the complex growth medium. After decreasing the amounts of phosphate and sulfate in the medium, sufficient activity was incorporated for measurement of distribution among the lipid components (see Table XIII). Of the ^{14}C-labeled precursors, mevalonate showed the highest degree of incorporation amounting to 8% of the tracer supplied; acetate and glycerol each showed about 2% incorporation.

The distribution of the supplied tracers among the lipid components is given in Table XIII. Almost half of the ^{14}C derived from mevalonate was associated with the major phospholipid, phosphatidylglycerophosphate (spot 6 Fig. 8), while the pigments and neutral lipids (spot 9, Fig. 8), the glycolipid sulfate (spot 2, Fig. 8), and phosphatidylglycerol (spot 8, Fig. 8) had much lower (and about equal) proportions of ^{14}C activity. In contrast, the highest proportions of ^{14}C derived from either acetate or

TABLE XIV

Distribution of ^{14}C among Moieties of Ether-Linked Polar Lipids[a]

Precursor	^{14}C-distribution, % of ^{14}C in total polar lipids[b]		
	Glycerol diether moiety		Water-soluble moieties[c]
	Phytanyl groups	Glycerol moiety	
1-^{14}C-Acetate[d]	99	trace	1
2-^{14}C-Mevalonate	98	trace	2
1,3-^{14}C-Glycerol	52	18	30

[a] Data of Kates *et al.* (1968b).

[b] Corrected for ^{14}C in pigments and squalenes according to data in Table XIII.

[c] Contains glycerol diphosphate and glycerophosphate derived from the diether phosphatides, and glucose, galactose, and mannose derived from the glycolipid sulfate.

[d] Traces (<0.3%) of ^{14}C-labeled fatty acids methyl esters were found after methanolysis of ^{14}C-acetate-labeled lipids with methanolic HCl.

glycerol were associated with the glycolipid sulfate, and progressively lower proportions appeared in phosphatidylglycerophosphate, phosphatidylglycerol, and the pigment + neutral lipid fraction. The phosphatidylglycerol sulfate (spot 4, Fig. 8) was relatively heavily labeled with acetate or glycerol but less so with mevalonate. The ^{14}C-labeled neutral lipids were later shown to be squalene, dihydro- and tetrahydrosqualenes, and carotenoid pigments (Tornabene *et al.*, 1969).

The incorporation of ^{32}P-orthophosphate closely paralleled the distribution of lipid-P among the components (Table VI). Thus, the major phosphatide, phosphatidylglycerophosphate, and the minor phosphatide, phosphatidylglycerol, accounted for 77% and 8%, respectively, of the total ^{32}P incorporated into the lipids, values which are almost identical to the respective proportions of lipid-P in these components. The other minor phosphatide, phosphatidylglycerol sulfate (spot 4), was relatively strongly labeled with ^{32}P as well as ^{14}C, and also contained some ^{35}S. It should be noted that the traces of ^{32}P associated with spots 1, 2, and 3 are probably due to streaking of the major phosphatide (spot 6).

When ^{35}S-sulfate was used as precursor, 80% of the radioisotope incorporated into the lipids appeared in the glycolipid sulfate ester component (spot 2, Fig. 8), and about 13% was associated with a slow-moving unidentified sulfolipid (spot 1, Fig. 8). The remaining 6% of ^{35}S was associated with the phosphatidylglycerol sulfate component (spot 4) mentioned above.

To determine the distribution of ^{14}C among the various ether-linked lipid moieties, the lipids were degraded as shown in Scheme I, and the ^{14}C-activity in the phytanyl groups, the glycerol moiety of the diphytanyl glycerol ether, and the water-soluble moieties of the phosphatides (glycerophosphate and glycerol diphosphate) and glycolipids (galactose, glucose, mannose) was determined (Table XIV). With either ^{14}C-acetate or ^{14}C-mevalonate as precursors, 98% or more of the activity in the polar lipids was associated with the phytanyl groups, but only traces were in the glycerol moiety of the diether and only a few percent in the water-soluble groups. With ^{14}C-glycerol as precursor, however, considerable proportions of ^{14}C were found in the sugar and glycerol mono- and diphosphate moieties of the lipids and in the glycerol portion of the diether; but the phytanyl groups still accounted for the major amount of ^{14}C.

These findings show that the predominant biosynthetic route for the hydrocarbon chains in *H. cutirubrum*, starting from acetate, is the mevalonate pathway, and that glycerol can enter both the mevalonate pathway for phytanyl groups (via acetate), and the glycolysis cycle for the synthesis of glycerol phosphates and sugars, as shown in Scheme VII.

Evidence supporting the participation of glycerol in the glycolysis cycle was obtained during a study of the utilization of glycerophosphate for phospholipid synthesis in *H. cutirubrum* (Wassef *et al.*, 1970a), in which a rapid conversion of glycerol to glycerophosphate in intact cells was observed. This conversion is catalyzed by the enzyme glycerol kinase, as demonstrated in cell-free homogenates of *H. cutirubrum;* the glycerophosphate formed by this enzyme was found to be the *sn*-3-isomer (Wassef

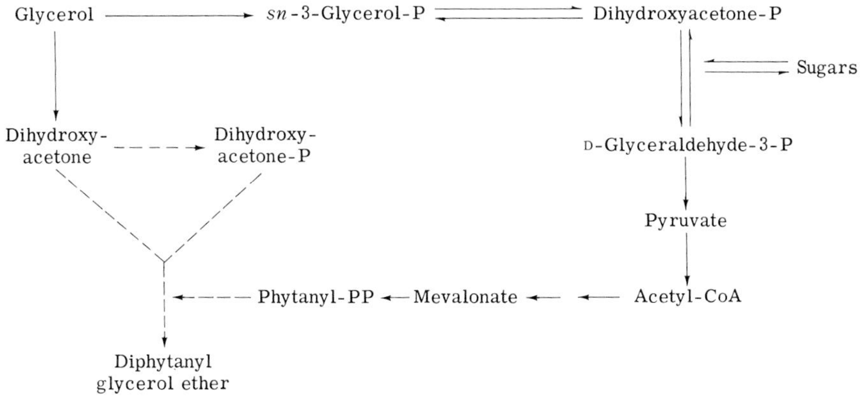

SCHEME VII. Metabolic pathways proposed for glycerol incorporation into lipids and and lipid moieties in *H. cutirubrum*.

TABLE XV

*Retention of ^3H in Lipids and Lipid Moieties of **H. cutirubrum** from Precursors 1(3)-^3H-Glycerol or 2-^3H-Glycerol*[a]

Lipid fraction or moiety	1(3)-^{3}H-glycerol percursor Retention of ^{3}H (%)		2-^{3}H-glycerol percursor Retention of ^{3}H (%)	
	Found	Calc.	Found	Calc.[d]
Total lipids	79 ± 1	—	5 ± 1	—
Water-soluble fraction	67 ± 1	—	6 ± 1	—
Sugars	53 ± 1	50[b]	3 ± 1	0
Diphytanylglycerol	70 ± 1	69	3 ± 1	0
Phytanyl groups	51 ± 1	54[c]	3 ± 1	0
Glycerol moiety	102 ± 2	100[d]	4 ± 1	0

[a] Data of Kates *et al.* (1970).
[b] Calculated on basis of conversion of glycerol to glucose via glycerophosphate by glycolytic pathway.
[c] Calculated on basis of conversion of glycerol to acetyl-CoA by glycolytic pathway, and then to phytanyl-PP by the mevalonate pathway.
[d] Calculated on the basis of complete dehydrogenation only at C-2.

et al., 1970b), thus showing that its stereospecificity is the same as in all other organisms studied. Further support for the participation of glycerol in glycolysis was provided by the demonstration of glycerophosphate dehydrogenase activity in cell-free homogenates of *H. cutirubrum*, and that this enzyme is also specific for the *sn*-3-glycerophosphate (Wassef *et al.*, 1970b).

A more detailed picture of the utilization of glycerol for synthesis of ether-linked lipids of *H. cutirubrum* was provided by studies with ^{3}H-labeled glycerol as precursor (Kates *et al.*, 1970). Cells were grown separately in the presence of (a) 1(3)-^{3}H-glycerol + 1(3)-^{14}C-glycerol and (b) 2-^{3}H-glycerol + 1(3)-^{14}C-glycerol. The lipids were then isolated and degraded as shown in Scheme I, and the $^3H/^{14}C$ ratios determined for the various lipid moieties. Striking variations in the $^3H/^{14}C$ ratios of the lipid moieties were found to depend on the position of the ^{3}H-label in the glycerol precursor supplied (Table XV). Thus, the total lipids and all lipid moieties from cells grown in the presence of the 2-^{3}H-glycerol were almost completely devoid of ^{3}H, whereas lipids from cells grown in the 1(3)-^{3}H-glycerol had a high retention of ^{3}H (79%). Furthermore, the phytanyl groups of the glycerol diether and the sugar moities of the glycolipids retained about 50% of the ^{3}H in the 1(3)-^{3}H-isomer, in good agreement with the values calculated for retention of ^{3}H in these moieties formed from glycerol via the glycolysis cycle and the mevalonate pathway, respectively (Table XV, Scheme VII). The complete absence of ^{3}H in these groups when the 2-^{3}H-isomer was used as precursor is also in accord with the pathway shown in Scheme VII, since dehydrogenation of glycerophosphate to dihydroxyacetone phosphate results in complete removal of ^{3}H at carbon 2, and hence no retention of ^{3}H can be expected in the sugars and the phytanyl groups.

Surprisingly, the glycerol moiety of the diphytanyl glycerol ether was found to retain 100% of ^{3}H from the 1(3)-3*H*-isomer. This finding, together with the fact that complete loss of ^{3}H occurred when the 2-^{3}H-isomer was used, shows that the glycerol moiety of the diether must undergo dehydrogenation at C-2 but not at C-1 (or C-3). These restrictions would eliminate participation of aldo–keto isomerizations, such as that between dihydroxyacetone phosphate and glyceraldehyde 3-phosphate catalyzed by triose phosphate isomerase. Thus, dihydroxyacetone phosphate could not act as precursor of the glycerol moiety of the diphytanyl glycerol ether, as it does for monoalkyl glycerol ethers (Snyder *et al.*, 1970; see Chapter VII), *unless* it were kept in a physically separate pool where it could not equilibrate with glyceraldehyde 3-phosphate (Scheme VII). Another possible precursor would be dihydroxyacetone itself, which is formed by the action of glycerol dehydrogenase, an enzyme known to be active in *H. cutirubrum* (Baxter and Gibbons, 1954). It should be emphasized that the steps leading from dihydroxyacetone or dihydroxyacetone phosphate to the diphytanylglycerol ether (Scheme VII) are completely unknown at present and probably represent a novel pathway for the biosynthesis of the ether linkages in a dialky glycerol ether.

Synthesis of the phytanyl group undoubtedly occurs from acetate via the pathway:

$$\begin{array}{c}\text{Acetate} \rightarrow \rightarrow \rightarrow \rightarrow \text{Mevalonate} \rightarrow \rightarrow \rightarrow \rightarrow \text{Isopentenyl pyrophosphate} \\ \downarrow\uparrow \\ \text{Phytanyl pyrophosphate} \leftarrow \leftarrow \leftarrow \leftarrow \leftarrow \text{Dimethyl allyl pyrophosphate}\end{array}$$

However, the individual steps in this pathway still have to be confirmed by studies with cell-free or purified enzyme systems.

The route from dimethylallyl pyrophosphate to the fully saturated C_{20} isoprenoid (phytanyl) group, about which nothing is known, is of particular interest. Presumably it is similar to the pathway, also unknown, for the biosynthesis of phytol in photosynthetic organisms. One might envisage a mechanism wherein reduction of the isoprenoid double bonds occurs at each step in the pathway:

Dimethylallyl pyrophosphate → Geranyl pyrophosphate →

Farnesyl pyrophosphate → Geranylgeranyl pyrophosphate

Alternatively, the four hydrogenation steps may occur entirely in the terminal geranylgeranyl pyrophosphate to form the fully saturated phytanyl pyrophosphate. However, all that can now be stated with certainty about this mechanism is that each of the reduction steps must be stereospecific and give rise to the D or *R* absolute configuration at the asymmetric carbon atoms 3, 7, and 11 in the phytanyl group (Kates *et al.*, 1967b).

B. Existence of the Malonyl-CoA Pathway for Fatty Acid Biosynthesis

As was mentioned earlier, very small amounts of fatty acids were detected in all of the extreme halophiles examined (Table I). Traces of fatty acids could have been derived at least in part from lipids in the culture medium (Kates *et al.*, 1966b), but the question still remained whether these organisms were capable of synthesizing fatty acids. Studies on the incorporation of ^{14}C-acetate into lipids of *H. cutirubrum* (Kates *et al.*, 1968b) showed a small (<0.3% of total incorporation) but definite labeling of fatty acids (Table XIV), suggesting that a pathway for fatty acid synthesis might be operative.

Studies were than carried out using a cell-free enzyme system and ^{14}C-malonyl-CoA as precursor (Pugh *et al.*, 1971). The soluble enzyme system was prepared by sonication of a suspension of *H. cutirubrum* cells followed by centrifugation at 40,000 *g* and precipitation of the enzyme from the supernatant with ammonium sulfate between 50% and 75% saturation. For comparison, a fatty acid synthetase preparation from *E. coli* was also examined. Using a standard assay system for fatty acid synthetase (Lennarz *et al.*, 1962) the activities of the fatty acid synthetase system of the halophile and *E. coli*, fortified with the respective heat-stable acyl carrier protein fraction, were determined as a function of increasing NaCl or KCl concentration (Table XVI).

The results showed clearly that *H. cutirubrum* does indeed contain a fatty acid synthetase system which is active only at very low salt concentrations, with an activity about 6% of that in *E. coli*. Even at moderate NaCl or KCl concentrations (1 *M*), the fatty acid synthetase of both *H. cutirubrum* and *E. coli* is inactivated to the extent of 77% and 99%, respectively; and at 4 *M* salt concentration inactivation reaches 92% and almost 100%, respectively. The halophile system, although strongly inhibited by high salt concentrations, appears to be less affected than the

TABLE XVI

Effect of Salt Concentration on Activity of Fatty Acid Synthetase in H. cutirubrum and E. coli[a]

NaCl or KCl concentration (M)	Specific activity[b]	
	H. cutirubrum	*E. coli*
0	1.3	22
1	0.3	0.1
2	0.3	<0.1
4	0.1	<0.1

[a] Data from Pugh *et al.* (1971).
[b] Moles $\times$ 10^{-10} malonyl-CoA incorporated/30 min/mg protein.

E. coli system. This may account for the slight incorporation of ^{14}C-acetate observed in cells of *H. cutirubrum* under normal conditions of growth.

C. Conclusions

The metabolic studies described have revealed some unusual pathways for synthesis of lipids in *H. cutirubrum* that differ from those in other microorganisms in several important respects:

(1) Hydrocarbon chains of polar lipids are synthesized exclusively by the mevalonate pathway for isoprenoid chains, which must be modified to include hydrogenation steps for formation of the saturated phytanyl groups; these enzymic systems are not inhibited by high salt concentration, but, on the contrary, probably *require* 4 M salt for full activity.

(2) The malonyl-CoA system for fatty acid synthesis does exist, but is operative at a very low level because of inhibition by the high intracellular salt concentration. As a result, only traces of fatty acyl ester lipids are formed.

(3) Glycerol incorporated into the polar lipids must pass through a dehydrogenation stage whereby the hydrogens at carbon 2 are removed; this unidentified dehydrogenated glycerol intermediate presumably acts as an acceptor of phytanyl pyrophosphate in a novel pathway for formation of the diphytanyl glycerol ether characteristic of this organism.

(4) A novel biosynthetic pathway must also exist for the diphytanyl ether analogs of phosphatidylglycerophosphate and phosphatidylglycerol, since both glycerol moieties in these lipids have the opposite configuration

to those in the corresponding diester analogs found in nonhalophiles. It should be noted that it is not simply a matter of the enzyme glycerophosphate:cytidine monophosphate phosphatidyl transferase being stereospecific for *sn*-1-glycerophosphate rather than the normal *sn*-3-glycerophosphate, since the latter isomer is the only one formed enzymically in *H. cutirubrum*, either by the action of glycerokinase or glycerophosphate dehydrogenase (Wassef *et al.*, 1970b).

V. Function of Diphytanyl Ether Lipids in Cells of Extremely Halophilic Bacteria

The presence of novel lipids in extremely halophilic bacteria raises the question as to their function in the cell and whether their unusual structures have a bearing on the ability of the extreme halophiles to cope with the high salt concentrations in their environment. Although few or no experimental data are available on the cellular function of these lipids, some aspects of this topic may be discussed at this time, in the hope of at least shedding some light on the problems involved (Kates, 1968).

As mentioned elsewhere, the lipids in extreme halophiles are associated almost entirely with the cell envelope (Kushner *et al.*, 1964), and most likely with the cytoplasmic membrane component of the envelope (Stoeckenius and Kunau, 1968). Furthermore, virtually all of the polar lipid components contain negatively charged groups (PO—O^- and SO_3^-) (Table VI). It is likely that the electrostatic forces between the negative charges on the lipid and protein components and the environmental cations would generally help to stabilize the membrane. In particular, interaction of Mg^{2+} with the acidic phosphatides and proteins has been found to be important in maintaining stability of the cell envelope (Kushner and Onishi, 1966). The strong affinity of the diether analog of phosphatidylglycerophosphate for Mg^{2+} ions (Rayman *et al.*, 1967) suggests that it is a major Mg^{2+} binding site in the membrane. Binding of phosphatides with protein in the membrane may also occur through magnesium chelation between basic amino acid residues (e.g., lysine, arginine) in the protein and the lipid phosphate groups, as well as by binding of the lipids to hydrophobic residues in the proteins (McClare, 1967).

The presence of diphytanyl glycerol ether-derived lipids in the membranes of extreme halophiles may also have a marked survival value for these organisms. The saturated phytanyl groups would impart stability of the membranes to peroxidation under the conditions of high aeration necessary for growth. Also, ether-linked alkyl groups would impart stability of the lipids to hydrolysis over a wide pH range such as might be encountered in their natural environment. Furthermore, the "unnatural" configuration of the glycerol moieties of the phosphatides makes them resistant to attack by the known plant, animal, and bacterial phospholipases (M. Kates, unpublished results). One other point might be mentioned in this connection, namely, that the carotenoid pigments present in the membrane have a definite survival value for extreme halophiles since they protect them from the harmful effects of direct sunlight to which they would naturally be exposed (Larsen, 1962).

Finally, the diphytanyl ether-linked phosphatides and the glycolipid sulfate may play a role in the selective ion transport of the membrane as a result of differential affinities for Na^+ and K^+. Vandenheuvel (1965) has constructed molecular models of the diphytanyl ether analog of phosphatidylglycerophosphate and found that the distance between the P=O groups fits very well into his model of a conductive cation–water monolayer, sandwiched between the lipid and protein layers, which would be highly selective for K^+. Further studies along these lines using lipid bilayers of the individual phosphatides of *H. cutirubrum* should be carried out to determine whether or not cation selectivity resides in the phosphatide components of the membrane.

REFERENCES

Abrahamsson, S., Ställberg-Stenhagen, S., and Stenhagen, E. (1963). *Progr. Chem. Fats Other Lipids* **7**, 1.

Ackman, R. G., and Hansen, R. P. (1967). *Lipids* **2**, 357.

Baxter, R. M., and Gibbons, N. E. (1954). *Can. J. Biochem. Physiol.* **32**, 206.

Bayley, S. T., and Kushner, D. J. (1964). *J. Mol. Biol.* **9**, 654.

Brown, A. D. (1963). *Biochim. Biophys. Acta* **75**, 425.

Burrell, J. W. K., Jackman, L. M., and Weedon, B. C. L. (1959). *Proc. Chem. Soc., London* p. 263.

Cho, K. Y., and Salton, M. R. J. (1966). *Biochim. Biophys. Acta* **116**, 73.

Christian, J. H. B., and Waltho, J. A. (1962). *Biochim. Biophys. Acta* **65**, 506.

Dawson, R. M. C. (1954). *Biochim. Biophys. Acta.* **14**, 374.

Faure, M., and Morelec-Coulon, M. J. (1958). *Ann. Inst. Pasteur, Paris* **95**, 180.

Faure, M., Marechal, J., and Troestler, J. (1963). *C. R. Acad. Sci.* **257**, 2187.

Faure, M., Marechal, J., and Troestler, J. (1964). *C. R. Acad. Sci.* **259**, 941.

Hancock, A. J., and Kates, M. (1972). *Chem. Phys. Lipids* **8**, 87.

Joo, C. N., and Kates, M. (1968). *Biochim. Biophys. Acta* **152**, 800.

Joo, C. N., and Kates, M. (1969). *Biochim Biophys. Acta* **176**, 278.

Joo, C. N., Kates, M., and Shier, T. (1968). *J. Lipid Res.* **9**, 782.

Kates, M. (1964). *Advan. Lipid Res.* **2**, 17.

Kates, M. (1968). *Abstr. Int. Congr. Biochem., 7th; 1967* Symp. I, Vol. 7, p. 3.

Kates, M., and Deroo, P. W. (1972). *Biochim. Biophys. Acta* (in press).

Kates, M., and Hancock, A. J. (1971). *Biochim. Biophys. Acta* **243**, 254.

Kates, M., and Hancock, A. J. (1972). *Biochem. J.* (in press).

Kates, M., Sehgal, S. N., and Gibbons, N. E. (1961). *Can. J. Microbiol.* **7,** 427.
Kates, M., Chan, T. H., and Stanacev, N. Z. (1963a). *Biochemistry* **2,** 394.
Kates, M., Sastry, P. S., and Yengoyan, L. S. (1963b). *Biochim. Biophys. Acta* **70,** 705.
Kates, M., Yengoyan, L. S., and Sastry, P. S. (1965a). *Biochim. Biophys. Acta* **98,** 252.
Kates, M., Palameta, B., and Yengoyan, L. S. (1965b). *Biochemistry* **4,** 1595.
Kates, M., Palameta, B., and Chan, T. H. (1966a). *Can. J. Biochem.* **44,** 707.
Kates, M., Palameta, B., Joo, C. N., Kushner, D. J., and Gibbons, N. E. (1966b). *Biochemistry* **5,** 4092.
Kates, M., Palameta, B., Perry, M. B., and Adams, G. A. (1967a). *Biochem. Biophys. Acta* **137,** 213.
Kates, M., Joo, C. N., Palameta, B., and Shier, T. (1967b). *Biochemistry* **6,** 3329.
Kates, M., Wassef, M. K., and Kushner, D. J. (1968a). *In* "Membrane Models and the Formation of Biological Membranes" (L. Bolis and B. A. Pethica, eds.), pp. 105–113. North-Holland Publ., Amsterdam.
Kates, M., Wassef, M. K., and Kushner, D. J. (1968b). *Can. J. Biochem.* **46,** 971.
Kates, M., Wassef, M. K., and Pugh, E. L. (1970). *Biochim. Biophys. Acta* **202,** 206.
Kates, M., Park, C. E., Palameta, B., and Joo, C. N. (1971). *Can. J. Biochem.* **49,** 275.
Kelly, M., and Liaaen Jensen, S. (1967). *Acta Chem. Scand.* **21,** 2578.
Kelly, M., Norgard, S., and Liaaen-Jensen, S. (1970). *Acta Chem. Scand.* **24,** 2169.
Kushner, D. J. (1964). *Exp. Chemother.* **2,** 114–168.
Kushner, D. J. (1968). *Advan. Appl. Microbiol.* **10,** 73.
Kushner, D. J. and Onishi, H. (1966). *J. Bacteriol.* **91,** 653.
Kushner, D. J., Bayley, S. T., Boring, J., Kates, M., and Gibbons, N. E. (1964). *Can. J. Microbiol.* **10,** 483.
Larsen, H. (1962). *In* "The Bacteria" (I. C. Gunsalus and R. Y. Stanier, eds.), Vol. 4, pp. 297–342. Academic Press, New York.
Larsen, H. (1967). *Advan. Microbiol. Physiol.* **1,** 97.
Lennarz, W. J., Light, R. J. and Bloch, K. (1962). *Proc. Nat. Acad. Sci. U.S.* **48,** 840.
McClare, C. W. F. (1967). *Nature (London)* **216,** 766.
Marshall, C. L. and Brown, A. D. (1968). *Biochem. J.* **110,** 441.
Michelson, A. M. (1959). *J. Chem. Soc., London* p. 1371.
Pugh, E. L., Wassef, M. K. and Kates, M. (1971). *Can. J. Biochem.* **49,** 953.
Rayman, M. K., Gordon, R. C., and MacCleod, R. A. (1967). *J. Bacteriol.* **93,** 1465.
Sehgal, S. N., Kates, M., and Gibbons, N. E. (1962). *Can. J. Biochem.* **49,** 69.
Siakotos, A. N. and Rouser, G. (1965). *J. Amer. Oil Chem. Soc.* **42,** 913.
Smithies, W. R. S., Gibbons, N. E., and Bayley, S. T. (1955). *Can. J. Microbiol.* **1,** 605.
Snyder, F., Malone, B., and Blank, M. L. (1970). *J. Biol. Chem.* **245,** 1790.
Stoeckenius, W., and Kunau, W. H. (1968). *J. Cell. Biol.* **38,** 337.
Stoeckenius, W., and Rowen, R. (1967). *J. Cell Biol.* **34,** 365.
Tornabene, T. G., Kates, M., Gelpi, E., and Oro, J. (1969). *J. Lipid Res.* **10,** 294.
Vandenheuvel, F. A. (1965). *J. Amer. Oil Chem. Soc.* **42,** 481.
Wassef, M. K., Kates, M., and Kushner, D. J. (1970a). *Can. J. Biochem.* **48,** 63.
Wassef, M. K., Sarner, J., and Kates, M. (1970b). *Can. J. Biochem.* **48,** 69.

CHAPTER XVI

THE SEARCH FOR ALKOXYLIPIDS IN PLANTS

Helmut K. Mangold

I. Introduction

For almost a quarter of a century it has been claimed that alkoxylipids are characteristic of animal tissues. Several investigators have reported that they are abundant in higher animals, scarce in lower animals, but are not present at all in bacteria (Hanahan and Thompson, 1963; Rapport and Norton, 1962; Thiele, 1964). Recently, however, alkyl and alk-1-enyl ethers of polyhydric alcohols have been identified in low forms of marine life (see Chapter XI), in protozoans (see Chapter XIII), and in bacteria (see Chapters XIV and XV). A few publications even describe the presence of alkoxylipids in plant tissues, but the validity of these reports has been subject to considerable doubt.

Plant materials have not, in the author's opinion, been investigated thoroughly for alkoxylipid content, so the question as to whether or not alkyl and alk-1-enyl ethers of polyhydric alcohols occur in plant tissues cannot yet be answered satisfactorily. This lack of information is due in large measure to the fact that alkoxylipids, if they do occur in plant tissues, are present only in very small proportions, and that the older methods used for the detection and identification of these compounds were rather unsatisfactory. However, with the help of new and improved techniques of fractionation and isolation, it should be possible to explore the problem systematically and comprehensively, and to arrive at more conclusive results. Detailed descriptions of various reactions and methods used in the analysis of alkoxylipids can be found in Chapter II.

II. Alkoxylipids in Plant Tissues

Nearly all the studies published so far regarding the occurrence of alkoxylipids in higher plants were carried out before efficient methods for the chromatographic fractionation of complex lipid mixtures had become available. However, work recently carried out on the lipids present in human and animal tissues has used chromatography, especially thin-layer chromatography (TLC), most advantageously for the detection and isolation of alkoxylipids. As a rule, neutral alkoxylipids can be detected in lipid extracts by adsorption chromatography. Thus, mixtures of alkyldiacylglycerols and alk-1-enyldiacylglycerols are well resolved as lipid classes by adsorption chromatography on thin layers of silica gel, both from each other and from triacylglycerols (Schmid and Mangold, 1966b). But it is advisable to analyze complex mixtures of neutral lipids by chromatography on both silica gel and magnesium oxide, since some diol lipids are not separated from glycerol-derived lipids on silica gel (Kaufmann *et al.*, 1971a).

Mixtures of alkylacyl, alk-1-enylacyl, and diacyl phosphoglycerides having the same base constituent are not easily resolved by the chromatographic techniques available at the present time (Renkonen, 1968). However, a great deal of information can be gained from studying the chromatographic behavior of the rather polar neutral lipids obtained from the ionic alkoxylipids through methanolysis and hydrogenolysis. The products of these reactions, alkyl and alk-1-enyl ethers of glycerol or other polyhydric alcohols, and methyl esters and dimethyl acetals, or long-chain alcohols, can be well resolved on thin layers of silica gel.

Although chromatography is certainly a great asset in detecting alkyl and alk-1-enyl moieties in a lipid fraction, rigid proof of the presence of these groups must be based on the results of chemical reactions and spectroscopic data.

A. Lipids Containing Alkyl Moieties

To the best of the author's knowledge, there are only three publications referring to alkylglycerols or their diacyl derivatives in plant tissues. Harrison and Hawke (1952a,b) reported the presence of alkylglycerols in the nonsaponifiable fraction isolated from the fat of Camelthorn (*Acacia giraffae*), a plant indigenous to South Africa. It was thought that 3% of the nonsaponifiable fraction of seed fat consisted of alkylglycerols (Harrison and Hawke, 1952a), as did 1.7% of seed pod fat (Harrison and Hawke, 1952b). However, identification of the alkylglycerols was based solely on the fact that the material found could be acetylated, and cleaved with periodate.

In 1958, Bodman and Maisin mentioned in a review article that they had detected alkyldiacylglycerols in Californian tung oil (*Aleurites fordii*), but experimental details were not given. A year later, in 1959, the author of this chapter tried to substantiate the finding, but TLC of the nonsaponifiable fractions isolated from several samples of tung oil did not reveal the presence of alkylglycerols. It must be noted that alkaline hydrolysis is a poor method when used to isolate alkylglycerols, because a considerable portion of nonsaponifiable material is likely to be lost in emulsions formed during extraction of the lipid hydrolyzate. However, alkylglycerols *can* be recovered in good yields from the products of hydrogenolysis, even if they are present in very small proportions (Schmid and Mangold, 1966a).

Recently, Polheim (1971) reported that he could not detect any trace of neutral or ionic alkoxylipids in baker's yeast (*Saccharomyces cerevisiae*). It is interesting that, according to Stowe (1960), *cis*-9-octadecenylglycerol (selachyl alcohol) exhibits growth-promoting activities in pea stem sections.

It is not an easy task to furnish proof of the presence of long-chain alkyl moieties in a lipid compound, and peralkylated substances are particularly difficult to detect and identify. The author considers the following criteria obligatory for the rigid proof of the presence of alkyl moieties in a lipid fraction: *Alkaline hydrolysis* with aqueous alcoholic potassium hydroxide, *alcoholysis* with methanol containing hydrogen chloride, and *hydrogenolysis* with lithium aluminum hydride followed by decomposition of the lithium alumino complexes with aqueous mineral acid must all produce lipid fractions whose *infrared spectra* exhibit a band at 1125–1120 cm^{-1} (9μ) and whose *cleavage with boiling hydroiodic acid* produces alkyl iodides. These iodides can be isolated by TLC, identified as a lipid class by their UV spectrum, and analyzed as a homologous series by gas–liquid chromatography. In addition, *mass spectra* of derivatives of alkyl ethers of polyhydric alcohols isolated from a natural source must agree with those of synthetic reference compounds.

B. Lipids Containing Alk-1-enyl Moieties

In 1949 Feulgen *et al.* reported that they had detected plasmalogens, i.e., lipids containing alk-1-enyl groups, in soybean (*Glycine soja*) and peas (*Pisum sativum*). According to these authors, dormant soybeans contained about 80 ng, and germinated soybeans over 400 ng of plasmalogens per bean. Evidence for the presence of the aldehydogenic lipids and the quantitative estimation of these compounds were based on the "plasmal" reaction, or Feulgen test: Alk-1-enyl ethers are cleaved with mercuric chloride, and the resulting aldehydes undergo a color-producing reaction with fuchsin-sulfurous acid (Feulgen and Grünberg, 1939) (see Chapter I).

A few years later, Lovern (1952) described the occurrence of plasmalogens in commercial groundnut (*Arachis hypogaea*) and soybean phosphatides. He, too, used the plasmal reaction as a test for aldehydogenic lipids. At about the same time, Thiele (1953) reported that he had subjected olive oil to methanolysis and had isolated dimethyl acetals from the products of this reaction. Thiele (1953) suggested, probably correctly, that these acetals must have been formed from plasmalogens. Using a solvent fractionation procedure, Wagenknecht (1957) isolated a fraction from the lipids of green peas that contained about 0.25% plasmalogens. Again, these compounds were detected and estimated by means of the "plasmal" reaction. Bergelson *et al.* (1966) presented evidence for the natural occurrence of a novel class of aldehydogenic lipids, the alk-1-enyl ether esters of ethanediol and other short-chain diols, or "neutral diol plasmalogens." These compounds were allegedly found in lipid fractions isolated from various animal tissues and also from corn (*Zea mays*) and sunflower (*Helicantus annus*) seeds as well as from soil yeasts (*Lipomyces* sp.). Bergelson *et al.* (1966) based their conclusions on aldehydogenic substances obtained from lipids that had been subjected to alcoholysis by treatment with a solution of potassium methylate in methanol. Acid methanolysis of these aldehydogenic substances yielded dimethyl acetals of long-chain aldehydes plus short-chain diols. Catalytic hydrogenation of the aldehydogenic compounds led to substances that did not release aldehydes on treatment with acid.

Quite recently, Albro and Dittmer (1970) indicated that they had found alk-1-enyldiacylglycerols ("neutral plasmalogens") in *Sarcina lutea*, a bacterium, and they suggested a pathway by which these compounds might be involved in the biosynthesis of long-chain hydrocarbons. However, the validity of their ideas has been questioned (Kolattukudy, 1970c).

Kaufmann *et al.* (1971b) have confirmed that, as claimed earlier, dormant green peas and soybeans as well as pea and soybean seedlings contain aldehydogenic lipids. Both the choline and ethanolamine phosphoglyceride fractions from these tissues exhibited the "plasmal" reaction. After reaction with lithium aluminum hydride and decomposition of the lithium alumino complexes with water (Schmid *et al.*, 1967), both types of phosphoglycerides yielded long-chain alcohols and a rather polar neutral fraction whose infrared spectrum exhibited a band at 1664 cm^{-1} (6.05 μ) considered characteristic of long-chain alk-1-enyl ethers (Norton *et al.*, 1962). Acid-catalyzed hydrolysis of these polar neutral fractions yielded mixtures of long-chain aldehydes, and catalytic hydrogenation led to 1-alkylglycerols. The various reactions used by Kaufmann *et al.* (1971b) in the identification of alk-1-enylacylglycerophosphorylcholine and alk-1-enylacylglycerophosphorylethanolamine are summarized in Fig. 1. Gas–liquid chromatographic analyses of the aldehydes derived from choline and ethanolamine phospholipids revealed that they were saturated and monounsaturated, and had from 14 to 20 carbon atoms (Hamza and Mangold, 1972). The chain-length distribution of the "free aldehydes" occurring in the surface wax of pea leaves is completely different; these

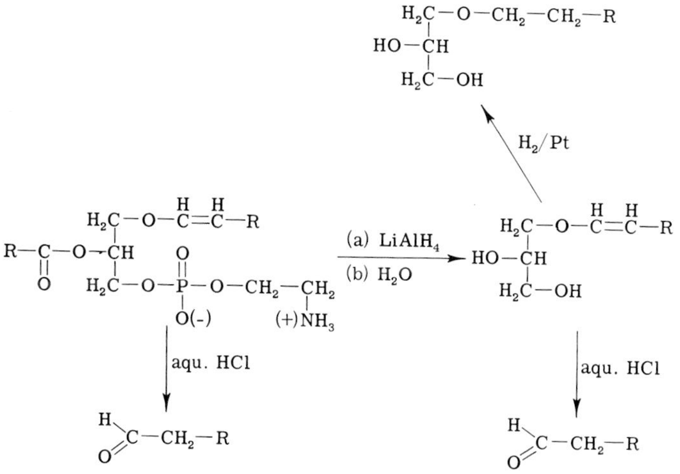

FIG. 1. Identification of alk-1-enylacylglycerophosphorylcholine and alk-1-enylacylglycerophosphorylethanolamine from green peas and soybeans (Kaufmann *et al.*, 1971b).

mixtures comprise aldehydes with 22 to 30 carbon atoms (Kolattukudy, 1970a). In this context, it is interesting that etiolated *Euglena gracilis* is able to reduce fatty acids or their esters to aldehydes and the latter to alcohols (Kolattukudy, 1970b). In animal tissues, alcohols serve as precursors of both the alkyl and alk-1-enyl moieties (see Chapter VII).

It is much easier to identify alk-1-enyl ethers than alkyl ethers, because of the greater reactivity of the former. However, the classic test for alk-1-enyl ethers, the plasmal reaction, is of limited value, in that autoxidation products can interfere with it.

In the author's opinion, rigid proof for the presence of alk-1-enyl moieties in a lipid fraction should be based on the following criteria: *Hydrogenolysis* with lithium aluminum hydride, followed by decomposition of the lithium alumino complexes with water, must yield a lipid fraction whose *infrared spectrum* exhibits a band of medium intensity near 1665 cm^{-1} (6.05 μ). *Acid-catalyzed hydrolysis* of this lipid fraction with aqueous hydrochloric acid in diethyl ether must yield aldehydes, and its *alcoholysis* with methanol-containing hydrogen chloride must produce dimethyl acetals. Both the aldehydes and the dimethyl acetals should yield 2,4-dinitrophenylhydrazones when reacted with an acidic solution of 2,4-dinitrophenylhydrazine in ethanol, and both aldehydes and dimethyl acetals can be analyzed by gas-liquid chromatography. Further, its *hydrogenation*, with platinum (dioxide) as catalyst, must lead to alkyl ethers.

III. Conclusions

The application of new and improved methods of analysis has brought to light more solidly based evidence that aldehydogenic lipids exist in higher plants, especially in developing plant tissues. In the opinion of the author, there is now no doubt that alk-1-enylacylglycerophosphorylcholine and alk-1-enylacylglycerophosphorylethanolamine do indeed occur in plant tissues, but the presence of alkylglycerols or their derivatives has yet to be proved. It will be a challenging task for years to come to systematically explore the distribution of alkoxylipids in the plant kingdom and in various parts of individual plants, to investigate the routes of biosynthesis and metabolism of these compounds, and to elucidate their functions in the living plant cell.

REFERENCES

Albro, P. W., and Dittmer, J. C. (1970). *Lipids* **5**, 320.

Bergelson, L. D., Vaver, V. A., Prokazova, N. V., Ushakov, A. N., and Popkova, G. A. (1966). *Biochim. Biophys. Acta* **116**, 511.

Bodman, J., and Maisin, J. H. (1958). *Clin. Chim. Acta* **3,** 253.
Feulgen, R., and Grünberg, H. (1939). *Hoppe-Seyler's Z. Physiol. Chem.* **257,** 161.
Feulgen, R., Feller, P., and Andresen, G. (1949). *Ber. Oberhess. Ges. Natur- Heilk.* **24,** 270.
Hamza, Y., and Mangold, H. K. (1972). *Chem. Phys. Lipids* (in press).
Hanahan, D. J., and Thompson, G. A., Jr., (1963). *Annu. Rev. Biochem.* **32,** 215.
Harrison, G. S., and Hawke, F. (1952a). *J. S. Afr. Chem. Inst.* **5,** 1.
Harrison, G. S., and Hawke, F. (1952b). *J. S. Afr. Chem. Inst.* **5,** 13.
Kaufmann, H. P., Mangold, H. K., and Mukherjee, K. D. (1971a). *J. Lipid Res.* **12,** 506.
Kaufmann, H. P., Hamza, Y. and Mangold, H. K. (1971b). *Chem. Phys. Lipids* **6,** 325.
Kolattukudy, P. E. (1970a). *Lipids* **5,** 398.
Kolattukudy, P. E. (1970b). *Biochemistry* **9,** 1095.
Kolattukudy, P. E. (1970c). *Arch. Biochem. Biophys.* **141,** 381.
Lovern, J. A. (1952). *Nature (London)* **169,** 969.
Norton, W. T., Gottfried, E. L., and Rapport, M. M. (1962). *J. Lipid Res.* **3,** 456.
Polheim, D. (1971). Private communication.
Rapport, M. M., and Norton, W. T. (1962). *Annu. Rev. Biochem.* **31,** 103.
Renkonen, O. (1968). *J. Lipid Res.* **9,** 34.
Schmid, H. H. O., and Mangold, H. K. (1966a). *Biochem. Z.* **346,** 13.
Schmid, H. H. O., and Mangold, H. K. (1966b). *Biochim. Biophys. Acta* **125,** 182.
Schmid, H. H. O., Baumann, W. J., and Mangold, H. K. (1967). *Biochim. Biophys. Acta* **144,** 344.
Stowe, B. B. (1960). *Plant Physiol.* **35,** 262.
Thiele, O. W. (1953). *Z. Gesamte Exp. Med.* **121,** 246.
Thiele, O. W. (1964). *Z. Klin. Chem.* **2,** 33.
Wagenknecht, A. C. (1957). *Science* **126,** 1288.

AUTHOR INDEX

Numbers in italics refer to the pages on which the complete references are listed.

C

H

L

N

O

T

U

V

Y

Z

SUBJECT INDEX

I

L

M

N

O

P

R

S